Physics and Chemistry of Finite Systems:
From Clusters to Crystals

NATO ASI Series

Advanced Science Institutes Series

*A Series presenting the results of activities sponsored by the NATO Science Committee,
which aims at the dissemination of advanced scientific and technological knowledge,
with a view to strengthening links between scientific communities.*

The Series is published by an international board of publishers in conjunction with the
NATO Scientific Affairs Division

A	**Life Sciences**	Plenum Publishing Corporation
B	**Physics**	London and New York
C	**Mathematical and Physical Sciences**	Kluwer Academic Publishers
		Dordrecht, Boston and London
D	**Behavioural and Social Sciences**	
E	**Applied Sciences**	
F	**Computer and Systems Sciences**	Springer-Verlag
G	**Ecological Sciences**	Berlin, Heidelberg, New York, London,
H	**Cell Biology**	Paris and Tokyo
I	**Global Environmental Change**	

NATO-PCO-DATA BASE

The electronic index to the NATO ASI Series provides full bibliographical references
(with keywords and/or abstracts) to more than 30000 contributions from international
scientists published in all sections of the NATO ASI Series.
Access to the NATO-PCO-DATA BASE is possible in two ways:

– via online FILE 128 (NATO-PCO-DATA BASE) hosted by ESRIN,
Via Galileo Galilei, I-00044 Frascati, Italy.

– via CD-ROM "NATO-PCO-DATA BASE" with user-friendly retrieval software in
English, French and German (© Springer Science+Business Media Dordrecht
1989).

The CD-ROM can be ordered through any member of the Board of Publishers or
through NATO-PCO, Overijse, Belgium.

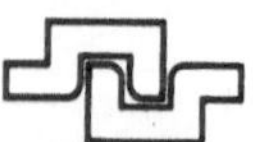

Series C: Mathematical and Physical Sciences - Vol. 374

Physics and Chemistry of Finite Systems: From Clusters to Crystals

Volume I

edited by

P. Jena
S. N. Khanna

and

B. K. Rao

Department of Physics,
Virginia Commonwealth University,
Richmond, VA, U.S.A.

Springer-Science+Business Media, B.V.

Proceedings of the NATO Advanced Research Workshop on
Physics and Chemistry of Finite Systems: From Clusters to Crystals
Richmond, VA, U.S.A.
October 8–12, 1991

Library of Congress Cataloging-in-Publication Data

```
Physics and chemistry of finite systems : from clusters to crystals /
  edited by P. Jena, S.N. Khanna, and B.K. Rao.
       p.    cm. -- (NATO ASI series. Series C, Mathematical and
  physical sciences ; vol. 374)
    "Published in cooperation with NATO Scientific Affairs Division."
    Includes index.
```

ISBN 978-94-017-2647-4   ISBN 978-94-017-2645-0 (eBook)
DOI 10.1007/978-94-017-2645-0

```
    1. Atomic structure--Congresses.  2. Electronic structure-
  -Congresses.  3. Crystals--Electrical properties--Congresses.
  4. Crystals--Optical properties--Congresses.  5. Cluster analysis-
  -Congresses.   I. Jena, P.  II. Khanna, S.N.  III. Rao, B. K.
  IV. Series: NATO ASI series.  Series C, Mathematical and physical
  sciences ; no. 374.
  QC172.P48  1992
  530.4'1--dc20                                          92-16307
```

ISBN 978-94-017-2647-4

This book contains the proceedings of a NATO Advanced Research Workshop held within the programme of activities of the NATO Special Programme on Chaos, Order and Patterns as part of the activities of the NATO Science Committee.

Other books previously published as a result of the activities of the Special Programme are:

ABRAHAM, N.B., ALBANO, A.M., PASSAMANTE, A. and RAPP, P.E. (Eds.) - *Measures of Complexity and Chaos* (B208) 1990 Plenum Publishing Corporation. ISBN 0-306-43387-7.

BUSSE, F.H. and KRAMER, L. (Eds.) - *Nonlinear Evolution of Spatio-Temporal Structures in Dissipative Continuous Systems* (B225) 1990 Plenum Publishing Corporation. ISBN 0-306-43603-5.

CHARMET, J.C., ROUX, S. and GUYON, E. (Eds.) - *Disorder and Fracture* (B235) 1991 Plenum Publishing Corporation. ISBN 0-306-43688-4.

MARESCHAL, M. (Ed.) - *Microscopic Simulations of Complex Flows* (B236) 1991 Plenum Publishing Corporation. ISBN 0-306-43687-6.

MacDONALD, G.J. and SERTORIO, L. (Eds.) - *Global Climate and Ecosystems Change* (B240) 1991 Plenum Publishing Corporation. ISBN 0-306-43715-5.

CHRISTIANSEN, P.L. and SCOTT, A.C. (Eds.) - *Davydov's Soliton Revisted: Self-Trapping of Vibrational Energy in Protein (B243) 1991 Plenum Publishing Corporation. ISBN 0-306-43734-1*

HOLDEN, A.V., MARKUS, M. and OTHMER, H.G. (Eds.) - *Nonlinear Wave Processes in Excitable Media* (B244) 1991 Plenum Publishing Corporation. ISBN 0-306-43800-3.

LING-LIE CHAU, NAHM, W. (Eds.) - *Differential Geometric Methods in Theoretical Physics: Physics and Geometry* (B245) 1991 Plenum Publishing Corporation. ISBN 0-306-43807-0.

ATMANSPACHER, H. and SCHEINGRABER, H. (Eds.) - *Information Dynamics* (B256) 1991 Plenum Publishing Corporation. ISBN 0-306-43912-3.

BABLOYANTZ, A. (Ed.) - *Self-Organisation, Emerging Properties, and Learning* (B260) 1991 Plenum Publishing Corporation. ISBN 0-306-43930-1.

PELITI, L. (Ed.) - *Biologically Inspired Physics* (B263) 1991 Plenum Publishing Corporation. ISBN 0-306-44000-8.

BISHOP, A.R., POKROVSKY, V.L. and TOGNETTI, V. (Eds.) - *Microscopic Aspect of Nonlinearity in Condensed Matter* (B264) 1992 Plenum Publishing Corporation. ISBN 0-306-44001-6.

JIMENEZ, J. (Ed.) - *The Global Geometry of Turbulence: Impact of Nonlinear Dynamics* (B268) 1992 Plenum Publishing Corporation. ISBN 0-306-44014-8.

MOSEKILDE, E. and MOSEKILDE, L. (Eds.) - *Complexity, Chaos and Biological Evolution* (B270) 1992 Plenum Publishing Corporation. ISBN 0-306-44026-1.

ROY, A.E. (Ed.) - *Predictability, Stability and Chaos in N-Body Dynamical Systems* (B272) 1992 Plenum Publishing Corporation. ISBN 0-306-44034-2.

BEN AMAR, M., PELCE, P. and TABELING, P. (Eds.) - *Growth and Form: Nonlinear Aspects* (B276) 1992 Plenum Publishing Corporation. ISBN 0-306-44046-6.

CONTENTS

STABILITY AND EVOLUTION

DYNAMICS

ELECTRONIC STRUCTURE

MAGNETISM

PREFACE

Recent innovations in experimental techniques such as molecular and cluster beam epitaxy, supersonic jet expansion, matrix isolation and chemical synthesis are increasingly enabling researchers to produce materials by design and with atomic dimension. These materials constrained by size, shape, and symmetry range from clusters containing as few as two atoms to nanoscale materials consisting of thousands of atoms. They possess unique structural, electronic, magnetic and optical properties that depend strongly on their size and geometry. The availability of these materials raises many fundamental questions as well as technological possibilities.

From the academic viewpoint, the most pertinent question concerns the evolution of the atomic and electronic structure of the system as it grows from micro clusters to crystals. At what stage, for example, does the cluster look as if it is a fragment of the corresponding crystal. How do electrons forming bonds in micro-clusters transform to bands in solids? How do the size dependent properties change from discrete quantum conditions, as in clusters, to boundary constrained bulk conditions, as in nanoscale materials, to bulk conditions insensitive to boundaries? How do the criteria of classification have to be changed as one goes from one size domain to another?

Potential for high technological applications also seem to be endless. Clusters of otherwise non-magnetic materials exhibit magnetic behavior when constrained by size, shape, and dimension. Nanoscale metal particles exhibit non-linear optical properties and increased mechanical strength. Similarly, materials made from nanoscale ceramic particles possess plastic behavior. Recent discovery of C_{60} buckyballs and micro-tubules of graphitic carbon and their ease of production show promise in a variety of industrial applications starting from drug delivery to lubricants and superconductivity. The strong dependence of cluster size on their reaction with reagent gases opens new doors for understanding catalysis. Since the properties of clusters can be drastically modified by changing only a few atoms, one can imagine having a three-dimensional periodic table of elements (cluster size providing the third dimension) from which new materials can be custom made. The field, therefore, represents a true interdisciplinary forum that needs coordinated efforts by physicists, chemists, material scientists, and engineers to tap its full potential in aiding our fundamental knowledge and quest for new materials.

The Physics and Chemistry of finite systems consisting of clusters, nano-structures, and quasicrystals was the subject of a recent international symposium held in Richmond, Virginia during the week of October 8-12, 1991. The purpose of the symposium was to bring together researchers from different disciplines including Physics, Chemistry, Mathematics, materials science, and Engineering to discuss the various challenging problems of both fundamental and technological importance. The symposium consisted of fourteen plenary sessions and two poster sessions. There were fifty two invited papers and 211 contributed poster presentations presented by nearly 300 scientists from 28 countries. The symposium concluded with a panel discussion on outstanding problems and future directions of the field. The proceedings of this symposium constitute this book.

The discussions and presentations included a number of topics:

-Preparation and characterization
-Atomic and electronic structure
-Stability and fragmentation
-Electronic, optical, magnetic, and thermodynamic properties
-Molecular dynamics simulation
-Melting
-Cluster reactions and catalysis
-Cluster support interaction
-Cluster assemblies
-Cluster materials involving C_{60} fullerenes
-Technological applications

The symposium was in the planning stage for nearly two years, and we owe our gratitude to a large number of colleagues and institutions who have assisted with the organization of the meeting. We are indebted to Professor Linus Pauling for serving as the honorary chairman of this conference and to members of the International Advisory Board for their help in selecting the topics and speakers. We are grateful to the members of the Local Organizing Committee for spending countless numbers of hours during the planning stage. We wish to thank the conferees for the high quality of their participation in the invited and poster presentations, and exchange of ideas. Our special thanks go to Ms. Barbara Martin for her tireless effort in looking after all the arrangements.

This symposium was made possible by generous grants from the North Atlantic Treaty Organization and Philip Morris, U.S.A. The symposium was also supported in part by grants from the National Science Foundation, Office of Naval Research, Oak Ridge Associated Universities, Army Research Office, and Extrel Corporation. We acknowledge, with gratitude, financial assistance from Virginia Commonwealth University, the symposium host.

P. Jena
S. N. Khanna
B. K. Rao

Richmond, Virginia
December 1992

CLUSTERS AS A KEY TO THE UNDERSTANDING OF PROPERTIES AS A FUNCTION OF SIZE AND DIMENSIONALITY

JOSHUA JORTNER
School of Chemistry
Tel Aviv University
Tel Aviv 69978, Israel

ABSTRACT. The exploration of size effects in conjunction with surface phenomena in clusters builds bridges between molecular, condensed matter and surface chemical physics. Cluster size equations are advanced, which quantify the "smooth" evolution of condensed matter properties and of microscopic surface characteristics with increasing the cluster size.

I. PROLOGUE

Structural, electronic and chemical characteristics of materials depend primarily on the state (phase) and degree (size) of aggregation. Clusters, i.e., finite aggregates containing 2-10^4 particles, exhibit unique physical and chemical phenomena [1-12], which provide ways and means to explore the gradual transition from molecular to condensed matter systems [13-15]. Furthermore, a large surface to volume ratio in clusters allows for the exploration of microscopic surface phenomena [16-31]. Clusters constitute bridges between molecular, surface and condensed matter chemical physics.

This paper will focus on some of the unique novel and basic features of clusters, which involve:

1. The size dependence of structural, energetic, electronic, electromagnetic, thermodynamic, dynamic and chemical properties of finite systems.
2. The evolution of macroscopic surface characteristics with increasing the cluster size.
3. The evolution of condensed matter properties with increasing the cluster size.

II. THE CLUSTER SIZE EQUATION

Fig. 1 portrays an "artist's view" of the general features of the dependence of a cluster property $\chi(n)$ (which corresponds to a structural, energetic, electromagnetic, etc. observable) on the number n of the cluster constituents. The size dependence of $\chi(n)$ falls into two distinct domains:

(I) Specific cluster size effects, which are exhibited for "small" clusters. In this size domain an irregular size dependence of structural, energetic, electronic and electromagnetic observables is exhibited, which is manifested most dramatically in the existence of "magic numbers" in $\chi(n)$ vs n, which reflect atomic [32-36], electronic [37-42] an Fermion [43-46] shell closure effects.

P. Jena et al. (eds.), Physics and Chemistry of Finite Systems: From Clusters to Crystals, Vol. I, 1–17.
© 1992 *Kluwer Academic Publishers.*

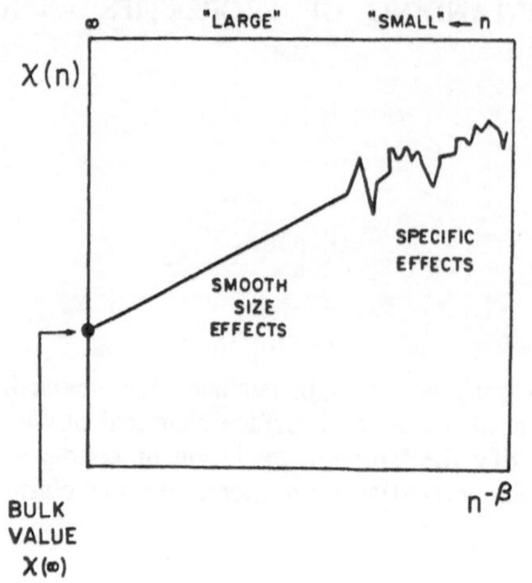

Figure 1. The cluster size dependence of a cluster property $\chi(n)$ on the number, n, of the cluster constituents. The data are plotted vs $n^{-\beta}$ where $0 \leq \beta \leq 1$. "Small" clusters reveal specific size effects, while "large" clusters are expected to exhibit for many properties a "smooth" size dependence of $\chi(n)$ which converges for $n \to \infty$ to the bulk value $\chi(\infty)$.

(II)   Smooth cluster size effects, which are revealed for "large" clusters. In this size domain $\chi(n)$ interpolates smoothly to the corresponding bulk value $\chi(\infty)$.

The cluster size equation (CSE) attempts to provide a quantitative description of the size dependence of a cluster property in range II, which is given by

$$\chi(n) = \chi(\infty) + An^{-\beta} \quad , \tag{II.1}$$

where A is a numerical constant and β the positive exponent in the range $0 \leq \beta \leq 1$. The features of the CSE are:

(1) $\chi(n)$ is expressed in terms of the corresponding bulk value and a correction term.
(2)   The value of the observable in the bulk, $\chi(\infty)$, is obtained either from experiment, which is usually the case, or from extremely reliable theoretical calculations or simulations.
(3)   The correction term

$$C(n) = An^{-\beta} \tag{II.2}$$

accounts for the modification of the bulk value in the cluster due to the excluded volume outside it.
(4)   The applicability range of the CSE. The correction term, C(n), is a "smooth" function of n. Accordingly, the cluster is sufficiently large, so that specific size effects are eroded.
(5)   The expression of the CSE in terms of the cluster radius $R_c = R_0 n^{1/3}$ can be readily obtained by setting

$$C(n) \equiv C(R_c) = A(R_c/R_0)^{-3\beta} \tag{II.3}$$

so that

$$\chi(R_c) = \chi(\infty) + (AR_0^{3\beta}) R_c^{-3\beta} \quad . \tag{II.4}$$

(6)   The generalization of the CSE. A more general form can appear as a polynomial

$f(x) = \Sigma A_j x^j$ in $x = n^{-\beta}$, where A_j are numerical constants. Thus $C(n) = f(n^{-\beta})$, so that

$$\chi(n) = \chi(\infty) + f(n^{-\beta}) \tag{II.5}$$

or

$$\chi(R_c) = \chi(\infty) + f[(R_c/R_0)^{-3\beta}] \quad . \tag{II.6}$$

The generalized CSE can originate from physical corrections, i.e., the incorporation of several distinct contributions to $\chi(n)$, or from the extension of the CSE [47-49].

Eqs. (II.1) and (II.5) constitute the CSE. The CSE is better than it appears at first sight for the quantification of the physical observables. All the complicated short-range effects, which are usually difficult to evaluate, are incorporated in $\chi(\infty)$ (often taken from experiment). What is explicitly evaluated is the correction term $C(n)$ arising from the excluded volume and determined by long-range effects, which are relatively easy to calculate in a reliable manner. The CSE constitutes a quantitative description of the evolution of bulk properties and can also be extended to describe the transition from microsurfaces to macrosurfaces (Chapter VI).

III. ENERGETIC CLUSTER SIZE EFFECTS

Several energetic attributes establish a quantitative "transition" from clusters to bulk condensed matter, which can be handled by the CSE.

III.A. The Cohesion Energy of He_n Clusters.

The ground state energy $E(n)$ of $(^4He)_n$ clusters at 0 K [46,50,51] was accounted for [50,51] by the liquid drop model [52]

$$E(n)/n = E_v + E_s n^{-1/3} + E_c n^{-2/3} \quad , \tag{III.1}$$

where E_v is the volume energy, E_s is the surface energy and E_c is the curvature energy. The calculated energies (Fig. 2) are well fitted by Eq. (II.1) with the energies being expressed in K units [50,51]

$$E(n)/n = -6.91 + 18.9n^{-1/3} - 12n^{-2/3} \quad . \tag{III.1a}$$

The limiting energetic result for the infinite cluster $[E(n)/n](n \to \infty) = -6.91K$, extrapolates well to the bulk liquid energy of $-7.21K$ per atom [50,51].

III.B. The Cohesive Energy of Metal Clusters.

The liquid drop model was applied for the cohesion energy per particle of Na_n and K_n clusters [53,54], which was expressed as a superposition of bulk and surface terms

$$E(n)/n = E_v + E_s n^{-1/3} \quad . \tag{III.2}$$

A good fit, according to this CSE, was achieved for atomization energies of Na_n ($n = 4$-100) [54] and K_n ($n = 3$-200) [53] clusters (see Fig. 3 for Na_n clusters [54]). Again, the CSE converges smoothly to the corresponding bulk heats of sublimation. From the analysis of the cohesion energies it is apparent that the liquid drop model corresponds to a CSE.

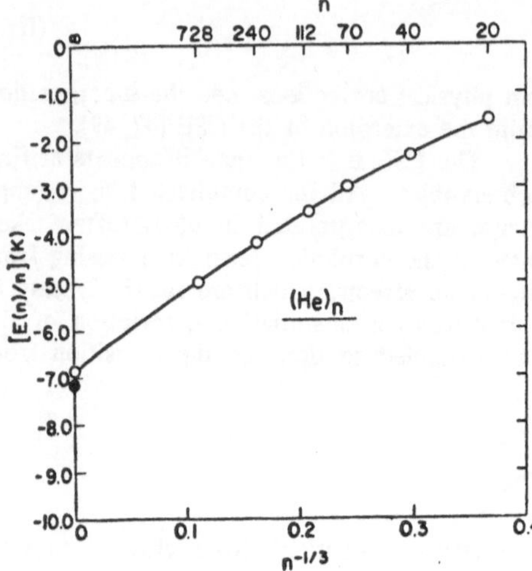

Figure 2. The size dependence of the energy per atom in $(^4\text{He})_n$ clusters. Open circles are calculated data [50,51] while the black point at $n = \infty$ corresponds to the experimental bulk value. The dependence of $E(n)/n$ on $n^{-1/3}$ is given in terms of the liquid drop (extended CSE) model, Eq. (III.1), represented by the solid line.

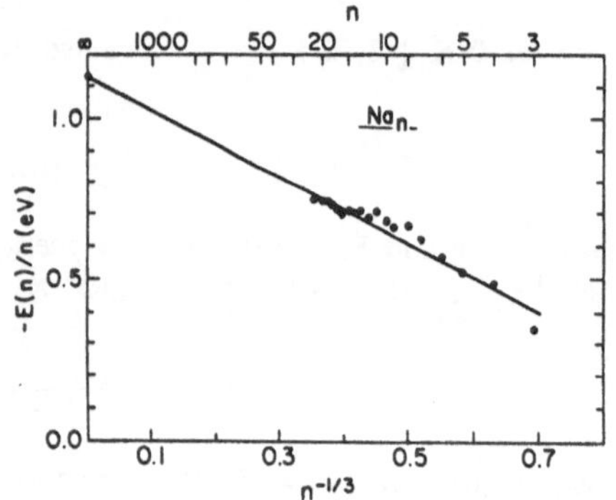

Figure 3. The size dependence of the experimental cohesive energy per atom for Na_n clusters [54]. The linear $n^{-1/3}$ plot, converging to the bulk value, is in accord with Eq. (III.2).

III.C. Ionization Potentials of van der Waals Clusters.

The simplest case involves an atomic internal impurity state in an internal position in a rare-gas cluster, e.g., $XeAr_n$ or $A\ell Ar_n$ [55]. The ionization potential of a doped rare-gas cluster, with the impurity occupying an interior location, can be expressed in the form

$$I(n) = I(\infty) + A\, n^{-1/3} \quad , \tag{III.3}$$

where

$$A = \frac{e^2}{2R_0}(1 - \epsilon^{-1}) \quad . \tag{III.4}$$

and ϵ is the cluster optical dielectric constant, which is taken to be equal to that of the bulk solid.

Eqs. (III.3) and (III.4) are applicable to (i) internal impurity states, (ii) a free hole with a narrow bandwidth and (iii) a trapped hole via two-center or multicenter trapping. Of course, the values of $I(\infty)$ are different for each case. We have calculated the slopes A of the CSE, Eq. (II.4), for the ionization potentials of rare-gas clusters (Table I) utilizing the experimental low frequency optical dielectric constants and took $R_0 = (3/4\pi\rho)^{1/3}$, where ρ is the solid density. Experimental clusters ionization potentials were experimentally reported by Hertel et al [56] and by Ding et al [57], with the latter data resting on an arbitrary energetic correction. The limiting values of $I(\infty)$, which correspond to the solid state photoionization thresholds, are experimentally known [58]. In Figures 4 and 5 we present the CSE plots according to Eq. (III.3) for the ionization potentials of rare-gas clusters together with the bulk data, which reveal that:
(1) The IP data of Ar_n [56] (Fig. 5), Kr_n [56,57] and Xe_n [57] (Fig. 4) for $n \geq 13$ extrapolate linearly to $I(\infty)$.
(2) A good agreement between experiment and theory is accomplished for the linear slopes, A, of the CSE for the rare-gas clusters Ar_n [56], Kr_n [56,57] and Xe_n [57] (Table I). This agreement strongly supports the physical significance of these CSE plots over their applicability domain.

TABLE I. CSE Data for Ionization Potentials of Rare-Gas Clusters.

Cluster	$R_0(\text{Å})$	ϵ [a,b] (calc)	A (eV) (expt)	A (eV)
Ar_n	2.08	1.66	1.37 0.65 [d,e]	1.20 [c]
Kr_n	2.21	1.88	1.52	1.10 [d]
Xe_n	2.40	2.22	1.65	1.41 [d]

(a) A.C. Sinnock and B.C. Smith, Phys. Rev. <u>181</u>, 1297 (1969).
(b) J. Kruger and W.J. Ambs, J. Opt. Soc. Amer. <u>49</u>, 1195 (1959).
(c) Reference 56 and Fig. 5.
(d) Reference 57 and Fig. 4.
(e) The A value is not reliable.

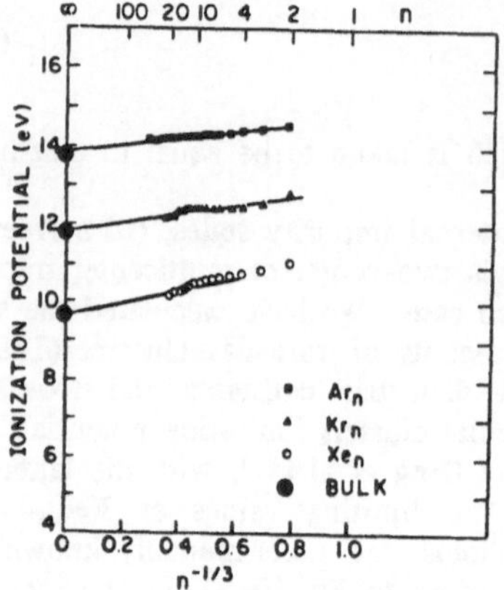

Figure 4. The size dependence of the ionization potentials of rare-gas clusters inferred by Ding et al [57] from their experimental data. Linear plots were drawn according to Eq. (III.3), which converge to the bulk experimental values [58].

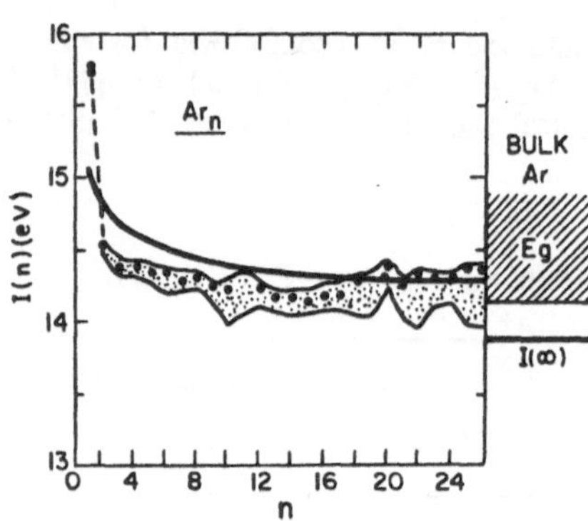

Figure 5. An analysis of the experimental ionization potential data of Hertel et al [56] for Ar_n clusters, according to the CSE. Experimental data are marked by black points while the shaded area represents the experimental uncertainty. The bare atom ionization potential is connected to the cluster I(n) data by a dashed line. The bulk solid band gap (lower limit of the conduction band) and the photoionization threshold I(∞) [58] are also marked. The ionization potential data are fit by the CSE, Eq. (III.3), with I(∞) = 13.9eV (the experimental value [58]) and A = 1.2eV (solid line).

III.D. Ionization Potentials of Metal Clusters.

Schumacher et al [59,60] advanced a simple description of the ionization potential of a metal cluster, which rests on the picture of the ionization of metallic spheres. The ionization potential (or electron affinity) of a cluster of charge Z is given for the metal droplet model [59-62]

$$I(R_c) = W + \left(Z + \frac{3}{8} \right) e^2/R_c \qquad (III.5)$$

while for the polarization model [63-65]

$$I(R_c) = W + (2Z + 1)\, e^2/2R_c \ .$$

(III.6)

These relations were widely applied for the ionization potentials of metal clusters, and, in general, reasonable agreement was accomplished. Notable in this context is the Schumacher plot [59,60] of ubiquitous information on metal clusters ionization potentials, however, the metal clusters used for this analysis might have been too small to warrant the applicability of the CSE. The most extensive information currently available involves the ns ionization potentials and nd inner shell ionization potentials of Cu_n^- clusters [66,67] (up to n = 400) and of Au^- clusters [67] (up to n = 200). The size dependence of the experimental ionization potentials [66,67] (Fig. 6) faithfully follows the artist's view of size effects (Figure 1). These results reveal that:

a) A smooth behavior of the ionization potentials is exhibited for n > 60. Thus the quantitative applicability of the CSE for smaller metallic clusters should be viewed with caution.

(b) For larger clusters the smooth CSE plots converge to the corresponding bulk values.

(c) The slopes, a, of the two CSE plots, Eqs. (III.5) or (III.6), for the ns and nd ionization potentials of Cu_n^- and Au_n^- are [67] $a \simeq -0.55$, which is close to $a = -1/2$ inferred from the polarization model, Eq. (III.6). Nevertheless, an unambiguous distinction between the 3/8 and the 1/2 plot cannot be made, in view of the uncertainty in the relation between n and R_c.

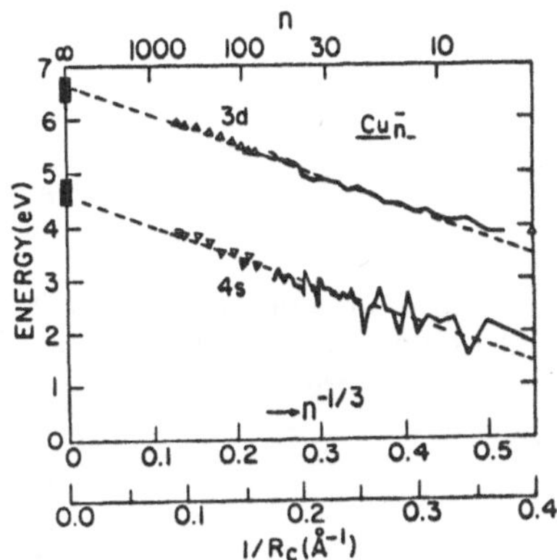

Figure 6. The dependence of the energies of the ns and nd bands of Cu_n^- (n = 1-411) clusters reproduced from Cheshnovsky et al [66], which reveal a CSE behavior according to Eq. (III.6) for n > 60, converging to the corresponding bulk energies.

The size dependence of the ionization potentials in range I provides information on specific cluster size effects on the electronic level structure, e.g., magic numbers, but does not elucidate the central issue of metal-nonmetal transition (MNMT) in clusters. Some information does, however, emerge on the size evolution of the electronic band structure of Hg_n clusters [68-72] (Fig. 7). For small n < 18 clusters very marked deviations of I(n) from Eq. (III.6) are exhibited [68]. A notable break at $n \simeq 18$ is revealed, while at large cluster sizes ($n \simeq 70$) the ionization potential data converge to the CSE, Eq. (III.6). The largest Hg_n cluster data (for $n \geq 70$) converge to the bulk value [68] with $a \simeq 1/2$ (Fig. 7). Small Hg_n (n < 18) clusters correspond to molecular van der Waals clusters. Around $n \simeq 18$ the overlap between electronic levels of s and p parentage sets in, with further increase of the cluster size resulting in the congestion of these s and p mixed energy

8

levels, with the gradual evolution of metallic properties. This mechanism of MNMT in clusters bears analogy to the MNMT in bulk mercury, which originates from the overlap of s and p bands [73]. The onset of cluster size behavior according to Eq. (III.6) (with the correct intercept W and with a = 1/2) marks the gradual onset of metallic properties in this system. However, the central issue of MNMT transition in finite systems with increasing the cluster size is still open.

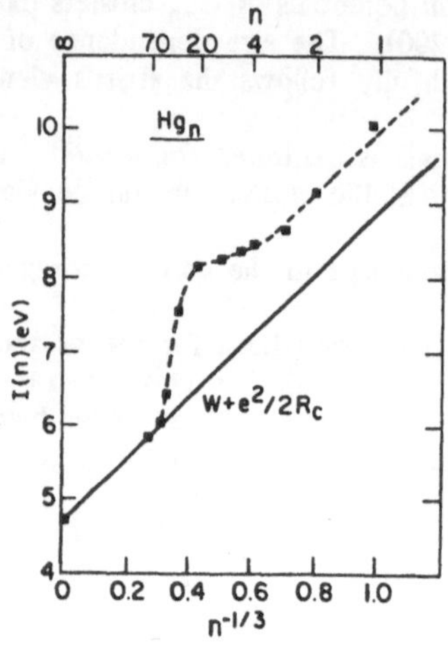

Figure 7. The size dependence of the ionization potentials of Hg$_n$ clusters [68], which reveal marked deviations from the CSE, Eq. (III.6). The dashed line is drawn through the experimental data for visual orientation. Only for n > 70 the CSE is obeyed with a $\simeq$ 1/2, converging to the bulk value.

IV.   A CLUSTER SIZE EQUATION FOR THE ENERGETICS OF WANNIER EXCITONS

Excitons in bulk solid and liquid rare-gases are described in terms of Wannier-Mott electron-hole pairs [74,75]. The energies E_N of these excitons in the bulk are based on the effective mass approximation, being given by a hydrogenic series [74]

$$E_N(\infty) = E_G - B/N^2 \quad , \qquad (IV.1)$$

where N is the principal quantum number, E_G is the gap energy, B is the effective Rydberg constant $B = \mu e^4/2\epsilon^2\hbar^2$, where ϵ is the static dielectric constant and μ is the reduced effective mass of the exciton $\mu^{-1} = m_e^{-1} + m_h^{-1}$, with m_e and m_h being the effective mass of the electron and hole, respectively. The exciton radius is $r_N = N^2 \epsilon \hbar^2 / \mu e^2$. This simple model was successful in describing the energies of N $\geq$ 2 exciton states in solid rare-gases and for the description of excitations from deep impurity states in rare-gas alloys [58,74].

In the description of excitons in semiconductor clusters [76-80], it was assumed that the boundary exerts localization of the electron and hole within the cluster, and the same description will be adopted for rare-gas clusters [28]. Two limiting situations can then be distinguished for an exciton characterized by a fixed N value [76].

(1) Strong confinement. $R_c \ll r_N$ for some N. This situation corresponds to individual electron and hole confinement within the well, with negligible spatial correlation [76,77,79]. The energy of the lowest excited state [77] does not correspond to a CSE.

This can be readily rationalized, as another limiting situation is encountered when R_c is increased

(2) Weak confinement. $R_c \gg r_N$ for some N. Now the character of the exciton as a pseudoparticle is preserved, while the translational degrees of freedom are confined within the cluster. The energy E_N of the Nth hydrogenic exciton within the cluster is [76,79]

$$E_N = E_N + \pi^2 \hbar^2 / [2(m_e + m_h) R_c^2] \ , \tag{IV.2}$$

where the second term corresponds to the confinement energy of a pseudoparticle with the mass $(m_e + m_h)$. Eq. (IV.5) constitutes a CSE. The CSE was expressed in the form

$$E_N(n) = E_N(\infty) + A\, n^{-2/3} \ , \tag{IV.3}$$

where $A = G/R_0^2$ and $G = \pi^2 \hbar^2 / 2(m_e + m_h)$. These relations imply that

(1)   The cluster size correction is $C(n) \propto n^{-2/3}$, revealing a distinct size dependence from that exhibited by the common $C(n) \propto n^{-1/3}$ relation.

(2)   The cluster size correction is independent of the exciton quantum number N.

These expectations are borne out by the experimental data of Möller et al [28]. Fig. 8 displays Möller's experimental data [28] for $N = 2$ and $N = 3$ excitons in Ar_n clusters (in the size domain which satisfies the validity condition for weak confinement), plotted according to relation (IV.3). The CSE is obeyed, converging to the experimental bulk $N = 2$ and $N = 3$ exciton energies (Fig. 8). The slope of the CSE, which is independent of N, as expected, is given by $G = 55\pm5$ eVÅ². This result implies an abnormally low hole effective mass [28], i.e., $(m_e + m_h) \simeq 2\mu$, which requires a reexamination of the physical picture of exciton confinement in rare-gas clusters.

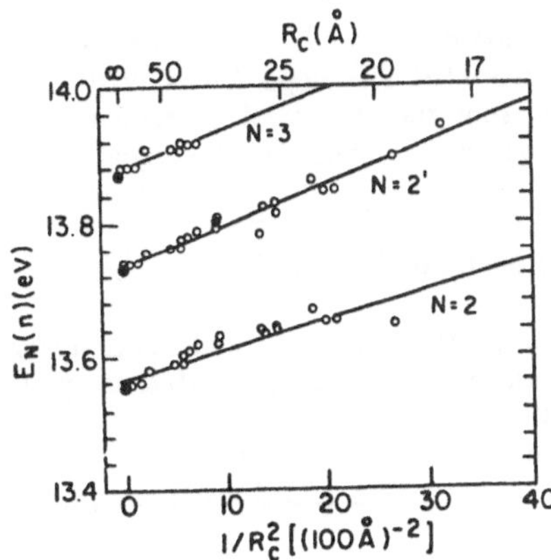

Figure 8. The size dependence of the energies of $N = 2$, $N = 2'$ and $N = 3$ Wannier-Mott excitons in Ar_n clusters. Experimental data (open points) from Möller [28]. The linear plots correspond to the CSE, Eq. (IV.3), which converge to the bulk data [58] represented by black points.

V. CLUSTER SIZE EFFECTS ON INTRAVALENCE ELECTRONIC EXCITATIONS

Van der Waals $M \cdot A_n$ heteroclusters, consisting of M bound to rare-gas atoms (A) and to molecular ligands [14,17,81-85] are of considerable current interest in relation to the microscopic exploration of solvation phenomena [84], and the elucidation of physisorption phenomena on microsurfaces [17]. Detailed information emerged from resonance two-color two-photon electronic spectroscopy of mass-selected $M{\cdot}A_n$ clusters [85-90] regarding the spectral shifts and the linewidths (FWHM) of the $S_0 \to S_1$ transition, the ionization potentials from the ground electronic state S_0 and the pure radiative lifetimes from the first spin-allowed electronic excitation S_1 of $M{\cdot}A_n$ heteroclusters. The size dependence of these spectroscopic observables provides information concerning the evolution of condensed matter properties with increasing the cluster size.

V.A. Spectral Shifts

The spectral shifts $\delta\nu$ for the $S_0 \to S_1$ electronic origin of $M{\cdot}A_n$ clusters (A = heavy rare-gas atom or nonpolar hydrocarbon) relative to the bare M molecule, originate essentially from dispersive interactions and are invariably expected to be negative (i.e., red spectral shifts). Quantitative information concerning the size effects on the spectral shifts was experimentally and theoretically obtained for heteroclusters of 9,10-dichloro-anthracene (DCA) with Ar (Fig. 9) [85]. The size dependence of the spectral shift (Fig. 10) indicates a weak size dependence. Moving towards very large clusters, $\delta\nu$ for $DCA{\cdot}Ar_n$ (n = 1000±500) was determined [91] and the spectrum of DCA in the Ar matrix, which corresponds to the infinite cluster limit, was also measured [92]. The cluster size equation for dispersive spectral shifts assumes the form [85]

$$\delta\nu(n) = \delta\nu(\infty) + An^{-1} \quad . \tag{V.1}$$

where $A \propto \rho$. In Figure 11 we plot the size dependence of the spectral shifts according to Eq. (V.1) for $DCA{\cdot}Ar_n$. The cluster size dependence of $\delta\nu(n)$ for $DCA{\cdot}Ar_n$ is (Fig. 11)

$$\delta\nu(n)/cm^{-1} = -610 + 4.2{\times}10^3\, n^{-1} \quad . \tag{V.2}$$

Thus the cluster size equation for $\delta\nu(n)$ is extremely well obeyed with: (i) $\delta\nu(\infty)$, which coincides with the experimental matrix value. (ii) The slope $A = 4.2{\times}10^3$ cm^{-1} is close to the value $A = 3.0{\times}10^3$ cm^{-1} estimated from the theory of dispersive spectral shifts together with ρ and R_0 estimates for the cluster taken to be equal to these attributes in the bulk.

V.B. Molecular and Electromagnetic Size Effects on Radiative Lifetimes

The pure radiative lifetime τ_r (i.e., the Einstein coefficient) for spontaneous emission from S_1 for the $M{\cdot}A_n$ herocluster is modified relative to the radiative lifetime τ_r^0 of the bare M molecule, due to intermolecular interactions between the multicenter transition monopoles of M and the rare-gas atom transition dipoles. The relative change $\Delta\tau_r/\tau_r^0 = (\tau_r-\tau_r^0)/\tau_r^0$ exhibits a marked structural sensitivity, serving as a useful spectroscopic probe for the interrogation of structure and isomerization dynamics of $M{\cdot}A_n$ heteroclusters [86,91].

Some experimental data [85,91,93] for $DCA{\cdot}Ar_n$ clusters (n = 1-54) and for n = 1000±500 [91] reveal that $\Delta\tau_r/\tau_r^0 > 0$ (Fig. 12). A cursory examination of these experimental results may lead us to the wrong conclusion that $\Delta\tau_r/\tau_r^0$ converges to a positive value in the bulk limit. In contrast, the experimental observation for DCA in solid Ar [92] reveals that $\Delta\tau_r/\tau_r^0 = -0.27$ in the matrix (Fig. 12). Clustering effects on pure radiative lifetimes fall into two distinct categories [86,91,93,94].

(A) The molecular domain. Small and many-atom clusters, whose size is considerably lower than the wavelength of radiation λ at the molecular frequency of M, where

$\Delta\tau_r/\tau_r^0 > 0$ [86,91,93,94].

(B)   The electromagnetic limit. Very large clusters whose size is comparable to or exceeds λ, where $\Delta\tau_r/\tau_r^0 < 0$ [86,94,95].

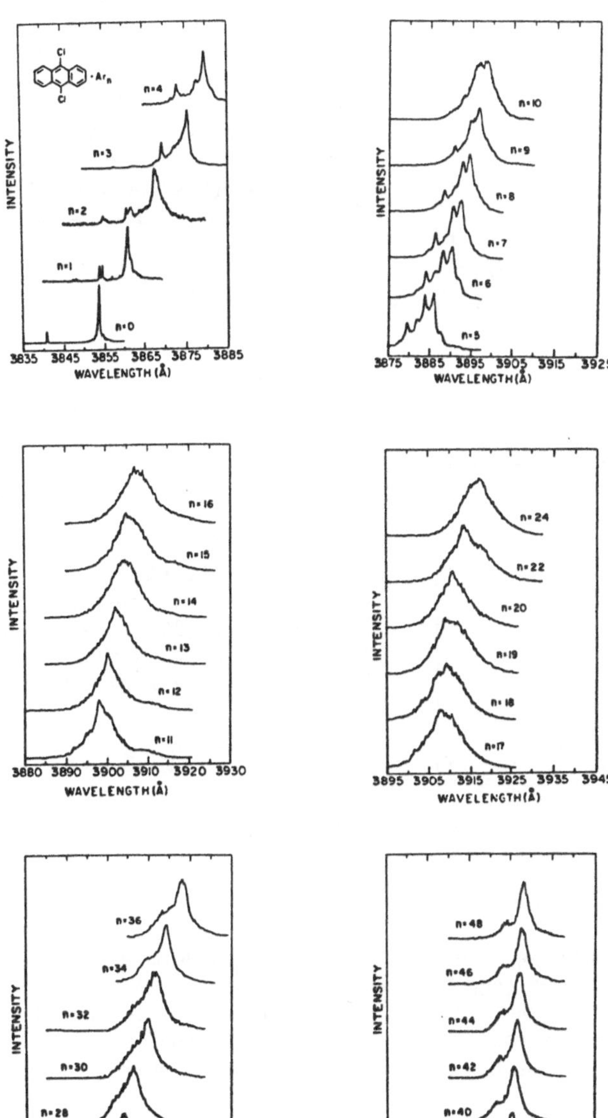

Figure 9. Resonance two-photon two-color ionization mass-resolved $S_0 \rightarrow S_1$ spectra of 9,10-dichloroanthracene·Ar$_n$ (n = 1-48) clusters [85].

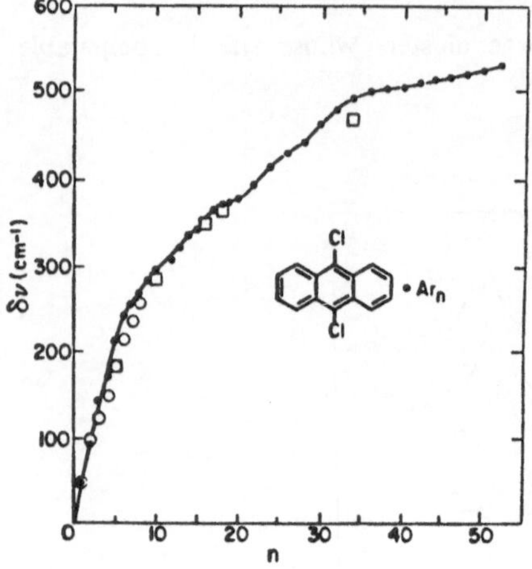

Figure 10. The size dependence of the red spectral shift of the main spectral feature of the $S_0 \rightarrow S_1$ electronic origin of DCA·Ar$_n$ (n = 1–55) clusters. Experimental data (•) from reference [85]. Calculated data from MD simulations of the spectra [85] are marked (o) at 3K and ([]) at finite temperatures (10K for n = 5, 12K for n = 10 and 30K for n = 16, 18 and 34).

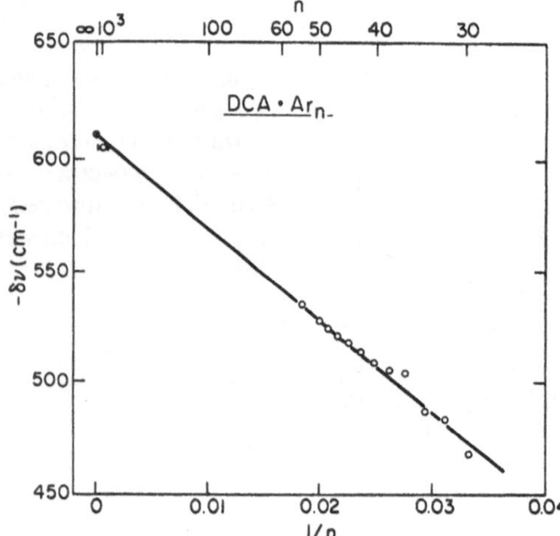

Figure 11. An analysis of the spectral shifts of DCA·Ar$_n$ according to the CSE, Eq. (V.1). The sources of experimental data are: n = 30–54 [85], n = 1000±500 [91] and Ar matrix data [92]. The straight line corresponds to Eq. (V.1).

It should be noted that r_r converges towards its bulk value only in range (B), where electromagnetic size effects dominate (Fig. 12). In contrast, other spectroscopic size effects, i.e., spectral shifts, homogeneous linewidths and ionization potential shifts of M in M·A$_n$ heteroclusters, have already converged towards their bulk value in range (B). An important conclusion emerging from this discussion is that distinct cluster size effects are not universal.

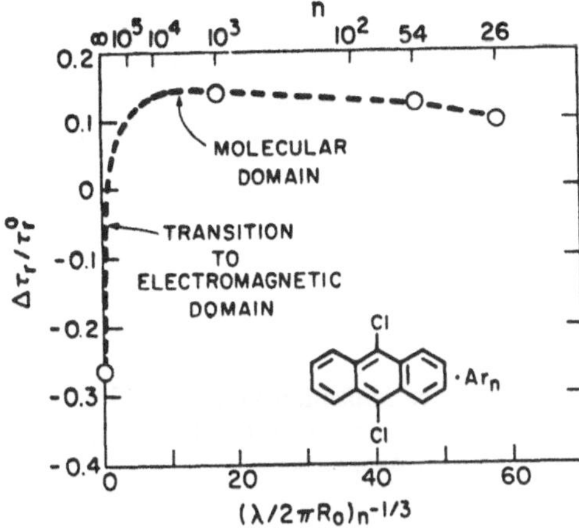

Figure 12. The size dependence of $\Delta\tau_r/\tau_r^0$ for DCA·Ar$_n$ (n = 30,54 [85], 1000±500 [91] and $n = \infty$, i.e., matrix [92]). The data in the range n = 30–1000 exhibit the molecular domain for this observable, while the transition to the electromagnetic domain is exhibited only for huge cluster sizes.

VI. SIZE EQUATIONS FOR MICROSURFACES

Surface states of excess electrons and of atomic and molecular impurities are ubiquitous in clusters. The formation of excess electron surface states [96-104] can be traced to the following causes:

(i) Strong short-range repulsive interactions between the excess electron and the cluster constituents, as is the case for electron–helium cluster interactions [101-103].

(ii) A large cluster reorganization energy for interior excess electron states, as is the case for $(H_2O)_n^-$ (n = 2-32) [14,96-98,100].

(iii) Structural effects which cause effect (ii), as is the case for surface states on $Na_{14}Cl_{13}^+$ clusters [104].

The formation of surface states of atomic and molecular impurities was experimentally demonstrated for SiF_4·Ar$_n$ by infrared vibrational spectroscopy [19] and for Xe·Ar$_n$ by electronic spectroscopy [28]. The spectral shifts for the electronic excitation of an impurity atom or molecule in moderately large clusters of benzene·Ar$_n$ (n = 30), benzene·$(N_2)_n$ (n = 40) and Aℓ·Ar$_n$ (n = 28) were found by Whetten et al to be definitely too small relative to the experimental bulk values [22]. These spectral shifts cannot be accounted for by the CSE, Eq. (V.1), for interior impurity states and a tentative possible assignment is to impurity surface states [22].

Cluster size equations can be advanced for the description of the ground-state and the excited-state energetics (and other properties) of cluster microsurfaces and their convergence with increasing the cluster size to the corresponding properties of the macrosurface. Two types of atomic or molecular surface impurity states were distinguished [19,28,31]:

(i) the impurity atom is located inside the cluster surface, i.e., within the first surface layer;

(ii) the impurity atom is located on the cluster surface.

For the surface state of molecules on heteroclusters a crossover from type (i) to type (ii) is expected with increasing the cluster size, provided that these surface states will be energetically or kinetically favored. Further experimental studies on the crossing cluster beams with impurity molecule beams, in conjunction with simulations are required. A rough description of the size dependence of an energetic property $\chi_s(n)$ of a surface state, e.g., the spectral shift or the ionization potential shift of an impurity molecule on a microsurface, can be obtained by considering the impurity located on a hemisphere of radius R_c. Under these circumstances the asymptotic observable for the macrosurface is $\chi_s(\infty) \simeq \chi(\infty)/2$, where $\chi(\infty)$ is the corresponding bulk value. Similarly the correction term originating from the contribution of the excluded volume will be of the form $A_s n^{-\beta}$ where $A_s \simeq A/2$, with A being the corresponding bulk value, while the exponent β is identical for the surface and for the interior state. Thus the approximate energetic size equation for surface states is

$$\chi_s(n) \simeq \chi(\infty)/2 + (A/2)\, n^{-\beta} \tag{VI.1}$$

with $\beta = 1$ for dispersive spectral shifts and $\beta = 1/3$ for ionization potential shifts. This relation provides a first step towards the quantitative elucidation of the energetics of excitation and ionization of molecular impurity states on microsurfaces.

VII. CONCLUDING REMARKS

The description of size effects on some ground state energetics, electronic excitations and electromagnetic attributes of clusters results in the quantification of two classes of phenomena. First, the gradual "transition" from the finite cluster to the infinite bulk system can be described in terms of the CSE. Second, the gradual "transition" from the cluster microsurface to a macrosurface can be described in terms of a surface CSE. Cluster size effects provide a unified (but not universal) description of the merging between the microscopic and macroscopic points of view for the description of large finite systems.

ACKNOWLEDGMENT

This research was supported by the Binational German-Israeli James Franck Program for Laser-Matter Interaction.

REFERENCES

1. Proceedings of the International Meeting on Small Particles and Inorganic Clusters. J. Phys. (Paris), Colloque C2, 38 (1977).

2. Proceedings of the Second International Meeting on Small Particles and Inorganic Clusters, Lausanne, 1980. Surface Sci. 106, 1-608 (1981).

3. Proceedings of Bunsengesellschaft Discussion Meeting on Experiments on Clusters, Koningstein, 1983. Ber. Bunsenges. Phys. Chem. 88 (1984).

4. Proceedings of the Third International Meeting on Small Prticles and Inorganic Clusters, Berlin, 1985. Surface Sci. 165, 1-1072 (1985).

5. Träger, F. and zu Putlitz, G. (eds.) (1986), Metal Clusters, Proceedings of an International Symposium, Heidelberg. Springer-Verlag, Berlin.

6. Jena, P., Rao, B. K., and Khanna, N. (eds.) (1986), The Physics and Chemistry of Small Clusters, NATO ASI Series. Plenum, New York.

7. Sugano, S., Nishima, Y., and Onishi, S. (eds.) (1987), Microclusters, Springer, Heidelberg.

8. Jortner, J., Pullman A., and Pullman, B. (eds.) (1987), Large Finite Systems, D. Reidel, Dordrecht.

9. Benedek, G., Martin, T. P., and Pacchioni. G. (eds.) (1988), Elemental and Molecular Clusters, Springer, Heidelberg.

10. Scoles, G. (ed.) (1990), The Chemical Physics of Atomic and Molecular Clusters, Proceedings of the International School of Physics "Enrico Fermi", Course CVII, North Holland, Amsterdam.

11. Proceedings of the Fourth International Symposium on Small Particles and Inorganic Clusters, Aix en Provence, 1988. Z. Phys. D12 (1989).

12. Echt, O. and Recknagel, E. (eds.) (1991), Proceedings of the Fifth International Symposium on Small Particles and Inorganic Clusters, Konstanz, 1990, Z. Phys. D19,20.

13. Friedel, J. (1977) J. Phys. (Paris) C2, 38, 1.

14. Jortner, J. (1984) Ber. der Bunsenges. Phys. Chem. 88, 188.

15. Bjørnholm, S. (1990) Contemporary Physics 31, 309.

16. Whetten, R. L., Cox, D. M., Trevor, D. J. and Kaldor, A. (1985) Phys. Rev. Lett. 65, 1494.

17. Leutwyler, S. and Jortner, J. (1987) J. Phys. Chem. 91, 5558.

18. Barnett, R. N., Landman, U., Cleveland, C. L., and Jortner, J. (1987) Phys. Rev. Lett. 59, 811.

19. Gu, X. J., Levandier, D. J., Zhang, K., Scoles, G., and Zhuang, D. (1990) J. Chem. Phys. 93, 4898.

20. Perera, L. and Amar, F.G. (1990) J. Chem. Phys. 93, 4884.

21. Adams, J. E. and Stratt, R. M. (1990) J. Chem. Phys. 93, 1332.

22. Xinling, Li., Hahn, M. Y., El-Shall, M. S., and Whetten, R. (1991) J. Phys. Chem. 95, 8524.

23. Perera, L. and Berkowitz, M. L. (1991) J. Chem. Phys. 95, 1954.

24. Stapelfeldt, J., Wörmer, J., and Möller, T. (1989) Phys. Rev. Lett. 62, 98.

25. Wörmer, J., Guzielski, J., Stapelfeldt, V., and Möller, T. (1989) Chem. Phys. Lett. 150, 321.

26. Wörmer, J., Guzielski, V., Stapelfeldt, V., Zimmerer, G., and Möller, T. (1990) Phys. Scr. 41, 490.

27. Wörmer, J. and Möller, T. (1991) Z. Phys. D20, 39.

28. Möller, T. (1991) Z. Phys. D20, 1.

29. Scharf, D., Jortner, J., and Landman, U. (1986) Chem. Phys. Lett. 126, 495.

30. Jortner, J., Scharf, D., Ben-Horin, N., Even, U., and Landman, U. (1990) reference 9, p. 148.

31. Scharf, D., Jortner, J., and Landman, U. (1988) J. Chem. Phys. 88, 4273.

32. Mackay, A. L. (1962) Acta Crystallogr. 15, 916.

33. Hoare, M. R. (1979) Adv. Chem. Phys. 40, 49.

34. Echt, O., Sattler, K., and Recknagel, E. (1981) Phys. Rev. Lett. 47, 1121.

35. Farges, F., de Feraudy, M. G., Raoult, B., and Torchet, G. (1983) J. Chem. Phys. 78, 5067.

36. Echt, O., Kandler, O., Leisner, T., Miehle, W., and Recknagel, E. (1989) Faraday Symposium No. 25 on Large Gas Phase Clusters.

37. Ekardt, W. (1984) Phys. Rev. B29, 1558.

38. Knight, W., Klemenger, K., de Heer, W., Saunders, W., Chou, M., and Cohen, M. L. (1984) Phys. Rev. Lett. 52, 2142.

39. de Heer, W., Knight, W., Chou, M., and Cohen, M. L. (1986) Solid State Phys. 40, 93.

40. Knight, W. G. (1990) reference 10, p. 413.

41. Pewerstorf, W., Bonacic-Keutecky, V., and Keutecky, J. (1988) J. Chem. Phys. 89, 5794.

42. Pantucci, P., Bonacic-Keutecky, V., and Keutecky, J. (1989) Z. Phys. D12, 307.

43. Stringari, S. (1985) Phys. Lett. A107, 36.

44. Stringari, S. and Treiner, J. (1987) J. Chem. Phys. 87, 5021.

45. Lewart, D. S., Pandharipande, V. T., and Pieper, S. R. (1988) Phys. Rev. B37, 4950.

46. Stringari, S. (1990) reference 10, p. 199.

47. Lang, N. D. and Kohn, W. (1970) Phys. Rev. B12, 4555.

48. Brechignac, C., Cahuzac, Ph., Carlier, F., and Leygnier, J. (1989) Phys. Rev. Lett. 63, 1368.

49. Cheshnovsky, O. (to be published).

50. Pandharipande, V. T., Zabolitzky, J. G., Pieper, S. C., Wiringa, R. B., and Helmbrecht, U. (1983) Phys. Rev. Lett. 50, 1676.

51. Pandharipande, V. R., Pieper, S. C., and Wiringa, R. B. (1986) Phys. Rev. B34, 4571.

52. Bohr, A. and Mottelson, B. R. (1975) Nuclear Structure, Vol. II, Benjamin, New York.

53. Brechignac, C., Cahuzac, Ph., Carlier, F., de Frutos, M., and Leygnier, J. (1990) J. Chem. Phys. 93, 7449.

54. Brechignac, C., Cahuzac, Ph., Carlier, F., de Frutos, M., and Legnier, J. (1991) Z. Phys. D19, 1.

55. Whetten, R. L., Schriver, K. E., Presson, J. L., and Hahn, M. Y. (1990) J. Chem. Soc. Farad. Trans. 86, 2375.

56. Kamke, W., de Vries, J., Krauss, J., Kaiser, E., Kamke, B., and Hertel, I. V. (1989) Z. Phys. D14, 339.

57. Gantefőr, G., Brőker, G., Holub-Krafe, E., and Ding, A. (1989) J. Chem. Phys. 91, 7972.

58. Schwentner, N., Koch, E. E., and Jortner, J. (1985) Electronic Excitations in Condensed Rare Gases, Springer Verlag, Berlin.

59. Herrman, A., Schumacher, E., and Wőste, L. (1978) J. Chem. Phys. 68, 2327.

60. Kappes, M. M., Schar, M., Radi, P., and Schumacher, E. (1986) J. Chem. Phys. 84, 1863.

61. Wood, D. M. (1981) Phys. Rev. Lett. 46, 749.

62. van Staveren, M. P. J., Bron, H. B., de Jongh, L. J., and Ishii, Y. (1987) Phys. Rev. B35, 7749.

63. Brus, L. E. (1983) J. Chem. Phys. 79, 556.

64. Perdew, J. P. (1988) Phys. Rev. B37, 6175.

65. Nitzan, A. (1987) reference 8, p. 333.

66. Cheshnovsky, O., Taylor, K. J., Conceicao, J., and Smalley, R. E. (1990) Phys. Rev. Lett. 64, 1785.

67. Taylor, K. J., Pettietle-Hall, C. L., Cheshnovsky, O., and Smalley, R. E. (in press) J. Chem. Phys.

68. Rademann, K., Kaiser, B., Even, U., and Hensel, F. (1987) Phys. Rev. Lett. 59, 2319.

69. Brechignac, C., Broyer, M., Chauzac, P., Delaretaz, G., Labastie, P., and Wőste, L. (1988) Phys. Rev. Lett. 60, 275.

70. Rademann, K. (1989) Ber. Bunsenges. Phys. Chem. 93, 653.

71. Haberland, H. and Langosch, H. (1991) Z. Phys. D19, 223.

72. Gracia, M. E., Pastor, G. M., and Bennemann, K. H. (1991) Z. Phys. D19, 219.

73. Mott, N.F. (1974) Metal-Insulator Transitions, Tayler and Francis, London.

74. Knox, R. S. (1963) Theory of Excitons, Academic Press, New York.

75. Rice, S.A. and Jortner, J. (1967) in F. Fox, M. Labes, and A. Weissberger (eds.), Physics and Chemistry of the Organic Solid State, Interscience Pulishers, New York, chap. 4.

76. Efros, Al. L. and Efros, A. L. (1982) Sov. Phys. Semicond. 16, 772.

77. Brus, L. E. (1984) J. Chem. Phys. 80, 4403; (1986) 90, 2555.

78. Nair, S. V., Sinha, S., and Rustagi, K. C. (1987) Phys. Rev. B35, 4098.

79. Kayanuma, Y. (1988) Phys. Rev. B38, 9797.

80. Rama Krishna, M. V. and Freiser, R. A. (1991) Phys. Rev. Lett. 67, 629.

81. Ondrechen, M. J., Berkovitch-Yellin, Z., and Jortner, J. (1981) J. Am. Chem. Soc. 103, 6586.

82. Even, U., Amirav, A., Leutwyler, S., Ondrechen, M. J., Berkovitch-Yellin, Z., and Jortner, J. (1982) J. Farad. Discuss. Chem. Soc. 73, 153.

83. Leutwyler, S. and Bosinger, J. (1990) Chem. Rev. 90, 489.

84. Amirav, A., Even, U., and Jortner, J. (1981) J. Chem. Phys. 75, 2489.

85. Ben-Horin, N., Bahatt, D., Even, U., and Jortner, J. (submitted).

86. Shalev, E., Ben-Horin, N., and Jortner, J. (1991) J. Chem. Phys. 94, 7757.

87. Even, U., Ben-Horin, N., and Jortner, J. (1989) Phys. Rev. Lett. 62, 140.

88. Hahn, M. M., and Whetten, R. L. (1988) Phys. Rev. Lett. 61, 1190.

89. Shalev, E., Ben-Horin, N., Even, U., and Jortner, J. (1991) J. Chem. Phys. 95, 3147.

90. Ben-Horin, N., Even, U., Jortner, J., and Leutwyler, S. (to be published).

91. Penner, A., Amirav, A., Jortner, J., and Nitzan, A. (1990) J. Chem. Phys. 93, 147.

92. Crepin, C. and Tramer, A. (1990) Chem. Phys. Lett. 170, 446.

93. Bahatt, D., Even, U., Shalev, E., Ben-Horin, N., and Jortner, J. (1991) Chem. Phys. 156, 223.

94. Gerstein, J. and Nitzan, A. (1991) J. Chem. Phys. 95, 686.

95. Konester, J. and Mukamel, S. (1989) Phys. Rev. A40, 1065.

96. Barnett, R. N., Landman, U., Cleveland, C. L., and Jortner, J. (1987) Phys. Rev. Lett. 59, 811.

97. Barnett, R. N., Landman, U., Cleveland, C. L., and Jortner, J. (1988) J. Chem. Phys. 88, 4429.

98. Barnett, R. N., Landman, U., Cleveland, C. L., and Jortner, J. (1988) Chem. Phys. Lett. 145, 382.

99. Barnett, R. N., Landman, U., Cleveland, C. L., Kestner, N. R., and Jortner, J. (1988) Chem. Phys. Lett. 148, 249.

100. Barnett, R. N., Landman, U., Scharf, D., and Jortner, J. (1989) Acct. Chem. Res. 22, 350.

101. Jortner, J., Scharf, D., and Landman, U. (1987) 'Excited State Spectroscopy', in V. M. Gassano and N. Terzi (eds.), Solids, Proceedings of the International School of Physics "Enrico Fermi", Course XCVI, 1985, North Holland, Amsterdam, p. 438.

102. Rama Krishna, M. and Whalley, K. (1988) Phys. Rev. B38, 11839.

103. Nubotovski, V. and Romanov, D. (1985) Sov. J. Low Temp. 11, 277.

104. Landman, U., Scharf, D., and Jortner, J. (1985) Phys. Rev. Lett. 54, 1860.

ATOMIC STRUCTURE OF CLUSTERS THROUGH CHEMICAL REACTIONS*

S. J. RILEY AND E. K. PARKS
Chemistry Division
Argonne National Laboratory
Argonne, IL, USA

ABSTRACT. Techniques for probing the structure of isolated metal clusters through adsorbate binding patterns will be described. The saturation of clusters with reagents such as ammonia and nitrogen provides information on the number of preferred binding sites for these reagents. The dependence of this number on cluster size can suggest particular structural themes. The equilibrium reaction with water can be used to identify cluster sizes having especially enhanced binding for the water molecule. Again, the sequence of cluster sizes showing such enhancement can point to specific cluster structure. The reaction with oxygen can identify cluster sizes having particularly high ionization potentials, and these can be compared to simple models for the electronic structure of metal clusters. Representative applications of these probes to iron, cobalt, nickel, and copper clusters will be discussed.

1. Introduction

This paper will be primarily concerned with chemical probes of the geometrical structure of isolated clusters of transition metal atoms. There is little question that the structure of a metal cluster can be very important in determining its chemical properties, so, conversely, it may be no surprise that the tables can be turned and chemistry can be used to determine structure. The detailed interactions between a cluster and an adsorbate molecule will depend on the microscopic arrangement of surface metal atoms in the vicinity of the binding site. With the use of appropriate adsorbate molecules and an understanding of how they bind to metal surfaces, it is possible to determine cluster structure from a careful investigation of the dependence of binding patterns on cluster size.

The traditional method of probing the structure (both geometrical and electronic) of isolated clusters is the observation of magic numbers in cluster mass spectra. These numbers correspond to cluster sizes having enhanced stability, hence enhanced intensity in the spectra. Classical examples include the alkali metals,[1] where closed electronic shells dominate cluster stability, and rare gas clusters,[2] where the closings of icosahedral geometrical shells and subshells produce clusters of enhanced stability. No such traditional magic numbers have ever been seen for transition metal clusters. This is because in order for cluster intensity to reflect stability, cluster growth must occur under equilibrium or quasiequilibrium conditions. The most common method for transition metal cluster generation is laser vaporization in a flow tube,[3] where cluster growth is kinetically,

P. Jena et al. (eds.), Physics and Chemistry of Finite Systems: From Clusters to Crystals, Vol. I, 19–28.
© 1992 *Kluwer Academic Publishers.*

rather than thermodynamically, controlled. This leads to cluster size distributions determined primarily by the statistical probability of cluster-atom collisions, and cluster stability plays little role in mass spectral intensity.

An alternative probe of cluster structure uses the dependence of cluster chemical properties on cluster size. For example, the number of binding sites for a given adsorbate molecule can be determined by counting the number of molecules that saturate the cluster. Systematics in this number may point to a particular structural theme. For cluster-adsorbate reactions that are at equilibrium, the relative amount of a reaction product will be a strong function of the cluster-adsorbate binding energy, and this may be quite sensitive to cluster structure. Even for reactions that are kinetically controlled, electronic closed shells might be identified by cluster sizes that are relatively inert. Examples of such chemical probes will be presented here, taken from our studies of the reactions of iron, cobalt, nickel, and copper clusters with ammonia, water, oxygen, and nitrogen.

2. Experimental Methods

Clusters are generated by pulsed laser vaporization of an isotopically enriched metal target that is located in a flow-tube reactor (FTR). The vaporized metal atoms are entrained and cooled by a continuous flow of helium carrier gas, leading to rapid cluster growth. At a point downstream of the target where cluster growth has terminated and clusters have cooled to the ambient temperature, reagent gas is introduced into the FTR. Since the pressure of helium is 20 Torr or more, the resulting reactions are usually of the association type, because the excess energy of the reaction complex can readily be dissipated by collisions with the helium. At the end of the FTR, the clusters and their reaction products expand through a nozzle into vacuum and are formed into a molecular beam. After traversing several stages of differential pumping, the clusters arrive at a time-of-flight mass spectrometer where they are ionized by a second pulsed laser. Typical experimental spectra are averaged over several thousand laser pulses to provide adequate signal-to-noise ratios.

In the early days of transition metal cluster chemistry research, the most common experiment was a determination of cluster reactivity as measured by the kinetics of reaction with a single (i.e., the first) reagent molecule. Study of the prototypical system, the reaction of iron clusters with hydrogen, showed that reaction rate constants can vary substantially with cluster size, changing by as much as two orders of magnitude with the addition of a single iron atom.[4] Similar behavior has since been reported for many other systems.[5] While such reactivity dependence indicates there might be substantial changes in cluster structure with size, in general these kinetics experiments do not provide any direct information on actual cluster structure, with the possible exception of electronic structure (see below).

More recently, a procedure for following the entire range of interaction between a cluster and an adsorbate has been developed.[6] It consists of determining the coverage, as measured by the average number $\bar{m}$ of adsorbate molecules on a cluster, as a function of the partial pressure of reagent in the FTR. A plot of $\bar{m}$ vs log (pressure), an *uptake* plot, provides a map of the coverage dependence of the cluster-adsorbate interaction. For example, at low coverages it may be that the cluster-adsorbate interaction is sufficiently strong that the reaction is kinetically controlled. This region is characterized by a rapid increase in $\bar{m}$ with log (pressure). At some point the adsorbate binding energy becomes weak enough that the reaction enters an equilibrium regime, in which the uptake is much more gradual. Then at higher pressure a plateau may appear, a region where coverage does not depend on pressure. This implies that the binding energy for the next site to be filled is significantly lower than that of the last site, meaning that the the primary binding sites have been saturated. Such a determination of cluster saturation can be an important clue as to cluster structure. If the preferred type of binding site for a given adsorbate is known, then saturation data provide a count of such sites. The dependence of this number on cluster size may display systematics that point to a particular structure, as will be shown below.

For reactions that are at equilibrium, the relevant quantitative measure of reactivity is the equilibrium constant K_{eq}. This can be calculated from the ratio of mass spectral intensities and the measured reagent partial pressure in the FTR. Since $RTlnK_{eq} = -\Delta G° = -\Delta H° + T\Delta S°$, variations in K_{eq} are directly affected by variations in $-\Delta H°$, i.e., the cluster-adsorbate binding energy. In fact, with a systematic determination of the temperature dependence of K_{eq}, $-\Delta H°$ and $-\Delta S°$ can be extracted directly via a van't Hoff analysis.[7] Again, the dependence of these thermodynamic quantities on cluster size can provide information on cluster structure.

3. Results and Discussion

Our studies have identified four reagents whose reactions with metal clusters can probe cluster structure, albeit in somewhat different ways. Ammonia and nitrogen are convenient reagents for cluster saturation: Since they are gases, high partial pressures can be established in the FTR and cluster saturation is easy to accomplish. The equilibrium reaction with water can be used to probe changes in cluster-adsorbate binding energies. The reaction with oxygen can, in some cases, provide information on cluster electronic structure. Examples of the use of each of these reagents will be presented.

3.1. AMMONIA

Surface science studies on the adsorption of ammonia on metals have shown that the ammonia dipole binds nitrogen-end-down with the molecular axis normal to the metal surface.[8] The binding is partly electrostatic, from dipole-induced-dipole forces, and partly due to σ donation of the nitrogen atom's lone pair electrons into the metal's d orbitals. The electrostatic forces are maximized over atop (single-atom) sites, since for these sites the positive charge of the metal atom nucleus is directly below the molecule, and the induced dipole is larger than for multiatom sites. Furthermore, the more exposed the metal atom, the stronger the metal-ammonia interaction, since electronic charge tends to be drawn away from such atoms into the surrounding surface. This attracts the ammonia dipole more strongly, and reduces the repulsive forces associated with the σ donation. Thus we assume that as a cluster adsorbs ammonia molecules, the sites that will be preferentially filled first are those on the most exposed metal atoms, i.e., apex atoms or atoms that have minimum metal-metal coordination. If an uptake plot shows a clear plateau, then the number of ammonia molecules at the plateau provides a count of the number of apex atoms. Knowing this number, we can begin to suggest possible cluster structures.

Examples of such data are shown in Figure 1 for Ni_{19}[9] and Co_{19} reacting with ammonia at 25 °C. It is immediately apparent that under these conditions the way these two clusters take up ammonia is quite different. Ni_{19} shows a plateau at $\bar{m} = 12$, while for Co_{19} there is a plateau at $\bar{m} = 6$. For a 19-atom cluster, there are two possible atomic configurations that provide a clear choice as to the number of apex atoms. These are the fcc octahedron and the double icosahedron. Models of such structures are shown in Figure 2. In this figure, the atoms having minimum metal-metal coordination are shown in gray. It is clear that the octahedron has six such sites, while the double icosahedron has twelve. This would suggest that a good candidate for Co_{19} would be the octahedron, and for Ni_{19} the double icosahedron. For the octahedron, the apex atoms have only four nearest neighbor metal atoms, while the other surface atoms have seven. This large difference in metal-metal coordination explains the extensive plateau in the Co_{19} uptake plot: The cluster-ammonia binding energy for a non-apex site is substantially lower than for an apex site. For Ni_{19}, the 12 atoms on the caps have six nearest neighbors, while the atoms around the "waist" have eight. Also, the 12 preferred sites are quite close together, resulting in higher repulsion between bound ammonia molecules than for the octahedron. This explains why a much higher ammonia pressure is required to saturate Ni_{19} than Co_{19}.

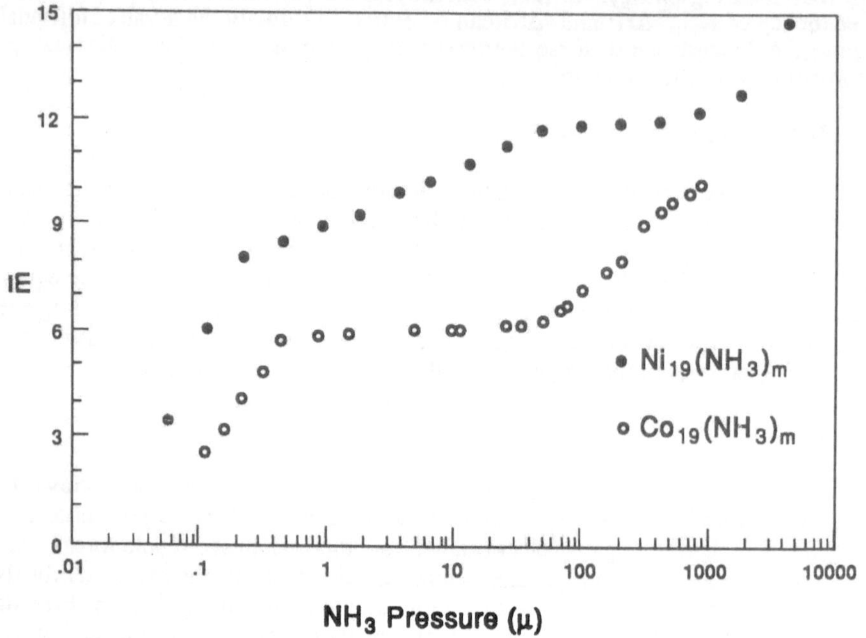

Figure 1. Uptake plots for the reaction of Ni_{19} and Co_{19} with ammonia. Nickel data adapted from Ref. 9.

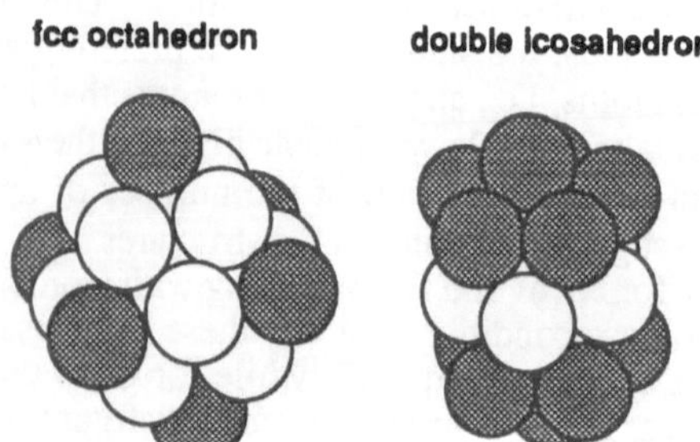

Figure 2. Models of the 19-atom octahedron and double icosahedron clusters. Primary ammonia binding sites are shown in gray.

Similar ammonia saturation experiments have been done for larger cobalt[10] and nickel[9] clusters, and for cobalt and nickel clusters saturated with hydrogen.[11] In many cases, for ammoniated clusters in the 50- to 200-atom size range there is evidence for clusters having icosahedral structure. The evidence comes from a sequence of magic numbers corresponding to the filling of icosahedral shells and subshells, just as for rare gas clusters. In the case of nickel and cobalt clusters the magic sizes are characterized by a *minimum* in the number of ammonia molecules bound at saturation, and often this minimum is 12. An inspection of models of these icosahedral closed (sub)shell clusters explains why. Clearly, the closed icosahedral shell clusters at 55 and 147 atoms have 12 apex atoms, and these clusters would be expected to preferentially bind 12 ammonia molecules. The addition of another metal atom to the closed shell will provide a 13th ammonia binding site, since the new atom will sit outside the closed shell and in fact will have only three nearest neighbor metal atoms. The removal of an atom from the closed icosahedron will most likely be from an apex position, since these atoms have the lowest coordination. Although this might appear to reduce the number of ammonia binding sites, in fact in most cases it increases it, because the "hole" left at the apex position now has enough room around it to accommodate two ammonia molecules, for a total once again of 13. Thus the closing of an icosahedral shell provides a local minimum (12) in the number of binding sites. A consideration of the arrangement of metal atoms on the surfaces of closed subshell icosahedral clusters shows that in many cases these clusters will likewise provide a local minimum, often 12, in the number of preferred ammonia binding sites. Thus the sequence of $\bar{m}$ minima for cobalt and nickel clusters, both hydrogenated and nonhydrogenated, points to cluster growth schemes based on icosahedral packing.

3.2. WATER

While the cluster-ammonia reactions provide evidence for icosahedral structure, there is certainly the possibility that saturation with ammonia might change cluster structure, i.e., that the bare clusters (or the purely hydrogenated clusters) might not be icosahedral in structure. This would be the cluster analog of surface reconstruction, and in fact there is evidence for such effects for iron, cobalt, and nickel clusters in the 19- to 34-atom size range.[10] There is, however, a less invasive structural probe, the reaction with water, that points to icosahedral structure for nonammoniated cobalt and nickel clusters in most of the size range covered by the ammonia probe. The significant aspect of the reaction of water with these clusters is that the cluster-water binding energy is sufficiently low that under normal experimental conditions the reaction is an equilibrium one. This means that changes in the water binding energy comparable to or greater than RT will result in significant changes in the relative amount of cluster hydrate product, since the equilibrium constant is exponentially dependent on $-\Delta H°/RT$ for the reaction.

That such large changes occur can be seen in Figure 3, which shows a portion of the mass spectrum recorded when hydrogenated nickel clusters are reacted with 0.1 mTorr of H_2O in the FTR. It is clear that certain cluster sizes, noted in the figure, have a dramatically enhanced propensity to bind water. In fact, an analysis[11] of data such as this indicates that these special cluster sizes can have $-\Delta G°$ values as much as 2 kcal/mol greater than other clusters. Furthermore, the sequence of the "magic" cluster sizes having water binding maxima is often found to comprise clusters with just one metal atom more than the cluster sizes having *minima* in ammonia binding; in other words, the sequence of icosahedral closed subshells. It would appear, then, that the pattern of water binding is also somehow related to the icosahedral sequence.

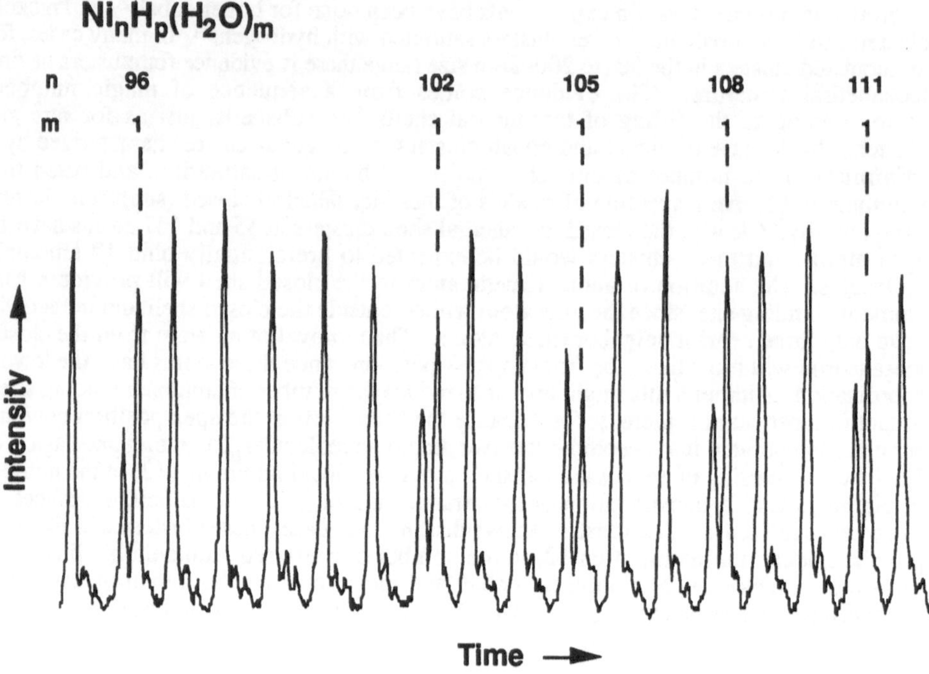

Figure 3. Portion of the mass spectrum recorded when hydrogenated nickel clusters are reacted with 0.1 mTorr water. The mass peaks corresponding to clusters $Ni_nH_p(H_2O)_m$ having enhanced binding for water are annotated with n, m values.

This relationship becomes obvious if we consider the nature of the cluster-water interaction. Like ammonia, water binds through the lone pair electrons, now on the oxygen, and is expected to prefer atop sites with metal atoms of lowest metal-metal coordination. As was pointed out above, a single atom outside a closed icosahedral shell or subshell will in fact have the lowest coordination of all atoms on the cluster surface, and in this case will provide the strongest binding site for the water molecule. Thus the sequence of "closed-subshell-plus-one" clusters having enhanced water binding is consistent with icosahedral structure for these clusters. This means that for the most part saturation with ammonia does not change cluster structure.

Studies[9-11] of the reactions of bare and hydrogenated cobalt and nickel clusters with water show that, just as for the ammonia probe, many clusters in the 50- to 200-atom size range are icosahedral in structure. There are notable exceptions, however, in particular the 147-atom nickel cluster, which does not appear to be the closed-shell icosahedron. In most cases the hydrogenated clusters show more pronounced icosahedral features in their binding patterns than the nonhydrogenated ones: They generally have deeper $\bar{m}$ minima and have more enhanced water binding maxima. This most likely reflects the fact that the hydrogen atoms, while not competing directly for binding sites with the ammonia or water, do weaken the relative binding strength to the non-apex sites on the faces and edges of the clusters. Interestingly, these two probes provide no evidence at all that iron clusters in the 50- to 200-atom size range are icosahedral.

3.3. OXYGEN

As a general rule, the oxidation of transition metal clusters is a process quite different from the simple surface chemisorption reactions with molecules such as ammonia and water. Oxidation can result in a complete reconstruction of the cluster, yielding what is essentially a cluster of a metal oxide, and can be so exothermic that even etching of the cluster can occur. Earlier studies[12] of transition metal cluster oxidation showed that a constant metal-oxygen ratio was established for remarkably small clusters, but provided no information on cluster structure. In fact, it is unlikely that oxidation would be sensitive to geometrical structure, since it is such a violent reaction. However, it would appear that in favorable cases cluster oxidation can be a probe of *electronic* structure.

This is illustrated by the oxidation of copper clusters. Figure 4 shows a portion of the mass spectrum recorded when copper clusters are reacted with 4.2 mTorr of O_2 in the FTR.[13] [Actually, to generate this spectrum 0.9 mTorr of water was also added to the FTR. Since the mass of four oxygen atoms is close to that of a copper atom, the cluster oxide spectrum is very congested and difficult to interpret. The addition of water has the effect of spreading the products out over a large number of $Cu_nO_m(H_2O)_p^+$ peaks, since the oxides readily adsorb water.] The peaks that stand out in Figure 4 represent those cluster sizes that are unreactive with oxygen. It is most likely that the reactivity in this system is dominated by cluster electronic properties, in particular, ionization potentials (IPs). The reaction is probably governed by an electron transfer from the cluster to the oxygen molecule, so that the higher a cluster's IP the lower its reactivity. Thus the pattern of unreactive clusters in Figure 4 reflects a pattern of clusters with maximum IPs.

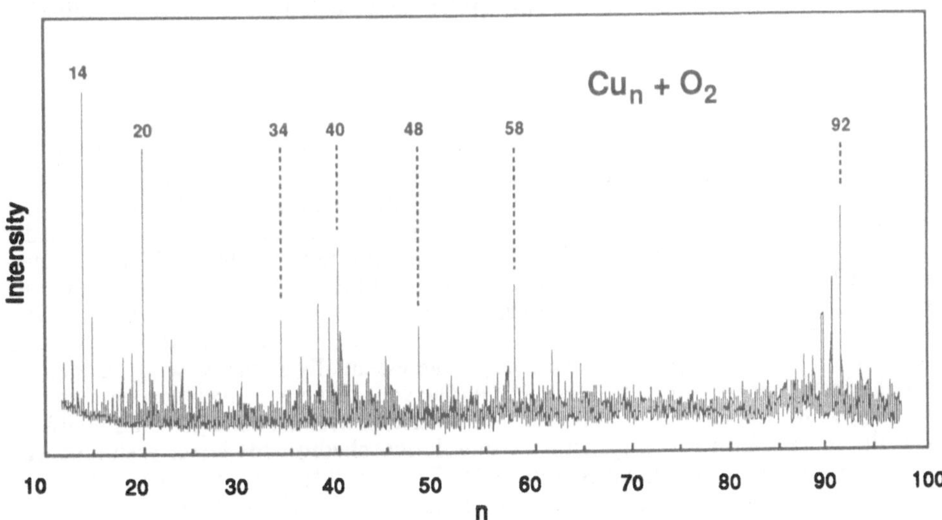

Figure 4. Mass spectrum resulting from the reaction of copper clusters with 4.2 mTorr O_2 and 0.9 mTorr H_2O. Labeled intense peaks correspond to unreactive bare clusters. Adapted from Ref. 13.

The well-known spherical drop model[14] has been quite successful in explaining the magic numbers seen in mass spectra of alkali metal clusters. Based on the idea of electrons moving in a spherical-well potential, it predicts electronic orbitals having filled shells for clusters with, among other numbers, 20, 34, 40, 58, and 92 electrons. Such clusters have higher IPs and enhanced stability compared to other cluster sizes. As can be seen in Figure 4, copper clusters with these numbers of atoms likewise have high IPs, suggesting that copper clusters can also be explained by the spherical drop model, assuming that copper atoms are monovalent (i.e., contribute one electron per atom to the cluster electronic orbitals). However, there are at least two clusters, Cu_{14} and Cu_{48}, showing low reactivity that cannot be explained by the simple spherical model. At this point, the question of geometrical structure becomes important.

For alkali metals, there is evidence that under most experimental conditions clusters are highly fluxional, if not actually liquid.[15] In principle the nuclei can adjust their geometry to accommodate closed electronic shells, if that results in the lowest total energy, and strong magic numbers are seen for the predicted spherical clusters. Copper clusters, however, have more rigid geometries. This is shown by the ammonia and water probe reactions. Water binding maxima are found for just those closed (sub)shell-plus-one clusters expected for icosahedral structure, at least in the 40- to 100-atom size range. This geometrical structure can impose a nonsphericity on the system that produces additional electronic subshell closings. For example, calculations[14] show that there is a subshell closing at 14 electrons associated with a spheroidal distortion. These calculations do not predict a closing at n = 48, so the relative unreactivity of this cluster may represent a different type of distortion, or may result from a more complex interaction between geometrical and electronic structure that affects chemical properties. The copper cluster system is the first one in which information has been provided on both electronic and geometrical structure. A better understanding of the interplay between these two structural properties will hopefully provide us with a fuller understanding of the basic relationships between cluster structure and chemical reactivity.

3.4. NITROGEN

Studies of the reaction of nitrogen with bulk nickel surfaces suggest that nitrogen chemisorption is nondissociative, and that the N_2 molecules are oriented perpendicularly to the metal surface and bind to single-atom sites.[16] Nitrogen also appears to react with small nickel clusters via a nondissociative chemisorption process. Cluster saturation studies suggest that the N_2 also binds to single atoms, and furthermore that each surface metal atom can accommodate a nitrogen molecule. This is illustrated by Figure 5, which shows portions of mass spectra recorded for nickel clusters reacting with 553 mTorr N_2 at 20 °C (upper spectrum) and with 985 mTorr N_2 at -40 °C (lower spectrum). At these pressures and temperatures, clusters in these size ranges are essentially saturated with nitrogen. As is highlighted in the figure, Ni_7 is seen to saturate with seven N_2 molecules, Ni_{13} with 12, and Ni_{19} with 17. Such saturation levels are consistent with the pentagonal bipyramid for Ni_7, the icosahedron for Ni_{13}, and the double icosahedron for Ni_{19}, assuming each surface metal atom can bind an N_2 molecule. This result for Ni_{19} is consistent with the ammonia probe, as discussed above. Notice also that if the 19-atom cluster were the octahedron (Figure 2) we would expect a saturation at 18 N_2 molecules, since the octahedron has 18 surface metal atoms. Furthermore, a detailed uptake plot of N_2 on Ni_{19} shows plateaus at both $\bar{m} = 12$ and $\bar{m} = 17$, suggesting that the first 12 sites, just those cap sites preferred by ammonia, are more strongly bound than the last five, waist-atom sites.

As can be seen in Figure 5, something interesting happens at Ni_{14}, which saturates at 14 N_2 molecules. If this cluster were simply the 13-atom one with an additional metal atom on the surface, it might be expected to bind 13 or fewer molecules, but not 14. The binding of 14 suggests that the 14-atom cluster, at least when saturated with N_2 molecules, adopts some completely different structure in which there is no central metal atom. Theoretical simulation studies[17] in our group indicate that Ni_{14} is substantially more fluxional than the icosahedral Ni_{13}. The exothermicity of the reaction with N_2 may well be sufficient to heat the cluster enough to allow a restructuring that exposes the central metal atom, allowing the binding of the maximum number of N_2 molecules. This type of restructuring might, in fact, be quite common for non-closed-shell clusters, and may play a dominant role in determining the actual structure of many metal clusters with adsorbates bound to their surfaces.

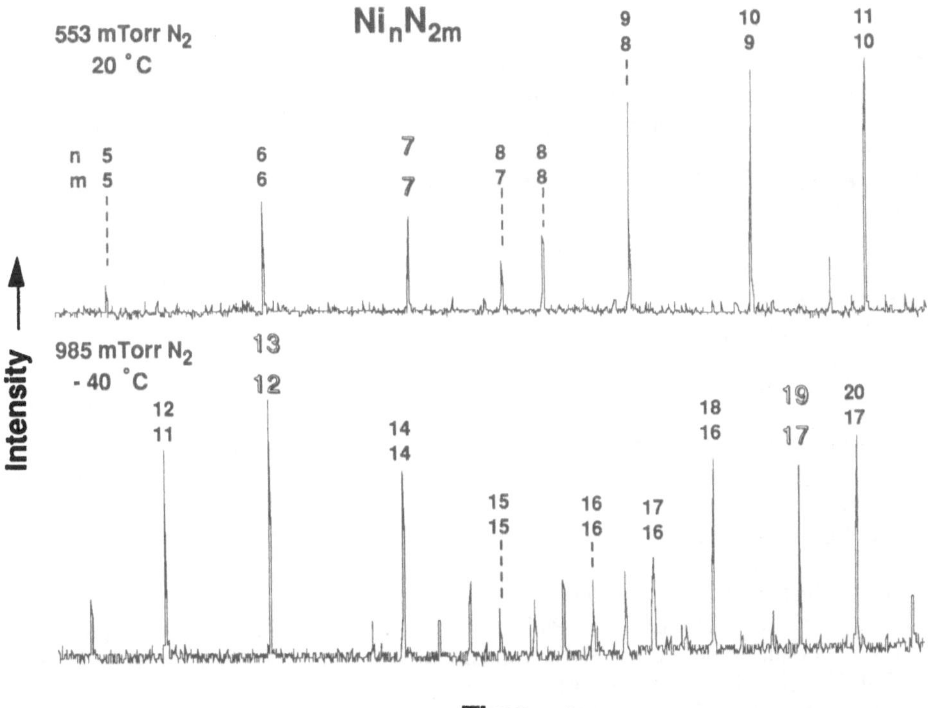

Figure 5. Mass spectra illustrating the saturation of small nickel clusters with nitrogen.

4. Conclusions

This paper has presented an overview of chemical probes of metal cluster structure. To date, these probes have been applied to iron, cobalt, nickel, and copper clusters, and have shown evidence for icosahedral packing in the latter three of these metals. There is every reason to expect these probes will work for other metals, particularly those of the second and third rows of the periodic table. However, it must be noted that probes such as these have yet to reveal binding pattern sequences characteristic of nonicosahedral structure. This may mean that the differences in the numbers of nearest neighbors between atoms in a closed (sub)shell and atoms outside it are not large enough in the case of nonicosahedral packing for adsorbate binding energy differences to be significant. Or, it may mean that for metals not showing icosahedral structure, such as iron, clusters have several isomers, effectively hiding magic numbers. Clearly additional work will be needed before we have a complete picture of the geometrical structure of isolated transition metal clusters.

[*] Work performed under the auspices of the Office of Basic Energy Sciences, Division of Chemical Science, US-DOE under contract number W-31-109-ENG-38.

[1] See, for example, M. L. Cohen, M. Y. Chou, W. D. Knight, and W. A. deHeer, J. Phys. Chem. **91**, 3141 (1987), and references therein.

[2] See, for example, W. Miehle, O. Kandler, T. Leisner, and O. Echt, J. Chem. Phys. **91**, 5940 (1989), and references therein.

[3] T. G. Deitz, M. A. Duncan, D. E. Powers, and R. E. Smalley, J. Chem. Phys. **74**, 6511 (1981).

[4] S. C. Richtsmeier, E. K. Parks, K. Liu, L. G. Pobo, and S. J. Riley, J. Chem. Phys. **82**, 3659 (1985); R. L. Whetten, D. M. Cox, D. J. Trevor, and A. Kaldor, Phys. Rev. Lett. **54**, 1494 (1985); S. J. Riley and E. K. Parks, in "Physics and Chemistry of Small Clusters," edited by P. Jena, B. K. Rao, and S. N. Khanna (Plenum, New York, 1987) p. 727.

[5] See, for example, A. Kaldor, D. M. Cox, and M. R. Zakin, Adv. Chem. Phys. **70**, 211 (1988).

[6] E. K. Parks, G. C. Nieman, L. G. Pobo, and S. J. Riley, J. Chem. Phys. **86**, 1066 (1987).

[7] B. J. Winter, T. D. Klots, E. K. Parks, and S. J. Riley, Z. Phys D **19**, 381 (1991).

[8] C. W. Seabury, T. N. Rhodin, R. J. Purtell, and R. P. Merrill, Surf. Sci. **93**, 117 (1980).

[9] E. K. Parks, B. J. Winter, T. D. Klots, and S. J. Riley, J. Chem. Phys. **94**, 1882 (1991).

[10] Unpublished results from this laboratory.

[11] T. D. Klots, B. J. Winter, E. K. Parks, and S. J. Riley, J. Chem. Phys. (in press).

[12] G. C. Nieman, E. K. Parks, S. C. Richtsmeier, K. Liu, L. G. Pobo, and S. J. Riley, High Temp. Sci. **22**, 115 (1986).

[13] B. J. Winter, E. K. Parks, and S. J. Riley, J. Chem. Phys. 94, 8618 (1991).

[14] W. de Heer, W. D. Knight, M. Y. Chou, and M. L. Cohen, Solid State Phys. 40, **93** (1987).

[15] E. C. Honea, M. L. Homer, J. L. Persson, and R. L. Whetten, Chem. Phys. Lett. **171**, 147 (1990).

[16] K. Horn, J. DiNardo, W. Eberhardt, H. -J. Freund, and E. W. Plummer, Surf. Sci. **118**, 465 (1982).

[17] J. Jellinek and I. L. Garzón, Z. Phys. D. **20**, 239 (1991); Z. B. Güvenç and J. Jellinek, these Proceedings.

STRUCTURE OF DECAGONAL QUASICRYSTALS

A.R. Kortan, R.S. Becker, F.A. Thiel and H.S. Chen
AT&T Bell Laboratories, Murray Hill, New Jersey

ABSTRACT. Recent discoveries made in synthesizing stable quasicrystals allows us to apply conventional techniques in growing macroscopic single grains and use conventional probes to study the complex structure of this novel form of matter. The decagonal quasicrystals in particular appears to have a far less complicated phase diagram and an extreme phase stability compared to their three dimensional icosahedral counterparts. We have grown several millimeter size single grains and studied the structure of stable $Al_{65}Co_{20}Cu_{15}$ Decagonal quasicrystals, using High Resolution X-ray Diffraction (HRXD) and Scanning Tunneling Microscopy (STM). Unlike diffraction probes which provide large volume reciprocal space information, STM is a local probe and provides direct real-space information with atomic resolution. The two techniques are therefore complimentary and when combined becomes very powerful in structure determinations of complex systems like quasicrystals. We have been able to prepare clean single grain surfaces in UHV and resolve individual atoms in quasiperiodic ordering with STM. Images of 10-fold quasiperiodic surfaces exhibit very well defined five sets of mass density lines, separated by multiples of 72 degree rotations. Near perfect quasiperiodic order seem to extend beyond thousands of angstroms, and can be modeled by a Penrose lattice. All atomic layers observed have identical local structure unlike the two dissimilar layers of the approximant $Al_{13}Fe_4$ phase. High resolution x-ray scattering from single grains finds 2000 angstroms size defect free quasicrystalline layers stacked periodically.

1. INTRODUCTION

Quasicrystalline form of matter was first discovered by Schectman[1] while studying an Al-Mn compound under a transmission electron microscope. For this particular compound the quasicrystalline icosahedral phase was metastable, and micron size grains contained large number of defects, which made detailed quantitative structural determinations impossible. Over the past several years we have witnessed the discovery of tens of more quasicrystalline compounds in other systems, some of which were thermodynamically stable and structurally near defect free[2-5]. In particular Al-Co-Cu and Al-Co-Ni decagonal quasicrystals are found to be remarkably stable and defect free. The decagonal quasicrystals first discovered by Bendersky[6], are described by periodically stacked layers of two dimensional quasicrystalline layers. Despite their reduced dimensionality structure of decagonal quasicrystals are far less understood than their three dimensional counterparts. Intensive studies[7-23] of several different decagonal systems, however, have not resulted in a widely accepted model for the decagonal quasicrystals.

On the theory side numerous models are developed to understand the quasicrystalline symmetry. An early model developed by Schectman et.al.[24] and Stephens et.al.[25] makes use of the random packing of icosahedral units that yields a highly defective structure. Pauling[26] on the other hand explains the quasicrystalline structure by multiply twinned small crystallites. Levine and Steinhardt[27] generate a deterministic structure by projecting from a six-dimensional hyper-cubic lattice onto three dimensions and provide a description for a defect free quasicrystal.

29

P. Jena et al. (eds.), Physics and Chemistry of Finite Systems: From Clusters to Crystals, Vol. I, 29–38.
© 1992 *Kluwer Academic Publishers.*

Henley et. al.[28] and Audier et.al.[29] derived the icosahedral structure from a crystalline neighbouring phase. Widom et.al.[30] and Strendburg et.al.[31] discovers the importance of the entropic terms in their recent Monte-Carlo simulations of the random tiling model. In light of the latest observations on perfect quasicrystals[32-34] the correct modelling will likely be provided by either the perfect[27] or random tiling models.

Our present knowledge of quasicrystals is limited to observations made by x-ray or electron diffraction techniques. These diffraction techniques probe the atomic structure through a large area Fourier transform and therefore particularly sensitive to regularities in the structure. In quasicrystals the issue often becomes the degree of perfection or the nature of disorder. In diffraction experiments the part of the intensity that contain this information is spread over a large volume in q-space which makes it very difficult to interpret. Because of the large sample volumes required, diffraction probes most often can not make a distinction between a multiply twinned structure and a true quasicrystalline structure. The scanning tunneling microscope (STM) on the other hand is a local real-space probe that provide atomic resolution surface imaging. Here we present single crystal x-ray diffraction measurements of the reciprocal space, and the high resolution STM images of the clean surface, for the single grain decagonal $Al_{65}Co_{20}Cu_{15}$ quasicrystal[16]. The (HRXD) measurements and the STM images provide strong evidence for the two-dimensional tiling model and rule out multiple twinning as the source for the surface features. Strikingly, all the evidence suggests that the surface structure is an exact termination of the bulk quasiperiodic structure. This is supported by the agreement of STM measurements with the HRXD and also by the chemical analysis of the surface layers by Auger Electron Spectroscopy (AES).

2. RESULTS

The decagonal phase of the $Al_{65}Co_{20}Cu_{15}$ compound is thermodynamically stable[14]. We employed conventional techniques to grow several millimeter size facetted single grains which exhibit a colomnar decagonal prismatic morphology with the colomn axis being parallel to the periodic direction. The x-ray sample was chosen to match the absorbtion length and the STM sample was oriented with x-rays, cut and polished to optical flatness to better than one degree to the ten-fold quasicrystalline planes.

2.1. SINGLE CRYSTAL X-RAY DIFFRACTION

These measurements are carried out on a triple-axes, four circle ganiometer equipped with a 12kW rotating anode generator. Resolution of 8×10^{-4} $\mathring{A}^{-1}$ (10^{-2} $\mathring{A}^{-1}$) in the longitutional and 3×10^{-5} $\mathring{A}^{-1}$ (5×10^{-3} $\mathring{A}^{-1}$) in the transverse directions are achieved by a pair of flat Ge(111) (Pyrolitic Graphite) monochromator-analyser crystals. A sample of approximately 1 mm^3 single grain with well defined facets is illuminated with monochromatic x-rays and the reciprocal space is explored to 8 A^{-1}. The reciprocal space is found to consist of planar quasicrystalline reflections which are stacked periodically with extinctions. The indexing procedure[28] employs least path criterion to eliminate alternate indexing yet makes use of all the five inplane unit vectors to preserve the symmetry. Figure 1 shows three radial scans within the quasiperiodic planes separated by 18 degree rotations. Remarkably, even on a logarithmic scale, scans made along two P directions separated by 36 degrees are identical indicating that there is a true 10 fold rotational symmetry. The scan shown in the middle is the D-axis scan and is precisely separated from the P-axis by a 18 degree rotation. All the peak positions shown in Figure 1 are uniquely indexed by the six Miller indices that describes the decagonal quasicrystals. High intensity x-ray source allows us to do these measurements over five orders of magnitude dynamic scale in intensity, and resolve the weak peaks that are not measurable by electron microscopies.

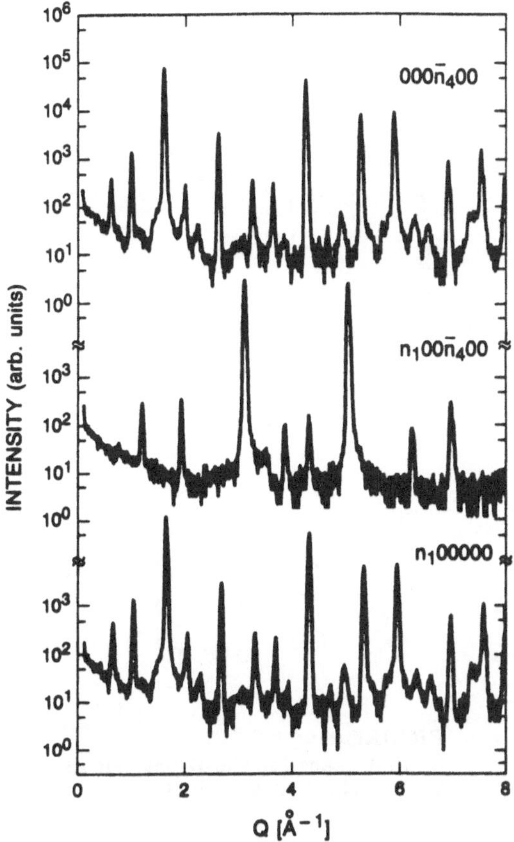

FIGURE 1.
Single crystal x-ray radial scans made along the three principal symmetry axis in the quasicrystalline plane. Scans are separated by 18 degree rotations. Miller indices in the n variable form denote the scan direction. The n1 and n4 scans are P-pattern scans and the center scan is for the D-pattern.

Single crystal scans made along the periodic axis are shown in Figure 2 and Figure 3 for D and P patterns respectively. The scans $00000n_6$ contains no quasicrystalline inplane components in both Figures. Other scans in each figure include an inplane component that increases by the golden section constant. As shown in Figure 2 the scans made in the D-pattern reveal a $8.26\mathring{A}$ periodicity whenever the scans include a quasiperiodic plane component as shown by the middle two scans in the figure. In the specular scan $00000n_6$ and in the P-pattern scans (Figure 3), only $4.13\mathring{A}$ periodicity is measurable, indicating that when viewed along the parallel planes the repeat distance of the structure becomes $4.13\mathring{A}$. Line width measurements made by high resolution on the peaks shown in Figure 1 indicate that the structure is near perfect decagonal with a longitutional correlation length of $2000\mathring{A}$.

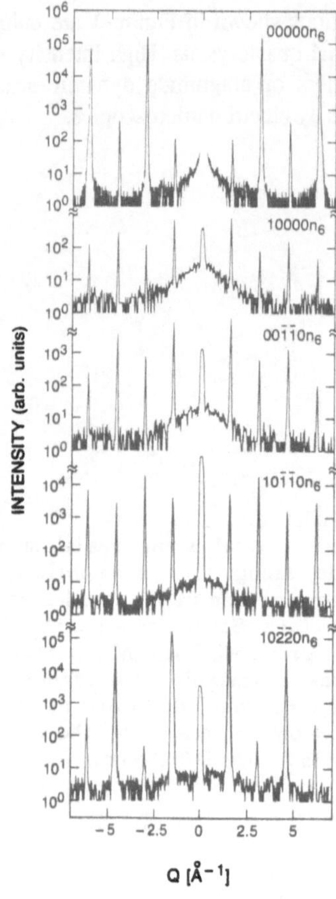

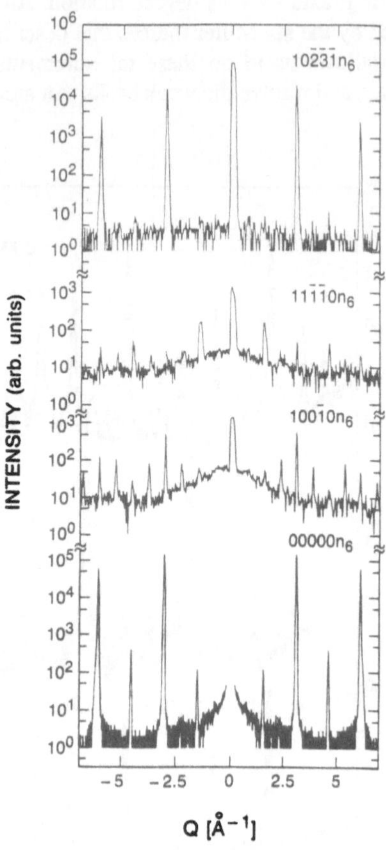

FIGURE 2.
Single crystal x-ray scans made along the periodic direction for the P-pattern.

FIGURE 3
Similar scans for the D-pattern revealing the 8.26 A periodicity.

2.2. SURFACE STRUCTURE BY SCANNING TUNNELING MICROSCOPY

The ultra-high vacuum (UHV) STM used in these experiments has been described elsewhere[35]. The UHV chamber includes facilities for sample characterization using low energy electron diffraction (LEED) and Auger electron spectroscopy (AES), along with an ion gun for surface cleaning by ion sputtering. The quasicrystal sample surface was cleaned by repeated cycles of ion bombartment (Ne+, 1.0 keV) and annealing at temperatures of up to 700 C for periods of 10 to 18 hours. AES analysis of the surface constituents after the STM measurements performed indicates that the surface stochiometry is essentially unchanged from the bulk composition. Figure 4 shows an unprocessed, constant current tunneling image (bias voltage (Vb) 100meV, demanded tunneling current 1nA) of a 150Å square region of the clean surface, at an inclined angle. The observed features on these apparently low free energy surfaces consist of terraces flat to +0.05Å, separated by 1.94+0.10 Å step heights. On these terraces point and ring like structures are visible. Four atomic steps are prominant, along with a network of mass density lines criss crossing the surface. The irregularly shaped ~10 angstroms size white features observed on some terraces are identified as residual oxide islands by AES. Presence of these oxide islands obscure the surface structure at few isolated locations. The local structure does not change from layer to layer. The mass density lines become highly visible when the unprocessed image is viewed at a glancing angle, as shown in Figure 4. Close examination shows that these mass density lines corresponding to atomic positions run along ten symmetry directions spaced at 36 degrees and the spacing between these lines are quasiperiodic. These lines are therefore analogs of the Ammann quasilattice lines that are used to decorate Penrose tiles to generate a linear quasiperiodic pentagrid[36]. Continuity of these lines therefore indicate a defect free quasicrystalline structure.

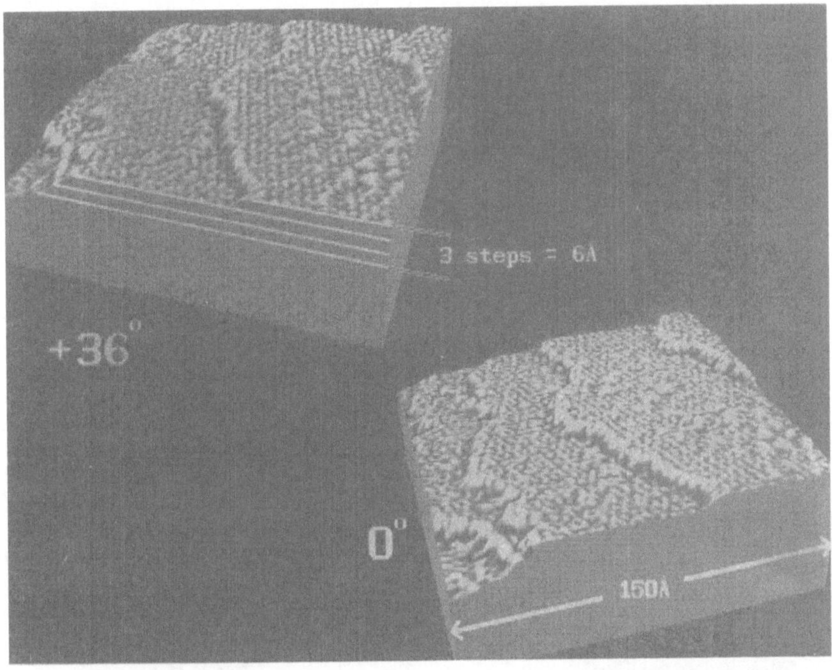

FIGURE 4
Unprocessed raw STM images of the 10-fold quasicrystalline surface of the Al-Cu-Co decagonal phase.

The Fourier transform of this image show 10 very intense peaks at $1.62+0.16 \text{ A}^{-1}$ corresponding to the mass density lines of Figure 4. This value is also in perfect agreement with the bulk x-ray measurement of 1.6489 A^{-1}. Figure 5 shows a Fourier filtered version of Figure 4. Here we amplified features close to 1.62 A^{-1} by a factor of two to enhance the image, without adding spurious features. In the filtered image the local atomic structural features of rings and points become more pronounced and the steps are preserved.

In order to understand the atomic structure, we examine a small region on the surface in Figure 6. A local STM image; Figure 6a can be fitted nicely with the solid circles of the model in (b), as shown in (c). The closest distance between these atoms (solid circles) is 3.99Å, as measured by both STM ans HRXD. This obviously is a very open structure and must be supported by another,

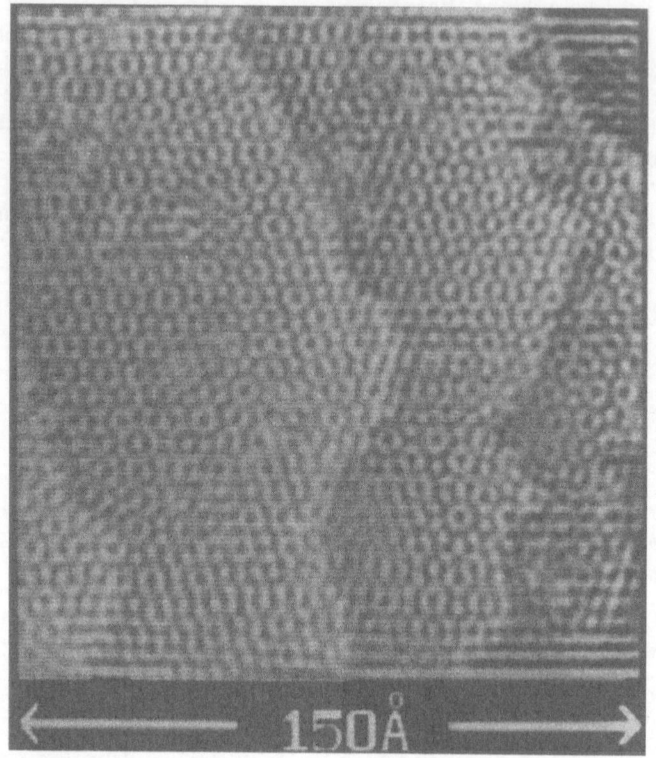

FIGURE 5
Fourier Filtered STM image (top view). Procedure enhances the images of the structural features.

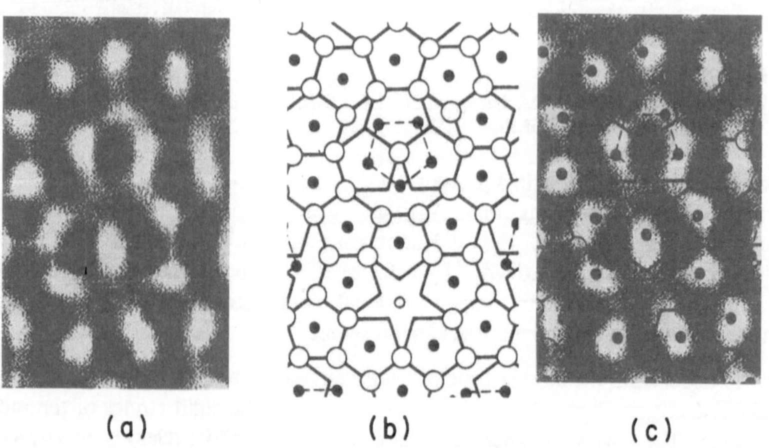

FIGURE 6
Structural model developed (b), for a local image (a) determines the atomic positions. Atoms marked by filled circles in (c) are imaged by STM.

almost co-planar layer of atoms. This layer of atoms can be constructed by placing atoms at the hollow sites. Figure 6b shows the resulting pentagonal network of atoms (open circles). The elevated layer of atoms imaged by the STM (solid circles) combined with the pentagonal network defines the unit layer or the unit step height of $1.94 + 0.1$Å. The atomic model developed also predicts very reasonable interatomic distances as we will explain later. A slight shift in the STM image from the ideal atomic positions can be seen in Figure 6c. This can be accounted for by the drift problems of the STM, which distort the image and cause an expansion along a prefered direction. The pentagonal network we just described is a perfect tiling quasilattice. First constructed by Penrose[37], this quasilattice requires six distinct tiles with appropriate matching rules. This is similar to the findings of Onada et. al.[38], who use eight distinct vertices in addition to matching rules to define the local arrangement of Penrose tiles and generate perfect quasicrystals. Imposing matching rules for tiles beyond the nearest neighbor tiles by vertex rules or going beyond the two rhombic tiles as Penrose did with six tiles, seem necessary for modeling perfect quasicrystals.

In the projection formalism[39], a tetragonal lattice with lattice parameters of a and c in 6D projects onto physical space as a decagonal lattice. The projected quasilattice constants a_R and c_R become, $a_R = a / 2.5$ and $c_R = c$. Identifying the fundamental vector (100000) as q = 0.6298 A^{-1}, we find $a_R = 3.9906$Å, and c=8.26Å Using this information, we draw the top view of the tiling on the magnified image. Figure 7 shows the real image and its selective Fourier filtered form along with the tiles. Few local distortions of the tiles are caused by the drift problem of the microscope and could not be systematically corrected. Aside from these small distortions, the proposed tiling model work properly and suggests an atomic decoration for the tiles that is asymmetric. The thick tile is decorated on all vertices except an acute vertex, and the thin tile decorated on all vertices except an obtuse vertex. Penrose's pentagonal quasilattice to a great extend correctly describes the STM images except inside the decagonal features. Here STM images show almost a complete ring of atoms (shown by dashed pentagons in Figure 7 whereas pentagon center decorated quasilattice would show three atoms only. Since STM resolves individual atoms with identical separation for three atoms case we believe ring like features are real and mostly involves five atoms. Along the periodic axis these tiles are parallelepipeds with four side faces parallel to the five-fold axis.

The STM measured uniform step height of $1.94 + 0.10$Å is approximately one-fourth of the 8.26 c-axis periodicity and one-half of the fundamental length of 4.13Å as suggested by the strong 00000(2n) reflections in Figure 2 and 3. The fundamental length of 4.13Å approximates the height of an icosahedron made up of two Al pentagons stacked together with a 36 degree rotation in between and two Co atoms capping the top and the bottom of these two Al pentagons. The most prominant feature of the STM images; the mass density lines extend uninterrupted across all steps in all five distinct directions, suggesting that the local atomic structures remain identical along the c axis from layer to layer. This condition is satisfied by positioning pentagons in adjacent layers to overlap but also rotate by 36 degrees to maximize the packing and match the measured step height. Starting from a pentagonal quasilattice layer and constructing next layer by 36 degree rotated pentagons one finds that the new quasi-epitaxial layer is also a pentagonal quasilattice in the PLI class, with a local structure identical to the starting layer. The two layers are in fact related by an inflation[40]. The positional invariance of ring like features from layer to layer can then be understood as periodically stacked icosahedrons on top of each other to give a chain structure along the c-axis. The quasicrystalline layers giving rise to steps shown in Figure 1 then consist of a mostly Al pentagonal quasilattice frame and Co and Cu atoms decorating this

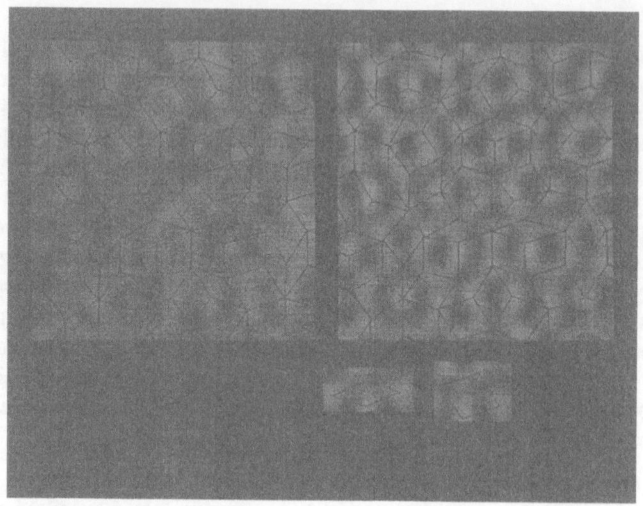

FIGURE 7
A local STM image and its Fourier enhancement decorated by Penrose tiles. Thin and thick rhombic tiles exhibit different decoration.

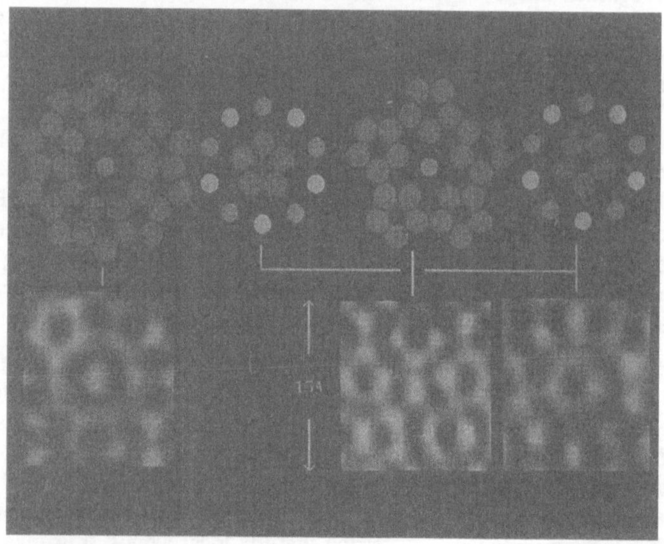

FIGURE 8
Local structural arrangements found very frequently in the STM images reveal the three dimensional stacking of quasicrystalline layers. Four layers displayed stack periodically to define the decagonal phase.

quasilattice. The edge length of the pentagons is 2.899Å, a very typical Al-Al distance[41-43].

The proposed stacking structure for the decagonal phase is shown in Figure 8 for a typical local pattern of a decagon and also compared to actual observed features in the STM images. The four layers shown correspond to the fundamental length of 4.13Å along the periodic axis.

3. DISCUSSION AND CONCLUSION

Aluminum rich compounds usually contain layers of Al atoms corresponding to some form of close packing of the Al atoms. The actual structure of the layers depends more on the restrictions imposed by the number of the transition metal atoms that are sandwiched between these layers. A well known example is the A-type layer that is very dominant in most Al-rich compounds, consists of linking square and triangular arrays of Al atoms. Structures of compounds like $Al_7CoCu_2, Al_9Co_2, Al_7Cu$ all consist of some form of stacking of A and B type of layers, where a B-type layer is a simple square array of transition metal atoms. The monoclinic phase[42] of $Al_{13}Co_4$ is also made up of two different types of layers that are stacked on top of each other. The layers here consist of linking decagons, pentagons, triangles and rectangles of mostly Al and some Co atoms. Overall such an inefficient packing seem to be tolerated at the expense of preserving the arrangement of Co-atom neighbours that dominate the pattern. Here Co atoms are coordinated with nine or ten (preferably ten) Al atoms only. This is also reflected in interatomic distances of the T(AlCoCu)[43] phase which is isomorphous to Al_7FeCu_2. Some local structural similarity exist between the crystalline $Al_{13}Co_4$ and our decagonal phase. Most importantly, the fragments of the Al quasilattice network exist in the form of edge sharing distorted pentagons and adjacent layers are arranged to place distorted Al pentagons on top of each other with a 36 degree rotation.

Models developed for the Al-Cu-Co decagonal phase utilized the possible similarity to the $Al_{13}Co_4$ system. Kumar et.al.[18], and Yamamoto et.al.[20-21] considered this similarity and build their models essentially by decorating Penrose sublattices for two significantly different quasicrystalline layers. Steurer et.al.[22] carried out a five-dimensional structure analysis and constructed the decagonal structure by periodically packing two similar quasiperiodic planes. Burkov[23] assumed that layers were similar and described the structure by large Penrose tiles. Our STM observations are in some agreement with the Steurer's findings, that all the layers are identical, however we additionally find that transition metal atoms are not co-planar with the rest of the atoms but rather occupy the center vacancy between two pentagonal rings.

The combined high resolution x-ray and STM measurements of the atomic structure of the decagonal phase finds a near-perfect quasicrystalline order and clearly rules out the twinning model.

REFERENCES

1. D. Schectman, I. Blech, D. Gratias and J.W.Cahn, Phys.Rev.Lett.53,1951(1984).
2. "Quasicrystals: The State of the Art", edited by P. Steinhart and
   D. DiVincenzo, World Scientific, Singapore, (1991).
3. "Quasicrystals and Incommensurate Structures in Condensed Matter", edited
   by M.J. Yacaman, World Scientific, Singapore, (1990).
4. "Introduction to Quasicrystals", edited by M.V. Jaric, Academic Press, San
   Diego, (1988).
5. "Physics of Quasicrystals", edited by P.J. Steinhardt and S. Ostlund, World

38

Scientific, Singapore, (1987).

6. L. Bendersky, Phys. Rev. Lett., 55,1461(1985).

7. T.C. Choy, J.D. Fitz Gerald and A.C. Kalloniatis, Phil. Mag.B.,58,35(1988).

8. S. Muller, Z. Phys. B. Cond. Matt.,74,369(1989).

9. T.-L. Ho, Phys. Rev. Lett., 56,468(1986).

10. D. Romeu, Materials Science Forum, 22,257(1987).

11. L.X. He, Y.K. Wu, X.M. Meng and K.H. Kuo, Phil. Mag. Lett.,61,15(1990).

12. X.D. Zou, K.K. Fung and K.H. Kuo, Phys.Rev.B.35,4526(1987).

13. S.H. Idziak and P.A. Heiney , Phil. Mag. B.

14. A.R. Kortan, R.S. Becker, F.A. Thiel and H.S. Chen, Phys. Rev. Lett. 64, 200(1990).

15. L.X. He, Y.K. Wu and K.H. Kuo, J. Mat. Sci.Lett.7,1284(1988).

16. A.R. Kortan,F.A.Thiel,H.S.Chen,A.P.Tsai,A.Inoue and T.Masumoto, Phys.Rev.B, 40,9397(1989).

17. S. Takeuchi and K. Kimura, J.of Phys.Soc.Jap.56,982(1987).

18. V. Kumar, D. Sahoo and G. Athithan, Phys. Rev.B.,34,6924(1986).

19. A.P. Tsai, A. Inoue and T. Masumoto, J. Mater. Trans. JIM 30,463(1989).

20. A. Yamamoto and K.N. Ishihara,Acta. Cryst.A44,707(1988).

21. A. Yamamoto, K. Kato, T. Shibuya and S. Takeuchi, Phys. Rev. Lett. 65,1603 (1990).

22. W. Steurer and K.H. Kuo, Phil. Mag. Lett. 62,175(1990), and Acta Cryst. B46, 703(1990).

23. S. Burkov, Phys. Rev. Lett. 67,614(1991).

24. D. Shechtman and I.A. Blech, Met. Trans. A,16,332(1985).

25. P.W. Stephens and A.I. Goldman, 56,1168(1986).

26. L. Pauling, Phys. Rev. Lett. 58,294(1987).

27. D. Levine and P.J. Steinhardt, 53,34(1984).

28. C.L. Henley and V. Elser, Phil. Mag.B 53,L59(1986).

29. M. Audier and P. Guyot, Phil. Mag.B 53,L43(1986).

30. M. Widom and D.P. Deng and C.L. Henley,Phys. Rev. Lett. 63,310(1989).

31. K.J. Strandburg, L.H. Tang and M. V. Jaric, Phys. Rev. Lett.63,314(1989).

32. A.P. Tsai, A. Inoue and T. Masumoto, J. Mat. Sci. Lett. 6,1403(1987).

33. A.P. Tsai, A. Inoue and T. Masumoto, Jap. Jour.of App. Phys. 27,L1587(1988).

34. A.P. Tsai, A. Inoue and T. Masumoto, Mat. Trans.,JIM, 30,463(1989).

35. R.S. Becker, B.S. Swartzentruber, J.S. Vickers, and T. Klitsner, Phys. Rev. B. 39, 1633(1989)

36. J.E.S. Socolar and P.J. Steinhardt, Phys.Rev. B. 34,617(1986).

37. R. Penrose, Inst. Math. and its Applications,July/Aug,266(1974).

38. G.Y. Onoda, P.J. Steinhardt, D.P. DiVincenzo and J.E.S. Socolar, Phys. Rev. Lett., 60, 2653(1988).

39. N.K. Nukhopadhyay, K. Chattopadhyay and S. Ranganathan, preprint.

40. R. Luck, Mat. Sci. Forum,22,231(1987).

41. P.J. Black, Acta Cryst.,8,175(1955).

42. R.C. Hudd and W.H. Taylor, Acta Cryst.15,441(1962).

43. M.G. Bown and P.J. Brown, Acta Cryst.9,911(1956).

ATOMIC STRUCTURE OF QUASICRYSTALS WITH AND WITHOUT PHASON DEFECTS

Ruizhong Hu, T. Egami
Department of Materials Science and Engineering
and Laboratory for Research on the Structure of Matter,
University of Pennsylvania, Philadelphia, PA19104-6272

A.-P. Tsai, A. Inoue, T. Masumoto
Institute for materials Research
Tohoku University, Sendai 980, Japan

J. C. Holtzer and K. Kelton
Department of Physics, Washington University
St. Louis, MO 63130

ABSTRACT. The atomic structures of Al-Ru-Cu, Al-Li-Cu and Ti-Mn-Si quasicrystals are discussed, in particular with respect to the origin of the phason defects. X-ray diffraction measurements, including anomalous x-ray diffraction studies at the Ru K-edge for Al-Ru-Cu, were carried out for these quasicrystals. Results are analyzed both in terms of 6-dimensional crystallography and 3-dimensional atomic pair distribution function (DDF). Structures were obtained by a direct projection of a 6-dimensional hypercubic lattice. Fundamental differences were observed among these quasicrystals in the nature of the projection window in the perpendicular space and the tendency of local cluster formation, which may be directly related to the density of phason defects.

1. INTRODUCTION

The atomic structure of quasicrystalline materials have been actively studied since the discovery of the first quasicrystal, Al-Mn [1]. The atomic structure of quasicrystals has been modelled preserving the long range icosahedral symmetry appeared in the diffraction pattern, for example, by the decoration of the 3 dimensional Penrose tiling (3-DPT) [2-3], or by the decoration of the 6-dimensional (6-D) hypercubic lattice projected onto the 3 dimensional (3-D) space [4]. However, these approaches require knowledge of the 6-dimensional structure, while we are provided only with the information in 3-dimensions. Thus a considerable amount of educated guesses are necessary for the structural determination. One of these guesses is to assume that the structure of a related crystalline compound can provide information regarding the perpendicular space. However, such a compound cannot always be found.

P. Jena et al. (eds.), Physics and Chemistry of Finite Systems: From Clusters to Crystals, Vol. I, 39–48.
© 1992 *Kluwer Academic Publishers.*

40

In this paper we show that by the combined use of both the 6-dimensional crystallography and the atomic pair distribution analysis the atomic structure can be determined even without knowing the structure of the related crystalline compound, and furthermore discuss the relationship between the atomic structure and the density of phason defects.

2. EXPERIMENTAL METHODS

X-ray diffraction measurements were carried out on powder samples of quasicrystalline $Al_{65}Ru_{15}Cu_{20}$ [5] and $Ti_{58}Mn_{37}Si_5$ [6], at the X-7A beamline of the National Synchrotron Light Source at the Brookhaven National laboratory. The synchrotron beam was monochromated by a channel-cut Si (111) crystal, and the scattered photons was detected by a solid state Ge detector. The sample was contained in a helium environment, and the scattering studies were made in the reflection geometry. The output from the solid state Ge detector was received by a CAMAC based multi-channel analyzer (MCA) operated on line by a micro-Vax II computer. The energy resolution of the detector allowed the separation of the K_α fluorescence from the diffraction intensity even when the incident energy is close to the K absorption edge in the anomalous x-ray scattering experiment. The intensity of the K_β fluorescence was calculated from the measured K_α intensity. The measured scattering intensity was corrected for the dead time of the detector, the K_β fluorescence, the background due to air scattering, absorption, multiple scattering and inelastic (Compton) scattering, and normalized to obtain the total elastic scattering intensity per atom $I_e(q)$. The total powder structure factor S(Q) is defined as;

$$S(Q) = I_e(Q) / \langle f(Q) \rangle^2 + [\langle f(Q)^2 \rangle - \langle f(Q) \rangle^2] / \langle f(Q) \rangle^2 \quad , \tag{1}$$

where Q is the scattering vector ($= 4\pi\sin\theta/\lambda$), $\langle...\rangle$ is the compositional average, and f(Q) is the atomic scattering factor. By integrating each peak and correcting for the multiplicity and Lorentz factor, we lobtain the square of the single crystal total structure factor,

$$|F_t(\vec{Q})|^2 = \frac{1}{N} \sum_{i,j} \frac{f_i(Q) f_j(Q)}{\langle f(Q) \rangle^2} e^{i\vec{Q}\cdot\vec{r}_{i,j}} \quad . \tag{2}$$

The purpose of x-ray anomalous scattering measurement is to single out the contribution of a particular element to the total scattering intensity by making use of the rapid change in the atomic x-ray scattering factor close to the absorption edge of an element. The total structure factor S(Q) can also be expressed as a weighted sum of compositionally resolved partial structure factors $S_{\alpha\beta}(Q)$ in the Faber-Ziman form [7],

$$S(Q) = \sum_{\alpha=1}^{n} \sum_{\beta=1}^{n} C_\alpha C_\beta \frac{f_\alpha(Q)\ f_\beta(Q)}{\langle f(Q) \rangle^2} S_{\alpha\beta}(Q) \quad , \qquad (3)$$

where C_α and f_α are the atomic concentration and the structure factor of the element α, respectively, and n is the number of elements in the material. The differential structure factor of an element A, defined below, can be obtained from the scattering intensities measured at two different incident photon energies below the absorption edge by;

$$S_A(Q) = [\sum_{\beta}^{n} C_\beta\, f_\beta(Q)\, S_{A\beta}(Q)]\, /\, [\sum_{\beta}^{n} C_\beta\, f_\beta(Q)]$$

$$= \Delta I_c(Q) / [2C_A \Delta f_A' \sum_{\beta}^{n} C_\beta Re(f_\beta(Q))] \quad , \qquad (4)$$

where ΔI_c and $\Delta f_A'$ are the differences in the elastic scattering intensity and the atomic scattering factor, respectively, which are obtained from the two separate scans at different incident x-ray energies. Here I_c is given by $S(Q) * \langle f(Q) \rangle^2$.

The total pair density function (PDF), $\rho(r)$, is obtained by Fourier transforming the total structure factor $S(Q)$,

$$\rho(r) = \rho_0 + \frac{1}{2\pi^2 r}\int Q[S(Q)-1]\sin(Qr)\,dQ \quad . \qquad (5)$$

The total PDF describes the distributions of interatomic distances weighted by compositions and structure factors of each elements in the material. It can also be expressed approximately as a weighted sum of partial pair distribution functions $\rho_{\alpha\beta}(r)$ as:

$$\rho(r) = \sum_{\alpha=1}^{n} \sum_{\beta=1}^{n} C_\alpha C_\beta [f_\alpha f_\beta / \langle f \rangle^2]\, \rho_{\alpha\beta}(r) \quad , \qquad (6)$$

where the partial pair density function $\rho_{\alpha\beta}(r)$ represents the α-β pair correlations which can be obtained by the Fourier transformation of the partial structure factor $S_{\alpha\beta}(Q)$ in the form of:

$$\rho_{\alpha\beta}(r) = \rho_0 + \frac{1}{2\pi^2 r}\int Q[S_{\alpha\beta}(Q)-1]\sin(Qr)\,dQ \quad . \qquad (7)$$

In practice, it is very difficult to determine all these correlation functions even for a binary alloy. However, the differential pair distribution function (DDF) [8] obtained by the Fourier-transformation of the differential structure factor, eq. (4), defined as,

$$\rho_A(r) - \sum_{\beta=1}^{n} \frac{f_\beta}{\langle f \rangle} \rho_{A\beta}(r) \quad , \tag{8}$$

offers an excellent intermediate step. This function describes all the atomic distances from the A element, weighted by the scattering factor.

3. RESULTS FOR Al-Ru-Cu

The i-phase $Al_{65}Ru_{15}Cu_{20}$ was prepared by casting and was annealed at 1130 °C for 12 hours. The anomalous X-ray powder diffraction measurement was carried out with the incident energies at 12.6 eV and 300 eV below the absorption edge which is at 22.1193 KeV. The scans were carried out up to $Q = 21$ ($Å^{-1}$) to ensure the PDF obtained from the Fourier transformation to be without substantial termination errors. The absolute value of the single crystal total structure factors of $Al_{65}Ru_{15}Cu_{20}$ are shown in Fig. 1 as a function of $Q_\perp$. The total structure factor is similar to the one obtained by Cornier-Quiquandon et al. for a single crystal $Al_{63}Cu_{25}Fe_{12}$ [9]. Since for the 6-dimensional simple hypercubic lattice

$$e^{i\vec{Q}\cdot\vec{R}_i} = e^{i\vec{q}_\parallel \vec{r}_{\parallel,i}} e^{i\vec{q}_\perp \cdot \vec{r}_{\perp,i}} = 1$$
$$\vec{Q} = \vec{q}_\parallel + \vec{q}_\perp , \quad \vec{R} = \vec{r}_\parallel + \vec{r}_\perp \quad , \tag{9}$$

where Q is the 6-dimensional reciprocal lattice vector and R is the 6-dimensional lattice vector, we obtain

$$\sqrt{I(Q)} = |F(q)| = \frac{1}{N}\sum_i e^{i\vec{q}_\parallel \vec{r}_{\parallel,i}} = \frac{1}{N}\sum_i e^{-i\vec{q}_\perp \cdot \vec{r}_{\perp,i}} \quad . \tag{10}$$

Therefore a plot of $|F_t(Q)|$ against $Q_\perp$ as in Fig. 1 gives the Fourier transform of the window function [4], and provides direct information about the shape of the projection window. Furthermore, from the differential structure factor for Ru, $S_{Ru}(Q)$, we obtain information regarding the Ru partial structure factor, and the sub-window for the Ru sites.

The diffraction peaks can be indexed by an icosahedral index scheme [10] by assigning (4,2,2,2,2,2) to the peak at $Q = 2.949$ $Å^{-1}$. This leads to the quasilattice parameter equal to 9.022 Å. This indexing assumes the NaCl type sublattices, to account for the f.c.c. symmetry deduced from the extinction condition [11]. If, for the time being, we neglect the chemical ordering, the quasilattice constant will be a half, or 4.511 Å. Thus the reciprocal quasilattice parameter is

0.696 Å⁻¹. If the structure is a standard 3-dimensional Penrose tile
(3-DPT) decorated by atoms, the size (average radius) of the window
should be about 40 % larger than this value. The results shown in Fig.
1, however, indicate that the size of the window in the perpendicular
space is larger than that of the standard and is close to the one
inflated by τ. In fact such a window, a 3-DPT window inflated by τ with
truncation along the 5-fold axis to avoid atomic overlaps, was studied
by Henley [12]. The original study by Henley was on the window deflated
by τ^2, but an inflation by τ^3 does not change the structure, but simply
rescales it, resulting in the window under discussion. This window is
known to produce a structure with the highest packing density. The PDF
calculated for such a structure shows already fairly good agreement with
the experimental PDF. However, a peak at 2.75 Å could not be reproduced
by this model. An additional window with the b.c.c. sites suggested by
Cornier-Quiquandon et al. [9] provides this interatomic distance. By
assigning Ru to one of the sublattices with the $r_\perp$ value within the
truncated 3-DPT window, and assigning Cu to the b.c.c. sublattice and
the truncated 3-DPT window of the opposite sublattice than Ru, the
calculated PDF and DDF show very good agreement with the experimental
results as shown in Figs. 2 and 3. More detailed descriptions of the
structure will be given elsewhere.

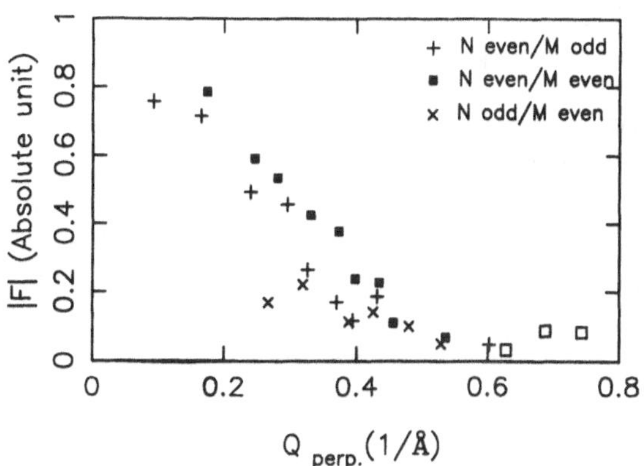

Fig. 1.   Total structure factor, $F_t(Q)$, of $Al_{65}Ru_{15}Cu_{20}$ plotted against
the perpendicular wave vector, $Q_\perp$. N and M refer to the Cahn index [9].

4. RESULTS FOR Al-Li-Cu AND Ti-Mn-Si

The x-ray and neutron PDF's obtained by the earlier study [13,14]
can be modelled well with the decoration scheme proposed by Henley and
Elser [3], as shown in Figs. 4 and 5. As noted by Henley and Elser, the
atomic decoration of the 3-DPT lattice (Al/Cu on vertices and edge
centers, Li on the two sites on the diagonal of the prolate rhombohedra)
leaves a half of Li atoms unaccounted for. In the structure of the
related rational approximant crystalline (R-) phase contains, in

addition to the decorated prolate and oblate rhombohedra, rhombic
dodecahedra having a hexagonal ring of Li atoms are found [3]. Thus it
is reasonable to assume that the i-phase also has these rhombic dodecahedra. The rest of the Li atoms can be accommodated to these sites by
slightly increasing the window size to include extra vertex points to
define the diagonal sites, and placing Li atoms to these diagonal sites
without placing additional Al/Cu to the extra vertex points. In
increasing the window size for the diagonal decoration, one has to make
sure that atoms are not too close to each other.

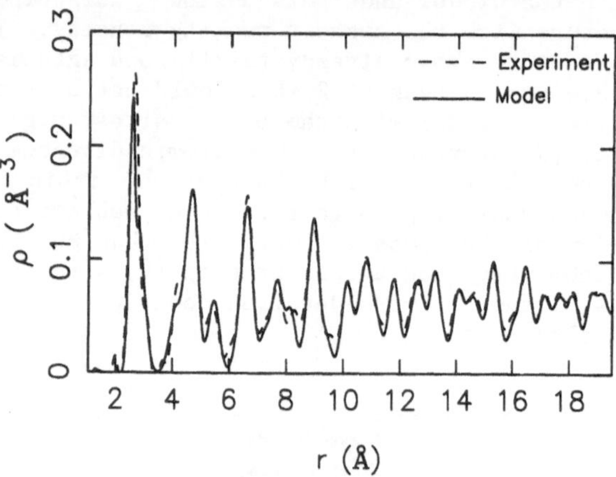

Fig. 2. Total x-ray atomic pair-distribution function for quasicrystalline $Al_{65}Ru_{15}Cu_{20}$, experimentally determined (dashed line) and calculated for the model (solid line).

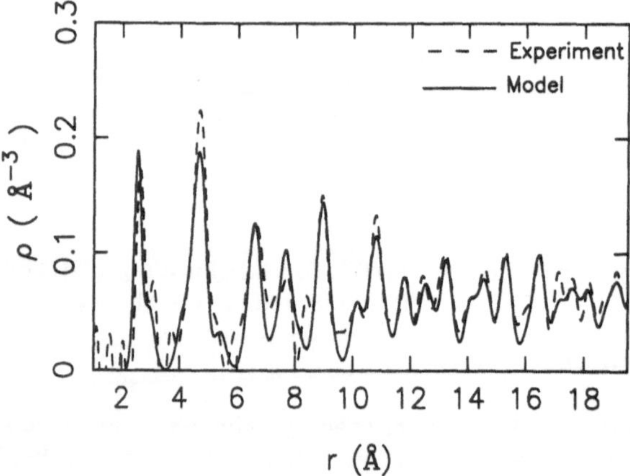

Fig. 3. Ru differential distribution function for $Al_{65}Ru_{15}Cu_{20}$, experimentally determined (dashed line) and calculated for the model (solid
line).

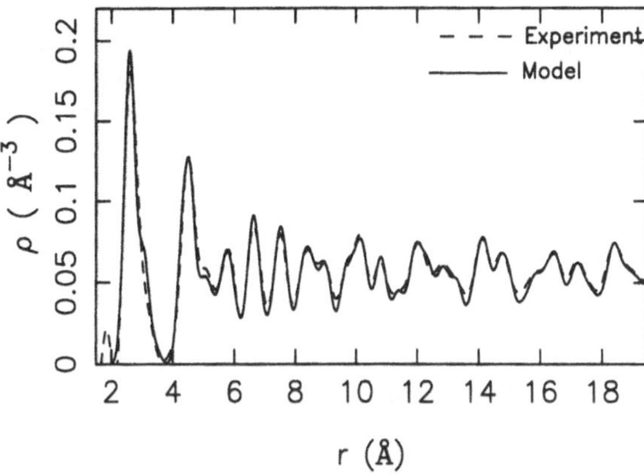

Fig. 4. X-ray total PDF for quasicrystalline $Al_{57}Li_{33}Cu_{10}$, experimentally determined (dashed line) [xx], and calculated for the model (solid line).

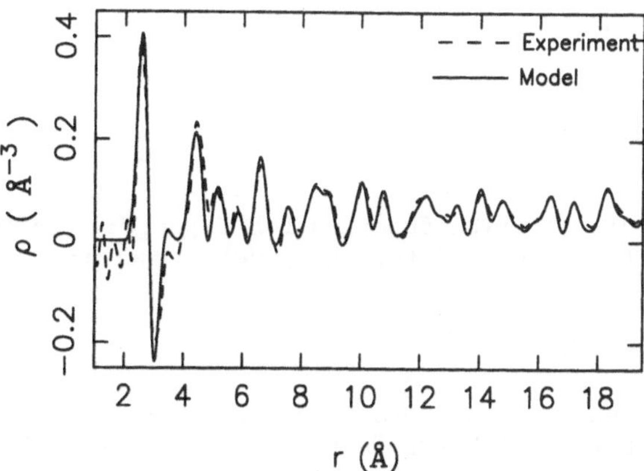

Fig. 5. Neutron total PDF for quasicrystalline $Al_{57}Li_{33}Cu_{10}$, experimentally determined (dashed line) [xx], and calculated for the model (solid line).

Although the diffraction pattern for Ti-Mn-Si is rather different from the one for Al-Li-Cu, mainly because the weak scattering by Li, their structures were found to be similar. The PDF calculated for the Al-Li-Cu structure, assuming that Ti and Mn randomly occupy the atomic sites, shows reasonable agreement with the experimental PDF. However, important discrepancies were found. In particular, prominent peaks at 5.3 Å and at 6.2 Å in the calculated PDF were not found in the experimental PDF, but were replaced by a broader peak at 5.6 Å. The PDF was

46

reproduced quite well as shown in Fig. 6, by expanding the window for
the edge-center sites. In the 3-DPT decoration scheme the projection
window for the edge-center sites is smaller than the 3-DPT window, since
the edge-center sites are defined by both of the nearest vertices being
within the 3-DPT window. In the present case, we chose the window for
the edge-center sites to be the 3-DPT window itself, thus allowing the
edge center positions for which one of the nearest vertices was not
within the 3-DPT window. This, however, resulted in many sites which
were too close to each other. These sites were rejected randomly to
avoid the atoms from overlapping.

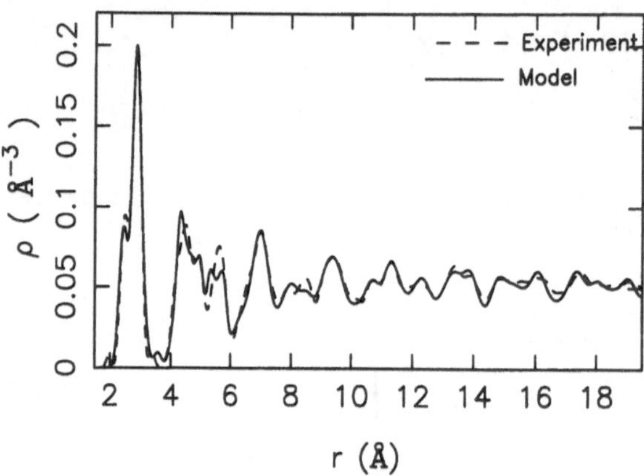

Fig. 6.   X-ray atomic pair-distribution function for $Ti_{58}Mn_{37}Si_5$ (solid
curve), compared with the PDF calculated for the model structure (dashed
curve).

5. DISCUSSIONS ON PHASON DENSITY

There are large differences in the phason defect density among the
three compounds discussed above. Al-Ru-Cu is practically phasonless,
Al-Li-Cu has some phasons, while Ti-Mn-Si has a high phason density
which results in diffuse scattering [6]. An obvious difference in the
structure is that the structure of Al-Ru-Cu is well defined by the
window alone, while others contained some ambiguities resulting in a
selection of sites to avoid atoms being too close to each other, which
can introduce some randomness in the structure. However, this may not
be the fundamental difference, since by defining the shape of the window
more carefully the element of randomness could be removed. Another
difference is that Al-Ru-Cu requires basically the decoration of the
vertex sites alone, while Al-Li-Cu and Ti-Mn-Si need edge center as well
as diagonal decorations, thus many sublattices have to be specified.
This certainly appears to be an important point, since having many
sublattices implies a larger probability of making mistakes, for
instance, due to thermal fluctuations. However, the need of the b.c.c.
decoration in Al-Ru-Cu considerably weakens this argument.

A more fundamental point appears to be the compatibility between the local structure and the global structure. In amorphous Al-Fe-Ce, an Fe atom is surrounded by Al neighbors, about 6 of them at 2.5 Å and about 4 ~ 5 at 2.8 Å [15]. Since the sum of the metallic radii for Al and Fe is 2.7 Å, the Al-Fe contact distance is significantly reduced compared to the distance expected for a normal free-electron-like metallic bonding, presumably due to the strong s-d hybridization [15]. The atomic environment in the glassy state most likely represents the most preferred one for the atom, because the topological constraint is absent in the glass. Now this preferred environment of Fe indeed is found in quasicrystalline Al-Fe-Cu, which is undoubtedly isomorphous to Al-Ru-Cu. In Al-Ru-Cu, however, the distances are slightly modified because of the larger atomic size of Ru (larger than Fe by 0.07 Å), and the Ru atoms have 6 Al at 2.6 Å and 4 ~ 5 Al at 3.0 Å. Thus this rather special environment of Fe or Ru atoms in the quasicrystalline structure is compatible with the atomic preference. This compatibility and the simple form of the projection window may be the source of an excellent long range order. The strong local interaction due to the s-d hybridization may be contributing heavily toward stabilizing this structure.

In Al-Li-Cu, such a perfect compatibility is not found. The structure is basically described as the 3-DPT lattice, decorated by Al, Cu and Li. Thus there must be some chemical force to prefer the 3-DPT window. However, a perfect decoration of the 3-DPT structure leaves a half of the Li atoms unaccounted for, and the placement of these extra Li atoms can lead to phason strains. To illustrate this situation let us consider building an Al-Li-Cu cluster from the origin. We assume that the center of the projection window is also at the origin. As the cluster grows with the decoration discussed above, it becomes necessary to place extra Li atoms to form a rhombic dodecahedron with a hexagonal Li ring. For the first group of extra Li sites, the additional vertex points which define the diagonal decoration are virtually on the surface of the window, thus the extra Li sites can be incorporated without a substantial change in the window. However, as the extra Li site becomes more removed from the origin, the $r_\perp$ value for such a vertex point moves further away from the 3-DPT window, and therefore it may become energetically unfavorable to place extra Li atoms, if there is a force to prefer the original 3-DPT window. This implies that every time the rhombic dodecahedra are formed, there will be a force to place them on the surface of the window, by relocating the center of the window. An atomic cluster around the lattice point which is located at the center of the window is highly symmetric, so this tendency could be interpreted as the propensity to produce such symmetric clusters. This will result in phason defects, as well as the tendency to form a crystalline R-phase. Thus in this case the tendency of forming large, well defined, symmetric local clusters may be the origin of phason defects.

As we discussed above the PDF calculated for Ti-Mn-Si based upon the Al-Li-Cu structure has a peak at 5.3 Å. This peak corresponds to the edge center-to-edge center distance within a rhombohedral unit cell. The fact that the experimental PDF does not have this peak means that the atoms do not like this local structure, and have a tendency of breaking up this unit cell. Consequently the quasicrystalline long range order is often disrupted by displacive defects which may result in the diffuse scattering of x-rays and electrons.

6. CONCLUSIONS

By determining the atomic structure of three quasicrystals by x-ray (anomalous) scattering and pulsed neutron scattering, differences in the propensity for local cluster formation were found. The preference of a local cluster which is inconsistent with the long range order (Ti-Mn-Si), or a well defined large cluster which can form a crystalline phase (Al-Li-Cu) can lead to phason formation. I-phase Al-Ru-Cu is exceptional in that the local environment Ru prefers is consistent with the quasicrystalline structure, and the window in the perpendicular space is contiguous, allowing little chance of phason formation.

Acknowledgment

This work was supported by the National Science Foundation through DMR90-01704. A part of this work was performed at the National Synchrotron Light Source of Brookhaven National Laboratory, operated by the Department of Energy, Division of Chemical Sciences.

References:

[1]   D. Shechtman, I. Blech, D. Gratias and J.W. Cahn, Phys. Rev. Lett., **53**, 1951 (1984).
[2]   V. Elser and C. L. Henley, Phys. Rev. Lett., 55, 2883 (1985).
[3]   C.L. Henley and V. Elser , Phil. Mag., **B53**, L59 (1986).
[4]   "Quasicrystallline Materials", edited by Chr. Janot and J.M. Dubois, (World Scientific, Singapore, 1988).
[5]   C.A. Guryan, A.I. Goldman, P.W. Stephens, K. Hiraga, A.P. Tsai, A. Inoue, and T. Masumoto, Phys. Rev. Lett., **62**, 2409 (1989).
[6]   C.Gibbons, K.F.Kelton, L.E.Levine and R.B.Phillips, Phil. Mag., **B59**, 593 (1989).
[7]   T.E. Faber and J.M. Ziman, Phil. Mag. **11**, 153 (1965).
[8]   P.H. Fuoss, P. Eisenberger, W.K. Warburton, and A. Bienenstock, Phys. Rev. Lett., **46**, 1537 (1981).
[9]   M. Cornier-Quiquandon, A. Quivy, S. Lefebvre, E. Elkaim, G. Heger, A. Katz and D. Gratias, Phys. Rev, **B44**, 2071 (1991).
[10]   V. Elser, Phys. Rev., **B32**, 4892 (1985).
[11]   S. Ebalard and F. Spaepen, J. Mater. Res., **4**, 39, (1989).
[12]   C.L. Henley, Phys. Rev., **B34**, 797 (1986).
[13]   W. Dmowski, T. Egami, Y. Shen, S.J. Poon, and G.J. Shiflet, Phil. Mag., **56**, 63 (1987).
[14]   Y. Shen, S.J. Poon, W. Dmowski, T. Egami and G.J.Shiflet, Phys. Rev., **37**, 1146 (1988).
[15]   H. Y. Hsieh, B. H. Toby, T. Egami, Y. He, S. J. Poon and G. J. Shiflet, J. Mater. Res., **5**, 2807 (1990).

STRUCTURE OF QUASICRYSTALS IN TERMS OF ATOMIC CLUSTERS

P. GUYOT, M. AUDIER and M. DE BOISSIEU
Laboratoire de Thermodynamique et Physico-Chimie Métallurgiques,
UA CNRS 29, INPG-ENSEEG, BP 75, 38402 St Martin d'Hères, France.

ABSTRACT. Experimental results showing that icosahedral atomic clusters, known to exist in complex crystalline phases, can be also considered in the structure of related quasicrystalline phases. Three parts are presented:
i) the observation of the existence of 5-fold polyatomic clusters in icosahedral quasicrystals by local spectroscopies and electron microscopy investigation techniques.
ii) the direct analysis of diffraction experiments in hyperspaces, and in particular the determination of the atomic surfaces, which are shown to restore after cut by the physical space several types of polyatomic clusters.
iii) the construction of atomic structure models of some icosahedral quasicrystals based on the connection of cluster units according to specific rules. Several ways of modeling are described. The atomic order of the models is analysed as well as their diffraction properties.

1. Introduction

The aggregation of poly-atomic clusters, which local symmetry prefigurates the forbidden crystallography symmetry of quasicrystals, is conceptually a seducting idea to model their structure, and there are several definite experimental signatures in favour of cluster models.

Many intermetallic compounds are long known to contain in their motives clusters of icosahedral or slightly distorted icosahedral order (for a review see ref. [1–3]). Interestingly, some of these compounds have been found to co-exist in as-cast products with identified quasicrystals, with two remarkable properties:

i) their structures are mutually oriented according to the subgroups of the icosahedral symmetry point group. For example, in the $AlMnSi$ and $AlLiCu$ alloys [3], the cubic crystal phases are $Pm\bar{3}$ and $Im\bar{3}$, and the crystallographic orientational relationship corresponds to the strict alignment of the $\langle100\rangle$ cubic axes with three orthogonal 2-fold axes of the icosahedral phase.

ii) in the diffraction patterns, the more intense reflections of the two phases are in near coincidence, inferring that common i-clusters modulate, as motives do via their form factor, the Fourier transform of the periodic and quasiperiodic cluster center structures [3, 4].

Still more convincing perhaps is the case of $AlFeCu$, where a trigonal phase is observed to transform by in-situ heating in an electron microscope, in a perfect i-phase, with the two above characteristics [5], as shown in Fig.1. Such near

49

P. Jena et al. (eds.), Physics and Chemistry of Finite Systems: From Clusters to Crystals, Vol. I, 49–59.
© 1992 *Kluwer Academic Publishers.*

crystals have been called "approximant" crystals, the larger the crystal unit cell the better being the approximation for a given space group. Within the cut and projection of hyperlattice representations, this is equivalent to replace the irrational number characteristic of the quasicrystal symmetry by a rational approximant which brings the strip in rational orientation with respect to the hyperlattice [6]: the closer to the irrational number is this approximant, the larger is the crystal unit cell, and the higher is the number of the atomic fragments (clusters or cluster groups) shared with the quasicrystal, which is nothing but a crystal of infinite unit cell. The unit cell lengths of the observed approximants are indeed quite large: about 1.4 nm for the cubic AlMnSi or AlMgZn and AlLiCu, 3.2 nm for the trigonal AlFeCu, 1.4 and 2.8 nm for the hexagonal Z-(AlLiCuMg), 1.4 and 8.3 nm for the tetragonal τ-(AlLiZnCu) [1, 3, 4], and of the same order in GaMgZn [7].

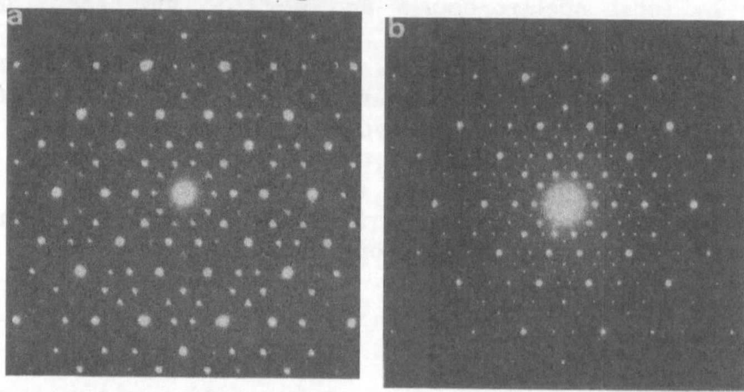

Figure 1. Electron diffraction patterns of 5-fold zone axes of (a) a trigonal phase, transformed upon heating at 860°C in (b) an icosahedral phase. The trigonal phase has in fact a microdomain structure, each domain having its ⟨111⟩ axis aligned with one of the ten 3-fold axes of the i-phase, in a stellate configuration [5].

These considerations open a strategy to build atomic structures of quasicrystals: once the known structure of the approximant has been analysed in terms of clusters of icosahedral symmetry, a scheme of connections of these clusters, going from periodic to quasiperiodic at long range, must be imagined. This transformation, supported by the tools of quasicrystallography, largely developed during the last few years (hyperlattice decoration, tilings, self-similarity...) leads however, most of the time, to only a 3-dimensional cluster backbone. An ultimate atomic filling and position adjustement of the connected outer-shells atoms of the clusters is required.

The approach is illustrated below for two well studied systems, AlMnSi and Al-Li-Cu, according to the several features previously outlined and supported by a variety of experimental data.

2 . Pioneer Clusters

Soon after the discovery by Shechtman, Blech, Gratias and Cahn [8] of the i-phase in AlMn, two icosahedral clusters, identified in cubic approximants, were considered as reasonable building units of related icosahedral quasicrystals. They are the so-called "Mackay" 54 atom cluster

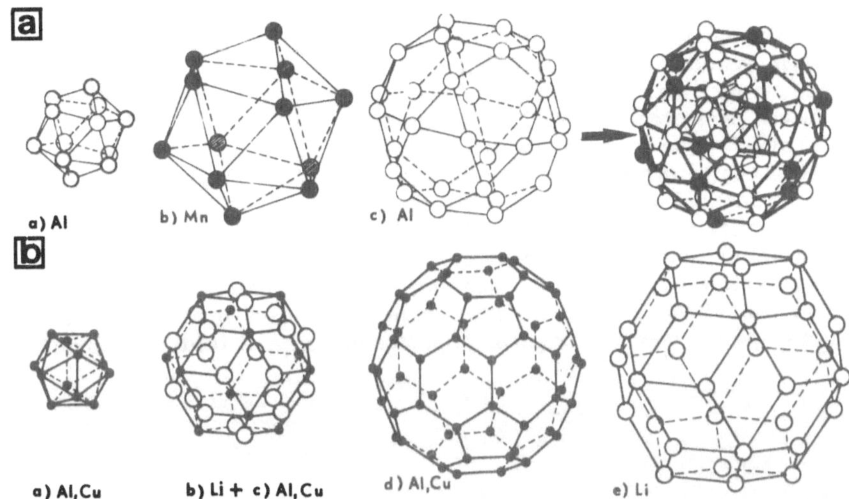

Figure 2. a) The three atomic shells constituting the "Mackay" icosahedron and b) the 5 shells-cluster (in the cubic phase the shell d, or "soccer-ball" exhibits a slight distortion with respect to the icosahedral symmetry; the shell b (Li codecahedron) + shell c (Al,Cu icosahedron) form a small triacontahedron).

double-icosahedron in Al Mn Si [6,9,10] and a cluster of 136 atoms distributed over five icosahedral shells in Al Mg Zn and Al Li Cu [11,12], (Fig. 2). They are both linked in the same way in the cubic structures: with the same orientation, first cluster neighbours are linked through the 3-fold cluster axes parallel to $\langle 111 \rangle$ axes, and second cluster neighbours, through the 2-fold cluster axes parallel to $\langle 100 \rangle$ axes. For the 5 shells-cluster, the external triacontahedron partly overlaps 8 first neighbours along $\langle 111 \rangle$.

3. Cluster units in icosahedral structures

Several experimental investigations techniques have been used to detect icosahedral short range order in quasicrystalline phases: NMR, Mossbauer spectroscopy, EXAFS, EELS, Transmission Electron Microscopy and more classically X-ray and neutron diffraction experiments, where the whole structure is expected to be determined.

3.1. DIRECT PROOFS OF THE EXISTENCE OF CLUSTERS (EXAMPLES).

i) Electron diffraction at the nanometric scale is allowed by electron beams with a diameter reduced down to about 1 nm at the specimen level. For sufficiently thin specimens, the information comes therefore from a volume of matter containing only a very limited number of presumed clusters. Such nanodiffraction patterns have been shown to own in Al Mn Si the characteristic symmetry features of the classical micronic electron diffraction patterns [13].

ii) Similar results have been obtained by analytical electron microscopy in the electron energy loss mode: EEL spectra recorded in STEM with a nanometric beam, in i-Al Li Cu and R-approximant, have near edge structures of Li_K and

52

Al $_{L2-3}$ - which result from topological and chemical short range order around Li and Al atoms- indeed very similar [14].

iii) High resolution electron microscopy (atomic resolution), in optimal conditions (very thin foils, Scherzer focus), gives directly interpretable images in terms of the projected potential approximation - the image consists of dark and white "atomic" regions corresponding to high and low atomic potential projected parallely to the electron beam. More generally, to take properly into account the electron dynamical scattering, a full image computer simulation of a given structure model is required for well defined experimental imaging conditions. This is certainly not achieved today for a full structure determination, but some semi-quantitative proofs of the existence of local clusters exist [13,15,16]. Two examples are given: i) Fig. 3 shows high resolution images of the Al Fe Cu icosahedral phase and the trigonal approximant in a 5-fold and pseudo 5-fold

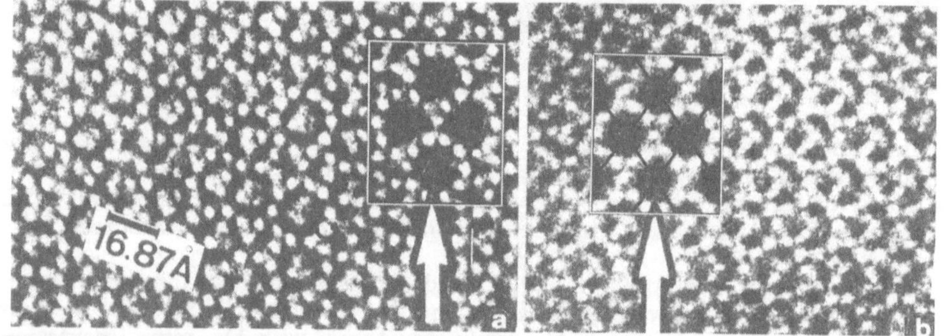

Figure 3. HREM images a) of the Al Fe Cu i-phase and b) of its trigonal approximant (note that a reversible transition has been observed between both these phases [5,19]).

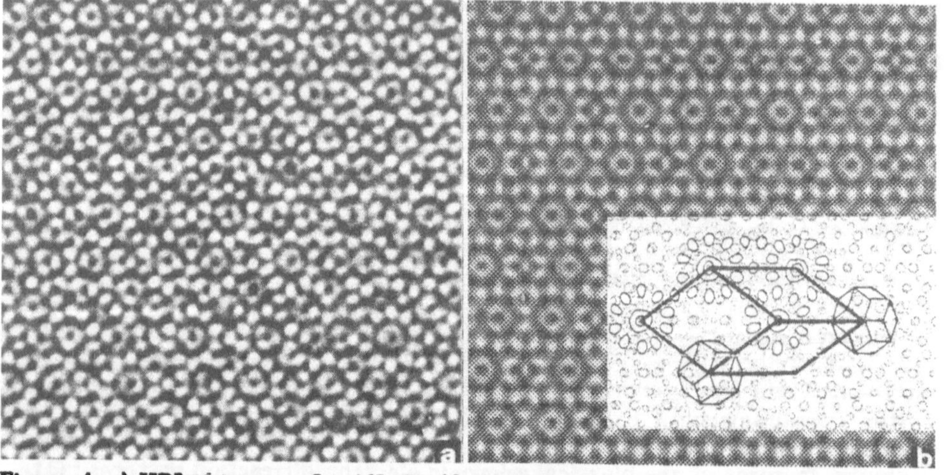

Figure 4. a) HREM image of i-AlLiCu (foil thickness < 5nm, Scherzer focus) [17], b) multi-slice image simulation of the cubic R-phase in a pseudo 5-fold orientation for a foil thickness of 6-7nm [18]; in insert commonly "decorated" rhombus of edge 1.2 nm.

orientation [5]; whereas ten white dots wheels of about 1.7 nm of diameter are reproduced periodically in the approximant, they are also observed, but with a lower number density and aperiodically spaced in the i-phase. ii) Fig. 4a is a high resolution image of i-AlLiCu in a 5-fold zone [17]; characteristic small rings surrounded by a decagonal wheel of white dots are distinguished. Fig.4b is an image simulation of the R-phase, calculated by the multislice technique near Scherzer focus for a thin foil [18]; one notes again the very close similitude between local white dots configurations in a) and b), extending up to 1.4 nm. One can deduce from the image simulation that the decagonal wheel corresponds to the projection of the small triacontahedron (Fig.2, shells b and c).

3.2. CLUSTERS RECONSTRUCTED FROM HYPERSPACE CRYSTALLOGRAPHY ANALYSIS.

A way for extracting structural informations from diffraction data of icosahedral quasicrystals, is to use a high dimensional approach. In that case, the periodicity property is recovered in 6-D space and thus usual tools of crystallography analysis, such as Patterson function, can be employed.

The 6-D cubic space decomposes in two orthogonal subspaces, the physical space or parallel space (E_{par}) and a perpendicular space (E_{perp}). In the cut scheme, the 6-cube (i.e. the unit cell) is decorated by a set of atomic surfaces which are 3-D objects lying in E_{perp}. The location of the atomic surfaces in the 6-D unit cell, but also their shape, has to be found in order to determine the atomic structure [see ref. 20 for more details].

3.2.1. *Icosahedral AlMnSi phase.*

The 6-D analysis has been first carried out by Gratias et al. [21], computing 6-D Patterson functions from both X-ray and neutron powder diffraction data. Two atomic surfaces have been found, a large one located at the origin and a small one at the body center of the 6-cube. The specification of the Al and Mn surfaces has been solved by neutron contrast variation techniques, substituting Mn isomorphically by a FeCr mixture. These studies [22,23] have shown that the Mn atomic surface lies at the origin of the 6-cube surrounded by an Al shell, the small body center atomic surface being occupied by Al atoms (Fig.5a).

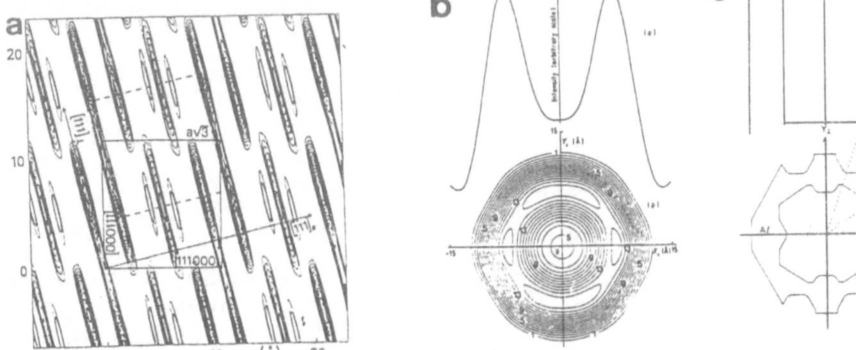

Figure 5. a) example of section of the 6-D atomic structure of i-AlMnSi, from ref.[22]; b,c) Al atomic surface at the origin of the 6-cube found experimentally and modelled by Duneau and Oguey.

The raw shape of the atomic surfaces is spherical but severe smoothing truncation effects are unavoidable [24]. After cut by E_{par}, only the two external Mn and Al shells of the Mackay icosahedron have been found, the small Al icosahedron being replaced by part of the vertices of a dodecahedron.

In that case, it has been interesting to make a comparison with the model of Duneau and Oguey [21], who have taylored the atomic surfaces in such a way that chemical composition and density are reproduced, short distances are avoided and the maximun number of Mackay generated. Their resulting atomic surfaces, quite similar to those found experimentally (Fig. 5b,c), exhibit as main difference some small Al surfaces lying on the mid-edges of the 6-cube and which have not been found experimentally. This means that such small Al surfaces could be characteristic of the small Al icosahedron belonging to the Mackay icosahedron. But this illustrates also the limits of a 6-D analysis for atomic positions which are dependent on weak atomic surface effects.

3. 2. 2. *Icosahedral AlLiCu phase.* A similar approach has been applied to i-AlLiCu but with more precision since large single grains could be grown [24], allowing both X-ray and neutron diffraction measurements. Using isotopic substitution of Cu and Li, it was first shown that Al and Cu atoms are randomly mixed. Neutron powder diffraction experiments with five different Li contrasts, together with previous X-ray and neutron single crystal diffraction results led to partial structure factors. Phases could be then reconstructed in an iterative way [25]. The 6-D structure is composed of three atomic surfaces with roughly the same size: two (Al,Cu) atomic surfaces lie at the origin and on the mid-edge, and one Li atomic surface is located on the body center. After cut by E_{par} the same icosahedral cluster as this existing in the crystalline R-phase [12] has been found, as examplified in figure 6, where two atomic planes are shown with respective traces of different shells.

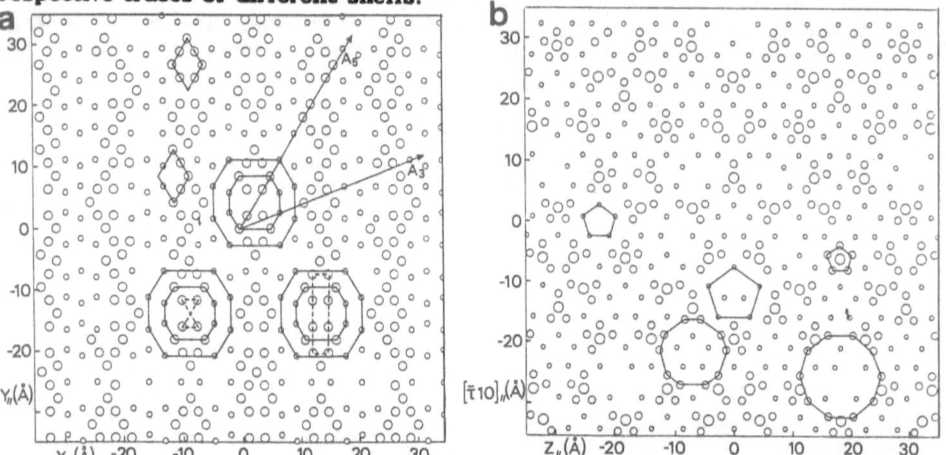

Figure 6. Cross sections of the 3-D atomic density corresponding to twofold (a) and fivefold (b) planes. Traces of the various cluster shells are outlined by thick lines. Al,Cu atoms correspond to large and medium circles and Li atoms to small circles (from ref. [25]).

Another point of interest is the comparison of these results with the early model of Henley and Elser [26], based on the decomposition of the crystalline

AlMgZn phase (nearly isostructural of the R-phase) into rhombic dodecahedron and prolate rhombohedra, which are decorated in a unique way. Yamamoto [27] has lifted the Henley-Elser model in the 6-D space, after some modifications to take into account the recent structure determination of the R-phase [12]. The result is quite similar to what is found directly from the experimental data, excepted for a small Al shell around the Li site. The agreement with diffraction data is similar in both cases. Differences between the two models concern mainly the outer shell of atomic surfaces and their precise shape, but obviously the same icosahedral clusters are generated.

Note that if the quasicrystalline structure may be considered as fully described in 6-D space with such a crystallographic approach, it appears, nevertheless, rather difficult to analyse what are the different structural characteristics of the icosahedral phase once it is projected in the physical space.

4 Cluster structure models

Cluster based quasicrystal models pack clusters as building units in parallel orientation (a necessary condition to enforce their own rotational symmetry propagation) and conveniently linked along high symmetry axes. See for example the aggregation of icosahedra face-to-face, i.e. connected along their threefold axes [28], used early for AlMn. Once the atomic short range order is thus, a priori, introduced in the structure, the long range order which will result on the cluster center positions has to be defined.

A simple aggregation rule which only avoids cluster overlap leads to a fully long range glassy structure [29], with an average i-symmetry. By reference to tilings, a quasiperiodic long range order may be insured by assigning the clusters to occupy a subset of the tiling vertices [3, 4, 30]. For the icosahedral phases, this subset may be an Ammann tiling expanded by τ^2 or τ^3 with respect to the elementary rhombohedral tiles of edge length a_R equal to about 0.4 to 0.5 nm.

At last, two properties of the models must be considered:

i) the degree of perfection of the long range order; a certain disorder may be incorporated by choosing for support a "random tiling", where there is a degeneracy in the ways of packing the clusters [31, 32], with resulting violations of the tile matching rules of the canonical tiling.

ii) the clusters form a backbone. In order to achieve the convenient density and chemical composition, additional so-called "glue" atoms must occupy the sites between clusters, another new source of "disorder".

We report now two examples of the cluster modelling procedure for i-AlMnSi and i-AlLiCu and compare the results with experimental data.

4. 1. i-AlMnSi

Several attempts have been made to model the structure of i-Al_4Mn in terms of Mackay cluster (MC) [4,6,33]. Whereas in ref. 4 the modelling was made at the a_R scale in 6-D space by the "cut-projection" method - by determining the acceptance domain of the MC - our approach was made directly in 3-D space by decoration of an Ammann tiling expanded by τ^2 ($\tau^2 a_R = 1.2$nm): an aggregate of 1095 MC face connected in first neighbouring positions was computer generated; half vertices of this Ammann tiling were MC decorated, and to half the number of prolate rhombohedra was added a supplementary MC using a graph algorithm. The structure projection shown in Fig. 7 displays a characteristic quasi-periodic

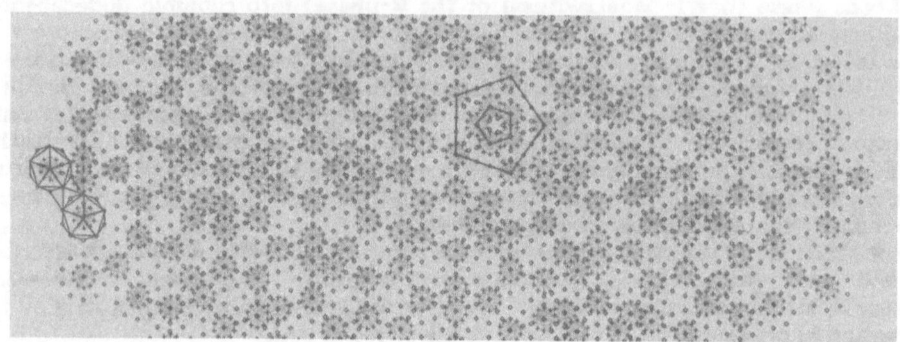

Figure 7. Projection of a MC i-backbone on a plane perpendicular to a 5-fold axis. The crosses stand for the Al atoms located at the vertices of the inner icosahedron, the circles stand from the Mn atoms on the vertices of the outer icosahedron. The 30 outer Al atoms are not shown. From ref. [33].

length scaling. The Fourier transform – which is simply the Fourier transform of the set of the decorated sites of the tiling times the MC structure factor – gives Bragg peaks positions fully consistent with the i-symmetry. But the calculated intensities were in relatively poor agreement with X-ray and neutron powder diffraction data [33].

The packing fraction of the model, expressed in terms of MC hard spheres, is somewhat small: 0.56 instead of 0.68 for the bcc approximant. Such a low cluster packing fraction is also shared by projection models [4,6,23]. It means that about 33% of the atoms do not belong to the clusters, instead of 22% for the α crystal.

4. 2. i-AlCuLi: A DENSE TWO-CLUSTER PACKING MODEL

A deterministic aggregation of two clusters has been proposed [34], in which the 2 basic rhombic triacontahedra of the R-phase, the small one (ST, shells a+b+c in Fig.2) and the large one (LT, shells a+b+c+d+e) decorate the Ammann rhombohedra expanded by τ^3 ($\tau^3 a_R = 21.39$ nm). In this cluster decoration, each vertex of the Ammann rhombohedra are decorated by a LT, while the ST have an internal densifying role, but also a face decorating one. Two classes of rhombohedra faces are defined, with consequent simple matching rules. The tiling is hierarchically generated by successive τ^3 inflation steps, the faces dichotomy and symmetry being restored at each step. Fig.8a shows the τ^3 inflated tiles (edge= $\tau^6 a_R$). Fig.8b represents typical ST aggregates surrounding a LT, which are located at the vertices of the τ^3 rhombohedra after achievement of the tiling. The last step of the modelling consists in adding (Al,Cu) atoms in the icosidodecahedral aggregates – between the LT and the 30 surrounding ST- to fit the model with the experimental density and chemical composition. However, though a comparison of the calculated structure factors with the experimental intensities for X-ray and neutron single grain diffraction has been quite satisfactory, such a model introduces a few half-indices reflections, typical of its hyper fcc structure, which were not observed experimentally [34].

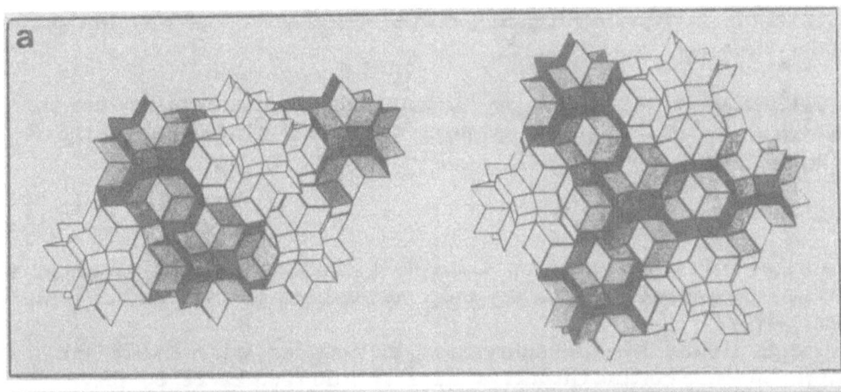

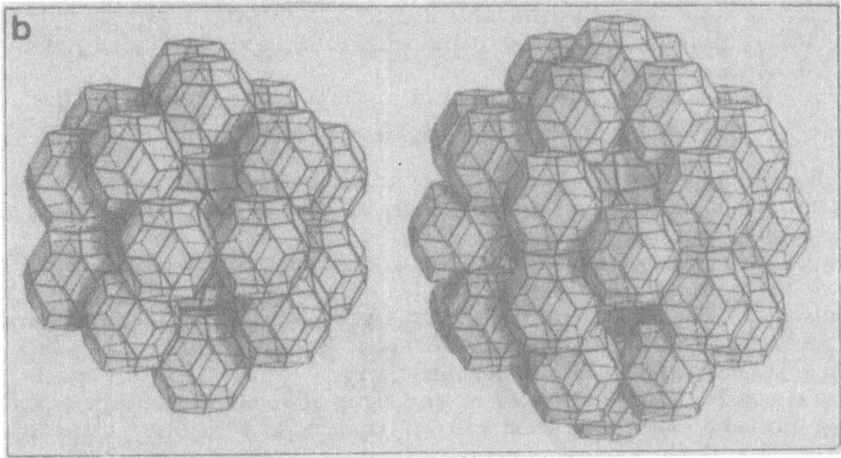

Figure 8. a) $\tau^6 a_R$ tiles decomposed in $\tau^3 a_R$ tiles; b) large aggregates of i-symmetry (dodecahedron of 20ST and icosidodecahedron of 30ST) generated in the dense two cluster model.

5 Conclusion

We have underlined in this paper several aspects of the atomic structure of quasicrystals in terms of the aggregation of polyatomic clusters. Two well studied icosahedral quasicrystals have been considered, but quasicrystals of lower symmetry, as the decagonal and the dodecagonal phases, are currently under investigation.

Several strong experimental evidences supporting such structures, have been reviewed, as well as some cluster structure modelling procedures based on the connection of clusters by convenient linkages on an appropriate quasilattice. Though each type of approach of the atomic structure of quasicrystals is based on different assumptions, all the results obtained so far allow to confirm that icosahedral atomic clusters defined in approximant periodic structures exist also in quasicrystalline phases. The clusters may involve up to more than hundreth atoms, confering to the structures a well defined "short and medium" range order.

At last, we believe that cluster structures offer a fairly well established support, to which the understanding of the stability and of physical properties of quasicrystals may be addressed.

Acknowledgements: we thank K. Hiraga for communication of the HRE micrograph of I-AlLiCu, and N. Tamura for his HR simulation of the R-phase prior to publication.

REFERENCES

1 Shoemaker D.P. and Shoemaker C.B.(1988) Icosahedral coordination in metallic crystals, Introduction to quasicrystals. Aperiodicity and Order, Vol.1, Academic Press, p.1-57.

2 Samson S. (1987) Fivefold aggregates in complex intermetallic compounds, Material Science Forum, Quasicrystals, K.H. Kuo (ed.) 22-24, 83-102.

3 Audier M. and Guyot P. (1989) Quasicrystal structure models related to crystalline structures, Extended icosahedral structures. Aperiodicity and Order, Vol.3, Academic Press, p.1-36.

4 Cahn J.W. and Gratias D. (1986) A structural determination of the Al-Mn icosahedral phase, Int. Workshop on Aperiodic Crystals, J. de Phys. France, C-3, 415-424.

5 Audier M. and Guyot P. (1989) Microcrystalline Al-Fe-Cu phase of pseudo-icosahedral symmetry, AAR Conf. Trieste, Quasicrystals, M.V. Jaric and S. Lundqvist (eds.), World Scientific, 74-91.

6 Elser V. and Henley C.L. (1985) Crystal and quasicrystal structures in Al-Mn-Si Phys. Rev. Lett., 55, 2883-2886.

7 Spaepen F., Chen L.C., Ebalard S. and Ohashi W. (1989) Microquasicrystals, superlattices and approximants, AAR Conf. Trieste, Quasicrystals, M.V. Jaric and S. Lundqvist (eds.), World Scientific, 1-18.

8 Shechtman D., Blech I., Gratias D. and Cahn J.W. (1984) Metallic phase with long-ranged orientational order and no translational, Phys. Rev. Lett., 53, 1951-54.

9 Mackay A.L. (1962) A dense non-crystallographic packing of equal spheres, Acta Cryst. 15, 916-918.

10 Guyot P. and Audier M. (1985) A quasicrystal structure model for Al-Mn, Phil. Mag. B52, L15-L19.

11 Bergman G., Waugh J.L. and Pauling L. (1957) The crystal structure of the metallic phase $Mg_{32}(Al,Zn)_{49}$, Acta Cryst., 10, 254-259.

12 Audier M., Pannetier J., Leblanc M., Janot C., Lang J.M. and Dubost B. (1988) An approach to the structure of quasicrystals: A single crystal X-ray & neutron diffraction study of the R-$Al_6 Cu Li_3$, Physica B,153, 136-142.

13 Guyot P. and Audier M. (1990) Electron microscopy and quasicrystals Geometry and Thermodynamics-Common problems of quasicrystals, liquid crystals and incommensurate systems, J.C. Tolédano (ed), NATO ASI Series, B: Phys. 229, Plenum Press, 89-107.

14 Sainfort P. and Guyot P. (1987) Low electron energy loss spectroscopy of quasicrystalline and crystalline AlLiCu alloys, Scripta Metall., 21, 1517-1522.

15 Hirabayashi M., Hiraga K. Shindo D. and Williams T.B. (1990) Icosahedral quasicrystalline Al-based alloys studied by HRTEM, Quasicrystals and Incommensurated Structures in Condensed Matter, J. Yacaman et al. (eds.), 262-267.

16 Beeli C., Nissen H.U. and Stadelmann, (1990) Comparison of HREM images and contrast simulations for decagonal Ni-Cr quasicrystals, J. de Phys. Paris. 51,

661-674.

17 Hiraga K., private communication.

18 Tamura N. and Verger-Gaugry J.L., (1991) HREM image simulations of the R-AlLiCu icosahedral approximant phase, in preparation.

19 Audier M., Guyot P. and Bréchet Y. (1990) High-temperature stability and faceting of the icosahedral AlFeCu phase, Phil. Mag. Lett., 61, 55-62.

20 Janot C., (1991) Recent developments in quasicrystalline materials, This conference.

21 Gratias D. Cahn J.W.and Mozer B. (1988) Six-dimensional Fourier analysis of the icosahedral $Al_{73}Mn_{21}Si_6$ alloy, Phys. Rev. B, 38, 1643-1646.

22 Janot C., Pannetier J., Dubois J.M. and de Boissieu M., (1989) Neutron diffraction approach of the atomic decoration of the AlMn icosahedral quasicrystal, Phys. Rev. Lett., 62, 450-453.

23 Duneau M. and Oguey C. (1989) Ideal AlMnSi quasicrystal: a structural model with icosahedral clusters, J. de Phys. Paris, 50, 135-146.

24 Dubost B., Lang J.M., Tanaka M., Sainfort P. and Audier M. (1986) Large AlLiCu single quasicrystals with triacontahedral solidification morphology, Nature (London), 324, 48-50.

25 de Boissieu M., Janot C., Dubois J.M., Audier M. and Dubost B. (1991) Atomic structure of the icosahedral AlLiCu quasicrystal, J. Phys.: Condens. Matter, 3, 1-25.

26 Henley C.l. and Elser V. (1986) Quasicrystal structure of $(Al,Zn)_{49}Mg_{32}$, Phil. Mag., 53, L59-L66.

27 Yamamoto A. (1991) Ideal structure of the icosahedral AlLiCu quasicrystal, preprint (submitted to Phys. Rev. B).

28 Audier M. and Guyot P. (1986) Al_4Mn quasicrystal atomic structure, diffraction data and Penrose tiling, Phil. Mag. 53, L43-L51.

29 Stephens P.W. and Goldman A.I. (1986) Sharp diffraction maxima from an icosahedral glass, Phys. Rev. Lett., 56, 1168-1171.

30 Henley C.L. (1986) Sphere packings and local environments in Penrose tilings, Phys. Rev., B34, 797-816.

31 Henley C.L. (1990) Random tilings and real quasicrystals, Quasicrystals and Incommensurated Structures in Condensed Matter, J. Yacaman, D. Romeu, V. Castano and A. Gomez (eds.), World Scientific, 152-169.

32 Shaw L.J., Elser V. and Henley C.L.(1991) Long-range order in a three-dimensional preprint (submitted to Phys. Rev.B).

33 Guyot P., Audier M. and Lequette R. (1986) Quasicrystal and crystal in AlMn and AlMnSi, model structure of the icosahedral phase, J. de Phys., C3, 47, 389-404.

34 Guyot P., Audier M. and de Boissieu M. (1990) Molecular structure of the icosahedral AlLiCu phase, Quasicrystals and Incommensurated Structures in Condensed Matter, J. Yacaman, D. Romeu, V. Castano and A. Gomez (eds.), World Scientific, 251-261.

STM STUDIES OF CLUSTERS

KLAUS SATTLER
University of Hawaii
Department of Physics and Astronomy
Watanabe Hall / 2505 Correa Road
Honolulu, Hawaii 96822, USA

ABSTRACT. The recent development in the research of small supported clusters and nanogranular thin films using a scanning tunneling microscope (STM) is reviewed. Metal clusters are imaged with atomic resolution on atomically flat graphite single crystal substrates. The adsorbed clusters tend to have their atoms situated above lattice sites of the graphite surface. Strong cluster-substrate interaction is observed yielding a broad variety of superstructures in regions close to the adsorption sites. The superstructure pattern are assigned to periodic charge density modulations (PCDM), and 15 observed pattern can be explained by just one type of charge density lattice superimposed on the graphite lattice with different phase adjustments.

1. Introduction

The formation of clusters on single crystal substrates plays an important role in nucleation and growth processes of thin solid films. Surface and interface research often has to take into account the microscopic features and morphology dynamics occurring in the first stages of film formation. In the submonolayer regime, adatoms can acquire various positions ontop of the substrate lattice and the question of commensurate or non-commensurate arrangements is of fundamental importance. Clusters, containing just a few atoms can be the nuclei for film growth and their initial structures can determine the final film quality. The adsorbed clusters can distort locally the surface charge density and may induce various types of surface reconstructions. Local surface tension and surface stress can be induced due to a mismatch of the cluster interatomic distances with the lattice constant of the substrate.

The scanning tunneling microscope (STM) is one of the main instruments for nanostructure generation and analysis. It is sensitive to extremely low coverages of admaterials in the far submonolayer range. It provides direct images of both periodic and nonperiodic features with atomic resolution. Besides the imaging of individual atoms it also gives electron density distributions not associated with atoms, like charge density waves (CDW) for low-dimensional solids, periodic charge density modulations (PCDM) at surface defects, or superperiodic charge density pattern with lattice constants very large compared to the atomic surface nets. Small clusters can be imaged atom by atom yielding cluster atomic structures, registry with the substrate lattice or dynamic cluster properties. STM-related methods like scanning tunneling spectroscopy (STS), STM-based local inverse photoemission spectroscopy (STM-IPES), or atomic force microscopy (AFM) can give further information about geometric, electronic and dynamic behavior of small supported clusters.

A few studies have been reported during recent years using STM for cluster-, small particle - and granular film research. Small clusters of Cu, Ag, Au, and Al were analyzed on highly-oriented pyrolytic graphite (HOPG), with the instrument operated either in air or in UHV, with static and

P. Jena et al. (eds.), Physics and Chemistry of Finite Systems: From Clusters to Crystals, Vol. I, 61–70.
© 1992 *Kluwer Academic Publishers.*

dynamic behavior being observed [1]. Small Si clusters where imaged on a Au(001)-(5x20) reconstructed surface and current-voltage characteristics were taken revealing the cluster energy band gap [2]. An STM/STS analysis of Fe particles on GaAs showed a size dependent metal-insulator transition [3]. Sb clusters on GaAs(110) where found to form for coverages in the range of a few % of a monolayer [4]. As the coverage was increased, the Sb islands were growing in size while maintaining a height corresponding to one monolayer. Then, for coverages close to 1.0 ML, the Sb islands formed a continuous monolayer. Planar clusters have also been imaged for Cr/GaAs(110) [5]. Preferential clustering along step edges was observed in this case. The mechanism for the formation of small clusters on GaAs(110) has been discussed under consideration of the adatom-substrate interaction, in particular the pairing of Au-adatoms [6]. A metastable 3D cluster phase was found for Ge on Si(001) with clusters having a {105} ordered structure [7].

In our laboratory we used graphite (HOPG) and mica as substrates because they are inert for most molecules, have micron-sized atomically flat regions with perfect crystal structure, and clean surfaces are easily prepared by cleavage. This guarantees that the clusters do not adsorb at edges or at lattice defects. However, as the studies showed, the cluster-substrate interaction is not negligible and periodic electronic modulations can be generated by adclusters which can have a feedback effect on the location and atomic structure of the clusters.

At low coverages, single metal adatoms as well as small clusters and particles can be observed in a broad size range. The cluster sizes can be controlled by the evaporation conditions and relatively narrow size distributions can be achieved. At high coverages, metals on graphite or mica tend to form small particles which agglomerate to larger islands. Such agglomerates can show ordered arrangements, with quite uniform spherical shapes and narrow size distributions for the particles. Photon emission from the STM due to electron tunneling probes elementary excitations and their relation to the morphology of the target. Simultaneously with the photon measurement a STM picture of the surface region can be obtained. We show the tunneling voltage dependence for photon emission with resonances at the energies for localized and extended surface plasmons of nanogranular gold films.

The STM used in our laboratory (Digital Instruments) is a single tube design. It is operated in both the variable current and the variable height mode. The x, y scan response of the piezo drive was calibrated before the experiments using graphite. In some of the experiments we prepared the samples in a high-vacuum system and then transferred them to the STM operating at atmospheric conditions. In other experiments we made the depositions in a preparation chamber (10^{-8} Torr) and subsequently transferred the samples to a UHV chamber (10^{-10} Torr) without breaking vacuum. In this case an STM in UHV was used (Mc Allister Technical Services with control unit from Digital Instruments). The supported cluster samples were prepared by cleavage of graphite or mica with subsequent deposition of metal vapor. Single adatoms diffuse at the surface of the substrate and form clusters by aggregation. We either used thermal evaporation from a filament of ion sputtering of a metal target. Both methods yield clusters which are well separated by large areas of clean graphite for the far-submonolayer coverage regime, typically 1% of a monolayer. Some of the images we analyzed by using a separate computer station equipped with a software package for image processing which we developed in our laboratory. This package allows us to load the coordinates of the substrate lattice and to map the adatoms and adclusters with great precision on the substrate atomic net. A separate computer program can be applied to find the center point for an individual atom and to give its coordinates relative to the coordinates of the substrate lattice. For the light emission experiments we enclosed the STM with a light-tight chamber and measured the photon intensity with a sensitive multiarray diode detector, cooled to -40C for noise reduction. The counting rate was very low, smaller than 400 counts per second which required a careful consideration of the noise signals. Also, during the long experimental runs (typically two hours for taking one data point) the surface structure was continuously monitored by the STM, where every 20 seconds a topographic picture was obtained.

2. Cluster Atomic Structures

Cluster imaging with the STM for obtaining atomic structures is mainly limited to two-dimensional structures because the STM is only sensitive to the first atomic layer of a surface. Atoms in the center of a cluster are hidden from view. However, for very small clusters containing up to about 10 atoms individual atoms can be imaged and bond lengths and bond angles can be determined. Such studies have been performed for various metal clusters; Cu, Ag, Au, Al [1], Pt [8,9] and Al [10].

Two examples of STM cluster images and the corresponding structural models are shown for

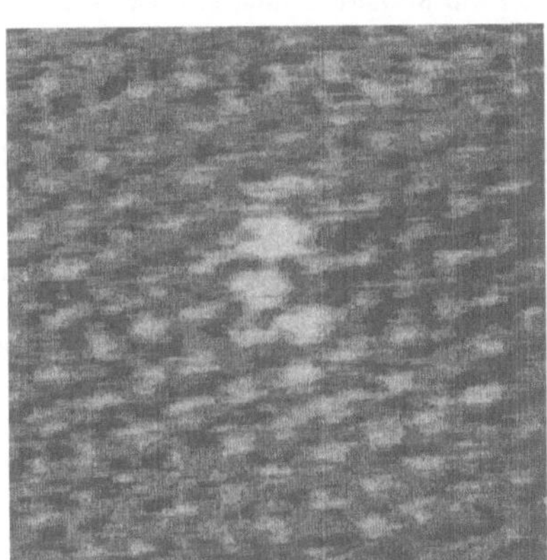
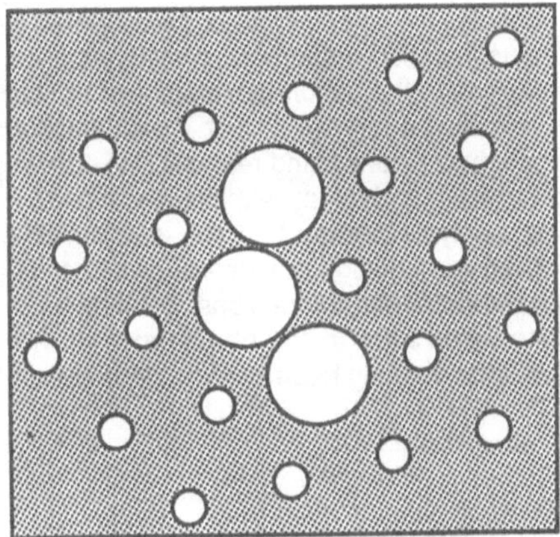

Fig. 1 $(2.6nm)^2$ STM image of a Pt_3 cluster on graphite, and structural model.

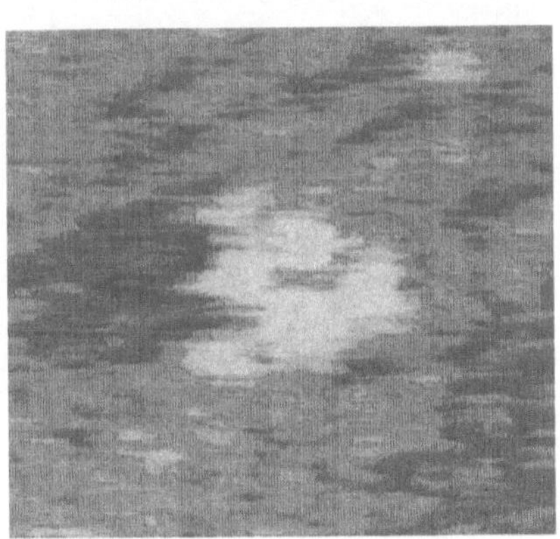
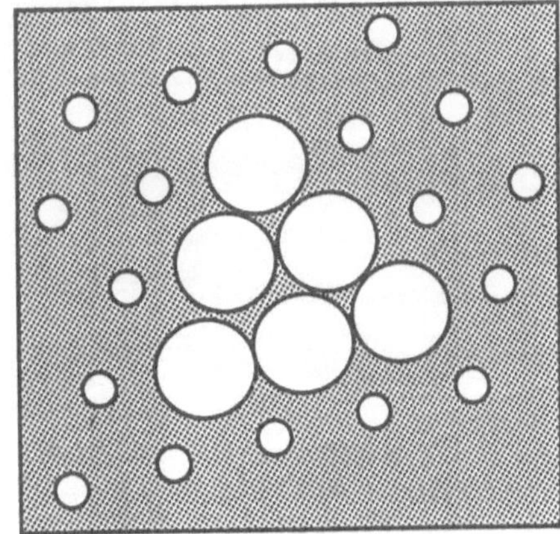

Fig. 2 $(1.9nm)^2$ STM image of a Pt_6 cluster on graphite, and structural model.

Pt$_3$ and Pt$_6$ on graphite in Figures 1 and 2, respectively. The atoms in the clusters are placed ontop of b-sites of the graphite lattice. Single, isolated ad-atoms of various metals also prefer b-sites [1,9]. Therefore, in a more general way we might say that the adatoms prefer areas at the substrate with enhanced electron density of states.

3. Cluster-Substrate Interaction

The interaction of an adsorbed cluster with the substrate can have two effects: (1) The cluster affects the substrate and/or (2) the substrate affects the cluster.

3.1 Cluster-Induced Periodic Charge Density Modulations (PCDM)

We consider the affect an adsorbed cluster can have on the substrate lattice. STM images of metal clusters adsorbed on graphite may show a variety of superstructures on the graphite surface in the vicinity of the adsorption sites. For Pt clusters of graphite, 10 different superstructures have been reported earlier [10]. For Co-clusters on graphite, the same types of structures and 5 additional types were found recently [11]. The superstructures are localized in small areas near the metal particles and decay within a distance of 2-5 nm into the graphite lattice. A two-dimensional Fourier analysis shows that the structures contain up to three pairs of first-order Fourier components in addition to those of graphite. These Fourier components have different intensities and phases in different directions, but they have the same period of 1.5a (a, the lattice constant of graphite, equivalent to the bb-distance in the surface net) and are rotated 30⁰ relative to the graphite lattice. It has been shown [10] that the observed structures are due to a charge density

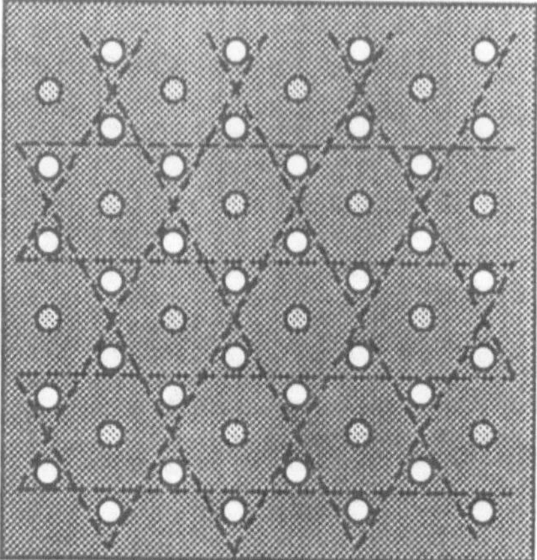

Fig. 3 (21nm)^2 STM image of a superstructure on graphite near an adsorbed Co-cluster and schematic model for the superlattice.

effect rather than to an atomic rearrangement. All of the 15 observed structures can be explained by periodic charge density modulations (PCDM) superimposed onto the graphite lattice.

Fig. 3 shows STM image and schematic model for a superstructure observed around Co-particles on graphite. It is one of the types of PCDM which was also found for adsorbed Pt-particles [10].

3.2 PCDM Feedback on Cluster Atomic Structures

The cluster-induced superstructures may have a decisive effect on the atomic arrangements in the clusters. The Pt particle in Fig.4 has induced one Fourier component of a PCDM yielding a periodic line-structure rotated 30^0 relative to the graphite lattice. Then, as a response, the atoms in the particle moved to adsorption sites given by the PCDM and the atomic lattice and the island shows a rectangular net.

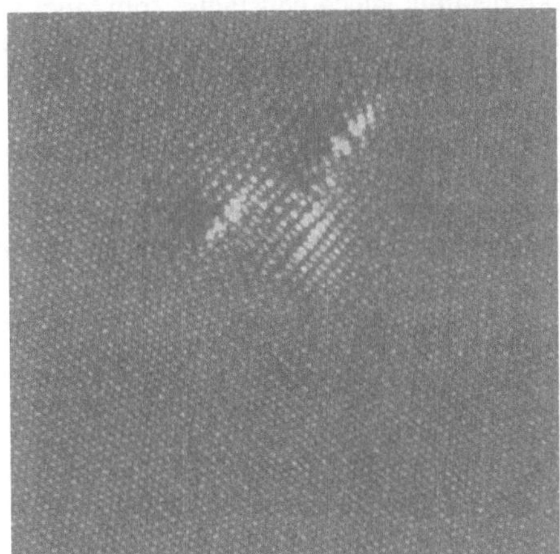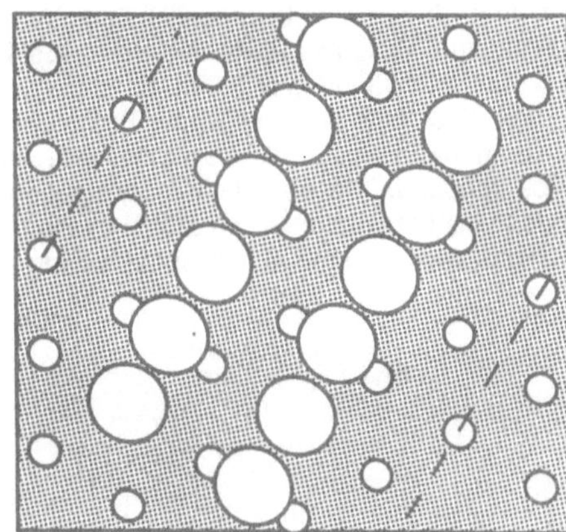

Fig.4   (18nm)^2 STM image of a Pt island structure, and structural model, considering adsorption of Pt atoms on the graphite surface net with a one-component 'periodic charge density modulation'.

We conclude that the lateral electron state distribution, as observed in the STM images, gives the substrate pattern for adsorption of the metal clusters. The atoms in the particles seem not to make any difference in choosing substrate atoms or PCDM-vertices. Therefore we might speculate that any enhanced lateral electron state density on a substrate could act as an adsorption site for adcluster atoms. Rectangular 2D-lattices as seen in Fig.4 have been found earlier for Au- and Ag-monolayer islands on graphite [12], and there might have been a correlation between PCDM and island structures in those experiments as well .

The origin of the PCDM is not well understood. One of the structures [(root3xroot3)R30^0] was found for defects on graphite or for adsorbed molecules [13]. It was suggested that the molecules perturb the surface charge density of graphite giving periodic oscillations similar to Friedel oscillations. Nakagawa et al. [14] ascribe the observed superstructures at defects to an interference of electron waves scattered at the defect. Both interpretation models however do not explain the large number of different structures which occur for metal cluster adsorption. The description of all the 15 observed types of pattern by only one PCDM wavelength and one angle of rotation relative to the atomic lattice indicates that a more general description is needed. The superimposed charge density pattern might be caused by a local distortion of the Fermi surface of the carbon toplayer produced by strain at the adsorption site due to a mismatch of the lattice parameters for adcluster and substrate.

4. Adsorption on Giant Lattice Systems (GLC)

Recently the observation of a giant lattice has been reported in STM studies of graphite [15]. Such lattices appear as superimposed on the atomic lattice having 2-D hexagonal symmetry and lattice constants of several tens of Angstroms. They can be found in STM images near long straight steps of the atomic lattice and they extend at least over micron dimensions without apparent decay. They have been explained by Moire fringes due to a small rotation of the first carbon layer relative to the second layer [15]. Depending on the angle of rotation different lattice constants for the giant lattices occur.

We have observed such giant lattices with three different lattice constants, 7nm, 4nm and 1nm. In all three cases hexagonal symmetry was found. We imaged the pattern together with the regular, atomic graphite lattice, atomically resolved. A giant graphite lattice with a lattice constant of 3.85 nm is shown in Fig.5. In Fig. 5a, a long straight step divides the image into two regions. On the right side of the step the giant lattice is observed, but it is missing on the left side of the step. If we zoom into a small area on the left side, we find the atomic graphite lattice with the .246 nm lattice constant. The giant lattice corrugation directly at the step appears enhanced which we relate to instrumental effects, caused by the response of the electronic feedback circuit when the tip approaches the step at high speed during scanning.

Fig.5   (143nm)^2 STM image of a giant graphite lattice (lattice constant 3.85nm) at a step in the atomic lattice, and (9.0)^2 zoomed image showing the atomic lattice of graphite together with the superlattice.

It is interesting to see how adsorbed clusters would respond to a giant lattice. The images of Fig. 6 show one and two Co-clusters adsorbed on a giant lattice substrate, respectively. We find that the clusters are situated above top sites of the giant lattice. The diameters of the clusters ($\sim$1nm; $\sim$Co$_{30}$) is about half the diameter of the vertex areas of the giant lattice, as seen from zoomed images.

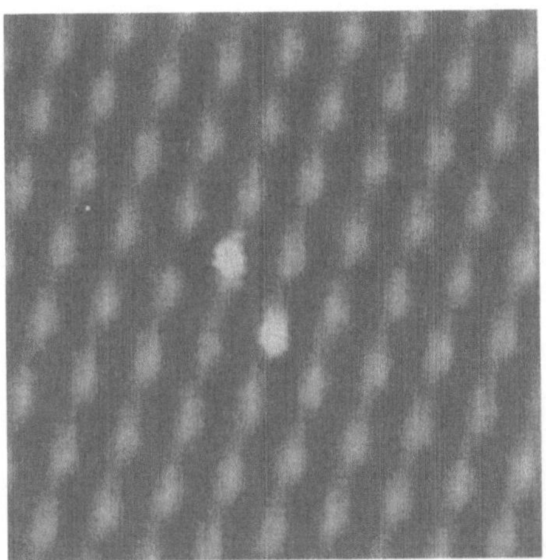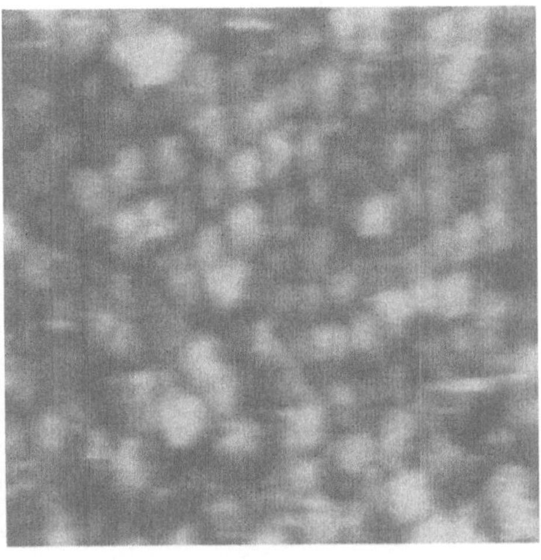

Fig.6 STM images of a giant lattice on graphite with one {(26.2nm)^2 image},
or two {(28.1nm)^2 image} cobalt particles adsorbed at top-sites of the giant lattice.

We believe that the appearance of the giant lattice is due to periodic electron density of states enhancement, which originally might (or might not) be generated by Moire pattern effects. Clusters on such a surface move to areas of enhanced density of states, the vertex points of the giant lattice. These vertex areas seem to provide higher bond strength for clusters on the substrate. The earlier results that single metal adatoms prefer top sites on the atomic lattice [1,9] and that the atoms in metal clusters and islands occupy the vertex points of PCDM's appears to be valid for metal adclusters on a giant lattice as well. In the first two cases single atoms or the atoms in a cluster are registered with the underlying lattice, which is found in the third case for the whole cluster. In this respect we might treat a substrate lattice merely as a periodic electron state density pattern independent if the pattern is produced by underlying atoms or not. Then the PCDM and GLS might be considered as excitations of the periodic state density associated with the atomic lattice. Atomic lattice, PCDM, and GLS as electron state density lattices become equivalent in respect to the adsorption of adatoms or clusters.

5. Cluster Lattice Systems (CLS)

Crystalline arrangements of clusters and small particles as a subgroup of cluster-assembled materials is highly interesting. Recently, STM images of close-packed ordered C_{60} films were reported [16].

In Fig.7 we show an STM image and a structural model for gold particles grown on a mica substrate. The diameter of the particles is about 280 Angstroms and they are arranged in a hexagonal lattice. The particle arrays are coherent in (2000A)^2 domains. Between the hexagonally ordered areas, the particles are showing directional order which extends at least over micron dimensions, the extend of areas which we have investigated.

Another phase for gold CLS can be observed as well. The film in Fig.8 is not as smooth as the one in Fig.7, showing roughness on the scale of the particle diameters. Such roughness in the particle films can be induced by annealing the films, without loosing their granular nature.

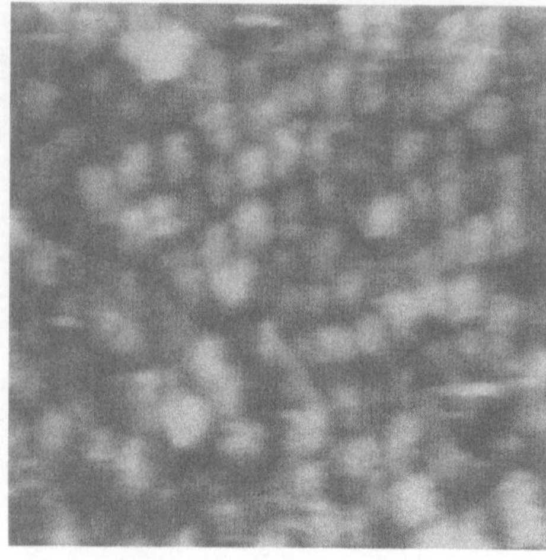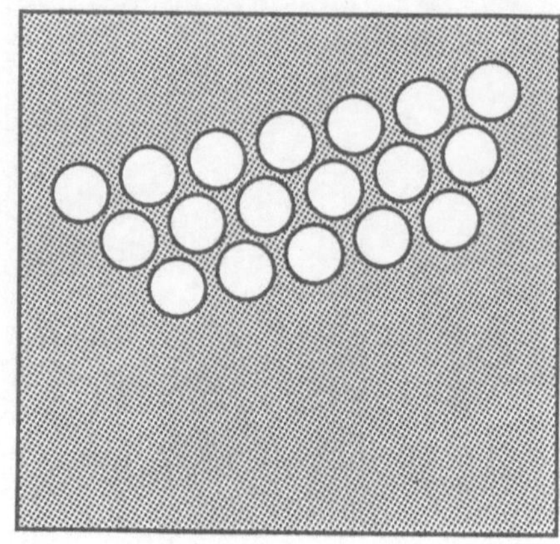

Fig.7 $(359nm)^2$ STM image and structural model for a film of 25-30nm gold particles showing a region of hexagonal lattice order (lattice constant a~30nm).

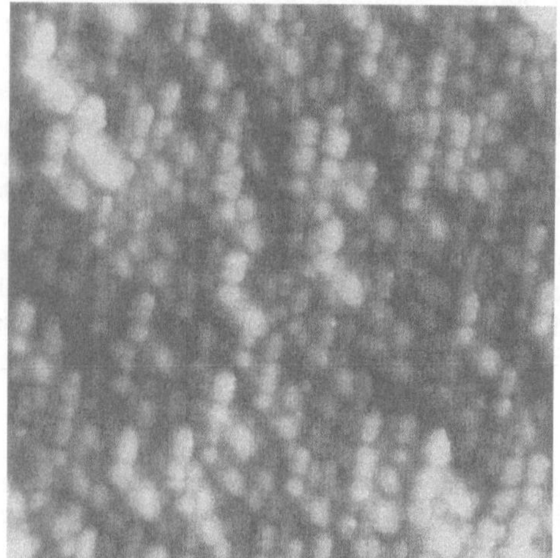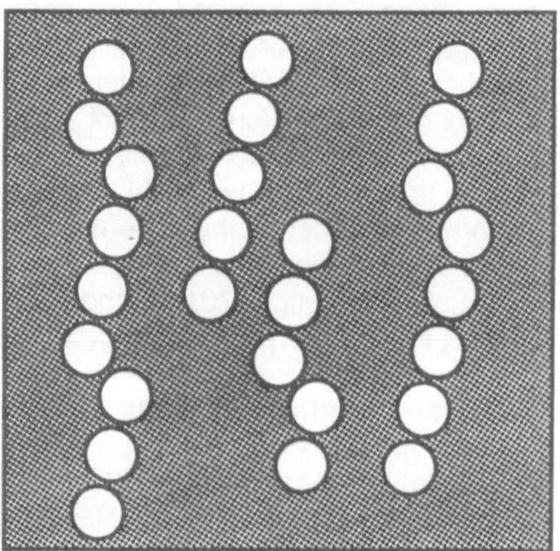

Fig.8 $(525)^2$ STM image of a film of ~15nm gold particles showing directional order with regions of 'herringbone' structure.

where arranged in pattern according to the 'herringbone' reconstructed surface. It might be that our observation of 'herringbone' order for gold particles is also originated by underlying lattices of zig-zag type reconstructed gold areas.

6. Localized and Extended Surface Plasmons on Granular Gold Films

Experiments on metal-insulator tunnel junctions showed photon emission due to electron tunneling [18], explained by the excitation of surface plasmons. The tunneling electrons are scattered in the surface region, exciting surface plasmons, and photons are emitted when the plasmons decay radiatively. For isolated gold particles of size 30nm a resonance in the emission spectrum was found at 1.9eV [19] which has been explained by localized surface plasmons [20]. On smooth surfaces the surface plasmon line is at 2.7eV.

Light emission from an STM has first been observed by Gimzewski et al. [21]. Using an STM as a tunnel junction allows to observe photon emission from small nanometer-scale areas at the surface. Simultaneously with the photon signals an image of the area can be recorded. Thus, the photon energy distribution can be related to the morphology of the surface area.

In Fig. 9 we show the dependence of the STM-induced photon intensity on the applied bias voltage, taken for a granular gold film. Due to the relatively low counting rate of less than 400 counts per second we restricted our step width to 0.25V. The data are the result of long experimental runs (several hours per data point). The images showed that the target was composed of assembled gold particles (see Figs.7 and 8).

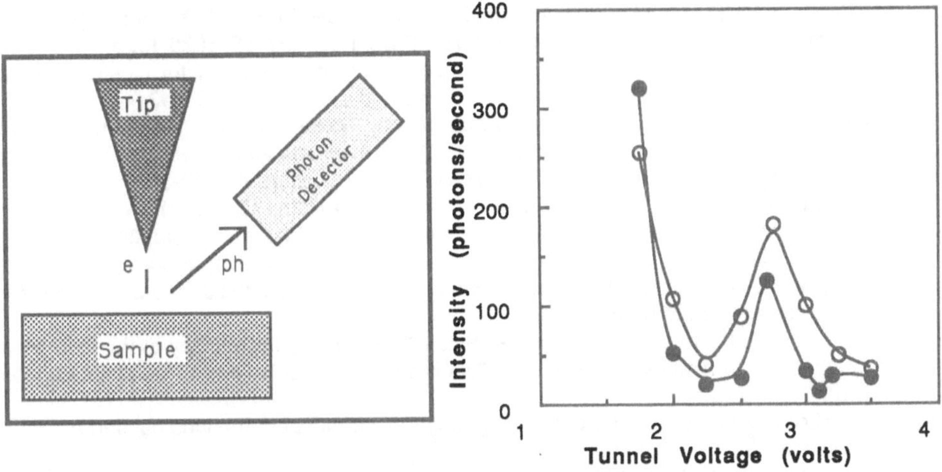

Fig.9   STM induced light emission from a granular gold film. The photon intensity is plotted versus the applied bias voltage.

Two experimental curves are displayed in Fig.9, taken in different runs for different samples. Both curves show the same behavior; enhanced photon intensity at about 1.9 eV and 2.7 eV. The two peaks can be assigned to the excitation of localized (1.9eV peak) and extended (2.7 eV peak) surface plasmons of gold. Further experiments are currently being performed in our laboratory using higher density of data points to obtain the resonance energies more accurately and to study in detail how the localized plasmon peak depends on the local surface morphology.

Acknowledgements

The author would like to thank his co-workers Maohui Ge, Narayana Venkatesvaran, and Jie Xhie for assembling the material for this review. Financial support from the Office of Technology Transfer and Economic Development (OTTED) and from the Research Council of the University of Hawaii is gratefully acknowledged.

References

1. Ganz E., Sattler K., and Clarke J. (1989) 'Scanning tunneling microscopy of Cu, Ag,Au, and Al adatoms, small cluster, and islands on graphite', Surf. Sci. 219,33-67
2. Kuk Y., Jarrold M.F., Silverman P.J., Bower J.E., and Brown W.L. (1989) 'Preparation and observation of Si_{10} clusters on a Au(001)-(5x20) surface', Phys. Rev. B 39, 11168-11170
3. First P.N., Stroscio J.A., Dragoset R.A., Pierce D.T., and Celotta R.J. (1989), 'Metallicity and gap states in tunneling to Fe clusters on GaAs(110)', Phys. Rev. Lett. 63, 1416-1419
4. Martensson P., and Feenstra R.M. (1988) 'Voltage-dependent imaging of antimony on a GaAs(110) surface', Journal of Microscopy, 152, 761-769
5. Trafas B.M., Hill D.M., Benning P.J., Waddill G.D., Yang Y.-N., Siefert R.L., and Weaver J.H. (1991) 'Clustering and reaction for Cr/GaAs(110): Scanning tunneling microscopy and photoemission studies', Phys. Rev. B 43, 7174
6. Allan G. and Lannoo M. (1991) 'Negative-U character of the adsorption on semiconductor surfaces: Application to metals on GaAs(110)', Phys. Rev. Lett. 66, 1209-1211
7. Mo Y.-W., Savage D.E., Swartzentruber B.S., and Lagally M.G. (1990) 'Kinetic pathway in Stranski-Krastanov Growth of Ge on Si(001)', Phys. Rev. Lett. 65, 1020-1023
8. Sattler K. (1991), 'Scanning tunneling microscopy and spectroscopy for cluster and small particle research' Z. Phys. D - Atoms, Molecules and Clusters 19, 287-292
9. Mueller U., Sattler K., Xhie J., Venkateswaran N., and Raina G. (1991), 'A scanning tunneling microscope study of single platinum atoms, platinum dimers and trimers on highly-oriented pyrolytic graphite', Z. Phys. D - Atoms, Molecules and Clusters 19, 319-321
10. Xhie J., Sattler K., Mueller U., Venkateswaran N., and Raina G. (1991) 'Periodic charge-density modulations on graphite near platinum particles' Phys. Rev. B 43, 8917-8923
11. Xhie J., Sattler K., Ge M. and Venkateswaran N. (1991), 'Scanning tunneling microscopy study of the interaction between adsorbed clusters and graphite substrates', Proceedings of this conference, to be published
12. Ganz E., Sattler K., and Clarke J. (1988) 'Scanning tunneling microscopy of the local atomic structure of two-dimensional gold and silver islands on graphite', Phys. Rev. Lett. 60, 1856-1859
13. Mizes H.A. and Foster J.S. (1989) , 'Long-range electronic perturbations caused by defects using scanning tunneling microscopy', Science 244, 559-562
14. Nakagawa Y. Bando H. Ono M. and Kajimura K.,unpublished
15. Kuwabara M., Clarke D.R. and Smith D.A. (1990), 'An anomalous super-periodicity in scanning tunneling microscope images of graphite', IBM Research Report (Materials Science RC 15344, #68325), 9 pages
16. Li Y.Z., Patrin J.C., Chandler M., Weaver J.H., Chibante L.P.F., Smalley R.E. (1991) ,'Ordered overlayers of C60 on GaAs(110) studied with scanning tunneling microscopy', Science 252, 547-548
17. Chambliss D.D., Wilson R.J., and Chiang S. (1991), 'Nucleation of Ordered Ni Island Arrays on Au(111) by Surface-Lattice Dislocations', Phys. Rev. Lett. 66, 1721-1724
18. Lambe J. and McCarthy S.L. (1976), 'Light emission from inelastic electron tunneling', Phys. Rev.Lett. 37, 923-925
19. Hansma P.K. and Broida H.P. (1978), 'Light emission from gold particles excited by electron tunneling', Appl. Phys. Lett. 32, 545-547
20. Rendell R.W., Scalapino D.J., and B. Muehlschlegel (1978), 'Role of local plasmon modes in light emission from small-particle tunnel junctions', Phys. Rev. Lett. 41, 1746-1750
21. Gimzewski J.K., Reihl B., Coombs J.H., and Schlittler R.R. (1988), 'Photon emission with the scanning tunneling microscope', Z. Phys. B 72, 497-501

STRUCTURE AND TRANSFORMATION. CLUSTERS AS MODELS FOR CONDENSED PHASES

LAWRENCE S. BARTELL, THEODORE S. DIBBLE, JAMES W. HOVICK
AND SHIMIN XU
Department of Chemistry
University of Michigan
Ann Arbor, MI 48109-1055

ABSTRACT. The special advantages and characteristic limitations of molecular clusters in investigations of the properties of condensed phases are evaluated. Clusters naturally lend themselves to the exploration of (1) the possible ways molecules can pack while undergoing various degrees of thermal agitation, and (2) the cooperative behavior of molecules in transitions between phases. Clusters can be studied experimentally, in expanding supersonic jets, or computationally, in molecular dynamics simulations. Of particular interest are clusters of polyatomic molecules because their structural diversity is much richer than that of atomic clusters. Novel phases not seen in the bulk have been generated, and explained. Extraordinarily rapid nucleation rates have been observed at the extremely high supercoolings that can be attained. Interfacial free energies of boundaries between liquid and solid phases have been inferred. Clusters are versatile and effective model systems of great promise in research on condensed matter.

Introduction

Clusters offer fresh insights into the nature of condensed matter. Much of the emphasis in cluster research to date has been concerned with the transition from molecular to bulk properties as aggregates increase in size or with the transition between disordered and ordered systems of atoms packed around a molecular chromophore [1-3]. Aims of the program at the University of Michigan are different. Although we seek to investigate transitions in clusters, the transitions are from one structural type to another (i.e., phase transitions) in one-component systems of polyatomic molecules. Such systems exhibit a richer diversity of structural types than do the more commonly studied monatomic systems [4] and, indeed, show an even richer diversity than have conventional crystallographic studies of bulk matter [5]. Observed transitions can be liquid-to-solid (freezing) or solid-to-solid. The techniques employed in our structural program are electron diffraction (ED) analyses of clusters generated in supersonic expansions (containing $\sim 10^3$ - 10^4 molecules) [6] and molecular dynamics (MD) simulations of clusters (~ 50 - 500 molecules) [5].

The special advantages of studying clusters as models of condensed systems in transition are:

1) the particularly large supercooling that can be attained,
2) the exceptionally fast transitions that can be induced,
3) the unusually clean measurements of nucleation rates possible, free from artifacts due to impurities,

71

P. Jena et al. (eds.), Physics and Chemistry of Finite Systems: From Clusters to Crystals, Vol. I, 71–76.

    4) the directness with which critical nuclei, which are themselves clusters, can be studied (by MD), and

    5) the facile generation of phases not ordinarily encountered in the bulk.

Advantage (1) is realized, in part, because of the small volume of clusters and, in part, because of the fast rate of cooling that can be imposed experimentally (in supersonic jets) and in computer simulations (via MD). Advantage (2) is a direct consequence of (1). Advantage (3) arises by virtue of the freedom of the clusters from any contact with a foreign surface or other phase than highly rarefied, inert gas. Concerning advantage (4), until MD simulations became feasible, all inferences about critical nuclei in phase changes had been indirect and conjectural. Current MD analyses can examine critical nuclei in detail [7]. Advantage (5) is aided in experimental work by the very fast measuring techniques possible with clusters which allow the observation of the metastable phases often formed under the extreme conditions that can be imposed. In the complementary computational research, this diversity owes to the great variety of starting configurations that can be explored in MD simulations.

Characteristic Aspects of Transitions in Clusters

It is instructive to consider the characteristic dimensions and time scales of observation of clusters, for these are responsible for both the special advantages and the natural limitations associated with the present approach to phase transitions. Nucleation rates, J, are defined in terms of the number N of critical nuclei formed per unit volume per unit time, or

$$J = (dN/dt)/V .\tag{1}$$

Under common conditions, a given cluster will change phase virtually as soon as a single critical nucleus is formed in it [8,9]. In supersonic jets, the timescale of observation may range from 10^{-6} to 10^{-4} s, and a very common cluster diameter is 100 Å (although larger and smaller clusters can be produced). In MD simulations the timescale is typically 10^{-10} to 10^{-11} s at a given temperature with a cluster diameter of perhaps 25 Å. For a given phase change to take place in the supersonic experiments, then, J must be of the order of 10^{28} m^{-3} s^{-1} or greater, which is a considerably greater value than that measured in conventional experiments. Transitions in MD computations ordinarily have to be 10^8-fold faster, yet, to be seen. Clearly, it cannot be taken for granted that equilibrium is established in either case. What is seen depends crucially upon what one starts with. That is a severe constraint but by no means a fatal one. Reproducible and informative phase changes are seen, both in supersonic jets [4,8,9] and in MD simulations [5,7]. Furthermore, adjustments of experimental conditions and computational parameters for some systems can narrow the gap between experiment and simulation enough for the same transition to be studied by both methods.

Despite the extreme requirements to be met if the time evolution of a phase transition in a cluster system can be observed, the special features of the two techniques can be turned to advantage in favorable cases. The potential advantages are outlined in the Introduction. The types of systems for which the cluster approach is most readily applicable are discussed in the following sections.

Illustrative Examples

Unless otherwise stated it will be assumed that the system variables in experiments are those corresponding to current practice at the University of Michigan [6,8,9] and that typical cluster sizes and observation times are those mentioned on the previous section. Nucleation rates are then restricted to within a few factors of ten of 10^{29} m^{-3} s^{-1}. Systems to be studied are evaporative ensembles whose temperatures are predetermined to within fairly narrow limits by the volatility of the substance under investigation [10-12]. Shortly after emerging from the nozzle, clusters reach a temperature at which their vapor pressure is of the order of 10^{-3} torr[6] unless they are cooled even more strongly by a concentrated, rapidly expanding carrier gas, or by a much more protracted period of evaporation [9]. For nonassociated species this temperature is related to the molar energy of vaporization approximately by [10-12]

$$T \approx 0.04 \ \Delta E_V / R$$

At such a temperature some substances are liquid and their clusters display no transitions. For many substances, however, T is far below the melting point. Whether clusters of such materials freeze before they are observed depends upon the rate of nucleation of the solid. A simple criterion [4] has been found to predict whether evaporatively cooled clusters can freeze by the time they are typically monitored. This criterion, corroborated by nucleation theory [13], is based on two dimensionless indices

$$R_1 = (T_b - T_m)/T_b$$

and

$$R_2 = 0.007 \ (\Delta S_m/R)^2$$

where T_m and T_b are the normal melting and boiling points and ΔS_m is the molar entropy of fusion. If the sum of the indices exceeds 0.3 the clusters can be expected to be liquid and, if not, solid. If the sum is approximately 0.3, it is likely to be possible to observe the time evolution of their freezing. Hence, one class of candidates for study has been identified.

Clusters of two examples of this class, carbon tetrachloride and methyl chloroform, have been studied in some detail [8,9]. In supersonic jets they freeze with nucleation rates of $\sim 10^{29}$ m^{-3} s^{-1}. These rates are enormously higher than any seen in previous studies of freezing and are very close to the maximum possible rates for such systems according to an application of classical nucleation theory. Although rates initially increase with increasing supercooling, they ultimately fall off when the liquid becomes too cold and viscous. The most significant information derivable from a measured nucleation rate is the interfacial free energy for the boundary between the solid and liquid. The interfacial free energies for CCl_4 and CH_3CCl_3, expressed in terms of energy per mole of molecules at the interface, have been found to be approximately one-third of the heat of fusion, in agreement with an idea proposed many years ago by Turnbull [14].

We have also seen comparable nucleation rates in transformations of solid clusters from one phase to another, for example, from bcc to monoclinic for SeF_6 [15]. In such a transition, however, the mechanism is markedly different from that in freezing and, as a consequence, the experimental rate referred to is very far from the maximum possible rate. No longer is the rate-limiting step a translational jump of molecules from a viscous liquid to a solid. It is now a 60 degree rotation of quasispherical molecules, only one-third of which need to change.[5,16,17] Because such rotations are not subject to the same activation barrier suffered by diffusing molecules in highly supercooled liquids, the observed solid-state transition is not so tightly limited. Indeed, the bcc to monoclinic transition for hexafluorides can be so much faster in highly supercooled systems that it can be readily followed even in MD simulations, as illustrated in Fig. 1.

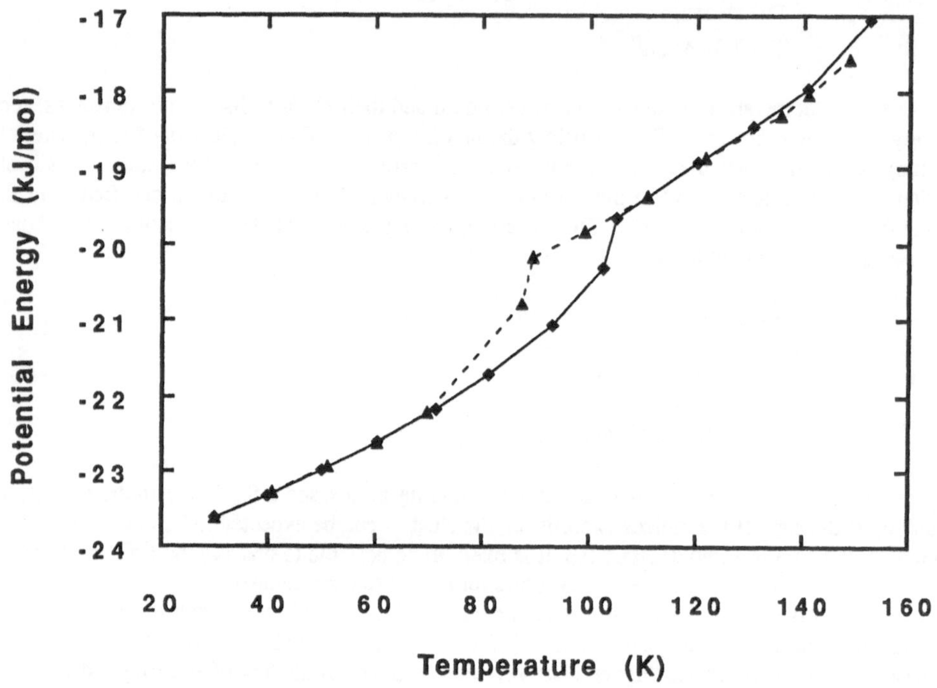

Figure 1. Variation of configurational energy of a 150-molecule cluster of SeF_6 with change in temperature according to a constant energy MD simulation. Dashed curve and triangles, transformation of bcc cluster to monoclinic by cooling. Solid curve and diamonds, warming of the monoclinic cluster so formed. It transforms back to the bcc phase. The hysteresis is due to nonequilibrium conditions arising during the high cooling/heating rates. Each point corresponds to a 5000-step run (timestep = 10^{-14} sec) with 1000 steps of rescaling of velocities in a heat bath followed by 4000 steps to average the potential energy and temperature. Atom-atom Lennard-Jones potential parameters for $\sigma(\text{Å})$ were 2.942 (FF), 3.348 (SeF), and 3.81 (SeSe), and for ε (kJ/mol) were 0.2184 (FF), 0.555 (SeF), and 1.41 (SeSe).

Generation of New Phases

New solid phases of a number of substances have first been seen in clusters [18-20]. A particularly illuminating series of compounds has proven to be that of the binary hexafluorides, AF_6. Although the molecules are quasispherical, they exhibit a completely different crystalline packing than do atomic solids such as Ar, and their clusters are strikingly different. Whereas solid clusters of argon are amorphous until they grow to approximately one thousand molecules [21], clusters of the hexafluorides can adopt crystalline structures when they contain as few as 50 molecules [5].

Submicroscopic crystals of the smallest hexafluoride, SF_6, have been found in a trigonal phase never seen in the bulk for reasons that are not yet clear [16,17]. However, another difference between clusters and the bulk has been explained by MD computations. It turns out that, in the bulk, the larger hexafluorides (A = Te, Mo, W) do not pack in the low temperature monoclinic structure found for SF_6 and SeF_6 but, instead, are orthorhombic when cold [22]. By contrast, clusters of these hexafluorides can easily be generated in the monoclinic phase [19-23]. Molecular dynamics simulations reveal that the monoclinic clusters are metastable with respect to the orthorhombic. They make it clear that it is the <u>kinetics</u>, not the thermodynamics of the transformation from the high temperature bcc phase that overwhelmingly favors the monoclinic form. An example of such an extraordinarily fast transition is shown in Fig. 1 for the case of SeF_6. The transition rates for SeF_6 and TeF_6 are similar, according to molecular dynamics simulations, and structural considerations readily account for the speed. The transformation from bcc to monoclinic described in the previous section is facile, unlike the major reorganization required for the transition to the orthorhombic structure. It is likely that monoclinic crystals will be detected even in the bulk if the solid is rapidly monitored during careful cooling of the bcc phase.

An additional crystalline phase of the transition metal hexafluorides that is unknown in the bulk has been seen in clusters under certain expansion conditions. An elucidation of its structure will depend upon simulations that screen the various ways such molecules can pack at different temperatures.

Clusters are providing a particularly fertile testing ground in explorations of how molecular systems can organize. As models of condensed phases they are able to demonstrate structural and dynamic properties in convenient ways. Moreover, they offer a more intimate view of molecular behavior, and can be subjected to nonequilibrium conditions unattainable in the bulk. The study of condensed matter along this avenue of approach is still in its infancy. Because of its natural advantages, it can be expected to mature rapidly.

Acknowledgment

This research was supported by a grant from the National Science Foundation. We gratefully acknowledge the indispensable contributions to this research made by Mr. Paul Lennon.

References

1. Jena, P., Rao, B.K., and Khanna, S.N. (eds.) (1987), Physics and Chemistry Small Clusters, Plenum, New York.
2. Chem Rev. 86, no. 3, Clusters thematic issue (1986).
3. Z. Phys. D 20, Cluster symposium thematic issue (1991)
4. Bartell, L.S., Harsanyi, L., Valente, E.J. (1989), J. Phys. Chem. 93, 6201.
5. Bartell, L.S. and Xu, S., J. Phys. Chem., in press.
6. Bartell, L.S. (1986) Chem. Rev. 86, 491.
7. Swope, W.C. and Andersen, H.C. (1990) Phys. Rev. B 41, 7042.
8. Bartell, L.S. and Dibble, T.S. (1991) J. Phys. Chem. 95, 1159
9. Dibble, T.S. and Bartell, L.S., J. Phys. Chem., submitted.
10. Gspann, J. (1976) in S. Datz (ed.), Physics of Electronic and Atomic Collisions, Hemisphere, Washington, D.C.
11. Klots, C.E. (1987) Nature 327, 222.
12. Klots, C.E. (1988) J. Phys. Chem. 92, 5864.
13. Bartell, L.S., J. Phys. Chem., in press.
14. Turnbull, D. (1950) J. Appl. Phys. 21, 1022.
15. Dibble, T.S. and Bartell, L.S., unpublished data.
16. Powell, B.M., Dove, M. T., Pawley, G.S., and Bartell, L.S. (1987) Mol. Phys. 62, 1127; (1988) *ibid.* 65, 353.
17. Pawley, G.S. and Dove, M.T. (1983) Chem. Phys. Lett. 99, 45.
18. Valente, E.J. and Bartell, L.S. (1983) J. Chem. Phys. 79, 2683.
19. Bartell, L.S., Valente, E.J., and Caillat, J.C. (1987) J. Phys. Chem. 91, 2498.
20. Harsanyi, L., Bartell, L.S., and Valente, E.J. (1988) J. Phys. Chem. 92, 4511; Bartell, L.S. and Powell, B.M. (1989) Mol. Phys. 67, 861.
21. Raoult, B., Farges, J., De Feraudy, M.F., and Torchet, G. (1989) Phil. Mag. B60, 881.
22. Bartell, L.S. and Powell, B.M., Mol. Phys., in press.
23. Hovick, J.W., Dibble, T.S., and Bartell, L.S., unpublished research.

STRUCTURE MODEL OF THE AL-CU-CO DECAGONAL QUASICRYSTAL

S. E. BURKOV

Cornell University, LASSP, Ithaca, NY 14853, USA and
Landau Institute of Theoretical Physics, Moscow, USSR

ABSTRACT. Structure model for decagonal quasicrystals AlCuCo and AlNiCo is proposed. The model agrees with available experimental data: the results of three-dimensional and five- dimensional Patterson analysis of X-ray diffraction data and with direct space atomic patterns found by high resolution electron microscopy. The model allows for a dualistic description: it can be viewed as both a set of specifically decorated overlapping decagonal clusters and as cutting and projecting only two simple "atomic surfaces".

1. Introduction

Among a dozen of experimentally discovered decagonal quasicrystals two alloys, $Al_{65}Cu_{20}Co_{15}$ and $Al_{70}Ni_{10}Co_{15}$, have attracted most attention. They deserve this attention by their outstanding properties: up to cm-size single crystals can be grown from melt; they are thermodynamically stable and exhibit extremely sharp diffraction peaks (the coherence length is about 2000 Å).[1-7] Moreover, the quality of the diffraction pattern improves upon annealing.[3,4] Both substances are believed to be isostructural with very close lattice parameters, for this reason no distinction between them will be made below. A decagonal quasicrystal has a layered structure, with every layer being a two-dimensional quasicrystal. The stacking of layers along the tenfold axis is usually presumed periodic. Although there are certain doubts in stacking periodicity[8] this question will not been addressed below, and the structure will be considered periodic with a period c of approximately 4.15 Å.[1-7] Doubling of this period reported in some samples[1,6,7] is not taken into account.

Determination of atomic structures of quasicrystals is not trivial, as standard crystallographic methods cannot be used straightforwardly because of the lack of periodicity. There are two general experimental approaches based on two theoretical perceptions of what quasicrystals are. One views a quasicrystal as a section of definite "atomic surfaces" in five-dimensional (5D) space, the coordinates of *individual* atoms being given by the cut-and-project method.[9] The atomic surfaces of a particular quasicrystal can be determined by measuring diffraction intensities, calculating Patterson function (electron density autocorrelation function) and deconvolving the Patterson function numerically. This method, previously used for treating icosahedral quasicrystals[10], has been recently applied to decagonal ones.[5,6] The other, competing, approach is based on the tiling and decoration concept[11], which views a quasicrystal as built of two elementary cells, or tiles, which fill the space quasiperiodically. The 5D space is used within this approach to refer to the coordinates of

77

P. Jena et al. (eds.), Physics and Chemistry of Finite Systems: From Clusters to Crystals, Vol. I, 77–83.
© 1992 *Kluwer Academic Publishers.*

the *tile vertices*, not individual atoms. Generally, coordinates of atoms inside tiles have nothing to do with higher dimensional space. Atoms are placed not at vertices of Penrose tiling, but in some positions providing reasonable local packing. The corresponding experimental method may be called "atomic modeling" or "playing sticks and balls". The idea is to invent some atomic decoration by guessing, making use of known rational approximants, common sense and quantum chemistry. There have been attempts to apply this approach to the decagonal phase. The two approaches are not mixed and sometimes considered as incompatible, which causes a sort of a dispute between adherents of the two methods. Neither of the two approaches works perfect. Direct determination of atomic surfaces by Patterson analysis relies on numeric structure refinement and gives only an *approximation* of atomic surfaces, which, after being cut and projected, often give unrealistic local environments in physical space. On the other hand, whereas any tiling model can be viewed as a cut-and-project one (the opposite is not always true), the number of atomic surfaces is, in general, equal the number of atoms in the tiles. In all known cases[5,6,10] the Patterson analysis shows only two or three very simple atomic surfaces. At a first glance, this rules out elaborate tiling models with many atoms in each tile which are usually introduced to achieve reasonable packing. One of the aims of this communicatoin is to present a structure which can be obtained by *both* cut-and-project *and* tiling-and-decoration methods, after certain restrictions are imposed on both ones. The model obtained this way appears to be a good model of AlCuCo/AlNiCo. It agrees with experimental data: the results of 5D- and 3D Patterson analysis of single crystal X-ray diffraction data [6,7] and with the direct space atomic patterns found by high resolution electron microscopy (HREM).[3]

2. Cluster packing description

There are two plane layers per vertical period in the proposed structure. Each has only five-fold symmetry, but the layer at $z = c/2$ is rotated by 36° with respect to the layer at $z = 0$ revealing overall decagonal symmetry. The atomic structure in the xy-plane can be presented by two methods, whose equivalence has been established analytically. The first, a modified tiling-and-decoration method, is formulated as follows. Our structure is built of *one* building block, a decagonal cluster (Fig.1). The particular atomic decoration of the cluster coincides 90% with the decoration inferred by Hiraga et. al. from the HREM images.[3] The same cluster was also obtained by three-dimensional (no 5D atomic surfaces) Patterson analysis[7]. However, it is not enough to determine cluster decoration: one must find out how clusters are assembled together. Since *tiling* of the plane by decagons is impossible, a *covering* of the plane is introduced. Two clusters are allowed to overlap, but only the two overlappings shown in Fig.1 are permitted. Note, that no "glue" atoms[11] are used when clusters intersect nor are substantial atomic displacements introduced: there is only one pairs of Al atoms (in each z-layer) which move from their original positions on the sides of internal, small, decagon when the latter share a side (as shown in Fig.2, right, central part, $3\pi/5$ overlapping).

Consider a covering of the plane by decagons, such that if any two clusters intersect then this is one of the two allowed intersections (multiple intersections are not excluded). First, one has to prove that such covering is possible. It has been done in[13] by proving that any such covering is equivalent to a *binary tiling*. A binary tiling (Fig.2) is a tiling by two Penrose rhombi such that in every vertex all angles are either all even or all odd (the angles $2\pi/5$ and $4\pi/5$ are called even, $\pi/5$ and $3\pi/5$ are odd.).[14] This restriction has a physical basis: an even and an odd angle cannot meet in one vertex, because atomic decorations of their interiors are different and cannot be matched. The basic building blocks, decagonal

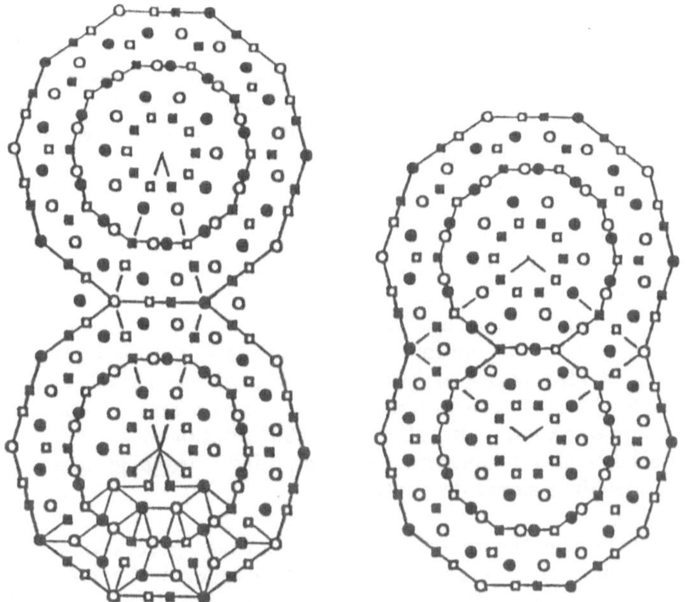

Fig.1 Two allowed overlappings of decagonal clusters and corresponding tiles. A portion of Penrose network hosting the atoms is shown. Circles - Al, squares - T-metal; open - z=0, full - z=c/2 layer.

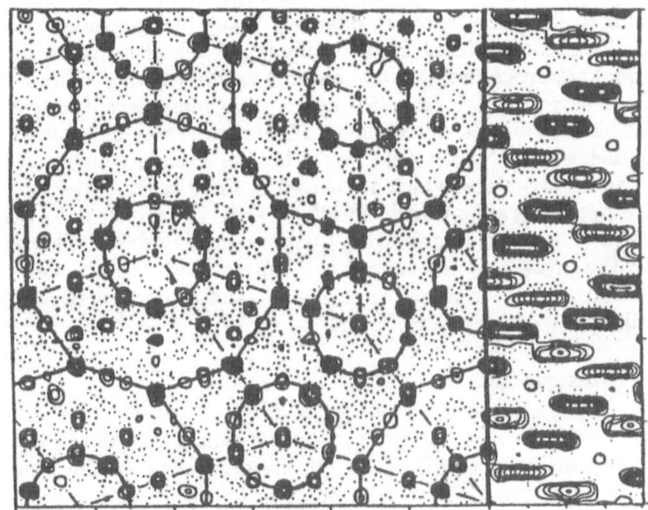

Fig.2 Atoms in the decagonal plane as obtained from 5D Patterson analysis.[6] The right insert is a section of atomic surfaces by a plane spanned by one parallel and one perpendicular basis vectors.

clusters of 10.3 Å radius, are centered at odd vertices, whereas even vertices are corners of decagons (Fig.1). The structure can also be viewed as built of smaller decagons (of τ^{-1} radius) centered at odd vertices and of 11-atom ring-like clusters of τ^{-3} radius centered at even vertices (do not mix them with 10-atom rings inside decagons which centers at odd vertices, Fig.1,2). Strictly speaking, the basic cluster has decagonal shape but only pentagonal symmetry, in particular, its corners are of two types ("colors"). When one goes around the cluster (i.e. an odd vertex) the colors of the decagon's corners (i.e. even vertices) alternate (Fig.1). This leads to a "colored" binary tiling, with all odd vertices being equivalent and two inequivalent types of even vertices, which alternate as described above. The selfconsistency of this alternation rule, not obvious from the definition, has been demonstrated in.[13] In terms of small decagons and 11-atom circular clusters a colored configuration looks as follows: all the small decagons are oriented the same way, including of coloring (i.e. all odd vertices are equivalent), whereas 11-atom ring-like clusters are in two inequivalent states (colors). To transform the latter cluster from a "white" to a "black" state, one has to apply a screw axis transformation (36° rotation plus half a period vertical shift, Fig.1). The scheme also demonstrates that imposing reasonable rules of overlapping inevitably leads to five-dimensional description.

There are two features making the present model distinct from a general tiling-and-decoration model. First, binary tiling emerges as a consequence of rules for cluster overlapping and the decoration of the Penrose tiles is deduced from the cluster decoration. Second, all atoms sit on a Penrose network (not a tiling), i.e. every two atoms can be connected by a path made of intervals of length a having ten decagonal orientations (a=2.44Å, whereas cluster radius is τ^3 a=10.33Å).

Before mapping the present tiling-and-decoration description to the cut-and-project representation one should realize that there are infinitely many binary configurations and all of them are different from conventional Penrose tilings. When all of these tile configurations are ascribed equal statistical weights one obtains random binary tiling. [14] When the binary tiling is viewed as projection from 5D space with some acceptance domain with sharp boundaries the binary tiling is called "ideal". There are some different choices of the acceptance domains, below one choice would be made: the domain consists of three "pancakes", a decagon [15] and two 5-stars, which are shown in Fig.3 as subdivisions of the whole polygons. The choice between random and "ideal" versions is equivalent to the choice between energy and entropy as a factor stabilizing quasicrystals.[2,3,4,5,12,14] The latter dispute is still unsettled and the present model cannot help to resolve it: the model may exist in the both frameworks. However, the mapping to the cut-and-project method is easier with the "ideal" binary tiling.

3. Cut-and-project description

To describe the same structure in a cut-and-project formalism one has to lift the coordinates of the individual atoms up to 5D space. If the decoration of the Penrose rhombi were to be of a general type, the lifting would create dozens of atomic surfaces. However, straightforward analytic calculations show that in the present model the atomic surfaces merge together, forming only two larger atomic surfaces (Fig.3). Thus, the second, equivalent, cut-and-project description is as follows: coordinates of all the *individual* atoms can be obtained by cutting and projecting 4 atomic surfaces. Two of them, shown in Fig.3, give the $z = 0$ layer. One surface, occupied by Al, is centered at -B, where B = (1,1,1,1,1)/5; the other surface, occupied mainly by transition (T) metal (Cu/Co or Ni/Co, which are mixed randomly), is centered at +2B. The $z = c/2$ layer is given by the same surfaces as in Fig.3, but rotated by 36° and centered at +B (Al) and -2B (T). Subdivisions of the atomic surfaces in Fig.3 reflect different atomic environments (the details will be presented elsewhere). The atomic surfaces are not of a general form: they are flat, have complicated boundaries (which were never guessed of in Patterson analysis) and are centered at special points of the 5D unit cell. The latter fact was noticed in virtually all experimental Patterson analysis works, which might mean that some analog of the present tiling-and-decoration and cut-and-project duality holds for icosahedral and other quasicrystals. Note, that shapes in Fig.3 have been obtained from exact geometric considerations, not from computer assisted refinement of experimental data, as in.[5,6]

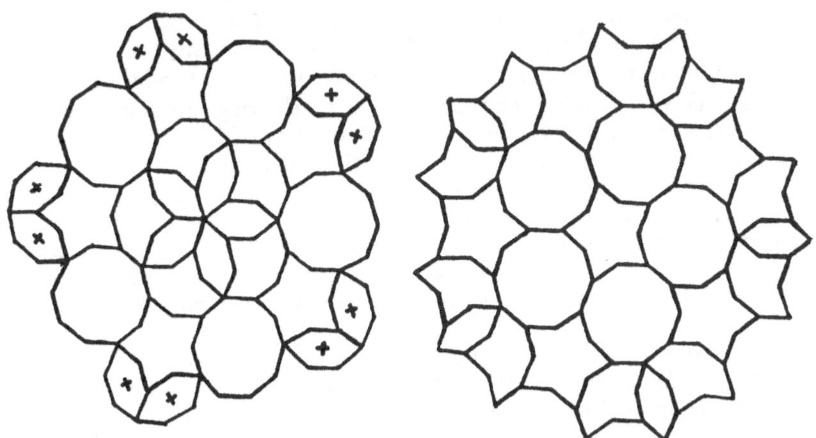

Fig.3 Atomic surfaces of "ideal" structure (no phason disorder). Right pancake is all Al, left is occupied by both T-metal and Al: the large central unmarked part is occupied by T-atoms, 10 crossmarked hexagon are occupied by Al. Decomposition to smaller polygons classifies local environments.

The atomic positions in Fig.2 have been obtained by Steurer and Kuo[6] by numeric 5D Patterson analysis of X-ray data. Their work had been published before Hiraga's clusters were announced.[3] For this reason the authors of[6] did not use the knowledge of cluster geometry in their work, they simply determined atomic surfaces in 5D space by direct numeric refinement. The atoms of Fig.2 were obtained by cutting and projecting these experimentally determined atomic surfaces. The lines showing the decagonal clusters and binary tiling have been added to the original figure of Ref.[6] by the author of the present paper. The agreement of Fig.2 with the cluster representation is fairly good but not perfect. This can easily be understood by realizing that the atomic surfaces in[6] are approximate: the authors restricted themselves to decagonal "pancakes" and conventional Penrose tiling, whereas real pancakes might have more intricate shape, such as in Fig.3, and the tiling is binary. Besides, the pancakes in Fig.3 correspond to the "ideal" binary tiling. If the real quasicrystal is stabilized by entropy or, at least, contains a substantial amount of phason disorder the boundaries of the atomic surfaces would be not as sharp as in Fig.3. Nevertheless, one can easily see in Fig.2 exactly the same decagonal clusters as observed in HREM images[3] and, moreover, the clusters overlap according to the rules shown in Fig.1.

So, Ref. [6] may be considered as a blind test of the cluster model. It also illustrates the equivalence of the cut-and-project and tiling-and-decoration description of the structure.

4. General properties

1. Space group is $P10_5/mmc$ (centrosymmetric), as verified by characteristic extinctions in both observed and calculated diffraction spectra.[16] 2. Kalugin's homology indices are $(1,1)$, i.e. the number density in a decagonal layer is expressed via basis reciprocal lattice vector Q_{10000} as:

$$n = \sqrt{5}sin(\pi/5)(Q_{10000}/2\pi)^2(N_1 + \tau N_2),$$

where N_1 and N_2, the homology indices, are both equal to 1. The latter fact means that from the homological point of view the atomic surfaces in Fig.3 are rather simple.[18] 3. The calculated density of 4.53 g/cm^3 agrees well with the measured density[6] of AlCuCo of 4.5 $\pm$ 0.05 g/cm^3. The interatomic distances are not shorter than 2.44Å, which is acceptable for an Al-T bond. The volume per atom is about 13.9 Å^3, showing sufficiently dense packing; there no big holes except for cylindric wells positioned at odd vertices. However, this is an experimental fact: both HREM[3] and X-ray[6,7] show that cluster centers are vacant. The structure described above is a sort of ideal, stochiometric structure; the real one may differ by elastic deformations, chemical disorder or, contrary, ordering of Ni, Co or Cu, deviation from stochiometry, doubling or tripling of the unit cell along the ten-fold axis, some puckeredness of the layers, etc. Moreover, there is a report [5] of a possible existence of another decagonal phase in AlNiCo with the space group $P10/mmm$.

The author is thankful to V. Elser, C. Henley, D. Mermin and P. Kalugin for useful discussions and to K. Hiraga, W. Steurer, S. Moss and Y. He for providing valuable experimental data. The work was supported by DOE grant #DE-FG02.89ER-45405.

REFERENCES

1. A. R. Kortan , F. A. Thiel, H. S. Chen, A. P. Tsai, A. Inoue and T. Masumoto , *Phys. Rev. B* **40**, 9397 (1989).
2. H. Chen, S. E. Burkov, Y. He, G. J. Shiflet and S. J. Poon, *Phys. Rev. Lett.* **65**, 72 (1990).
3. K. Hiraga, W. Sun, F.J. Lincoln, *Material Transactions JIM*, **32**, 308 (1991); *Jpn. J. Appl. Phys.*, **30**, L302 (1991).
4. Y. He, H. Chen, X. F. Meng, G. J. Shiflet and S. J. Poon, *Phil. Mag. Lett.*, **63**, 211 (1991).
5. A. Yamamoto, K. Kato, T. Shibuya and S. Takeuchi, *Phys. Rev. Lett.* **65** 1603 (1990).
6. W. Steurer and K.H. Kuo, *Acta Cryst.*, **B46** 703 (1990).
7. S. Hashimoto, K. Hiraga, X. Jiang, M. Kaneko, J. Kulik, Y. Matsuo and S.C. Moss, Texas AM Univ. Preprint (1991).
8. S.E. Burkov, *J. Stat. Phys.* **65**, 395 (1991).
9. P. A. Kalugin , A. Yu. Kitaev and L. S. Levitov, *JETP Lett.* **41**, 145 (1985); *J. Phys. (Paris)* **46**, L601 (1985); M. Denau and A. Katz, *Phys. Rev. Lett.* **54**, 2688 (1985); P. Bak, *Phys. Rev. Lett.* **56**, 861 (1986).
10. D. Gratias, J. W. Cahn and B. Mozer, *Phys. Rev.* **B 38**, 1638 (1988); *Phys. Rev.* **B 38**, 1643 (1988); S. Y. Qui and M. V. Jaric, in *Quasicrystals*, edited by T. Fujiwara and T. Ogava, (Springer-Verlag, Berlin, 1990), p.48; M. Cornier-Quiquandon, A. Quivy, S. Lefebvre, E. Elkaim, G. Heger, A. Katz and D. Gratias, preprint 1991.
11. V. Elser and C. L. Henley, *Phys. Rev. Lett.* **55**, 1883 (1985).

12. A. R. Kortan, R. S. Becker, F. A. Thiel, and H. S. Chen, *Phys. Rev. Lett.* **64**, 200 (1990).
13. S. E. Burkov, Cornell Univ. Preprint (1990).
14. F.Lancon, L.Billard and P.Chaudhari, *Europhys. Lett.* **2**, 625 (1986); F. Lancon, L. Billard, *J. Phys. (Paris)* **49**, 249 (1988). M. Widom, K. Strandburg, and R. H. Swendsen, *Phys. Rev. Lett.* **58**, 706 (1987).
15. E. Cockayne, private communication.
16. D. Mermin applied the results of Ref. [17] to this particular structure.
17. D. A. Rabson, N. D. Mermin, D. S. Roksar and D. C. Wright, Cornell Univ. Preprint (1990).
18. P. A. Kalugin, *Europhys. Lett.* **9**, 545 (1989).

Infrared Spectroscopy of SF$_6$ Attached to Classical and Quantum Clusters (finite size particles attaining bulk-like properties)

S. Goyal, D.L. Schutt , and G. Scoles
Department of Chemistry
Princeton University
Princeton, NJ 08544 USA

Abstract

Using a new thermal detection photodissociation spectrometer, we have measured the infrared spectra of the ν_3 vibrational mode of SF$_6$ seeded <u>in</u> clusters composed of argon, krypton, and xenon, and deposited <u>on</u> clusters composed of helium, neon, argon, and krypton. From the spectra of SF$_6$ in argon and krypton, it was possible to determine the approximate size, or number of atoms, which is necessary for a finite sized aggregate of rare gas atoms to change its structure from icosahedral to the bulk phase face-centered cubic structure. Clusters composed of neon and xenon were unable to solvate SF$_6$ for any size.

Introduction

The study of atomic and molecular clusters has been, and continues to be, a fruitful area of research. The field of clusters has developed and matured primarily to study the properties of matter in finite systems, and to try and understand the evolution of bulk-like properties in these systems as a function of their size. Cluster structure is one important area which has demanded the attention of many investigations. This has been especially true for clusters composed of rare gas atoms. The ease of producing these clusters in a supersonic expansion and the simple, well known interaction potentials of the rare gas systems has provided the impetus for many extensive investigations from both an experimental and theoretical perspective. It has been well established that small noble gas clusters are stable as polyicosahedra, medium clusters as Mackay icosahedra, and large clusters must eventually attain an fcc structural configuration as in the bulk phase solid. However, the cluster sizes where these transitions take place, especially the one from Mackay icosahedra to the fcc, are far from being precisely known. The experimental evidence in the smaller cluster regime has been in the form of magic numbers (greater stability for certain cluster sizes) observed in the mass spectra.[1] For the intermediate size and the larger cluster limit, the evidence has been through the electron diffraction measurements pioneered by the Orsay group.[2] They determined that the transition from the Mackay icosahedra to the fcc takes place at a size of 750 argon atoms. Similar observations of Lee and Stein[3] have placed that number between 1500 and 3500 atoms. In addition to experiments, theoretical attempts have also been made with the help of computer simulations to predict the cross-over point. Lee and Stein[3] have also performed energy minimization calculations for Lennard-Jones systems, and have predicted a transition at a size of roughly 3000 atoms. The molecular dynamics simulations of Andersen and Honeycutt[4] determined a cross-over point at approximately 5000 atoms for

85

P. Jena et al. (eds.), Physics and Chemistry of Finite Systems: From Clusters to Crystals, Vol. I, 85–91.
© 1992 *Kluwer Academic Publishers.*

argon. Northby et. al.[5] have reported, using a Lennard-Jones (6-12) potential, that the icosahedral structure is lower in total energy for a cluster composed of less than 14 shells (10179 atoms), but on applying the Aziz-Chen pair potential, the corresponding result was 13 shells (8217 atoms). More recently, van de Waal[6] has calculated the energy difference between the multi-shell cuboctahedra and icosahedra for a given size up to a maximum of 5083 atoms. On extrapolating his results to larger cluster sizes, he estimated that the smallest cluster with a cuboctahedron structure which has a higher binding energy than an icosahedral cluster has 10179 atoms. Quite clearly from this brief sampling of the literature, it is obvious there is no real consensus, experimentally or theoretically, on the critical size at which an aggregate of rare gas atoms will change their structure from polyicoshedral to the bulk phase face centered cubic structure.

The bulk of our knowleage for the rare gas systems has been derived through the use of classical mechanics. This is a reasonable approach for clusters composed of the larger rare gas atoms, which have been the systems of choice thus far. However, recently there has been an increased interest in systems where the clusters are fluid and where the dynamics and energetics of the cluster dictate the need for a fully quantum mechanical description. The most obvious example of such a study is clusters of helium. As in the bulk, helium clusters will remain fluid under all conditions of formation, and because of helium's small mass and weak interatomic potential, quantum effects will become significant. Consequentially, helium clusters will exhibit several unusual properties. In particular, the existence of magic numbers can not be predicted on the basis of sphere packing arguments, nor can the atoms be as localized as in clusters of the larger rare gases. Additionally, there exists the exciting possibility that clusters of helium may exhibit the analogous behavior of superfluidity as in the bulk phase. This question alone has prompted many theoretical investigations using computational techniques ranging from phenomenolgical approaches, studying the dynamics of aggregates of helium atoms, to *ab initio* calculations. Two of these studies[7,8] directly show that clusters of approximately seventy atoms manifest certain features characteristic of superfluidity in the bulk phase. Although some experimental work to probe the existence of superfluidity in confined geometries has been carried out using either small bubbles (<110Å) of helium trapped in copper foil[9] or atomic scattering experiments with helium cluster beams,[10] experimental evidence of superfluidity in finite aggregates of ^{4}He is at best inconclusive. Other experiments involving the possibility of superfluidity in helium clusters have also begun including the capturing of impurities by the clusters.[11]

With the questions of cluster structure and superfluidity unanswered, we decided to conduct a series of experiments based on the use of infrared spectroscopy of an impurity molecule embedded in a host cluster to investigate these areas. We selected this method because the vibrational spectrum is sensitive to the molecule's local environment. This technique provides the possibility of probing the cluster's structure and in the case of helium clusters perhaps superfluidity. The latter is an open question, since at present a theory which can fully explain the behavior of a vibrationally excited guest molecule interacting with a superfluid is not available.[12]

Experimental

The principle of our experimental technique has been published before[13,14] and is described here only briefly. The supersonic cluster beam was produced by expanding gas through a sonic nozzle into a primary chamber pumped by a 32,000 liter sec^{-1} diffusion pump. The nozzle was cooled using a closed cycle helium refrigerator and the temperature of the nozzle stabilized by a temperature controller utilizing a silicon diode sensor. For the production of neon, argon, krypton, and xenon clusters, a 50 μm diameter nozzle was used; while a 10 μm nozzle was used for helium clusters. The center portion of the beam was skimmed via a 390 μm skimmer and allowed to pass into a high vacuum chamber pumped by a 5000 liter sec^{-1} diffusion pump equipped with a water cooled baffle. The infrared active molecule was introduced in the cluster either by coexpanding a very dilute (typically 0.025%) mixture of SF_6 in the rare gas, or alternatively, producing neat clusters and attaching the chromophore by collisions with SF_6 contained in a pick-up cell located in the experimental chamber. Downstream of the pick-up cell, the cluster beam was crossed at right angles by the modulated output of a line tunable CO_2 laser. The spectrum of SF_6 in the clusters was recorded using one of two bolometers and a lock-in amplifier. The first bolometer was an annularly shaped bolometer positioned such that the primary beam passed through the center hole, while the second bolometer was a circular bolometer positioned three centimeters behind the first and placed directly on line with the primary cluster beam. The annular bolometer was used to detect fragments from the clusters as a result of the chromophore absorbing a photon and transferring the energy into the phonon modes of the cluster leading to evaporation. The second bolometer was used either to detect the fragmentation process as above (however, detecting a loss of particles from the main beam rather than an increase in particles as before) or to detect the energy of an absorbed photon which had not relaxed into the cluster prior to detection. The two types of signals could be easily differentiated from one another by simply monitoring the phase of the signals with respect to the incident scattered laser light. To compensate for the fact that CO_2 lasers are not continuously tunable, the number of points at which spectral intensities were measured was increased by employing three CO_2 isotopic combinations as lasing media, namely, $^{12}C^{16}O_2$, $^{13}C^{16}O_2$ and $^{12}C^{18}O_2$. The infrared laser beams were made to cross the molecular beam at a carefully controlled point, so that after normalization to laser power, no scaling of the data was necessary. This was demonstrated by measuring the SF_6 dimer spectrum which yielded a smoothly connected spectral envelope in agreement with previously published data.[15,16] From one laser scan to another, the intensities could be reproduced within 5%. Gaussian curves were fitted to the data for a more precise determination of the spectral features.

Results and Discussion

Previous investigations in our laboratory have focused on establishing the position (solvated vs. surface site) of different solute molecules in a cluster of variable size. Using mainly argon as a solvent, molecules such as SF_6, CH_3F, SiF_4, and CF_3Cl were studied.[17] With the help of computer simulation,[18] semiquantitative criteria were established for choosing the solute/solvent diameter and well depth ratios so as to favor either solvation or exclusion of the chromophore by the host cluster. In the work reported here, we focus instead on a single molecule (SF_6) while systematically changing the nature of the solvent. Using the whole noble gas family from Xe to He, we aim at spanning the full range of behavior from mainly classical to fully quantum mechanical.

Before beginning a discussion of this work however, it is instructive to take a brief look at the spectroscopy of SF_6 (gas phase absorption at 948 cm^{-1}) in rare gas matrices.[19,20] When SF_6 is initially deposited in a matrix of argon, two primary absorptions are observed, one at 937.9 cm^{-1} and another at 938.6 cm^{-1}. Upon annealing, the absorption at 937.9 cm^{-1} disappears and the intensity of the 938.6 cm^{-1} peak increases. The thermally labile absorption at 937.9 cm^{-1} has been attributed to SF_6 residing in a site which locally has a icosahedral structure and the second absorption to SF_6 in a fcc lattice site. This interpretation is entirely consistent with the fact that the packing density of the icosahedral structure is higher and therefore results in a larger red shift than for the less dense fcc lattice. Spectra of SF_6 in matrices of the remaining noble gases (except helium) were also measured and similar behavior to that of argon was observed. We have used these two absorptions as a signature of cluster structure and have been able to observe the change in packing of rare gas atom clusters as a function of their size. One final note is necessary prior to a detailed discussion. Since cluster size is critical to our interpretation of the data, it is essential that we have a realistic way of estimating the size. Barring effects due to nozzle shape (probably rather small), we believe that using the electron diffraction data of Torchet[21] combined with the scaling laws first introduced by Hagena,[22] provides a reasonable estimate of cluster size can be made.

The first system studied was SF_6 solvated by argon. Although this system had been previously investigated, we extended the study to include a much larger range of cluster sizes and have reported these results in a recent paper.[23] To avoid unnecessary duplication, only a brief summary of the findings will be recalled here. Figure one displays a series of spectra taken of SF_6 in argon clusters for a range of cluster sizes. It was found that clusters containing up to about 1000 atoms of argon show a spectral feature at 937.9 cm^{-1} characteristic of an SF_6 molecule residing in a tightly bound monomeric site. This absorption is red shifted 0.7 cm^{-1} with respect to that of SF_6 in a well annealed argon matrix. This is consistent with the existence of the unannealed matrix peak (with the same frequency) and the knowledge that small argon clusters are known to have a polyicosahedral structure and are more tightly packed than the bulk face centered cubic (fcc) lattice. As the cluster size is increased, the feature corresponding to the tightly bound monomeric sight disappears and absorptions indicative of a chromophore residing near the surface of the cluster emerge. However, on producing still larger clusters, a new absorption develops located at the same frequency, 938.6 cm^{-1}, as the primary absorption of SF_6 in a well annealed matrix of argon. This transition from the tightly bound monomeric absorption to a well annealed matrix-like absorption is located at approximately 2000 atoms per cluster and is likely related to the transition of the cluster's structure from polyicosahedral to an fcc structure.

To establish if this transition is a general property of the rare gas systems, the study was extended to include the other noble gases. The results for SF_6 solvated by krypton are shown in figure two. These spectra demonstrate a remarkable similarity in behavior with argon once the cluster size is taken into account. In particular, for krypton there is an absorption at 935.9 cm^{-1} which dominates the spectrum for the small clusters. With an increase in cluster size, this peak decreases in intensity as two new peaks located at 939.5 cm^{-1} and 941.0 cm^{-1} emerge. By means of pick-up experiments, these two peaks have been assigned to SF_6 residing near the surface of the cluster. Finally, as the cluster exceeds approximately 2000 atoms in size, an absorption at 935.3 cm^{-1} appears. Like argon, this final absorption is located at the same frequency as the bulk phase matrix spectrum for SF_6 solvated by krypton.[19]

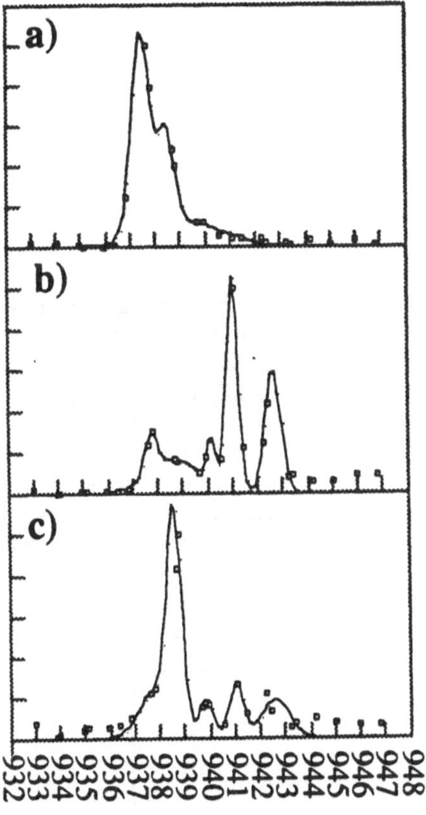
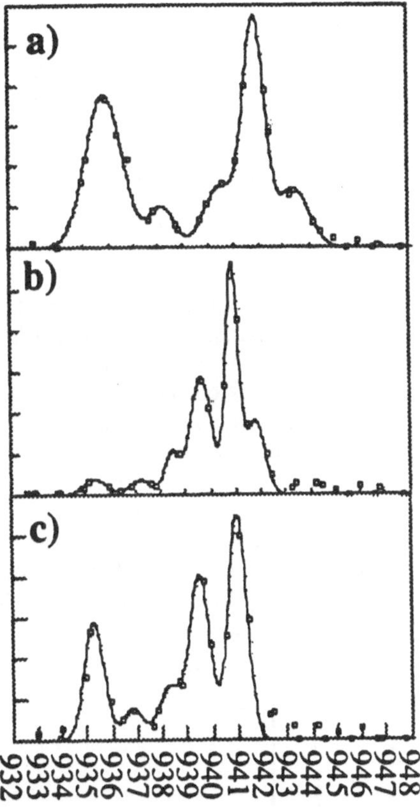

frequency (cm⁻¹)

figure 1. Spectra of SF_6 in clusters of argon for several stagnation pressures a)100 psi b) 400 psi c) 1200 psi

figure 2. Spectra of SF_6 in clusters of krypton for several stagnation pressures a)80 psi b)200 psi c) 500 psi

Studies were also conducted for SF_6 in clusters of xenon. Xenon however, was never found to solvate SF_6 regardless of cluster size. What was observed was a series of five peaks (937.5 cm⁻¹, 938.7 cm⁻¹, 940.0 cm⁻¹, 941.0 cm⁻¹, and 942.0 cm⁻¹) whose intensity varied only slightly with a change in cluster size. The five peaks were located between the gas phase absorption and the matrix, 931.2 cm⁻¹, absorption. Although pick-up experiments were not conducted for the Xe/SF_6 system for reasons of cost, we are confident that these peaks can be assigned to an SF_6 molecule residing near the surface of the cluster. Clusters of neon were also produced. For technical reasons however only the pick-up technique was used at this time. Two absorptions located at 943.4 cm⁻¹ and 944.5 cm⁻¹ were observed. These absorptions are tentatively assigned to SF_6 being near the surface of the clusters, again based on the fact that the absorptions are located between the gas phase and the observed absorption in a neon matrix (942.45 cm⁻¹).[20] We note that interpreting the xenon and neon

90

spectra in this way is entirely consistent with the model developed by Amar[18] to explain the argon results. Indeed the very large difference in the hard sphere diameter between the Ne and SF_6 and the relatively small difference in well depth between Xe and SF_6 locates these two systems in regions of the solvation diagram (see ref. 18) where solute expulsion by the cluster is possible. We note that the calculations by Amar have been carried out with a spherical SF_6-rare gas potential. When calculations are carried out with a more realistic potential, it is entirely possible that the boundary between the solvation and expulsion zones will vary . Therefore, too precise of an agreement at this time is unnecessary. What matters is that when going from Ar to Kr to Xe, the mixture's coordinates move towards the expulsion region.

While the detection of quantum effects in the SF_6/Ne cluster spectra may have to wait quantitative comparison with theory, it is possible that the spectrum of a molecule in contact with a He cluster will show qualitative features which may indicate a decidedly quantum behavior. This was our motivation for obtaining the spectrum of SF_6 after being picked-up by a helium cluster. The spectrum shown in figure 3 is that of SF6 attached to a helium cluster of approximately 10000 atoms.[24] The peaks located at 945.8 cm^{-1} and 946.4 cm^{-1} are due to an isolated SF_6 molecule in the cluster. Quite likely, the SF_6 is residing on the surface of the cluster as indicated by the lifting of the three-fold degeneracy of this level induced by the surface site's lack of spherical symmetry resulting in a doublet rather than a single absorption. The intensity of the two peaks agrees with this conclusion, since their 2:1 intensity ratio may be due to the larger shift of the singly degenerate parallel (to the surface normal) transition and the lesser shift of the doubly degenerate, and therefore more intense

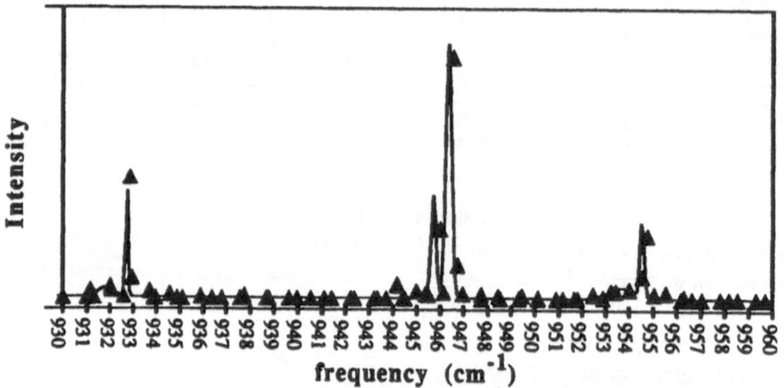

figure 3. Spectrum of SF_6 picked-up by helium clusters generated by a 1500 psi expansion from a 16K nozzle.

perpendicular transition. Although we do not have a calculation modeling this exact system to verify this conclusion, Whaley[12] has conducted a similar calculation for H_2 in helium clusters. In that study, H_2 was found to have a slightly higher probability of being found towards the surface rather than in the interior of the cluster. Extending this to a large molecule such as SF_6, it is very reasonable to assume that SF_6 is residing on the surface of a helium cluster. The remaining two peaks at 932.9 cm^{-1} and 954.7 cm^{-1} can be attributed to SF_6 complexing with itself to form a dimer. It is remarkable that these absorptions have a width which is less than 0.2 cm^{-1} FWHM (^{4}He clusters are predicted to be at about 0.4K[25]). This is in contrast to the 5-6 cm^{-1} wide absorption for the same dimer produced in a molecular beam expansion[15,16] where the temperature is estimated to be several kelvin. If the width of these absorptions can be found to be due to slightly hindered rotations or librations of the dimer on the cluster surface and not to cluster size or site inhomogeneity, these or other similar spectra may provide the first direct route to the measurement of a cluster's temperature. Furthermore, if indeed the peak widths indicate a temperature of ~0.5 K and the calculations of Klein[7] and Whaley[8] are correct, and provided the presence of the impurity does not destroy the existance of a superfluid phase[26], these clusters are superfluid. Presently we are expending a great deal of effort to extend these studies to include rotationally resolved spectra, by trying to use a smaller molecule and a continuously tunable infrared laser.

Summary

In this paper, we have demonstrated that the infrared spectrum of an impurity molecule located in a host cluster is a sensitive probe of the cluster's structure. In particular, using clusters of argon and krypton, we were able to determine the crossover point from an icosahedral structure to a face centered cubic structure for rare gas atomic clusters to be about 2000 atoms. Furthermore, we were also able to further demonstrate the validity of the previously developed model rationalizes the behavior of an impurity in a cluster of rare gas atoms. Finally, we have observed the vibrational spectrum of SF_6 in helium clusters which is, to the best of our knowledge, the first of its kind for a molecule in contact with liquid helium. It is our hope that, with additional analysis of the results coupled with continued experiments, finite size effects on the superfluid phase of liquid helium may be better understood.

References

1 Atomic and Molecular Beam Methods, Chapter 15, ed. G. Scoles, Oxford University Press(1988)
2 J. Farges, M. F.deFeraudy B. Raoult, and G. Torchet, *J. Chem. Phys.*, **84**, 3491 (1986).
3 J.W. Lee and G.D. Stein, *Surf. Sci.*, **156**, 112 (1985).
4 H.C. Andersen and J.D. Honeycutt, *J. Phys. Chem.*, **91**, 4950 (1987).
5 J.A. Northby, J. Xie, D.L. Freeman, and J.D. Doll, *J. Chem. Phys.*, **91**, 612 (1989).
6 B.W. van de Waal, *Z Phys. D.*, **20**, 349 (1991).
7 M.L. Klein and D.M.Ceperley, *Phys. Rev. Lett.*, **63**, 1601 (1990).
8 K.B. Whaley and M.V. Rama Krishna *Phys. Rev. Lett.*, **64**, 1126 (1990).
9 E.G. Syskakis, F. Pobell, and H. Ullmaier, *Phys. Rev. Lett.*, **55**, 2964 (1985).
10 J. Gspann, and H. Vollmar, *J. Chem. Phys.*, **73**, 1657 (1980).
11 A. Scheidemann, J.P. Toennies, and J.A. Northby, *Phys. Rev. Lett.*, **64**, 1899 (1990).
12 R.N. Barnett, and K.B. Whaley, K. B. private communication
13 D.J. Levandier, M. Mengel, R. Pursel, J. McCombie,and G. Scoles, *Z. Phys. D*, **10**, 337 (1988).
14 D.J. Levandier, S. Goyal, J. McCombie, B. Pate, and G. Scoles, *J. Chem. Soc. Faraday Trans.*, **86**, 2361 (1990).
15 T.E. Gough, D.G. Knight, P.A. Rowntree, and G. Scoles, *J. Phys. Chem.*, **90**, 4026 (1986).
16 B. Heymen, A. Bizzari, S. Stolte, and J. Reuss, *Chem. Phys.*, **132**, 331 (1989).
17 X.J. Gu, D.J. Levandier, B. Zhang, G. Scoles,and D. Zhuang, *J. Chem. Phys.*, **93**, 4898 (1990)
18 L. Perera, and F. Amar, *J. Chem. Phys.*, **93**, 4884 (1990).
19 B.I. Swanson and L.H. Jones, *J. Chem. Phys.*,**74**, 3205(1981).
20 B.I. Swanson and L.H. Jones, *Chem. Phys. Lett.*,**80**, 51 (1981).
21 J. Farges, M.F. deFeraudy, B. Raoult, and G. Torchet, *J. Chem. Phys.*, **78**, 5067 (1983).
22 O.F. Hagena, and W. Obert, *J. Chem. Phys.*, **56**, 1793 (1973).
23 S. Goyal, G.N. Robinson, D.L. Schutt, and G. Scoles, *J. Phys. Chem .*, **95**, 4186 (1991).
24 H.Bauer, M.Beau, B.Friedl, C.Marchand, K.Miltner, and H.J.Reyher, *Phys. Lett.*, A, **146**, 134 (1990).
25 D.M. Brink and S. Stringari, *Z. Phys. D - Atoms, Molecules and Clusters*, **15**, 257 (1990).
26 H.Buchenau, E.L.Knuth, J.Northby, J.P.Toennies, and C.Winkler, *J. Chem. Phys.*, **92**, 6875 (1990).

MORPHOLOGY AND STRUCTURE OF ULTRAFINE GOLD PARTICLES

JITENDRA KUMAR and ANU GUPTA
Indian Institute of Technology
Materials Science Programme
Kanpur-208016, India

ABSTRACT. Morphology and structure of ultrafine gold particles dispersed over alumina support films have been studied by transmission electron microscopy. It is shown that gold particles generally correspond to f.c.c. structure, but, also assume a distorted cubic phase in some situations. The unit cell becomes smaller on one side by about 5% and thus cubic symmetry is destroyed. The resulting structure is tetragonal with a ~ 0.407 nm or $a/\sqrt{2}$, c ~ 0.387 nm. On exposing the particles to hydrogen at temperatures 200 - 500°C for various lengths of time, an overall increase in particle size is found to occur. No evidence is obtained to support the particle migration, collision and subsequent coalescence for growth. Instead, Ostwald ripening mechanism is operative, whereby, the coarsening results due to transfer of atoms from smaller to larger particles via the substrate and/or by jump process. Gold particles show morphological variations, e.g. from irregular shapes to spherical or well defined such as hexagonal, pentagonal, etc.

1. Introduction

Ultrafine metal particles show unusual physical and chemical properties. Consequently, they have found immense applications in catalysis, photographic processes, etc. [1-4]. Also, metal particles incoporating a few hundred atoms represent a transition state of matter lying in between the single atom and the bulk. In-depth studies of small metal particles have, therefore, been undertaken in recent past with increasing vigour, principally, to understand the origin and nature of their characteristics. Metals such as platinum, gold, palladium, nickel, iron etc. supported on carbon, silica, or alumina have been chosen for case studies [2-5]. Of these, gold is probably the only metal which is least influenced by the environment because of its limited affinity for gas absorption [6]. Hence,it is not only simple but also unique system for understanding the behaviour of small metal particles without substantial interference from the gaseous atmosphere. It is with this broad objective, the present investigation was undertaken, in which, gold particles on alumina support have been dispersed and studied as such and after heat treatment in hydrogen atmosphere.

P. Jena et al. (eds.), Physics and Chemistry of Finite Systems: From Clusters to Crystals, Vol. I, 93–98.
© 1992 *Kluwer Academic Publishers.*

94

2. Experimental

Support films of alumina of thickness ~ 15 nm have been prepared by anodization in an electrolyte of 3 wt % tartraic acid in distilled water at pH 5.5, and transferred onto gold grids (see [6] for details). Gold thin films of mean thickness 0.4nm were then deposited on alumina supports by evaporation in vacuum ~ 10^{-5} torr at substrate temperature of 125°C to obtain reasonable dispersion of particles. Some samples were subsequently subjected to heat treatment in hydrogen at a flow rate of 45 to 50 ml/min. at several temperatures for various lengths of time. A Philips EM 301 transmission electron microscope (TEM) was used at 100 kV to obtain information about the morphology, dispersion and structural crystallography of particles.

3. Results and Discussion

3.1. EMERGENCE OF PARTICLES

The dispersion of gold particles over alumina supprot films was achieved by evaporation of metal under vacuum (~10^{-5} torr) on substrate held at elevated temperatures. Trial runs revealed that both mass thickness and substrate temperature influence the nature, size and distribution of particles. The deposition of gold corresponding to a mean thickness of 0.4 nm at substrate temperature of 125°C was found to yield reasonable dispersion of particles (Fig. 1a). The average projected diameter ranges between 7 to 9 nm depending upon the substrate conditions. Also, irregular shape particles including island of gold are noticed. On careful examination, it is realized that alumina substrate prepared from commercial grade aluminium foil display non-homogeneity character. Therefore, only those samples which had reasonable dispersion of particles both in terms of size distribution and shapes were selected for carrying out further study. The analysis of the electron diffraction pattern (Fig. 1b) suggests the presence of a f.c.c. phase of gold with a ~ 0.407nm. Notice that rings are somewhat diffuse perhaps due to the presence of fine particles.

3.2. EVIDENCE FOR A NEW PHASE

When gold particles are heated in hydrogen at 300°C just for 20 minutes they show a curious phenomenon (Fig. 2). The particles assume regular shapes and even exhibit faceting. More importantly, the nature of diffraction pattern has also changed. This suggests that gold particles no longer have the usual f.c.c. structure with lattice parameter a = 0.40783 nm. There exists no evidence in the literature to show any reaction of hydrogen with gold. It is therefore believed that particles are still of gold though new phase is perhaps stabilized by hydrogen. Our results also indicate that gold particles after heating in the temperature range 200 - 500°C for longer periods invariably exhibit f.c.c. structure. Therefore the new phase seems to appear during the transition from predominantly island - type crystallites to three dimensional particles. The survey of the interplanar spacings (Table 1)

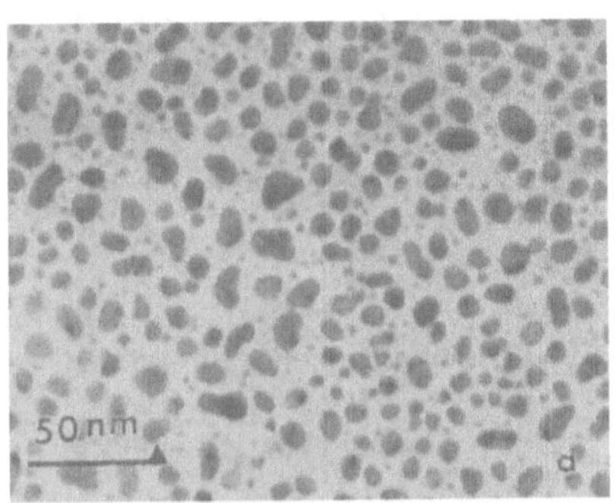
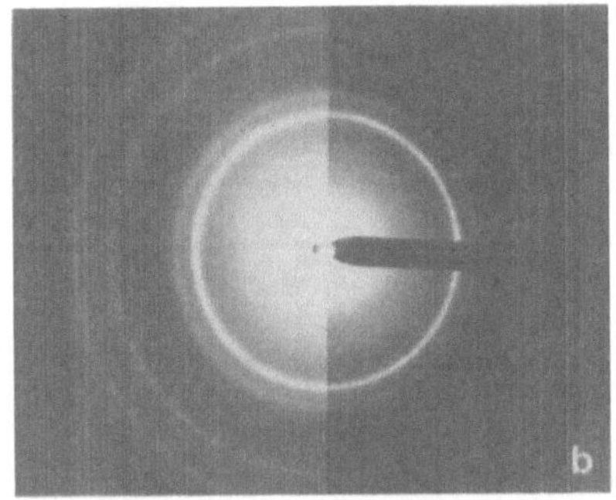

Figure 1. Electron micrograph a) and diffraction pattern b) of particles of gold dispersed over alumina at 125°C (mean thickness 0.4 nm).

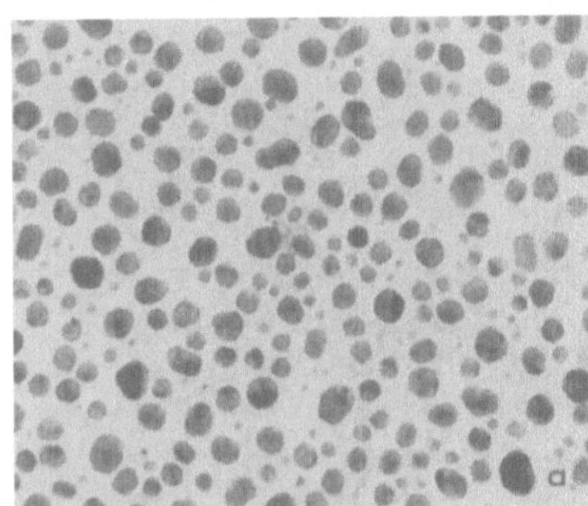
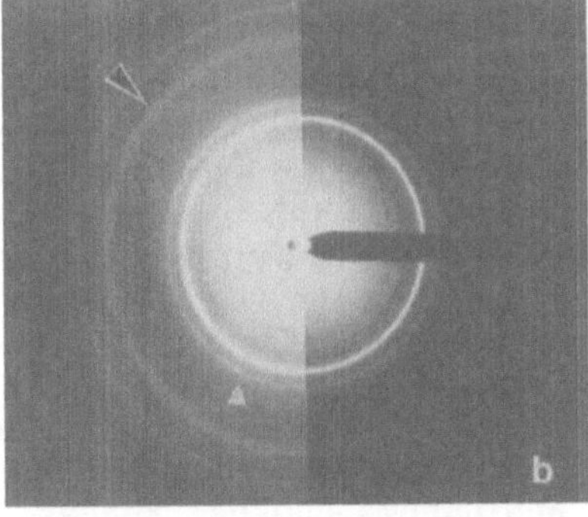

Figure 2. Electron micrograph a) and diffraction b) of gold particles after heat treatment in hydrogen at 300°C for 20 minutes.

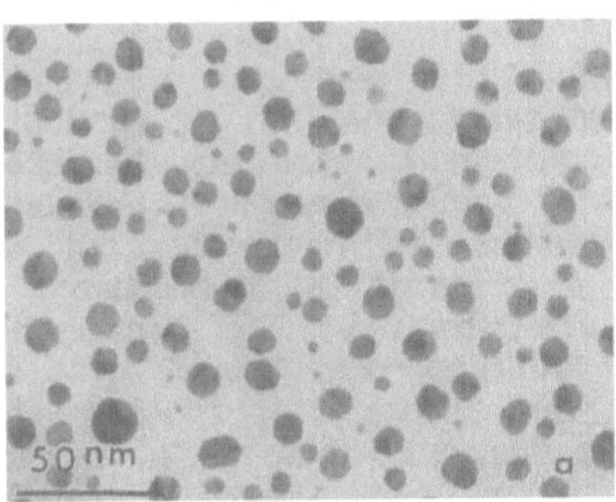
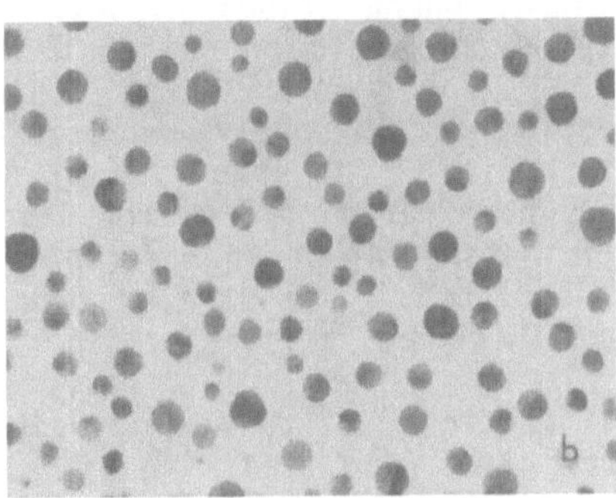

Figure 3. Gold particles exhibiting faceting and spherical shapes after heat treatment in hydrogen at 500°C for a) 16h and b) 64 h.

Table 1. Interplanar spacings and reflection hkl for gold particles after heating in hydrogen at 300°C for 20 min.

Line No.	Interplanar spacings (nm)		hkl*	
	observed	calculated*	a)	b)
1	0.384	0.3870	001	001
2	0.287	0.2878	110	100
3	0.230	0.2309	111	101
4	0.203	0.2035	200	110
5	0.190	0.1935	002	002
6	0.179	0.1820	201	111
7	0.144	0.1439	220	200
8	0.140	0.1402	202	112
9	0.122	0.1221	311	211
10	0.115	0.1155	222	202

*based on tetragonal unit cell a) with $a=0.407$ nm, $c=0.387$ nm or b) with $a'=a/\sqrt{2}=0.288$ nm, $c=0.387$ nm.

suggests that particles assume a distorted f.c.c. structure involving contraction of unit cell in only one dimension by about 5%. If the diffraction pattern is examined carefully, it appears that some lines are split into two while others displaced somewhat. Specifically, rings corresponding to planes (200) and (220) are affected most. There are two rings now for both 200 and 220 type reflections. It means that the cubic symmetry is destroyed. Further, 111 reflection continues to be stronger though its d-value is changed marginally. The unit cell becomes tetragonal with $a{\sim}0.407$ nm or $a/\sqrt{2}$, $c{\sim}0.387$ nm. This result is not unexpected as small metallic particles can display different atomic packing and assume unusual shapes. Atom by atom growth of metallic particles during vacuum deposition invariably form 13 atoms icosahedron, having a number of five fold rotation axes. Also, it has 42 nearest neighbour bond contacts in comparison to 36 in a cluster exhibiting f.c.c. structure [7]. It implies that an icosahedron has only (111) type closed packed surface while f.c.c. crystallites expose both (111) and higher energy (100) planes. The interfacial energies calculated purely on the basis of broken bonds for low index planes (100), (110) and (111) for gold come out to be 2395, 2126 and 2074 ergs cm^{-2}, respectively. Obviously, an icosahedron corresponds to a lower energy configuration. The stability of the icosahedral shape can, therefore, be thought of arising from the lowering of surface energy of the system by eliminating higher energy (100) like surface and having more and more (111) surfaces. The strains produced as a result of this change get accommodated by distortion in the particle itself. Icosahedron particles due to absence or decrease of (100) surfaces should give rise to diffraction pattern having weak 200 reflections – an experimentally observed fact (Fig. 2b). The diffraction rings are found to be somewhat broad possibly due to (i) particles being of very small size, and/or (ii) the presence of different degree of distortion in different regions. The conversion of icosahedral packing into f.c.c. is possible when bulk strain energy introduced into the particle overcomes its surface energy advantages. The tetragonal phase involves a 5.5% reduction in volume of the bulk f.c.c. unit cell. This result is not unusual as previous reports [2,8-10] do indicate contraction in lattice parameter of small metal particles, e.g., the lattice parameter decreases from 0.2 to 0.4% of the bulk value for spherical gold particles of diameter 12.5 to 3.5 nm [9]. A decrease of lattice parameter from 0.28 to 0.04 % of bulk value is also found in gold platelet of thickness 8.6 to 40nm [8]. The contraction is believed to be due to the compressive nature of surface stress felt by the body atoms.

3.3. FACETING IN PARTICLES AND COARSENING MECHANISM

Another important feature noticed in gold particles is faceting i.e. instead of assuming spherical shapes so as to reduce the total surface area, they prefer to have well defined morphologies such as hexagonal, pentagonal, etc. (Fig. 3). These observations are quite consistent with previous reports on morphology, mechanism of formation and stability of particles formed by vacuum deposition of f.c.c. metals on non-metallic substrates [11-16]. Such unusual shaped clusters are termed composite structures or multiply twinned particles (MTP's). They are made-up of basic polyhedra unit (decahedra or icosahedra) formed by the combination of regular tetrahedra bounded by (111) faces. While decahedran is generated by five tetrahedra with a common edge, icosahedra is formed by twenty tetrahedra with common vertex. To realize filling of space without gaps and loss of coherency at interfaces, the constituent tetrahedra undergo strains. (i.e distortions). Indeed, distortion accompanying slight bending of lattice images [15] or changes in symmetry and lattice parameters (Section 3.2) have been observed. These faceted particles are bounded by (111) planes and correspond to minimum energy configuration with elastic strains in the bulk, as explained above. On heating in hydrogen for longer duration, coarsening of particles occurs and is accompanied by overall decrease in surface area and emergence of higher interfacial energy surfaces. These,in turn, overcome the surface energy advantages and contribute to lowering of bulk strain energy. Consequently, faceted particles can not possibly be of minimum energy configuration and so begin to assume spherical shapes exhibiting f.c.c. structure. The present observations indeed match to this description (Fig. 3b).

Basically, two mechanisms are used to explain the growth of particles. In the first, particles can grow by migration followed by interparticle collision and coalescence. It implies that one should encounter particles touching each other quite frequently in the intermediate stages of heat treatment. Also, this mechanism predicts [4] a log normal distribution of particle size. The shape of distribution function contain a tail on the large diameter side. This means that percentage of particles approach zero asymtotically with increase in particle diameter. None of the micrographs and size-distribution curves show these features. In the second (i.e. Ostwald ripening process), movement of individual atoms take palce from smaller particles to larger ones. As a consequence, bigger particles grow further while smaller ones gradually disappear [16]. It implies that one should observe particles of small sizes in the vicinity of big ones. The micrographs indeed show this behaviour. Further, this mechanism predicts a negatively skewed distribution curve comprising of a tail towards the smaller diameters and a clear cut off at some size beyond the peak. The distribution curves do suport this characteristics. Therefore it is beieved that Ostwald ripening mechnanism is responsible for growth of particles in the present case.

References

1. Hamilton, J.F. and Baetzold, R.C. (1979) "Catalysis by small metal clusters" Science 205, 1213-1220.
2. Poppa, H. (1984) "Model studies in catalysis with UHV-deposited metal particles and clusters", Vacuum 34, 1081-1095.
3. Hayashi, C. (1987) "Ultrafine particles", J. Vac. Sci. Technol. A5, 1375-1384.
4. Granqvist, C.G. and Buhrman, R.A. (1976)"Ultrafine metal particles", J. Appl. Phys. 47, 2200-2218.
5. Kumar, J. and Palanisamy, R. (1987) "Formation of small particles of gold on alumina support films and their behaviour in oxygen and hydrogen atmospheres", Applied Surface Science 29, 256-270
6. Schwark, J. (1983) "Catalytic gold - application of elemental gold in heterogeneous catalysis", Gold Bulletin 16, 103-110.
7. Burton, J.J. (1975) " Structure and thermodynamic properties of microclusters, in G.C. Kuczynski (ed.), Sintering and Catalysis, Plenum Press, New York, pp. 17-27.
8. Smart, D.C., Boswell, F.W. and Corbett, J.M. (1972) "Lattice spacings of very thin gold platelets", J. Appl. Phys. 43, 4461-4465.
9. Mays, C.W., Vermaak, J.S. and Kuhlmann-Wilsdorf, D. (1968) "On the surface stress and surface tension II. determination of the surface stress of gold", urface Science 12, 134-140.
10. Moraweek, B., Clugnet, G. and Renouprez, A.J. (1979) " Contraction and relaxation of interatomic distances in small platinum particles from extended x-ray absorption fine structure (EXAFS) spectroscopy", Surface Science 81, L631- L634.
11. Ino, S. and Ogawa, S. (1967) " Multiply twinned particles at earlier stages of gold film formation on alkalihalide crystals", J. Phys. Soc. Japan 22, 1365-1374.
12. Darby T.P. and Wayman, C.M. (1975) " Nucleation and growth of gold films on graphite", J. Cryst. Growth 28, 41-52; 53-67.
13. Yagi, K., Takayangi K., Kobayashi, K. and Honjo, G. (1975) " In-situ observations of growth processes of multiply twinned particles", J. Cryst. Growth 28, 117-124.
14. Gillet, M., (1977) "Structure of small metallic particles", Surface Science 67, 139-157.
15. Komoda, T. (1968)" Study on the structure of evaporated gold particles by means of a high resolution electron microscope", J. Appl. Phys. 7, 27-30.
16. Geus, J.W.(1975) "Preparation and thermostability of supported metal catalysts", in G.C. Kuczynski (ed.), Sintering and Catalysis, Plenum Press, New York 29-61.

MAGIC NUMBERS OF BIMETALLIC CLUSTER IONS CONSISTING OF GROUP-III METALS: THE ICOSAHEDRIC 13 ATOM CLUSTERS

A. NAKAJIMA, T. SUGIOKA, T. KISHI, and K. KAYA
Department of Chemistry, Keio University, 3-14-1 Hiyoshi
Kohoku-ku, Yokohama 223, JAPAN

ABSTRACT. Bimetallic negative cluster ions consisting of group-III metals (In, Al, and B) were generated by two independent laser-vaporization sources. Icosahedron structure consisting of 13 atoms could be ascertained by their magic numbers of $In_{12}Al_1^-$ and $Al_{12}B_1^-$

1. INTRODUCTION

Since metal clusters play important roles in the process of chemisorption and catalysis [1], the properties of metal clusters have increasingly intrigued investigators in a past decade. Many investigators have developed the method of the production, the detection, and the characterization of the metal clusters experimentally [1,2]. In a mass spectrum, it has been found that neutral or charged metal clusters with certain numbers of atoms are more abundant than others. These "magic numbers" have been explained by using various models for different systems; electronic shell structure especially for alkaline metal clusters [3] and the microcrystal model for semiconductor clusters [4].

Recently several groups have reported the geometric structure of metal clusters by using various techniques. Parks et al. used a chemical reaction of nickel clusters with ammonia and water to determine their geometric structures [5], and have found that nickel clusters consisted of 50-100 atoms take the structures arising from closings of shell and sub-shells of icosahedral structures which has been suggested by Mackay [6]. Martin et al. observed the intensity periodicity in mass spectra [7].

We have recently developed a new method for producing a binary cluster containing two different elements [8,9]. Applying this method, we tried to infer geometric structures of metal clusters, especially clusters of grroup-III metals (In, Al, and B). Experimentally and theoretically, it has been reported that Al_{13}^- should be stabilized both by the electronic shell closing of $2p$ shell (40 electrons) and by the geometric structure of icosahedron [10-15].

2. EXPERIMENTAL SECTION

Details of the experimental apparatus have been reported previously [8,9]. In the generation of binary cluster ions, two target rods of aluminum (Nilaco Corp. 99.99 %) and indium (Nilaco Corp. 99.99 %) or boron nitride (Denka, N-1 type 99.2 %) were vaporized by two different pulsed YAG lasers (532 nm, Quantaray DCR-2A and DCR-2) in a He carrier gas (3-5 atm, 99.9999 %). The merit of this method is that the mixing ratio of two metal elements can be controlled easily by the fluence of individual lasers. The laser fluence for a major element was typically 6-10 mJ/pulse and that for a minor element was 2-4 mJ/pulse. The individual lasers were tightly focused on the metal rods by quartz

P. Jena et al. (eds.), Physics and Chemistry of Finite Systems: From Clusters to Crystals, Vol. I, 99–104.
© 1992 *Kluwer Academic Publishers.*

lenses. After thermalization with He, the reaction products was mass-analyzed by a time-of-flight (TOF) mass spectrometer equipped with reflectron. Negative ions were formed in the nozzle and no special secondary ionization was required. The ions were extracted by applying a fast high voltage pulse (Velonex Model 350 and V1742) and detected by a dual multi-channel plate. The signal was amplified and stored in a transient oscilloscope (LeCroy 9400A, 100 MHz sampling rate) coupled with a microcomputer (NEC PC-9801). The repetition rate was 7-8 Hz and a mass spectrum was typically obtained by averaging 700-800 outputs.

3. RESULTS & DISCUSSION

3.1. Mass Distributions of $Al_nIn_m^-$ clusters mainly composed by Al atoms

Figure 1 shows the TOF mass spectrum of $Al_nIn_m^-$ clusters (n=3-40, m=0-2). The spectrum was measured at appropriate voltages for a series of deflection plates adjusted for medium-sized clusters around n=10. As reported by Gantefor et al. [10] and Taylor et al. [11], the electron affinity of Al_n^- is positive at n≥1, but in the present case, the source conditions and the detection conditions were optimized for the measurements of the medium-sized clusters, then smaller clusters (n≤2) were not detected. Besides the series of Al_n^-, the binary clusters containing indium atoms were observed as shown in fig. 1. To show the magic numbers clearly, the intensity distributions in Al_n^-, $Al_nIn_1^-$, and $Al_nIn_2^-$ are separately shown in fig. 2. The intensity distributions were determined by averaging several mass spectra, and their uncertainty was about ± 3% of the value.

When one pays attention to the intensity distribution of Al_n^- in fig. 2, there is one prominently intense peak at n=13. The pattern of the obtained mass spectra was almost the same as that obtained by Gausa et al. [16] and that reported previously [9]. The magic number was observed as the result of the fragmentation of hot clusters and intensity anomalies reflect the stability of the decay products. To observe the magic number clearly, it was preferable to produce hot clusters under high laser fluence (9-10 mJ/pulse) and low density of He carrier gas (stagnation pressure of 3 atom).

The stability of neutral or charged aluminum clusters has been examined experimentally [10-12] and theoretically [13-15] using the electronic shell model. As for aluminum negative cluster ions, Al_n^-, their reactivities toward oxygen and their

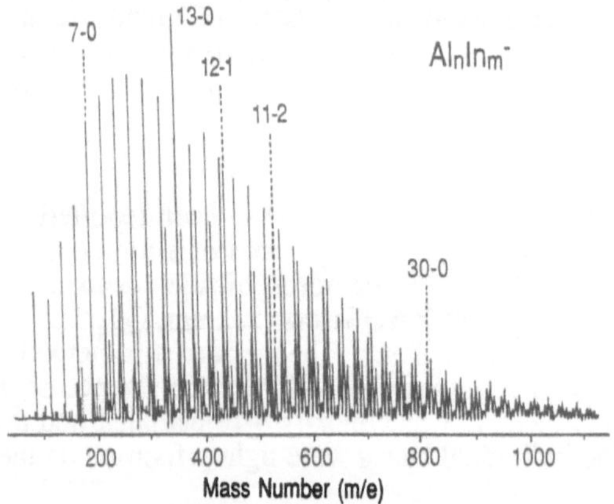

Figure 1 Time of flight mass spectrum of $Al_nIn_m^-$ (n=3-40, m=0-2). Several magic numbers are observed as a prominent peak or a intensity step; Al_{13}^-, $Al_{12}In_1^-$, and $Al_{11}In_2^-$.

photoelectron spectra have been reported [10-12]. Leuchtner et al. have studied the reactivities of Al_n^- clusters with oxygen [12], in which Al_{13}^- and Al_{23}^- have been reported to be less reactive compared with others and they have explained their low reactivities in terms of the stabilization by electronic shell structures. The numbers of valence electrons of Al_{13}^- and Al_{23}^- are 40 and 70, respectively, and these numbers strictly satisfy the closing of $2p$ and $3s$ shells, respectively. On the other hand, the photoelectron spectra of aluminum clusters ($2 \leq n \leq 25$) have been measured [10,11], and the change of their electron affinity shows irregularities at n = 6, 13, and 23; there are local maxima in the electron affinity as a function of number of atoms n. The irregularities can be attributed to the change of electronic structures which are related to the electronic shells. As mentioned above, the total number of valence electrons in Al_{13}^- and Al_{23}^- clusters strictly satisfy the shell closing. Either experimental result has been explained only by the electronic shell closings.

Figure 2 shows the intensity distribution of $Al_nIn_1^-$ and $Al_nIn_2^-$ clusters from n=3 to n=26. Surprisingly, the intensity distributions of $Al_nIn_1^-$ and $Al_nIn_2^-$ are almost the same with that of Al_n^- and the magic numbers appear at n=12 and 11, respectively. Furthermore, the peak of $Al_{10}In_3^-$ could also be observed as a magic number, although the intensity distribution of $Al_nIn_3^-$ is not shown in the figure. This result indicates that indium atoms can be substituted for aluminum atoms up to three atoms at least, as the 13 atoms cluster keeps its stability attributed to the Al_{13}^- cluster. Namely, the valence electrons of the indium atom can contribute to the formation of the $2p$ electronic shell.

As discussed previously, the geometric structure of the Al_{13}^- cluster can be presumed as an icosahedron structure [9], which is a close-packing structure. Since the Al_{13}^- cluster is also stabilized geometrically, it is figured out that the $Al_{12}In_1^-$, $Al_{11}In_2^-$, and $Al_{10}In_3^-$ clusters also take the icosahedron structure. Without the icosahedron structure, they could not be observed as magic numbers in the mass spectrum.

Mackay has pointed out that the icosahedron structure consisting of 13 atoms is made up by 20 distorted tetrahedron [6] ; the bond length between two surrounding atoms is 1.05 times larger than that between the center atom and the surrounding atom. The spheroidal closed shell is, therefore, slightly extended in order to include the atom at the

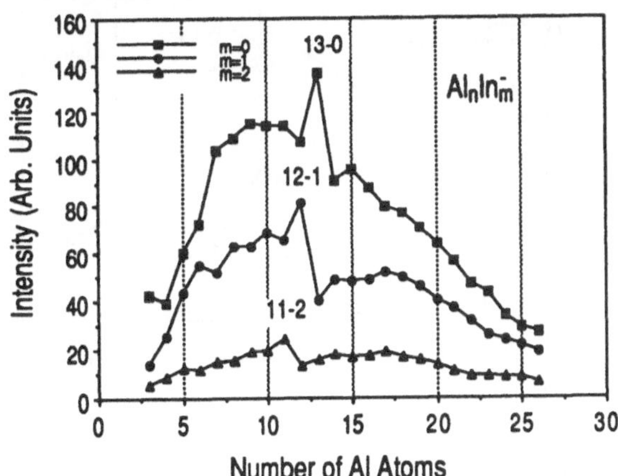

Figure 2 Intensity distribution of Al_n^-, $Al_nIn_1^-$, and $Al_nIn_2^-$ clusters. In each series, one magic number appears at n+m=13, and they can be explained by the electronic structure and the geometric structure (see text). The uncertainty was about ± 3 %.

center. The diameters of aluminum and indium atoms are 2.36 and 2.88 Å, respectively, and the diameter of indium atoms is larger than that of aluminum atoms. Therefore, indium atoms should be contained as surrounding atoms in the icosahedron structure, because the icosahedron structure can ill afford to include a larger diameter atom at the center. Moreover, larger diameter atoms prefer to be contained as surrounding atoms because the extension of the bond length in the spheroidal closed shell can be solved by the substitution into larger diameter atoms.

3.2. Mass Distributions of $Al_n In_m^-$ clusters mainly composed by In atoms

Figure 3 shows the intensity distributions in In_m^-, $Al_1 In_m^-$, and $Al_2 In_m^-$ ($7 \leq m \leq 16$) separately. The intensity distributions were determined by averaging several mass spectra, and their uncertainty was about ± 3% of the value. The spectrum was measured at appropriate voltages for a series of deflection plates adjusted for medium-sized clusters around n=10. When one pays attention to the intensity distribution of In_m^- in fig. 3, there is one intensity step between n=13 and 14. The pattern of the obtained mass spectra was almost the same as that obtained by Gausa et al. [16]. Compared with the magic number of Al_{13}^- as shown in fig. 2, the magic number of In_{13}^- never appears so clearly as that of Al_{13}^- and this reason is discussed below.

The stability of neutral or charged indium clusters has been examined experimentally [16,17] using the electronic shell model. Schriver et al. have measured ionization potentials of In_n clusters and have reported that the neutral In_n clusters (n<80) form electronic shells as trivalent atoms as similar as the Al_n clusters [17]. As for indium negative cluster ions, In_m^-, their photoelectron spectra has been measured in the range of n=2-15 [16], and the change of their electron affinity shows irregularities at n = 7; there is local maximum in the electron affinity as a function of number of atoms n. The irregularities can be attributed to the change of electronic structures which are related to the $1p$ electronic shell, and they have concluded that In_n^- clusters have an alkali-like character to some extent in such a small cluster-size.

Therefore, the magic number of In_{13}^- can be attributed mainly to the geometric stability, although the electronic shell partly contributes to the stabilization of the cluster.

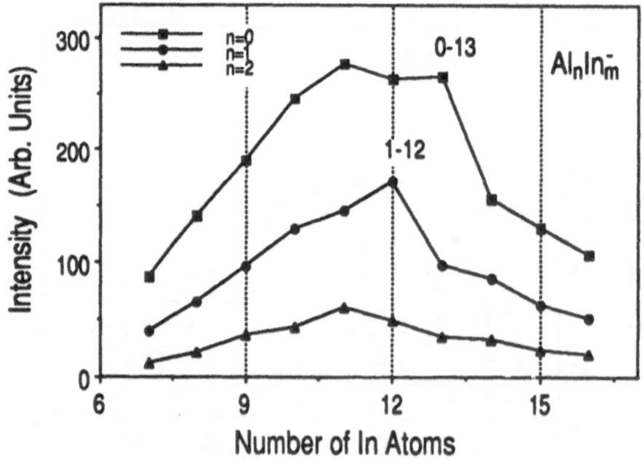

Figure 3 Intensity distribution of In_m^-, $Al_1 In_m^-$, and $Al_2 In_m^-$ clusters. In series of In_m^-, an intensity step appear at m=13, and in series of $Al_1 In_m^-$, an intensity peak appears at m = 12. These features can be explained by the geometric structure of icosahedron. The uncertainty was about ± 3 %.

Indeed, the mass peak of the In_{13}^- cluster never appears so clearly as that of the Al_{13}^- cluster. By an analogy of Al_{13}^-, it is reasonable that the geometric structure of the In_{13}^- cluster is the icosahedron structure. Figure 3 shows the intensity distribution of $Al_1In_m^-$ and $Al_2In_m^-$ clusters from m=7 to m=16. The intensity distribution of $Al_1In_m^-$ is almost the same with that of In_m^- and the magic numbers appear at m=12. However, there is no magic numbers in the intensity distribution of $Al_2In_m^-$ cluster. By an analogy of In_{13}^- and $Al_1In_{12}^-$, the mass peak of the $Al_2In_{11}^-$ cluster should be observed as a magic number, which is attributed to the geometric stability. The loss of the magic number of $Al_2In_{11}^-$ can be explained as follows; under the icosahedron structure, it is possible to substitute one atom into a smaller diameter atom because the smaller diameter atom can be contained in the cluster as the center atom. However, this structure can ill afford to substitute two atoms into the smaller diameter atoms because either atom must be located at the surface at least. As mentioned above, the spheroidal closed shell is slightly extended in order to include the atom at the center. Therefore, the surrounding atom of the smaller diameter atom should lead to a distortion of the spheroidal closed shell and the $Al_2In_{11}^-$ cluster loses the geometric stability. Then, the structure of $Al_1In_{12}^-$ can be presumed as follows; in the icosahedron structure the aluminum atom is located at the center of the cluster and is surrounded by twelve indium atoms (fig. 4a).

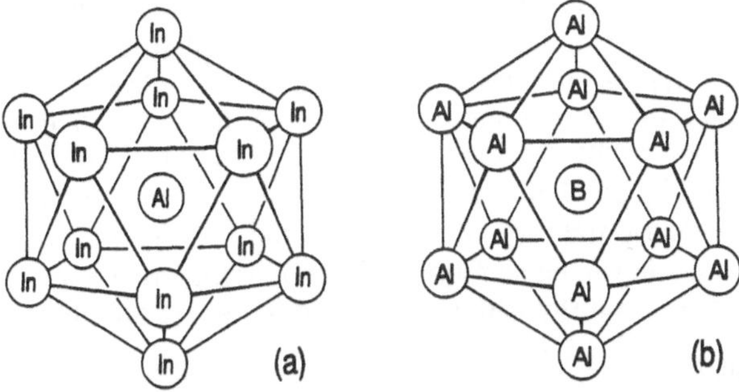

Figure 4 The most probable structures of (a) $Al_1In_{12}^-$ and (b) $Al_{12}B_1^-$. In the icosahedral structure, Al (B) atom is located at the center and is surrounded by twelve In (Al) atoms.

3.3. Mass Distributions of $Al_nB_m^-$ clusters mainly composed by Al atoms

Figure 5 shows the intensity distributions in Al_n^-, $Al_nB_1^-$, and $Al_nB_2^-$ separately. The intensity distributions were determined by averaging several mass spectra, and their uncertainties were about ± 3% of the values. The pattern of the intensity distributions is slightly different, especially around n=23, from that shown in fig. 2. This discrepancy seems to be due to the source conditions, for example, the temperature of the clusters. The intensity distribution of $Al_nB_1^-$ is almost the same as that of Al_n^- and the magic numbers appear at n=12 and 22, where the number of aluminum atoms decreases by one atom compared to the magic number of Al_n^- clusters. This result indicates that a boron atom can be substituted for an aluminum atom in Al_n^-. Namely, the valence electrons of the boron atom can contribute to the formation of the electronic shell and the shell closings of $2p$ shell and $3s$ shell can also be satisfied even in $Al_nB_1^-$ clusters. The diameters of boron and aluminum atoms are 1.64 and 2.36 Å, respectively. Irrespective of the large difference in their atomic diameters, magic numbers also holds at the same cluster size.

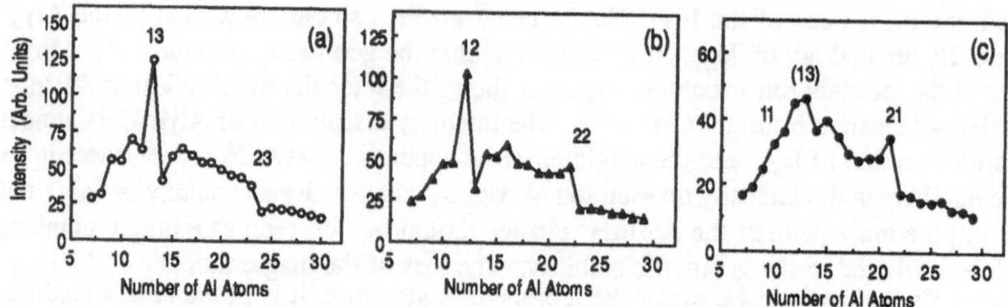

<u>Figure 5</u> Intensity distributions of (a) Al_n^-, (b) $Al_nB_1^-$, and (c)$Al_nB_2^-$ clusters. In series of Al_n^-, an intensity peak appears at n=13, and in series of $Al_nB_1^-$, an intensity peak appears at n = 12. but the mass peak of $Al_nB_2^-$ never appears as a magic number. These features can be explained by the geometric structure of icosahedron. The uncertainty was about ± 3 %.

As shown in fig. 5c, however, the mass peak of $Al_{11}B_2^-$ never appears as a magic number. This result cannot be explained by the electronic structure because $Al_{11}B_2^-$ cluster should be observed as a magic number only as a consequence of the $2p$ shell closing. The disappearance of the magic number of $Al_{11}B_2^-$ can be explained by the geometric structure as similar as the $Al_2In_{11}^-$ cluster. Namely, under the icosahedron structure, it is possible to substitute one smaller diameter atom in the center of the cluster, but this structure can ill afford to substitute two atoms into the smaller diameter atoms. The stable spheroidal shell cannot be constructed by one boron atom and eleven aluminum atoms so that the $Al_{11}B_2^-$ cluster loses the geometric stability. As similar as the $Al_1In_{12}^-$ cluster, the structure of the $Al_{12}B_1^-$ cluster can be presumed as follows; in the icosahedron structure the boron atom is located at the center of the cluster and is surrounded by twelve aluminum atoms (fig. 4b).

4. REFERENCES

[1] *Elemental and Molecular Clusters* ed. by G.Benedek, T.P.Martin and G.Pacchioni Springer-Verlag 1987
[2] *Microclusters* ed. by S. Sugano, Y. Nishina, and S. Ohnishi Springer-Verlag 1987.
[3] W. D. Knight, K. Clemenger, W. A. de Heer, W. A. Saunders, M. Y. Chou, and M. L. Cohen, Phys. Rev. Lett. 52 (1984) 2141.
[4] O. Echt, K. Sattler, and E. Rechnagel, Phys. Rev. Lett. 47 (1981) 1121.
[5] E. K. Parks, B. J. Winter, T. D. Klots, and S. J. Riley, J. Chem. Phys. 94 (1991) 1882.
[6] A. L. Mackay, Acta. Cryst. 15 (1962) 916.
[7] T. P. Martin, T. Bergmann, H. Gohlich, and T. Lange, J. Phys. Chem. 95 (1991) 6421.
[8] S. Nonose, Y. Sone, K. Onodera, S. Sudo, K. Kaya, J. Phys. Chem. 94 (1989) 2744.
[9] A. Nakajima, T. Kishi, T. Sugioka, Y. Sone, and K. Kaya, Chem. Phys. Lett. 177 (1991) 297.
[10] G. Gantefor, M. Gausa, K. H. Meiwes-Broer, H. O. Lutz, Z. Phys. D 9 (1988) 253.
[11] K. J. Taylor, C. L. Pettiette, M. J. Craycraft, O. Chevnovsky, and R. E. Smalley, Chem. Phys. Lett. 152 (1988) 347.
[12] R. E. Leuchtner, A. C. Harms, and A. W. Castleman, Jr., J. Chem. Phys. 91 (1989) 2753.
[13] M. Y. Chou, M. L. Cohen, Phys. Lett. 113A (1986) 420.
[14] M. P. Iniguez, M. J. Lopez, J. A. Alonso, and J. M. Soler, Z. Phys. D 11 (1989) 163.
[15] L. G. M. Pettersson and C. W. Bauschlicher, Jr., J. Chem. Phys. 87 (1987) 2205.
[16] M. Gausa, G. Gantefor, H. O. Lutz, and K. H. Meiwes-Broer Int. J. Mass Spectros. & Ion Processes 102 (1990) 227.
[17] K. E. Schriver, J. L. Persson, E. C. Hoena, and R. L. Whetten, Phys. Rev. Lett. 64 (1990) 2539.

SIGNIFICANT STRUCTURE THEORY OF MOLECULAR CLUSTERS

T. NGUYEN and K. COX
Phillip Morris USA Research Center, PO Box 26583, Richmond, VA, USA

D. D. SHILLADY
*Department of Chemistry, Virginia Commonwealth University, Richmond,
Virginia 23284-2006, USA*

ABSTRACT. The Significant Structure Theory of Eyring and Jhon, originally developed for bulk liquids, can be applied to <u>molecular clusters in the gas phase</u>. The canonical ensemble treatment of small clusters is reported using *ab initio* 6-31G vibrational frequencies for the Gibbs Free Energy and compared to the phenomonological equations of the Classical Theory and the Lothe-Pound formulation for water.

1. Introduction

Molecular cluster research [1] offers the prospect of materials with properties different from small molecules or bulk quantities of the same substance. A theory is needed for prediction of the properties of such clusters; for aerosols [2] and supersonic jet spectroscopy [3] of molecular clusters. It is shown here that a method developed by Eyring and Jhon [4] can be extended with one additional term to the formally exact treatment of molecular clusters. At low temperatures and low pressures one only needs to account for the vapor pressure of the monomer [5] to apply the Eyring-Jhon formalism to molecular clusters. This paper is a <u>proof-of-concept</u> of the method.

2. Method

We assume that a supersaturated vapor of monomer species is available prior to forming clusters and that this vapor can be treated as an ideal gas. In addition, there are aerosol clusters with g monomers interspersed in the monomer vapor. The canonical ensemble for each collection of g-mer clusters is given by [6,7] equation (1) where μ is the chemical potential and Q is the partition function for the gas of clusters.

$$\bar{Q} = \sum_{N(g)} Q(N(g), V, T) \ \exp\{\frac{\mu N(g)}{K_B T}\} \tag{1}$$

$$Q = Q(N(g), V, T) = \frac{Z(g)^{N(g)}}{N(g)!} \tag{2}$$

From ensemble theory, the average population of clusters which contain g monomers N(g) is given by equation (4) and equations (5) and (6) follow. Using the chemical potential for the g-mer as g times the chemical potential of a monomer, one obtains equations (7) and (8); the <u>concentration</u> C(g) is then

P. Jena et al. (eds.), Physics and Chemistry of Finite Systems: From Clusters to Crystals, Vol. I, 105–110.
© 1992 *Kluwer Academic Publishers.*

$$\bar{Q} = \sum_{N(g)} \frac{Z(g)^{N(g)}}{N(g)!} \exp\left\{\frac{\mu N(g)}{K_B T}\right\} = \exp\left\{\exp\left(\frac{\mu}{K_B T} Z(g)\right)\right\} \tag{3}$$

defined in equation (9). Taking the natural logarithm of equation (9) leads to the value of the Gibbs free energy of formation in equation (10) in terms of the total canonical ensemble partition functions for the g-mer, Z(g), and the monomer, Z(1). Note that Z(g) can be factored as in (11)

$$\bar{N}(g) = K_B T \frac{\partial \bar{Q}}{\partial \mu} \tag{4}$$

$$\bar{N}(g) = \exp\left\{\frac{\mu}{K_B T}\right\} Z(g) \tag{5}$$

$$\exp\left(\frac{\mu}{K_B T}\right) = \frac{\bar{N}(g)}{Z(g)} \tag{6}$$

$$\frac{\bar{N}(g)}{Z(g)} = \left[\frac{N(1)}{Z(1)}\right]^g \tag{7}$$

$$\frac{\bar{N}(g)}{V} = \frac{N(1)}{V} \exp\left[Ln\left[\frac{Z(g)}{V} \frac{\frac{N(1)^{g-1}}{V}}{\frac{Z(1)^g}{V}}\right]\right] \tag{8}$$

$$\bar{C}(g) = \frac{\bar{N}(g)}{V} = \frac{N(1)}{V} \exp\left[-\frac{\Delta G(g)}{K_B T}\right] \tag{9}$$

$$\Delta G(g) = -K_B T \left[Ln\frac{Z(g)}{V} - g\, Ln\frac{Z(1)}{V} + (g-1)\, Ln\frac{N(1)}{V}\right] \tag{10}$$

with the usual meanings for the translational, rotational and vibrational partition functions in equations (12), (13) and (14) with De as the binding energy of a given vibrational well including the zero-point energy with an electronic partition function of 1 for closed-shell clusters.

$$Z(g) = Z_{tr}(g)\, Z_{rot}(g)\, Z_{vib}(g) \tag{11}$$

$$Z_{tr}(g) = V\left[\frac{2\pi m K_B T}{h^2}\right]^{3/2} \tag{12}$$

$$Z_{rot}(g) = \frac{\pi^{1/2}}{\sigma} \left[\frac{T^3}{\Theta_A \Theta_B \Theta_C}\right]^{1/2} \tag{13}$$

$$Z_{vib} = \sum_{j=1}^{3n-6} \frac{\exp(-\theta_{v_j}/2T)}{1-\exp(-\theta_{v_j}/T)} \omega_{e_1} \exp\left[\frac{D_e}{K_B T}\right] \tag{14}$$

The key steps for this derivation [8] are given by equations (15), (16) and (17) which use the concept of a supersaturated vapor of monomers as is usual in aerosol research; this permits equation (18) to be written in terms of the bulk vapor pressure and the supersaturation S. The logarithmic term of the g-mer can thus be written in equation (19). Once the geometry of a given cluster has been obtained, the moment-of-inertia tensor

$$\begin{aligned} Ln\frac{Z(g)}{V} = {} & \frac{3}{2} Ln\left(2\pi m K_B \frac{T}{h^2}\right) + Ln\left(\frac{\pi^{1/2}}{\sigma}\left(\frac{T^3}{\Theta_A \Theta^B \Theta^C}\right)^{1/2}\right) \\ & + \frac{D^e}{K^B T} + Ln\omega_{el} - \sum_{j=1}^{3n-6}\left[\frac{\theta_{v_j}}{2T} + Ln\left(1-\exp\left(-\frac{\theta_{v_j}}{T}\right)\right)\right] \end{aligned} \tag{15}$$

$$S = \frac{P(1)}{P(\infty)} \tag{16}$$

$$P(1)V = N(1)K_B T \tag{17}$$

$$\frac{N(1)}{V} = \frac{P(1)}{K_B T} = \frac{SP(\infty)}{K_B T} \tag{18}$$

$$\frac{\Delta G(g)}{K_B T} = -Ln\frac{Z(g)}{V} + gLn\frac{Z(1)}{V} - (g-1)Ln\frac{P(\infty)S}{K_B T} \tag{19}$$

can be diagonalized to obtain the principle-axis moments for the rotational partition function and the translational contribution is easy to compute. While many past treatments have simply neglected the "extra" vibrations or estimated them by an empirical formula [9], in this work the HONDO5 [10] electronic structure program was used to <u>exhaustively</u> optimize the geometry of two forms of the dimer and a symmetrical form of the trimer. The C_3 trimer was arbitrarily restricted to conserve computer time, but still required over 20 CPU hours for the frequency calculations on the VCU IBM 3081K mainframe. Previous work in this laboratory by Dwyer [11] showed that vibrational frequencies can be calculated to within 2 /cm of the experimental frequencies using a minimal basis provided special attention is paid to optimizing the molecular geometry. In this work the standard [12] 6-31G basis set was used with repeated gradient optimization until the roundoff limit of the program was reached. Thus, these calculations were exhaustively optimized to better than 0.000001 hartree/bohr to ensure the geometry was very near all the vibrational minima. A step size of 0.001 bohr was used in the energy derivative for the force constant matrix without further scaling [13].

3. Results

Table 1 shows the calculated vibrational results for four "molecules"; the moments-of-inertia used in the rotational partition function were also very close to known values [15].

TABLE 1. Vibrational Frequencies for Water Clusters in 1/cm

Monomer	Open-Dimer	Closed-Dimer	C3-Trimer	
1678	132	28	(2)	243
(1595*)	165	103		274
3809	170	163	(2)	438
(3652*)	179	225		488
3951	364	457	(2)	646
(3756*)	696	662	(2)	766
	1683	1685		869
	1718	1690		1067
	3732	3790	(2)	1819
	3816	3802		1877
	3919	3941		3527
	3953	3944		3594
				3651
				3755
				3934
(* experimental Values [14])				3937

The geometries are very close to structures which have been reported for the dimers [16]; the "extra" frequencies reported here are the only such values known to us.

Once the vibrational frequencies and moments of inertia are available, one can use equation (19) to directly compute the Gibbs Free Energy of formation of the g-mer from the saturated vapor of the monomer using [8] equation (20). This quantity has

$$\log_{10}P(\infty) = 7.991 - \frac{1687}{T-43} \ mm \ Hg \tag{20}$$

also been derived classically by two previous methods [17], the so-called Classical Equation and the Lothe-Pound formulation. Since both of the phenomonological expressions depend on data derived from the bulk material, it is not clear that they have any meaning when extrapolated to the dimer and trimer level, while the Significant Cluster calculations are specifically for this case. The two smooth curves drawn in Figure 1 are extrapolated to the monomer. Only three (block) points result from the lengthy calculations necessary to obtain the vibrational frequencies, but it is clear that the ab initio significant cluster values are in very good agreement with the Classical Method. The previous objection to using the Classical Method for small clusters is that the equation depends on the bulk surface tension which might not be expected to apply to the case of small clusters. However, one of the successes of the Eyring-Jhon Significant Structure Theory was the excellent prediction of surface tension [18] for light and heavy water!

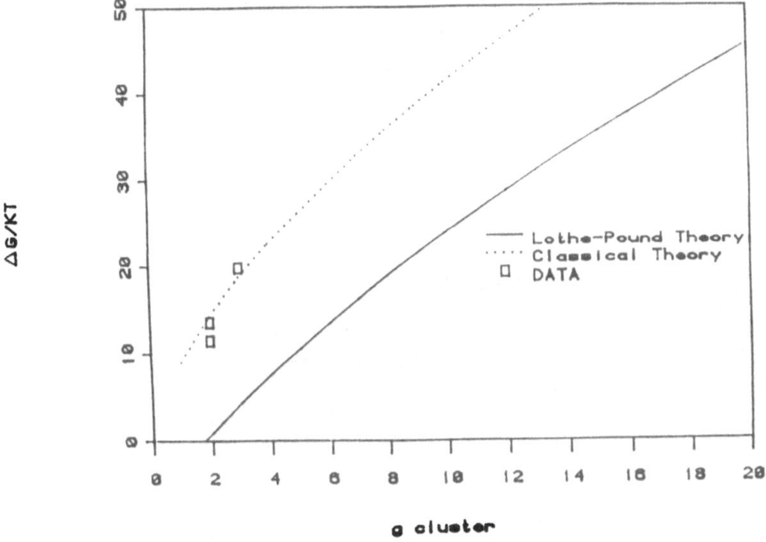

Fig. 1   Free energy of formation as a function of cluster size (g).

Comparison of this work can be made to the recent study of Argon clusters by Li and Sheraga [19] where a similar ensemble equation was derived for Argon clusters in equilibrium with saturated Argon vapor. The interactions were modeled by a Lennard-Jones potential instead of a Hartree-Fock-Roothaan SCF calculation and the cluster vibrations were not treated specifically. The Gibbs Free Energy points obtained by Li and Scheraga [19] for Argon clusters agreed less well with the Classical Equation than this work, so it is clear that the ensemble average method is capable of treating weak interactions and needs vibrational frequencies.

4. Conclusions

Geometry optimization of molecular clusters at the Hartree-Fock-Roothaan SCF level and harmonic frequencies obtained from exhaustively optimized structures give good agreement with the extrapolated Classical Equation for Gibbs Free Energy of Formation. The Lothe-Pound equation seems inappropriate for small clusters. This preliminary work constitutes a proof-of-concept and offers a straightforward way to calculate properties of molecular clusters.

5. Acknowledgements

We wish to thank the Computer Center Staff at VCU for the use of the IBM 3081K and Mr. Loyde Jones for assisting with the data input. Dr. Nguyen wishes to thank Dr. Cliff Lilly and Dr. Jerry Whidby of the Philip Morris USA Research Center for support of this public domain study as part of the requirements for the Ph.D. in Chemical-Physics from Virginia Commonwealth University.

110

6. References

[1]   P. Jena, B. K. Rao and S. Khanna, Eds., Proceedings of the International Symposium on the Physics and Chemistry of Small Clusters, (B158), Plenum Press, 1987.

[2]   G. M. Hidy, Aerosols, An Industrial and Environmental Science, Academic Press Inc., New York, 1984.

[3]   L. S. Zheng, P. J. Brucet, C. L. Pettiette, S. Yang and R. E. Smalley, J. Chem. Phys., 83, 4273 (1985).

[4]   H. Eyring and M. S. Jhon, Significant Liquid Structures, John Wiley and Sons, Inc., New York, USA, 1969.

[5]   G. W. Castellan, Physical Chemistry, Third Edition, Addison-Wesley Publishing Co., Reading, Mass. USA, 1983.

[6]   D. A. Mcquarrie, Statistical Thermodynamics, Harper and Row, New York, USA, 1973.

[7]   T. L. Hill, An Introduction to Statistical Thermodynamics, Addison-Wesley Publishing Co., Reading, Mass., USA, 1960.

[8]   T. T. Nguyen, Ph.D. Thesis, Virginia Commonwealth University, 1990.

[9]   M. Kulmala, M. Olin and T. Raunema, J. Aerosol Sci., 18, No. 6, 615, 1987.

[10]   M. Dupuis, J. Rys and H. F. King, Quantum Chemistry Program Exchange, No. 403, Bloomington Indiana, USA.

[11]   R. W. Dwyer, Ph.D. Thesis, Virginia Commonwealth Univ. 1975.

[12]   (a) W. J. Hehre, R. Ditchfield, and J. A. Pople, J. Chem. Phys., 56, 2257 (1972);
       (b) J. D. Dill and J. A. Pople, ibid., 62, 2921 (1975);
       (c) J. S. Binkley and J. A. Pople, ibid., 66, 879 (1977).

[13]   B. A. Hess, L. J. Schaad. P. Carsky and R. Zahradnik,. Chem. Rev., 86, 709, 1986.

[14]   J. C. Brand and J. C. Speakman, Molecular Structure, The Physical Approach, Edward Arnold Publishers Ltd., London, 1961, p152.

[15]   S. M. Blinder, Advanced Physical Chemistry, The Macmillan Company, London, 1969, p456.

[16]   H. Popkie, H. Kistenmacher and E. Clementi, J. Chem. Phys., 59, 1325 (1973).

[17]   G. S. Springer, Advances in Heat Transfer, Homogeneous Nucleation, Academic Press Inc., New York, 1978.

[18]   M. S. Jhon, E. R. Van Artsdalen, J. Grosh and H. Eyring, J. Chem. Phys., 47, 2231 (1967).

[19]   Z. Li and H. A. Scheraga, J. Chem. Phys., 92, 5499 (1990).

On the Origin of the Dendritic Shapes in Goethite Particles

M. RÖSLER[1], **H. HOFMEISTER**[1], **E. HELD**[2]
[1]*Institute of Solid State Physics and Electron Microscopy*
P.O. Box 250
4010 Halle
[2]*BASF Aktiengesellschaft*
6700 Ludwigshafen
Federal Republic of Germany

ABSTRACT. Electron microscopy is used to characterize particles of Goethite and the whole variety of their dendritic shapes as {021} and {061} twins and combinations of both resulting in 120°-, 60°-, three- and six-fold star-shaped dendrites, resp. Investigating the initial stages of the Goethite growth from the solution it becomes obvious that the interaction between the growing Goethite particles and some intermediate phases such as iron hydroxide and magnetite is responsible for the dendrite formation. The threefold epitaxial positioning of the orthorhombic Goethite lattice on the {0001} plane of the iron hydroxide or the {111} plane of magnetite favours the twinning along the {021} and {061} planes and yields the dendritic shapes.

1. Introduction

Particles of iron oxohydroxide are widely used as precursors of acicular particles of maghemite and iron generated by subsequent dehydration, oxidation and reduction processes. For magnetic recording purposes the final particles should exhibit well-defined sizes, shapes and crystal orientations in order to guarantee the ferri- or ferromagnetic behaviour and certain values of the crystalline and shape anisotropy. Therefore, for maghemite needle-shaped particles of 200–400 nm in length, 50–150 nm in thickness and of a needle axis orientation parallel to the [110] direction are preferred. This determines the requirements of the Goethite particles: needle-shaped crystals having an axis orientation parallel to [001] and sizes in the same order of magnitude as for maghemite.

P. Jena et al. (eds.), Physics and Chemistry of Finite Systems: From Clusters to Crystals, Vol. I, 111–118.
© 1992 *Kluwer Academic Publishers.*

Unfortunately, during the precipitation process of Goethite particles from the Fe(II) containing solutions the needle shape widely varies, frequently forming "dendritic shapes". These consist of needle segments intergrown in such a way, that the needle axes are inclined to each other by multiples of 60° /1/. One of them - the 120° needle axis inclination - was analysed and described by MAEDA /2/ as a {021} twin in the Goethite lattice. The reasons for such a twinning have not yet been clear. Therefore in real ensembles of Goethite particles the particle shape often deviates from the ideal acicular one. This effect influences also the particle shapes of the subsequent transformation steps up to the iron particles. The use of such particulate systems for advanced recording media will depress theire recording performance (the high frequency output, the noise level, etc.). Therefore the motivation for studying the dendritic shapes of Goethite is to minimize or to avoid their occurence.

The present paper i) describes the whole variety of the dendritic shapes arising in Goethite particles , ii) analyses the crystallographic origin of the needle axis deviations, and iii) gives experimental evidence of the formation mechanism of the dendritic shapes during the synthesis of the Goethite particles.

2. Experimental

Acicular particles of Goethite are prepared by the coprecipitation method under alkaline conditions. The starting materials are solutions of 0.63m $FeSO_4 \cdot 7H_2O$ (1.5l) and of 3.4m KOH (2.7l) in deionized water. At temperatures between 35 and 90 °C both solutions are mixed together by simultaniously stirring (300 - 400 rot/min) for a time of 30 min. After that air is bubbled (100 - 500 l/h) uniformly through the mixed solutions in order to promote the oxidation reaction.

The particles grown in the solution are washed to neutral pH, filtered, catched, and dried. Electron microscope investigation of these particles was carried out by dispersing a small amount of the powder in alcohol by ultrasonic treatment and dropping this dispersion onto a carbon covered microgrid. Particles obtained from not fully oxidized samples are carefully prevented from further oxidation during preparation and transfer to the microscope by covering them with a thin carbon shell. These specimens are investigated by transmission electron microscopy (TEM) in a JEM 100C microscope. A combination of bright- and dark field imaging, lattice fringe imaging and selected area diffraction (SAD) was used to determine the shape of the particles, the orientation of their lattice with respect to the needle axis and the interface boundaries inside individual particles. This procedure was carried out for as-grown particles catched at different times during the oxidation stage of the synthesis.

3. Results

3.1. MORPHOLOGY OF THE GOETHITE PARTICLES

Goethite particles catched at different times during the oxidation procedure are aciculare, varying only in lengths and aspect ratios. The needle axis of monocrystalline undisturbed particles is always parallel to the [001] direction of the orthorhombic unit cell (a=0.464nm, b=1.0nm, c=0.303nm /3/). The outer surface of these particles is smooth, with except of a few notches.

Bundles of intergrown needles exhibit a common zone axis, also in [001] direction. The outer surface of the bundles is stepped, because of the different length of the needles incorporated in the bundle. Fig.1 shows these both types of particles.

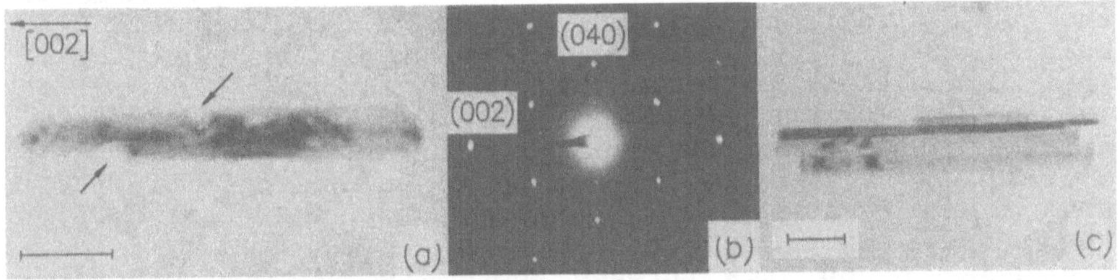

Fig.1: Electron micrographs of an isolated, single-crystalline Goethite needle (a), its SAD pattern (b), and a bundle of needle shaped α-FeOOH particles (c); /marker 100nm/. The arrows mark the typical surface irregularities on the needle (notches).

The occurence of dendritic shaped Goethite particles is obvious in all specimens investigated. Electron microscopy examination of the final (fully oxidized) ensemble yields the following types of dendritic shapes represented in fig. 2: -particles with a 120° needle axis inclination, each segment exhibits a needle axis in [001] direction (fig.2a),

-particles with a 60° needle axis inclination, each segment exhibits a needle axis in [001] direction (fig.2b),

-particles with combined 120° and 60° needle axis inclination, resulting in planar three- or sixfold starlike geometry (fig.2c,d), all of the segments exhibit needle axes in [001] directions; often the segments show different lengths.

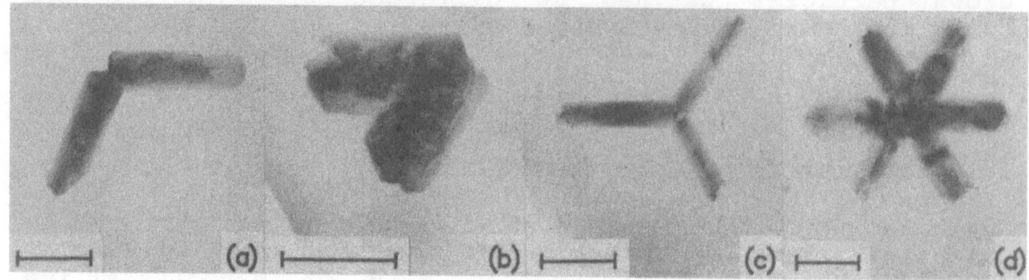

Fig.2: Electron micrographs of Goethite particles exhibiting dendritic shapes: 120° inclination(a), 60° inclination (b), combinations of both previous types (c,d);/ marker 100 nm /.

3.2. CHARACTERIZATION OF OF THE TWIN STRUCTURES

Electron diffraction patterns taken from both segments of the 120° inclined dendritic particles indicate, that there are monocrystalline segments with needle axis parallel to the [001] direction. The direction of the incident electron beam is always [100] with respect to the Goethite unit cell. The diffraction patterns of the disturbed region of inclination of the needles exhibit a sixfold symmetry, arising simply from the superposition of two 120° rotated diffraction patterns of the segments. In this case the (021) spot of the first segment is superposed on the (02$\bar{1}$) spot of the second segment. That means the {021} lattice planes continuously can grow through whole the inclined particle, whereas all the other lattice planes are "reflected" at the {021} twin plane. In the case of a 60° inclined needle shape the diffraction pattern of the disturbed region is composed of the patterns of the two segments rotated by 60°. Here the (061) spot of the first and the (06$\bar{1}$) spot of the second segment are superimposed. The {061} planes continuously run through whole the particle and all other lattice planes are folded at the {061} twin plane.

A proof of this is given in fig.3. Here both segments of the particle of fig.3a are visible in the dark field image (fig.3b) taken with the superimposed spots of the (021) and (02$\bar{1}$) planes of the diffraction pattern (fig.3c).

Bright field imaging of those particles enables to determine the location of the interface between the two segments. From the direction of the (020) lattice planes running parallel to the needle axis, the angle between the needle axis and the interface was determined to 60° for the

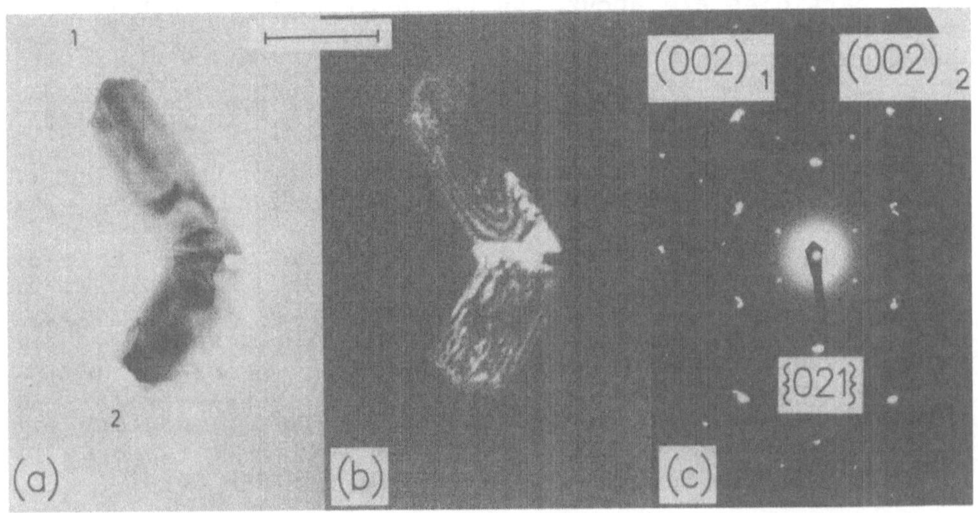

Fig.3: Bright field image of a 120° inclined particle (a) with the segments 1 und 2 and corresponding dark field image (b) taken with the {021} reflection of the SAD pattern (c); / marker 100 nm /.

120°-needle inclinations and 30° for the 60° needle-inclinations. Calculating the theoretical angle between the [001] direction and the {021} and {061} lattice planes in the Goethite lattice with 58.78° and 28.81°, respectively, the following conclusion can be made: The 120° inclined particles have to be interpreted as {021} twins and the 60° inclined particles as {061} twins.

All the other more complicated variations of the dendritic shapes are combinations of {021} and {061} twinning interfaces. In these combined forms of dendritic shaped particles frequently micropores with sizes less than 10 nm are found in the {061} twin boundaries.

3.3. THE FORMATION MECHANISM OF THE DENDRITIC SHAPES

For the examination of the early stages of formation of the dendritic shapes samples were obtained directly from the suspension after the homogenization of the KOH solution and the Fe(II)-salt solution without the following oxidation step. From these suspensions the particles are immediatly extracted and analysed. Using SAD the most particles were identified by their hexagonal lattice structure with d-values of 0.25 nm and their hexagonal platelike shape as distorted iron(II)hydroxide. The outer surfaces of these particles are attacked by limited oxidation owing to the presence of soluted oxygen in the suspension. The sizes of these particles are about 50 nm in diameter. Furthermore a small amount of cubo-octahedral particles of a spinel phase was found (probably magnetite).

Due to the oxygen soluted in the suspension Goethite particles are present from the beginning. These Goethite particles with lengths of less than 200 nm are small in comparison to the particles obtained after the oxidation step, but nevertheless all types of dendritic shapes are found.

Examination of the particle ensemble reveals that in this early stage of the process of Goethite formation frequently the dendritic Goethite particles are intergrown with other species of particles (iron hydroxide plates and spinell cubooctahedra). Fig. 4 shows a hexagonal shaped plate of the iron hydroxide which acts as substrate for the epitaxially oriented Goethite needles. The SAD pattern and dark field images of these connected particles reveale the following crystallographic relationship between both:

```
Goethite needle segment 1    [001] axis  ||  [110]   Fe-hydroxide
                             [010] axis  ||  [110]   Fe-hydroxide
                             (100) plane ||  (001)   Fe-hydroxide

Goethite needle segment 2    [001] axis  ||  [100]   Fe-hydroxide
                             [010] axis  ||  [120]   Fe-hydroxide
                             (100) plane ||  (001)   Fe-hydroxide
```

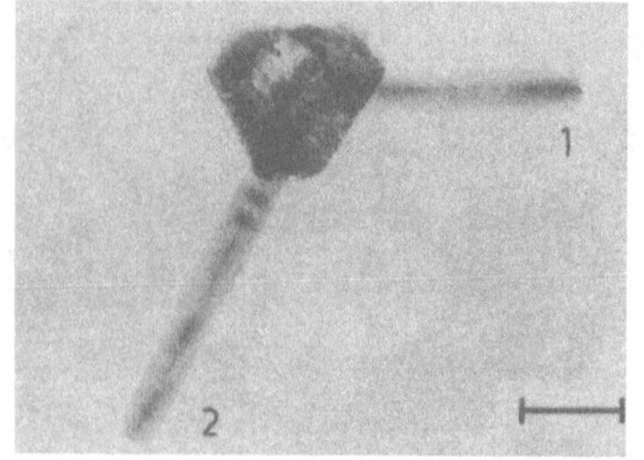

Fig.4: TEM image of the epitaxially grown FeOOH needle segments 1 and 2 on the (0001) plane of iron hydroxide /marker 50nm/.

Concerning the interaction of the Goethite particles with the magnetite, it was found, that the Goethite needles grow especially on the {111} faces of the cubooctahedra. Thereby the axes of the needle segments of Goethite are oriented in the same directions as on the (0001) plane of the iron hydroxide. Often magnetite particles are incorporated with their edges into the surface of Goethite needles, creating thereby the typical geometry of the notches found in the final particles. Both intermediate phases, the iron hydroxide and the magnetite were not found in samples taken from the solution after the oxidation step. Nevertheless also after growing the particles to the length of some 100 nm the typical features of the previous interaction with the hydroxide and the magnetite are present.

4. Discussions

The dependence of particle sizes, the particle shapes and the crystallinity of the Goethite-needles on the experimental conditions (temperature, pH, etc.) was intensively investigated by SCHWERTMANN and coworkers /4/.
   The orientation of the needle axis parallel to the [001] directions, the existence of 120° inclined dendritic shapes of Goethite and their interpretation as {021} twin are in agreement with previous results of VAN OOSTERHOUT /3/ and MAEDA /2/. The finding of an additional dendritic shape with a 60° inclination and its interpretation as {061} twin complete the basic structures required for explaining the more complicated dendritic shapes as composed shapes. With respect to the hexagonal close packed oxygen sublattice of the Goethite both twin planes, the {021} and the {061} planes are prismatic faces of the first and the second kind, respectively.
   The formation mechanism of these twins and therefore the origin of the dendritic shapes is found to be the epitaxial growth of the Goethite lattice on the particles of the intermediate phase Fe(II)hydroxide and magnetite. There the similarity between the hcp oxygen sub-lattice of the Goethite on the one hand and the oxygen sub-lattices of the (0001) Fe(II)hydroxide planes and {111} magnetite planes, respectively, on the other hand is responsible for the threefold positioning of the growing Goethite particles.
   This is demonstrated in fig.5, showing the view on the oxygen-sublattice of the Fe-hydroxide along the [001] direction and the [100] projections of the overgrown orthorhombic Goethite unit cells in three different, but equivalent positions. The oxygen sub-lattice of both the substrate and the deposit is nearly the same, resulting in a very small misfit between both sub-lattices. Only the distribution of the Fe-ions is changed between the octahedral positions of the oxygen lattices. If the Goethite lattice of the nuclei is continued in their respective growth directions [001] the growing particles reach the typical dendritic shapes with the {021} and/or {061} twins.
   During the oxidation step of the synthesis the intermediate phases are dissolved or removed, so that in the final particle ensemble the particle composites Goethite/Fe-hydroxide and Goethite/magnetite cannot be investigated.

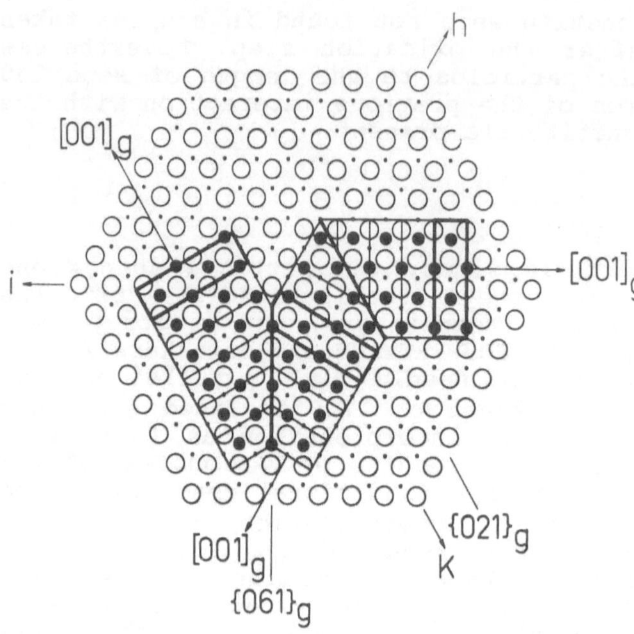

Fig.5: Sketch of the plane of $Fe(OH)_2$ and the epitaxial grown unit cells of α-FeOOH in threefold positioning; white and black circles - oxygen atoms of the first and the second layer, resp., points - octahedral positions.

In the first stage the synthesis of Goethite in strongly alkaline solutions of Fe(II) should be regarded as a combined process of solid-state transformation from the iron hydroxide to Goethite nuclei and an oxidation of $Fe(OH)^{3-}$ forming by dissolution of the hydroxide particles . With increasing dissolution of the hydroxide and increasing sizes of the Goethite nuclei the synthesis is governed by the growth of the α-FeOOH particles from $Fe(OH)^{3-}$ species via the solution.

5. REFERENCES

/1/ Cornell, R.M., Mann, S., and Skarnulis, A.J.(1983), 'High-resolution Electron Microscopy Examination of Domain Boundaries in Crystals of Synthetic Goethite', J.Chem.Soc.,Faraday Trans. 1, 79, 2679-2684

/2/ Maeda, Y., and Hirono, S.(1981), 'Electron Microscopic Observations of the Dendrites of Synthetic α-FeOOH Particles', Jpn.J.Appl.Phys.20, 1991-1992

/3/ Oosterhout, G.W.Van (1960), 'Morphology of synthetic submicroscopic crystals of α- and γ-FeOOH and γ-Fe_2O_3 prepared from FeOOH', Acta Cryst.13, 932-935

/4/ Schwertmann, U., Cambier, P., and Murad, E.(1985), 'Properties of Goethites of varying crystallinity,'Clays and Clay Minerals 33, 369-378

STRUCTURAL FLUCTUATION OF Au₅₅ AND Au₁₄₇ CLUSTERS

[1]S. Sawada and [2]S. Sugano
[1]Fundamental Res. Labs., NEC Corporation,
 34 Miyuki-ga-oka,Tsukuba 305, Japan
[2]Fuculty of Science, Himeji Institute of Technology,
 Kamigori-cho, Ako-gunn 678-12, Japan

Abstract

Structural fluctuation of Au₅₅ and Au₁₄₇ clusters with and without the substrate interaction are examined using model potentials and the transition state theory. It is shown that for Au₅₅ both with and without the substrate interaction, the cuboctahedral structure (COCT) has a lifetime too short to be observed with an electron microscope while the icosahedral structure (IC) has a lifetime long enough. On the other hand, for Au₁₄₇ without the substrate interaction, both COCT and IC have lifetimes long enough but the lifetime of IC is too long for both of them to be observed in an observation period, say several minutes. However, it is so much reduced by the substrate interaction that both the structures can be observed in an observation period. These results are compared to the experimental facts.

1. Introduction

Recently, small gold particles have been found to fluctuate between various structures under an intense electron beam in an electron microscope [1-4]. Each structure has a lifetime of the order of 0.1 sec. What is the source of the fluctuations is still an open question though several different ideas have been proposed.

In this paper, assuming the source of this phenomena to be the thermal effect, we examine the substrate effects on the structural fluctuations of Au₅₅ and Au₁₄₇. We calculate energies of the clusters in the icosahedral (IC) and cuboctahedral (COCT) structures and also energy barriers between them using the model potential, which reproduces bulk properties of gold very well. Then we calculate transition probabilities between two structures according to the transition state theory to discuss a possibility of the observation of the structural fluctuations with an electron microscope.

P. Jena et al. (eds.), Physics and Chemistry of Finite Systems: From Clusters to Crystals, Vol. I, 119–124.

120

2. Model and method

The Gupta's potential is used for potential energies between gold atoms [7 -10]:

$$V = (1/2) \cdot U \sum_j \{A \sum_i \exp(-p(r_{ij}-r_0)) - [\sum_i \exp(-2q(r_{ij}-r_0)]^{1/2}\},$$

$$(2.1)$$

where r_{ij} is the distance between the i-th and j-th atoms and r_0 is the nearest neighbor distance of the bulk crystal, being chosen to be equal to the experimental values 2.88 Å [11]. The values of U and A are determined by fitting the experimental values of the bulk cohesive energy per atom E_{bulk}= 3.81 eV [11]. We use $p \cdot r_0$=10.3 and $q \cdot r_0$=4.0, which reproduce the experimental values of bulk modulus and elastic constants very well [6].

We use the semi-infinite continuum model for a substrate and assume that the interaction energy between cluster atoms and the substrate is given by

$$V_a = U_a \cdot \sum_i [\exp(-2 \cdot \lambda \cdot z_i) - 2 \cdot \exp(-\lambda \cdot z_i)], \qquad (2.2)$$

where z_i is the z-coodinate of the i-th atom. (The z-axis is perpendicular to the substrate surface.) We use λ = 5.0 Å^{-1}, which is chosen somewhat arbitrarily to represent short-range nature of the interaction. The value of U_a is estimated to be 0.230±0.037 eV for the graphite substrate from the adhesion energy of the Au (111) face on the graphite (00.1) face, 514±82 erg·cm^{-2} [12]. This value of U_a is 0.06 ~ 0.07 times of E_{bulk}. We vary the ratio $\gamma = U_a/E_{bulk}$ in the numerical calculations to see the substrate effects.

In order to calculate transition probabilities between IC and COCT we use the transition state theory [13]: the transition probability from the structure A to B is given by

$$\Gamma_A = \nu_A^* \cdot \exp(-\Delta V_A/k_B T) \qquad (2.3)$$

with the frequency factor

$$\nu_A^* = (1/2\pi) \cdot \left(\prod_{\alpha=1}^{3N-6} \omega_\alpha \Big/ \prod_{\alpha=1}^{3N-7} \omega_\alpha' \right), \qquad (2.4)$$

where T is a temperature, k_B the Boltzmann's constant and ΔV_A a energy difference between the structure A and the

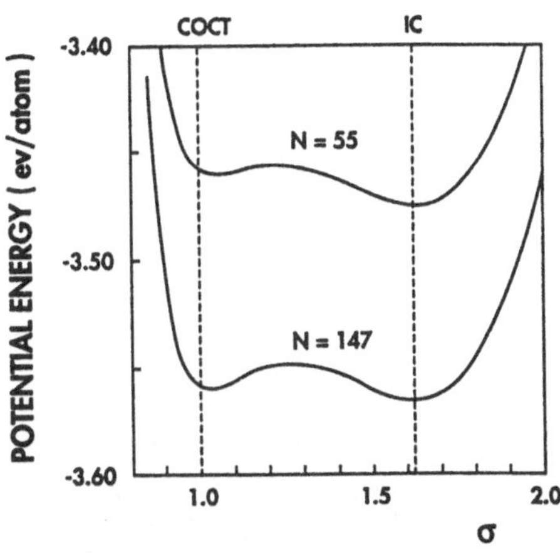

Fig.1. The potential energy variation (in eV per atom) along the transformation path between IC (σ = 1.62) and COCT (σ = 1) for Au55 (the upper curve) and Au147 (the lower curve) clusters.

transition state. The ω_α and ω'_α are eigenfrequencies in the structure A and at the transition state, respectively. They are defined by $\omega_\alpha = (\lambda_\alpha/m)^{1/2}$, where m is the mass of a gold atom and λ_α are positive eigenvalues of Hessian matrices (second derivative matrices of the potential).

3. Au55 clusters

First we calculate lifetimes of IC and COCT for a free Au55 cluster. There is a continuous transformation between IC and COCT specified by one parameter σ [14,5], which is defined only mathematically. The potential energy variation along this transformation is shown by the upper curve in Fig.1. This path is not necessarily the reaction path but must be very close to it. Starting with the maximum-energy point between IC and COCT on this path and using the Newton-Raphson method, we obtain the exact transition state and then calculate eigenvalues of Hessian matrices in IC, COCT and the transition state. In this way, we calculate lifetimes defined by the inverse of the transition probabilities given by (2.3). The results are shown in Fig.3 by the solid line (Y = 0) for COCT and the broken line (Y = 0) for IC as functions of the temperature. As seen in this figure, IC has lifetimes larger than 0.1 sec below 400K. On the other hand, COCT has lifetimes too short to be observed in an electron microscope around and above room temperature.

We now consider the substrate effects. COCT is assumed to be in contact with the substrate on one of its square faces. On the other hand, IC is on one of its triangle faces. COCT has a larger number of atoms in the contact face than IC so that the interaction with the substrate favors stabilization of COCT. In this case, we can also construct a mathematically defined continuous transformation between them by optimizing a rotation angle of the cluster relative to the substrate

122

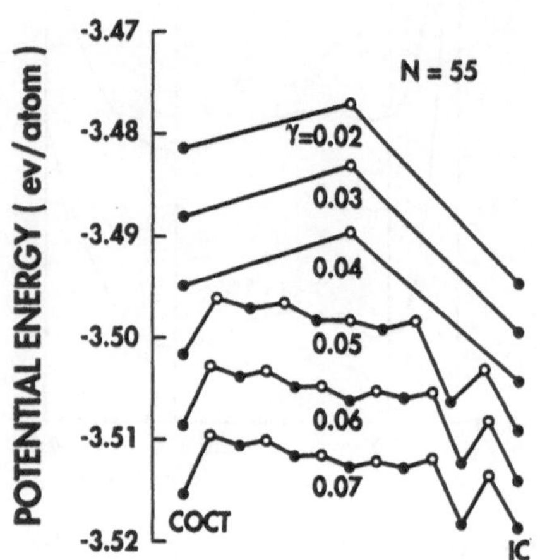

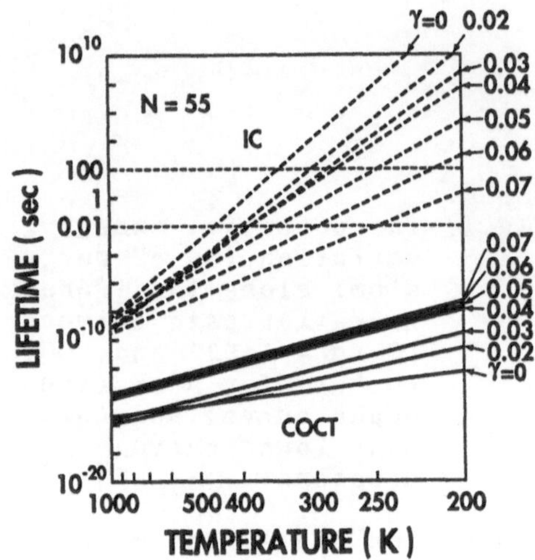

Fig.2   Schematic plots of
        the reaction paths
        between IC and COCT
        for Auss. The numbers
        beside the lines
        indicate the values
        of the ratio γ.

Fig.3   The lifetimes (in sec)
        of IC (the broken
        lines) and COCT (the
        solid lines) for Auss
        as functions of the
        temperature (in K).

surface and a distance between the center of mass of the
cluster and the substrate surface for each value of the
parameter σ. In this way, we obtain a maximum-energy point
between IC and COCT on the path and determine a transition
state by the Newton-Raphson method. However, we have found
that it is not always the transition state between COCT and
IC for large values of the ratio, $\gamma = U_a/E_{bulk}$, i.e. the
steepest descent path from it gets to a local minimum other
than IC and COCT in general. In order to obtain the reaction
paths among new local minima, IC and COCT, we use the
Czerminski-Elber methods [15]. Consequently, we have obtained
minima and transition states for the various values of γ. The
results are shown schematically in Fig.2, where the black
circles represent the minima and the white circles the transition states.

As seen in this figure, the larger the values of γ is, the
more minima exist on the paths between IC and COCT. According
to the "unified statistical theory" [16], the transition
probability for these cases is given by $t \cdot \Gamma_A$, where Γ_A is
calculated by (2.3) with the highest-energy transition state.
The "transmission coefficient", t, is calculated from the
other transition states and usually of the order of unity. We
omit the transmission coefficient since our purpose is just a
qualitative or semi-quantitative discussion.

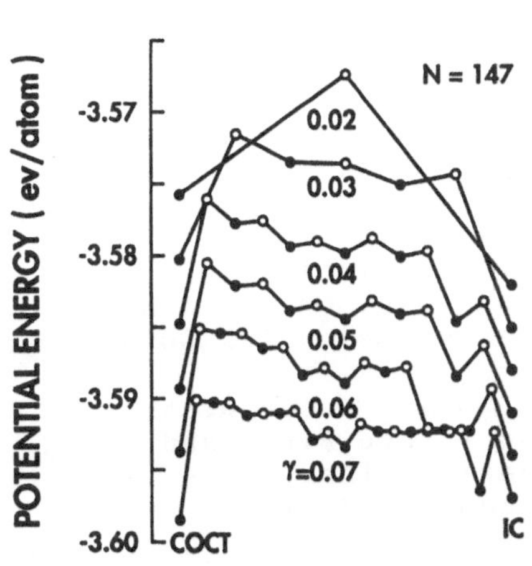

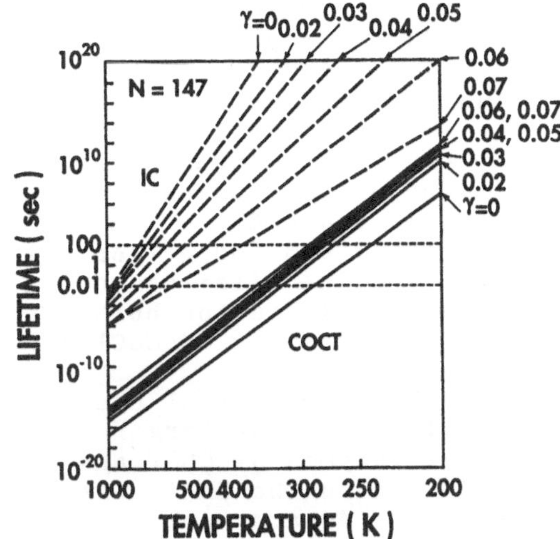

Fig.4. The same as Fig.2 except for Au₁₄₇.

Fig.5 The same as Fig.3 except for Au₁₄₇.

The results of the lifetimes of COCT and IC for various values of γ are shown by the solid lines for COCT and the broken lines for IC in Fig.3 as functions of the temperature. As seen in these figures, the substrate interaction stabilizes COCT. However, the lifetime of IC is long enough and that of COCT still too short to be observed with an electron microscope around room temperature even for $\gamma = 0.07$.

3. Au₁₄₇ clusters

Similarly to the case of Au₅₅, we first calculate lifetimes of IC and COCT for a free Au₁₄₇ cluster. The potential energy variation along the mathematically defined continuous transformation is shown by the lower curve in Fig.1. We obtain the exact transition state and then calculate the lifetimes of IC and COCT. The results are shown by the solid line ($\gamma = 0$) for COCT and the broken line ($\gamma = 0$) for IC in Fig.5 as functions of the temperature. As seen from this figure, COCT has a lifetime of the order of 1 sec around room temperature. On the other hand, IC has a lifetime larger than 10^{20} sec below 400K. This lifetime of IC is so long that both IC and COCT can be almost unobserved in an observation period, say several minutes.

Next we consider the substrate effect. The reaction paths are shown schematically in Fig.4, where the black circles represent the minima and the white circles the transition states. The results for the lifetimes are shown by the solid lines ($\gamma \neq 0$) for COCT and the broken lines ($\gamma \neq 0$) for IC in Fig.5 as functions of the temperature. As seen in this figure, while the lifetime of COCT is not changed very much even for $\gamma = 0.07$ around room temperature, the lifetime of IC is so much reduced that both IC and COCT can be observed in an observation period.

4. Discussions

We compare the present results with the experimental facts: both Au_5 and Au_{147} have been observed to fluctuate between IC and COCT on the graphite (00.1) surface above 300 K (for Au_5) and 500 K (for Au_{147}) [17]. The theoretical results for the lifetimes of IC are consistent with these experimental facts for both Au_{147} and Au_5. On the other hand, the lifetimes of COCT are underestimated for both Au_{147} and Au_5.

Though the potential energies of COCT are comparable to those of IC for the larger values of Y (see Fig.2 and Fig.4), the lifetimes of COCT are much smaller than those of IC. This is due to the fact that the frequency factors for COCT given by (2.4) are much larger than those for IC. The formula (2.3) with (2.4) is derived in harmonic approximation for multi-dimensional integrals [13] and includes the ratio of two terms of huge multiplication. The accumulation of the small error in the harmonic approximation for each dimension might cause the large error in the frequency factors. The improvement of this point is left for future works.

References

1. Iijima, S. and Ichihashi, T.: Phys. Rev. Lett. 56, 616 (1986)
2. Smith, D.J., Petford-Long, A.K., Wallenberg, L.R. and Bovin,J.-O.: Science 233, 872 (1986)
3. Mitome, M., Tanishiro, Y. and Takayanagi, K.: Z. Phys. D-Atoms, Molecules and Clusters 12, 45 (1989)
4. Ajayan, P.M. and Marks, L.D.: Phys. Rev. Lett. 63, 279 (1989)
5. Sawada, S. and Sugano, S.: Z. Phys. D-Atoms, Molecules and Clusters 14, 247 (1989)
6. Sawada, S. and Sugano, S.: Z. Phys. D-Atoms, Molecules and Clusters 20, 259 (1991)
7. Ducastelle, F.: J. Phys. (Paris) 31, 1055 (1970)
8. Gupta, R.P.: Phys. Rev. B 23, 6265 (1981)
9. Tomanek, D., Mukherjee, S. and Bennemann, K.H.: Phys. Rev. B 28, 665 (1983)
10. Rosato, V., Guillope, M. and Legrand, B.: Phil. Mag. A 59, 321 (1989)
11. Kittel, C.: 'Introduction to solid state physics', 6th Edn. New York: John Wiley 1986
12. Heyraud, J.C. and Metois, J.J.: J. Cryst. Growth 50, 571 (1980)
13. Vineyard, G.H.: J. Phys. Chem. Solids 3, 121 (1957)
14. Ogawa, T. and Nara, S.: Z. Phys. B-Condensed Matter and Quanta 33, 69 (1979)
15. Czerminski, R. and Elber, R.: J. Chem. Phys. 92, 5580 (1990)
16. Miller, W.H.: J. Chem. Phys. 65, 2216 (1976)
17. Takayanagi, K.: private communication.

A FACE CENTERED ICOSAHEDRAL APPROXIMANT PHASE WITH LOCAL ICOSAHEDRAL SYMMETRY

J.E. Shield, M.J. Kramer,
R.W. McCallum and A.I. Goldman
Ames Laboratory and Iowa State University
Ames, IA 50011

ABSTRACT. An approximant phase in the Al-Cu-Ru system, isostructural with α–AlMnSi, is described and compared with the face-centered icosahedral phase. The existence of this phase confirms the notion that the structures of simple and face-centered icosahedral alloys are closely related, differing perhaps by the atomic decorations of the quasicrystalline lattice.

INTRODUCTION

With no translational periodicity, a direct description of the atomic structure of quasicrystalline alloys in three dimensions is extremely difficult. Probably the best way to gain an understanding of the atomic structure of quasicrystals is to consider crystalline phases that contain atomic configurations which may possibly approximate the atomic structure in the quasicrystalline phase. Many quasicrystalline systems contain these "approximant" structures, and in fact quasicrystalline structures have been proposed based on the building blocks, or local atomic arrangements, found in the crystalline phases α-AlMnSi [1-3] and AlMgZn [4]. These building blocks, arranged neatly into periodic structures in the crystalline phases, are assembled quasiperiodically in quasicrystals. For example, α–AlMnSi and AlMgZn can be described as an arrangement of rhombic dodecahedron (RD) and prolate rhombohedron (PR) [2], which then can be rearranged into a quasicrystalline "lattice." The RD and PR are also common to three dimensional Penrose tiling models of the quasicrystalline structure [2]. For α-AlMnSi, the structure can also be described as a bcc packing of Mackay icosahedron (MI) connected along their 3-fold axes by "glue" atoms. MI present in quasicrystal structures could account for their 5-fold symmetry [1].

More recently, icosahedral alloys with a high degree of both structural and chemical order were discovered in the Al-Cu-Fe [5] and Al-Cu-Ru systems. The higher degree of structural order in these alloys is demonstrated by the presence of resolution-limited x-ray diffraction peak widths [6,7]. The diffraction patterns from these alloys can be indexed to a face-centered hypercubic lattice [8,9], indicating the possible presence of chemical order. Chemical ordering in these face-centered icosahedral (FCI) alloys is also suggested by the existence of antiphase boundaries [9].

The exact nature of the chemical ordering in the FCI alloys is not well understood. No approximant phases have been described in detail for any system containing the FCI phase. However, a cubic approximant has recently been identified in the Al-Cu-Ru system [10], and is isostructural with α-AlMnSi [11]. Other reports of approximant phases related to α–AlMnSi have also been reported in FCI alloy systems [12]. Here, we provide a more complete description of the cubic approximant phase discovered in Al-Cu-Ru and relate it to the FCI phase. A complete structural analysis will provide a better understanding of the ordering present in FCI structures.

P. Jena et al. (eds.), Physics and Chemistry of Finite Systems: From Clusters to Crystals, Vol. I, 125–129.
© 1992 *Kluwer Academic Publishers.*

126

EXPERIMENTAL PROCEDURES

An ingot of nominal composition $Al_{60}Cu_{25}Ru_{15}$ was alloyed by arc melting high purity (<99.99 percent) elemental Al, Cu and Ru. Rapidly solidified flakes were produced from this ingot by melt spinning at a wheel speed of 30 m/s in an argon atmosphere. Annealing was completed by wrapping the flake in Ta foil and sealing in quartz capsules under high purity Ar. Samples were quenched in water from the annealing temperatures.

X-ray diffraction was done on powdered flakes using a Philips PW1700 diffractometer with Cu K_α radiation. Transmission electron microscopy (TEM) samples were thinned by ion milling at 3 kV on a cold stage. TEM was done on a Philips CM30, equipped with a Link Analytical energy dispersive x-ray detector, operating at 150 kV.

RESULTS AND DISCUSSION

As previously reported, annealing $Al_{60}Cu_{25}Ru_{15}$ at 800°C from 24 to 96 hours produces both the FCI phase and an approximant phase [10]. After 24 hours, a mixture of FCI, tetragonal and Al_2Cu exists, with the majority of the material being the FCI phase (Figure 1a). Further annealing, at times greater than 96 hours, results in a change of structure (Figure 1b), which has been identified as a simple cubic structure with a=12.38 Å which belongs to the Pm3 space group [11], strongly suggesting that it is isostructural with α–AlMnSi. Hence, we will call this phase α-AlCuRu. Nearly all of the lines of Figure 1b were indexed to α–AlCuRu. The rhombohedral edge length a_r is 4.50 Å in α-AlCuRu, which can be found from the relation $a_r=(4 + 8/\sqrt{5})^{-1/2}a$ [2]. This is in comparison with α-AlMnSi, where a_r is 4.60 Å.

The transformation sequence in this alloy is also quite interesting. As noted, short anneals at 800°C produce a phase mixture which includes the FCI phase, which might be expected based on the phase diagram [10]. However, longer anneals produce the α–AlCuRu phase. In order to ensure that no significant change in composition occurred during the long-term anneal, which could explain the formation of α-AlCuRu far off the nominal composition, energy dispersive x-ray spectra were compared for the samples annealed for 24 hours and 8 days. Virtually no change in composition resulted from the annealing. The transformation to α–AlCuRu may be hindered by a lack of a significant driving force arising from the nearness in composition and structure of the FCI and α-AlCuRu phases.

In comparing the two x-ray diffraction patterns of Figure 1, it is interesting to note the many similarities which exist. For instance, many of the diffraction peaks of the α phase correspond closely to FCI peaks, with similar relative intensities. The close relationship between the two phases is also apparent when examining selected area diffraction (SAD) patterns from the two phases. Figure 2a shows the 2-, 3- and 5-fold patterns from the FCI phase. Similarly, Figure 2b shows the 2-, 3- and 5-fold directions from the α phase, which from the crystalline perspective are the <001>, <111> and <1τ0> zone axes, respectively. The presence of internal icosahedral order in the α phase is evident from the intensity modulations of Figure 2b. The angles between the 2-, 3- and 5-fold directions in the α phase also correspond to icosahedral point group symmetry.

The MI in α-AlCuRu, as in α-AlMnSi, are connected at their faces along the <111> direction, giving this direction 3-fold rotational symmetry. In comparison to the FCI phase (Figure 2a), the intensity modulations in the SAD patterns are nearly identical, both with respect to intensity and d-spacing. It is also interesting to note that the icosahedral intensity modulations in the α phase are approximately reated by τ. For instance, in the 5-fold pattern of Figure 2b the ratios of the spacings of icosahedral reflections are very close to τ. Also evident in this pattern are some shifts in the icosahedral diffraction spots, which in quasicrystalline alloys is evidence of phason disorder.

The similarities between the FCI and α-AlCuRu diffraction patterns indicate structural similarities between the two phases. Thus, it appears that the building blocks for the two structures, which for α–AlCuRu can be described in terms of RD and PR or MI [2], are very similar and only the way that they are assembled differ, as was found for icosahedral AlMnSi and a-AlMnSi [1-3] From this, it is apparent that the structures of simple and face centered icosahedral alloys may be nearly identical, except perhaps for the atomic decorations of the building blocks.

In considering possible ordering schemes, it is helpful to compare α-AlCuRu with α-AlMnSi. To begin with, the nominal composition of our specimen is similar to the composition of α-AlMnSi [13]. Grouping the Al and Cu (or Si) atoms together, we have $(Al,Cu)_{85}Ru_{15}$, as compared with $(Al,Si)_{84.4}Mn_{15.6}$. Chemical ordering of Al and Si has been proposed as an explanation of the effect of Si additions on the AlMn quasicrystalline structure [14]. However, Al and Si sites cannot be distinguished by x-ray diffraction, so the Al and Si sites were not identified in α–AlMnSi [13]. The nature of the chemical ordering in FCI alloys has been speculated to involve primarily Al and Cu [15]. If the strong scatterer in α–AlCuRu, Ru, is placed on the Mn sites of α-AlMnSi, it is very reasonable to assume ordering between the Al and Cu atoms on the Al(Si) sites of the α structure. Full profile x-ray analysis on α–AlCuRu is under way and should provide information on the specific Al and Cu sites.

CONCLUSIONS

In this paper, we have described α-AlCuRu, a 1/1 approximant to the FCI phase isostructural with α-AlMnSi. From x-ray and electron diffraction patterns, we have shown that the structural building blocks of the FCI and α-AlCuRu phases are nearly identical, which has previously been shown for simple icosahedral (SI) alloys and α-AlMnSi [1-3]. These results indicate that the structures of FCI and SI alloys are based on identical building blocks, and that some form of ordering occurs in the FCI alloys which distinguish them from SI alloys.

ACKNOWLEDGMENTS

This work was performed at Ames Laboratory, Iowa State University and was supported by the Director of Energy Research, Office of Basic Sciences, U.S. Department of Energy under contract No. W-7405-ENG-82.

REFERENCES

1. P. Guyot and M. Audier, Phil. Mag. B **52**, L15 (1985).
2. V. Elser and C.L. Henley, Phys. Rev. Lett. **55**, 2883 (1985).
3. M.Audier and P. Guyot, Phil. Mag. B **53**, L43 (1986).
4. C.L. Henley and V. Elser, Phil. Mag. B **53**, L59 (1986).
5. A.P. Tsai, A. Inoue and T Masumoto, J. Mater. Sci. Lett. **7**, 322 (1988).
6. C.A. Guryan, A.I. Goldman, P.W. Stephens, K. Hiraga, P.P. Tsai, A. Inoue and T. Masumoto, Phys. Rev. Lett. **63**, 2409 (1989).
7. P. Bancel, Phys. Rev. Lett. **63**, 2741 (1989); **64**, 496 (1990).
8. S. Ebalard and F. Spaepen, J. Mater. Res. **4**, 39 (1989).
9. J. Devaud-Rzepski, A. Quivy, Y. Calvayrac, M. Cornier-Quiguandon and D. Gratias, Phil. Mag. B **60**, 855 (1989).
10. J.E. Shield, C. Hoppe, R.W. McCallum, A.I. Goldman, K.F. Kelton and P.C. Gibbons, manuscript submitted to Phys. Rev. B.
11. J.E. Shield, L.S. Chumbley, A.I. Goldman and R.W. McCallum, in preparation.
12. S. Ebalard and F. Spaepen, J. Mater. Res. **6**, 1641 (1991).
13. M. Cooper and K. Robinson, Acta Cryst. **20**, 614 (1966).

14. C.H. Chen and H.S. Chen, Phys. Rev. B **33**, 2814 (1986).
15. S. Ebalard and F. Spaepen, J. Mater. Res. **5**, 62 (1990).

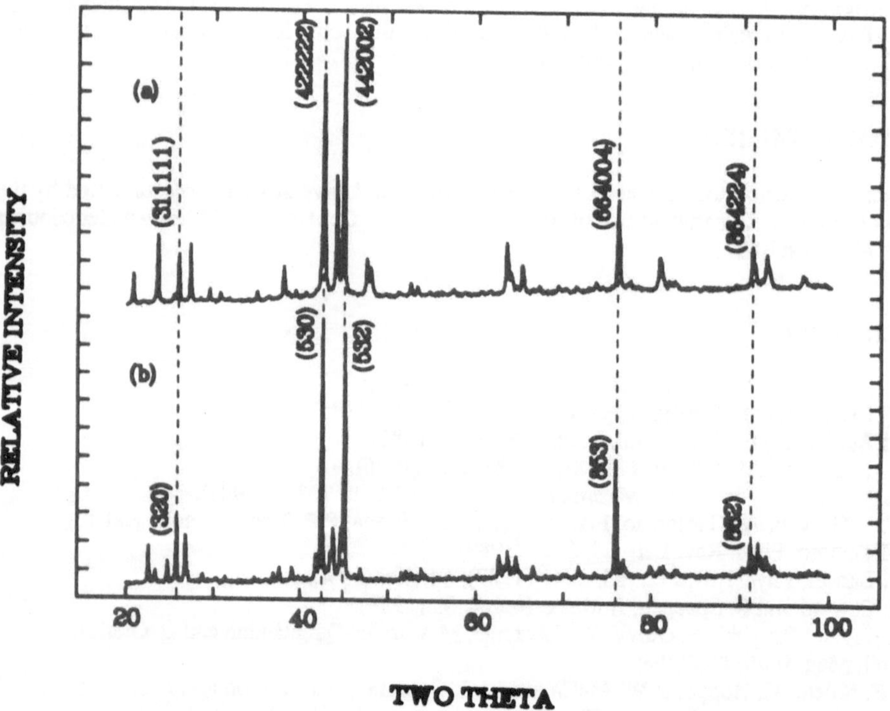

Figure 1. X-ray diffraction scans of $Al_{60}Cu_{25}Ru_{15}$ annealed at 800°C for (a) 24 h, which shows a majority of FCI phase present, and (b) 96 h, which is nearly all α-AlCuRu. Corresponding peak positions for the two phases are marked by dashed lines. FCI indices are from reference [6].

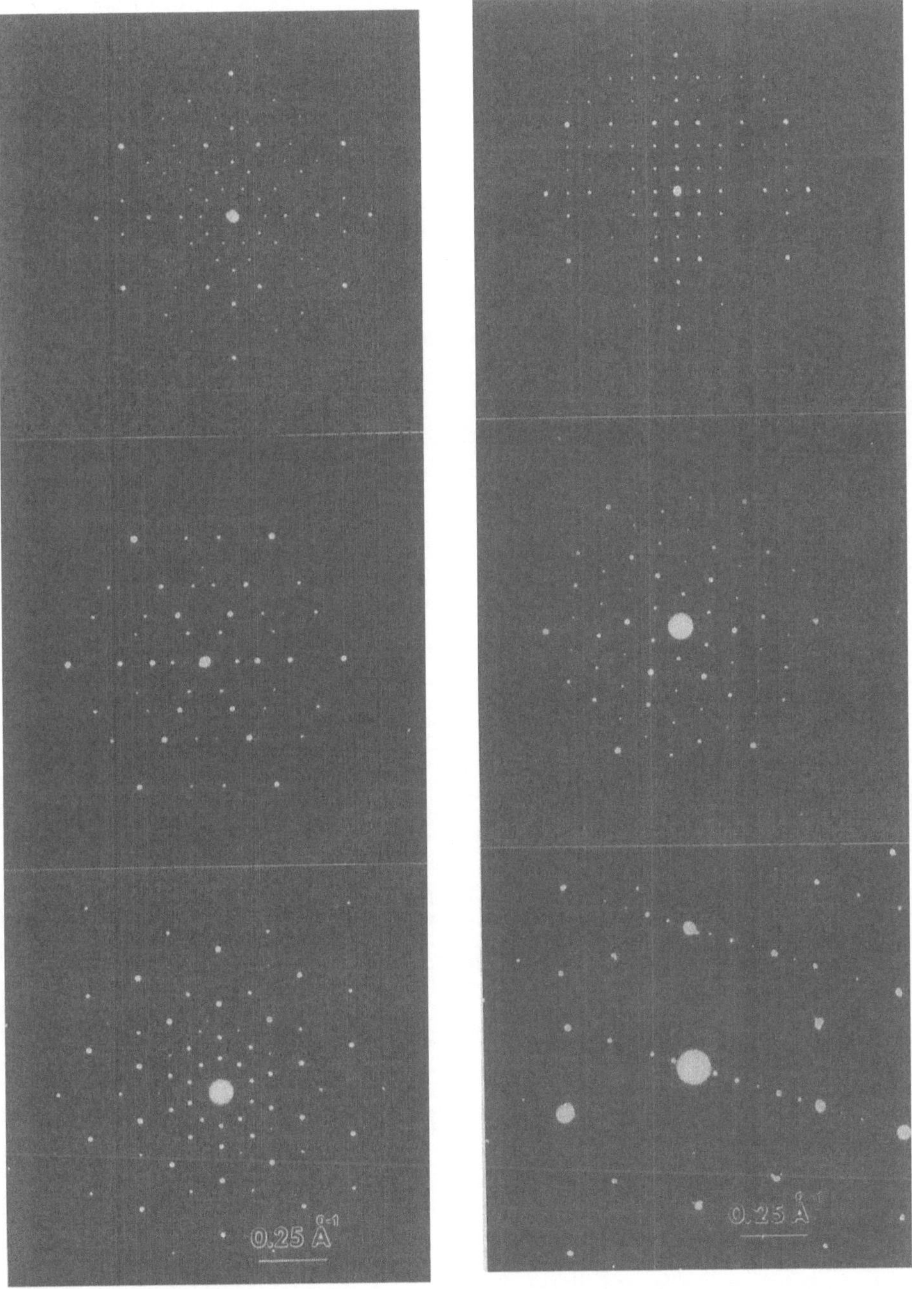

(a)   (b)

Figure 2. SAD patterns of (a) the 2-, 3- and 5-fold directions in the FCI phase, and (b) the 2-, 3- and 5-fold directions of the Mackay icosahedron in α–AlCuRu. These correspond to the <001>, <111> and <1 τ 0> directions of the crystalline unit cell.

DEFECT-FCC STRUCTURAL MODELS FOR ARGON CLUSTERS Ar$_N$, N $\gtrsim$ 500

BENJAMIN W. VAN DE WAAL
Department of Physics CT1324
University of Twente
P.O.Box 217
7500 AE Enschede
The Netherlands

ABSTRACT. In order to find support for the hypothesis that the preference of the havier rare gases for the *fcc* crystal – structure originates in the potential of the *fcc* structure to accommodate crossing stacking-faults, interference-functions for Ar$_N$ clusters with different internal structures have been calculated and compared with available electron-diffraction patterns. It was found that the introduction of two pairs of crossing stacking-faults in perfect *fcc* crystals of Argon results in simulated patterns that compare favourably with the observed curves, at least for clustersizes N $\gtrsim$ 500.

INTRODUCTION

One of the objectives of cluster-research is to clarify the origin of crystal-structures. The idea is that, whereas it is very difficult - if not impossible - to understand or predict, from first principles, the bulk crystal-structure of a substance, the explanation of the mutual arrangements of atoms or molecules in small solid clusters, consisting of only a few building-units, might be within reach. Once small clusters have been dealt with, the bulk crystal-structure could be envisaged as resulting from a plausible growth-scheme, possibly accompanied by size-dependent structural transitions.

The crystal-structure problem is most intriguing in the case of the (heavier) rare gases, appearing deceivingly simple in view of the isotropic van der Waals interactions between spherical atoms on one hand, and being intractable in view of the essentially infinite number of equally favourable close packings (including *fcc* and *hcp*) on the other hand.

In the case of Argon very accurate electron-diffraction intensities have been obtained from clusters in the size-range $20 \lesssim N \lesssim 3000$ by Farges *et al*, produced in their well-known experiments on homogeneous gas-phase nucleation by free jet expansion[1-3]. The precision of the diffraction-data allows the unambiguous assignment of the observed diffraction-lines to the *fcc* structure for the largest clusters seen, suggesting that the smaller cluster-structures could provide clues to the origin of the bulk *fcc* crystal-structure of Argon.

Such a clue has not been found. Rather, the smaller clusters had to be modelled after icosahedral motifs, in order to explain the observed diffraction-patterns, with no clear connection with the bulk structure.

It can not be ascertained, however, that the smaller clusters are actually the structural precursors of the larger clusters and crystals, the more so since they are produced in different experiments. Indeed, although Mackay-icosahedra[4] clearly form a *sequence*, no simple growth-mechanism can explain the evolution of one member into the next-larger, nor can the appearance of the *fcc* structure be connected to the icosahedra by a plausible structural transition.

The present paper describes a different approach, in which first a *growth-sequence* is *postulated*,

131

P. Jena et al. (eds.), *Physics and Chemistry of Finite Systems: From Clusters to Crystals, Vol. I*, 131–137.
© 1992 *Kluwer Academic Publishers.*

whose simulated diffraction-patterns are subsequently confronted with the experimental observations. We will concentrate on the region $500 < N < 3000$, in which the observed patterns suggest a transition from the icosahedral to the *fcc* arrangement, and present a sequence based on a single model, with no transitions involved.

As has been pointed out in a previous paper[5], the crystal-structure problem of the rare gases may have, despite its venerable reputation[6], a very simple solution: the *fcc* arrangement is preferred because it is the only close packing that permits lattice-defects that are essential to its growth. It was shown, that two crossing stacking-faults suffice to stimulate uninterrupted isotropic *fcc* growth, and to prevent new stacking-faults. It will be shown here, that a sequence, based on this growth-mechanism, does not conflict with experimental evidence.

DIFFRACTION PATTERNS

Electron-diffraction patterns for Argon clusters have been reported for $20 < N < 50$[1], $50 < N < 750$[2], $N \approx 1000$ and $N \approx 3000$[3], and also for $50 < N < 1500$[7] and $1500 < N < 3000$[8]. The experimental curves of Farges *et al*, with stagnation pressures 7 and 15 bar, have been reproduced in Fig.1. The estimated average clustersizes are $\overline{N} \approx 600$ and $\overline{N} \approx 3000$, respectively. The most important changes in the patterns with increasing size can be summarized as follows (the reader should consult the original papers for a detailed discussion): (1) All peaks narrow, except the first peak, near $s = 2$ Å^{-1}. (2) The broad peak near 4 Å^{-1} splits into two peaks, 220 and 311, according to the *fcc* labeling scheme (c.f.Fig.2.). (3) The feature near 6 Å^{-1} gets resolved into 422 and 333.(4) A shoulder on the right side of the first peak announces the splitting into *fcc* 111- and 200-peaks. The 111-intensity becomes intermediate between 220 and 311. (5) Other *fcc* lines become distinguishable as well: 533,551,642,553, and, as small shoulders, 400 and 440.

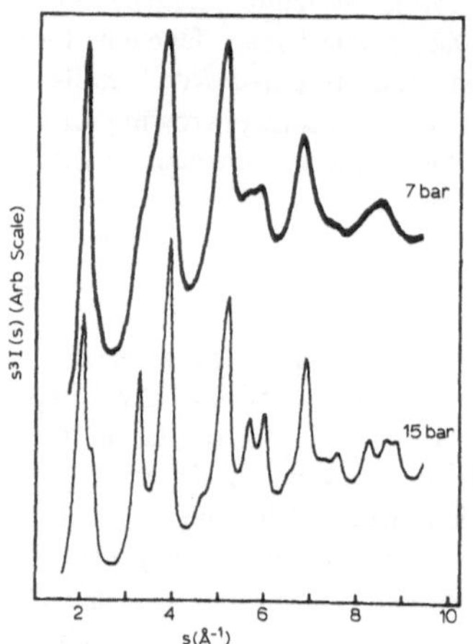

Fig.1.Experimentally observed electron-diffraction patterns of Argon clusters at different inlet pressures[2-3]. The estimated cluster sizes are $N \approx 600$ (top) and $N \approx 3000$ (bottom). (Reproduced with permission).

Fig.2 compares calculated interference-functions for three different model-sequences: (perfect) *fcc*-octahedra, defect-*fcc* octahedra, and icosahedra, respectively, with an increasing number of shells. The defect structures (*d-fcc*) have been obtained from the perfect crystals by introducing 4 crossing stacking-faults, as will be discussed in more detail below. Structural relaxation was applied to the icosahedra, but not to the *fcc* structures, having only negligible effect (other than a small shift of all peaks to the right, consistent with a slightly reduced lattice-constant) on the interference-patterns. The interference-functions were calculated in the same way as by Farges *et al*[1], to facilitate a direct comparison with their experimental results. The proportion t_A of free atoms in the scattering volume was taken zero. An overall temperature factor $\exp(-\frac{1}{2}u^2s^2)$ was assumed, with $u=0.20$Å.

It is clear that the introduction of crossing stacking-faults in perfect *fcc* crystals has a pronounced effect on the interference-functions: the splitting of the first peak in 111 and 200, already visible in *fcc* clusters of some 200 atoms, is postponed to much larger sizes, and may require some 10^4 atoms to get resolved. The same applies - although to a lesser extent - to the 220/311 splitting, whereas the

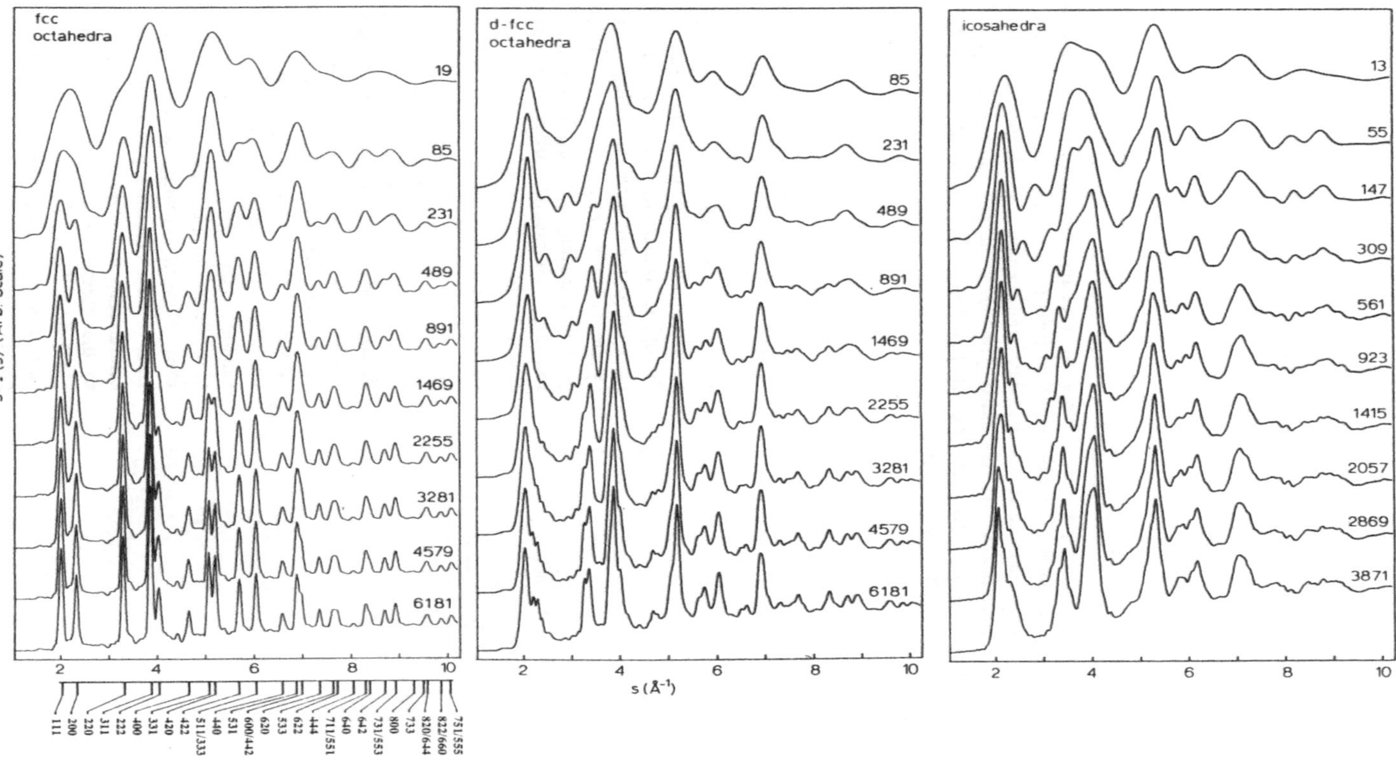

Fig.2. Calculated interference-functions for (from left to right) perfect *fcc* octahedra, defect-*fcc* octahedra, and multishell icosahedra, of increasing size (from top to bottom; number of atoms as indicated). The nature of the defects is shown in Figs.3*d* and 4. Only the icosahedra have been structurally relaxed. The *fcc* labeling scheme is included for reference. The *d-fcc* and icosahedral patterns are similar, but markedly different from those of perfect *fcc* crystals.

broadening of the peak near 5Å^{-1} (anticipating the 331/420-splitting) turns into a narrowing. On the other hand the *d-fcc* patterns have much in common with those of the icosahedra: the peaks are approximately in the same positions, and have comparable intensities. However, at equal sizes, the *d-fcc* peaks are narrower and get better resolved with increasing size (c.f. the peaks near 4 and 7 Å^{-1}).

Apparently, the *d-fcc* sequence reproduces, at least qualitatively, the observed evolution of the diffraction patterns with cluster-size.

DEFECT-*FCC* STRUCTURAL MODEL

Fig.3 shows some examples of structural models discussed in this paper: a 561-atom icosahedron (*a*), and three 489-atom *fcc* octahedra with 0, 1, and 2 pairs of crossing stacking-faults, respectively (*b,c,d*). The perfect octahedron (*b*) exposes only close-packed faces (as does the icosahedron), which complicates further growth. The faulted structures, on the other hand, exhibit non-integral surface-steps that can act as surface-nucleation centers. As discussed previously[5] these centers can not be exhausted by occupation and ensure correct *fcc* stacking of new layers. The crossing of faults is essential to provide *all* close-packed faces with steps.

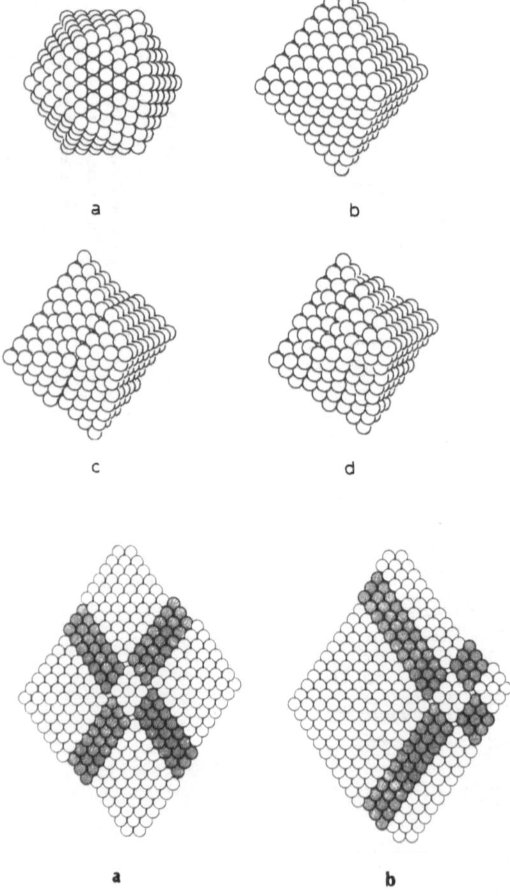

Fig.3. Some examples of cluster structures discussed in this paper: *a*: 561-atom multishell Mackay icosahedron; *b*: perfect *fcc* 489-atom centered octahedron; *c*: as *b*, but with *one* pair of crossing stacking-faults, approximately passing through the center; *d*: as *b*, but with *two* pairs of crossing stacking-faults, in a symmetric arrangement. The growth-promoting, stacking-fault resisting, non-integral surface-steps are clearly visible.

Fig.4. Symmetric (*a*) and asymmetric (*b*) arrangements of two pairs of crossing stacking-faults in a 3281-atom octahedral *fcc* cluster. Different parts of the crystal, that are separated by stacking-faults, are distinguished by shading; the view is down a [110]-direction, i.e. along the lines of intersection of crossing stacking-faults. Both crystals contain a part (around the central unshaded region) that is identical to the cluster of Fig.3d. The two (111)-faces facing the reader contain 4 surface-steps each, the other faces contain 2 steps.

Preliminary calculations showed that a single pair of crossing faults (Fig.3*c*) could not account quantitatively for the observed deviations from the *fcc* pattern of the diffraction-functions, in particular not for the height of the first maximum. Better results could be obtained by the introduction of a second pair (Fig.3*d*).

The arrangement of stacking-faults is more clearly visible in Fig.4*a*, where the view is down a [110]-direction, i.e. parallel to one of the edges of the octahedron. Parallel stacking-faults have been chosen three layers apart (in both directions), to fit in a small cluster, on one hand, but to avoid structural instability from the interaction of crossing-channels on the other hand. The crossing-region may be shifted to other parts of the crystal (c.f.Fig.4*b*), but was initially kept fixed, in the centre, to calculate the interference-functions of Fig.2. The stacking-faults divide the cluster in 9 regions of perfect *fcc* crystal, only 4 of which can grow in 3 dimensions. Since the fragments are parallel shifted with respect to each other, the defects would not be detectable by single crystal diffractometry   on macroscopic crystals (as twinning on (111) would be). As is apparent from Fig.4*b*, the relative sizes of the fragments may differ widely if the crossing-region is far from the centre, which may influence the interference-function significantly.

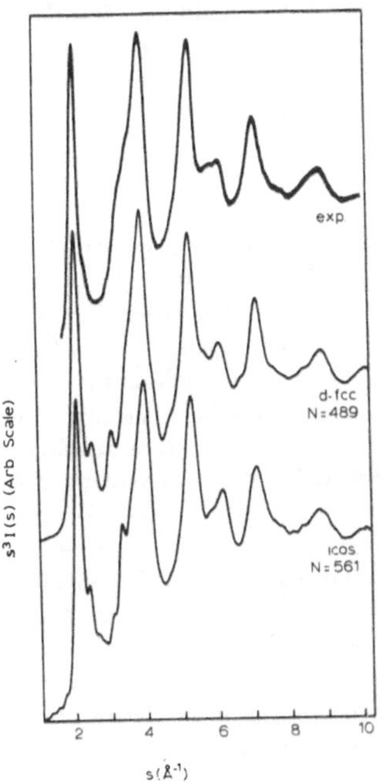

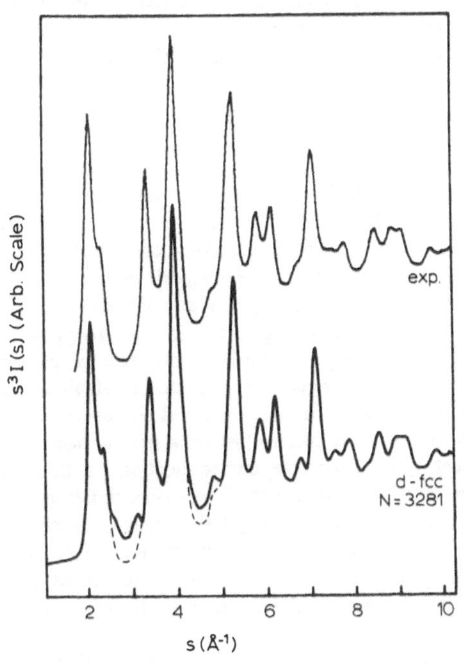

Fig.5. Comparison of two simulated interference-functions (lower curves) with the experimentally observed 7-bar diffraction-pattern (upper curve). The *d-fcc* 489-atom pattern (calculated for the structure of Fig.3*d*) and the 561-atom icosahedral pattern (Fig.3*a*) compare virtually equally well and have similar deficiencies.

Fig.6. Comparison of a simulated 3281-atom   *d-fcc* diffraction-pattern with the experimentally observed 15-bar curve ('exp'). Broken lines indicate main deficiencies. The calculated pattern is based on the structure of Fig.4*b*.

136

Calculated diffraction-functions for 489- and 3281-atom defect-*fcc* octahedral clusters are compared with the experimental curves of Fig.1, in Figs.5 and 6, respectively. The clusters have been structurally relaxed (E/N = 6.47 and 7.47 r.u., respectively, under a LJ-potential), and were assigned equal thermal factors u=0.18 Å. Unlike the smaller cluster (shown in Fig.3*d*) the larger cluster (Fig.4*b*) has an asymmetric fault-structure, which enhances the *fcc* character of the diffraction-pattern (c.f. corresponding curve in Fig.2). Finally, the calculated patterns have been smoothed by a (rather crude) procedure, to allow for electron-beam divergence and wavelength dispersion: F(s) is replaced by $[F(s)+F(s+\alpha s)+F(s-\alpha s)]/3$, with α = 0.015 and 0.008 in Figs.5 and 6, respectively. A simulated pattern for the 561-atom icosahedral cluster, produced in the same way (but with t_A = 0.23), has been included in Fig.5 for comparison. It is clear that both simulated curves compare equally well with the 7 bar curve, being deficient in the same regions: between the first two peaks, and near s=6Å^{-1}.

Discrepancies between observed and simulated 3000-atom patterns are near the main minima, as indicated in Fig.6; apparently the model produces too much diffuse scattering. Although the 311 line appears to be too high (as is 533), it is encouraging to find $I_{311}>I_{111}>I_{220}$, as observed. The experimental observation is of particular interest, since it suggests that the 3000-atom cluster is far from perfect (at no size the same order of intensities is found in perfect *fcc* octahedra, c.f.Fig.2), although nearly all *fcc* lines are visible (including 200, if not resolved). From Fig.4*b* it is clear, however, that a large proportion of the model cluster consists of perfect *fcc* crystal, the adjoining stacking-faults apparently being able to affect the *fcc* diffraction-pattern considerably.

DISCUSSION

These preliminary results suggest that the experimentally observed diffraction by Ar clusters in the size-range $500 \lesssim N \lesssim 3000$ can be accounted for by a single model. Although a number of discrepancies between observed and calculated curves remain, the model has sufficient flexibility to allow further investigation, e.g. by experimenting with the number, mutual separation, and orientation of stacking-faults. It is not yet clear whether the *fcc* diffraction-pattern is modified as it is by the perturbed atomic arrangements in the faults, or that the effects originate mainly in the crossing-regions. Substantially different contributions would necessitate the consideration of other stacking-fault arrangements, e.g. stacking-fault tetrahedra with 6 (rather than 4) crossings of 4 faults. It can be shown[9] that an increased frequency of occurrence of the distance $2\sqrt{(2/3)}$ (in units of the nearest-neighbour distance) between next-nearest neighbours across a stacking-fault has a pronounced effect on the interference-function (In this respect a stacking-fault is twice as effective as a twin-boundary, with the stacking-sequence *CBABA*, as compared to *CBABC*). Since this distance is also prominent in small icosahedral clusters, notably in the 13-atom cluster, similarities between *d-fcc* and icosahedral diffraction-patterns are to be expected. Also, the atomic arrangements near stacking-fault crossings are similar to those in 13-atom icosahedra: if the latter are considered as fused pentagonal bipyramids sharing an apex-atom and with their parallel pentagonal rings in a staggered orientation, the two pentagonal rings have the same spatial orientation in the *d-fcc* cluster and are deficient of one atom each. This similarity may not only be reflected by the diffraction-patterns, but may also be of significance in connection with the origin of the stacking-fault crossings.

Although the 489-atom *d-fcc* octahedral cluster of Fig.3*d* may be the precursor of the 3281-atom cluster of Fig.4*b* (through asymmetric growth on all but two faces), no evidence for precursors of the smaller cluster can be found in the observed diffraction-patterns, suggesting that the structural-transition problem is not removed but transferred to a smaller scale. However, it should be realized that, even if it were possible to conceive of a plausible icosahedral $\rightarrow$ *d-fcc* transition at size N $\approx$ 500, the genealogy of the icosahedral cluster would still remain obscure.

The *d-fcc* clusters, although mechanically stable, have not the highest possible binding-energy: a *perfect fcc* cluster of the same size is clearly more favourable. Multishell icosahedral clusters are the most favourable arrangements only for 'magic' numbers of atoms, but probably not in between. Since it is not very likely that a growing cluster switches from one sequence to another, whichever is the most favourable for the size at hand, other criteria may be helpful in finding possible cluster-structures. From the above discussion it will be clear that the potential to grow fast may be crucial in this connection. Apparently, *fcc* clusters have this property, thanks to their ability to accommodate a few, but essential lattice-defects. The view that this is actually the reason for the bulk crystal-structure of Argon to be *fcc* appears to be supported by experimental evidence.

REFERENCES

[1] Farges,J.,de Feraudy,M.F.,Raoult,B.,Torchet,G.:*J. Chem. Phys.* **78**,5067(1983)

[2] Farges,J.,de Feraudy,M.F.,Raoult,B.,Torchet,G.:*J. Chem. Phys.* **84**,3491(1986)

[3] Farges,J.,de Feraudy,M.F.,Raoult,B.,Torchet,G.:in *Large Finite Systems.* Jortner,J. et al (eds.),pp 113-119. D.Reidel Publishing Company 1987

[4] Mackay,A.L.:*Acta Cryst.* **15**,916(1962)

[5] van de Waal,B.W.:*Z. Phys. D.* **20**,349(1991)

[6] Niebel,K.F.,Venables,J.A.:in *Rare Gas Solids.* Klein,M.L., Venables,J.A. (eds.), 1st Edn.,pp.558-589.London,New York, San Francisco: Academic Press 1976

[7] Kim,S.S.,Stein,G.D.:*J. Coll. Interf. Sci.* **87**,180(1982)

[8] Lee,J.W.,Stein,G.D.:*J. Phys. Chem.* **91**,2450(1987)

[9] van de Waal,B.W.:to be published

BEYOND THE ICOSAHEDRAL GLASS MODEL, THE INTERPENETRATING PSEUDO - ICOSAHEDRAL CLUSTERS (IPIC) MODEL.

J.L. VERGER - GAUGRY
LTPCM - ENSEEG - INPG (CNRS UA 29)
BP 75 Domaine Universitaire
38402 Saint Martin d'Hères
France

ABSTRACT. We analyze the diffraction properties of infinite "formal" quasiperiodic, resp. periodic, packings of atoms in 3 dimensional Euclidian space that are considered as quasiperiodic, resp. periodic, distributions of identically oriented pseudo - icosahedral clusters which all have a finite spatial extension and interpenetrate. Restoring the icosahedral local order in the formalism leads to count atomic sites several times, when taken into account from the center of one cluster or another. Scattering functions are treated in perturbation over site multiplicity and icosahedral symmetry. The periodic case corresponds to approximant crystals, while the quasiperiodic one to icosahedral quasicrystals: the rank of the Z - module generated by the centers of clusters is then 3, resp. m, with m>3. A periodization of the quasicrystal can be performed in R^m. Pseudo - inflation rules arise from the geometry of an average cluster. This formalism is new and provides a "continuous link" between classical crystallography and quasi-crystallography.

1. INTRODUCTION

Icosahedral quasicrystals are now classified into four main classes (Henley, 1990 [1]): 1) the Al-TM class, where TM is one or several transition metals, such as i-AlMn or i-AlCuFe, 2) the AlMgZn class, such as i-AlLiCu or i-GaZnCu, 3) the PdUSi class with i-$Pd_{58}Si_{21}U_{21}$ [2], 4) the Ti-TM class, such as TiCo, TiMn [3 - 6]. On the other hand, many of these i-phases have known approximant crystals, such as α-AlMnSi, β-AlMnSi or μ-AlMn for i-AlMnSi [7 - 9]. The discriminator of the classification is the geometry of the icosahedral cluster (or its equivalent in terms of atomic surface in the hyperspace approach) as in the "nearest" approximant crystal or as determined by a Patterson analyzis from qc single crystals; it is limited to a few icosahedral shells, defined up to small deformations: the sequences of sites polyhedra, from the center, are either ico - icosi - ico (Double Mackay Icosahedron, case 1) or ico - dodeca - ico (Bergman unit, cases 2 and 4) with

P. Jena et al. (eds.), Physics and Chemistry of Finite Systems: From Clusters to Crystals, Vol. I, 139–145.
© 1992 *Kluwer Academic Publishers.*

centers which are empty (cases 1 and 2) or occupied (cases 1 and 4). Case 3) is somewhat different because of steric effects [8] (as usual, ico, icosi, dodeca stand for icosahedron, icosidodecahedron, dodecahedron respectively).

One possible key for modelling the i - phase consists in guessing new linkages between the atomic icosahedral clusters that we can observe in the approximant crystals in the same system, and which then give rise to quasiperiodic assemblies of atoms with <u>only one</u> common orientation for the icosahedral clusters: we call such an infinite collection of identically oriented clusters an orientational family in the following. However, it is a fact that approximant crystals possess in general <u>several</u> orientational families, as it was shown for instance by Tamura et al [5, 6, 9] for β - AlMnSi (with two families) and the isomorphous crystalline compounds η - Ti_2-TM

(space group $Fd\overline{3}m$) where TM = Ni, Fe, Co, Mn, Cr. In this latter case, eight orientational families do exist in real space for which, because of icosahedral extinctions due to superimposition effects on the reciprocal lattice, only four of them appear in reciprocal space. By this, we mean that four pseudo - icosahedral clusters of intense peaks (centered at the central spot) can be observed, for which some Bragg peaks (vertices) are common to several clusters. Duality relations (by Fourier transform) between the orientational families in real space and the pseudo - icosahedral icosahedral clusters of intense peaks in reciprocal space are totally satisfied [3, 5]. These clusters are centered at the TM(e) atomic sites. Of course, any investigation of reciprocal space (by any diffraction experiment) leads to sectioning such a 3 dimensional information, and we observe merely "rugous" sections of pseudo - icosahedral polyhedra, where all the spots to be considered for reasons of symmetry lie in a planar section having a certain (small) thickness.

Besides such examples of approximant crystals where several orientational families coexist, but where clusters are limited to a few layers [6, 9], we have the example of the α - AlMnSi phase (space group $Pm\overline{3}$) for which there is only one orientational family of pseudo - icosahedral clusters, but where the icosahedral shells extend over very large distances [9, 10], up to 4 nm and more, much larger than the lattice parameter, a = 1.268 nm, and therefore which interpenetrate a lot.

The purpose of this note is to report on some recent model of interpenetrating pseudo - icosahedral clusters (we call it "IPIC model") in a periodic context, that is to say approximant crystals, and in an aperiodic one, ie icosahedral quasicrystals (Verger - Gaugry et al., 1991 [11, 12]). Here, for simplicity's sake, only one orientational family is assumed to exist and pseudo - icosahedral clusters are all bounded in size.

2. SOME REMARKS ON THE DIFFRACTION PROCESS FROM "IPIC" IN APPROXIMANT CRYSTALS

The two main building blocks for cases 1) and 2) are the Double Mackay Icosahedron and the Bergman Unit, respectively constituted with 54 atoms and 44 atoms. Few researchers have investigated the external layers of these clusters [13] in approximant crystals (by this, we mean up to 20 or more layers) since there is a tacit agreement to consider that it is not of special interest to do so, one reason coming from the use of the decoration procedures of tiles (prolate and oblate elementary tiles, or more complicated tiles [14]) in quasiperiodic tilings, which actually utilize only a few shells of these icosahedral clusters and which force the study of the connections between the atomic units (matching rules and vertex-, edge-, face- packings [15]); another reason arises from the difficulty of making simple models of quasicrystals with interpenetrating pseudo - icosahedral clusters, where sites are belonging at the same time to several layers of several IPIC. This difficulty is a problem of conception and of handling on a computer since it is necessary to avoid different types of atoms at the same site (problem of coloration unambiguity).

However, assume now that a crystal of given space group and position parameters is composed of a finite number of pseudo - icosahedral clusters which spread over large distances and interpenetrate inside the unit cell and therefore everywhere, by periodicity (an accurate axiomatization of a G - approximant crystal, with G, a non -crystallographic point group, is given in [11]). An incident plane wave will "see" independently each pseudo - icosahedral cluster, in its totality, and will give rise to a diffracting contribution. Each cluster of the orientational family contributes equally in the global response to the incident wave, by the principle of superposition, provided that the atoms in the intersected regions are counted with their multiplicities and that a phase is introduced due to the relative positioning of the clusters centers, in the structure factor of the crystal [10, 11].

The notion of interpenetration is not classical in crystallography since the easiest way to build a crystal consists in reproducing a motif at each node of a lattice. Here, we voluntarily aim at investigating the consequences of complex overlaps of motives when some parts of them follow the icosahedral symmetry and can be coherently defined (propagation of the icosahedral symmetry) over many motives around the central one, and therefore everywhere by periodicity. In this sense, an approximant crystal is more than a "crystallographically defined" crystal since it contains extra informations on highly symmetric (empty or occupied) sites as centers of pseudo - icosahedral clusters and the spatial extents of these clusters around them. This information has to be found by hand or on the computer [12], but it is clear that classical crystallography cannot provide it.

As an example, let us consider the α - AlMnSi phase. Table 2 and Figure 2 in [10], reported from [9], show that a description of the structure in terms of icosahedral atomic units which do not overlap, necessarily implies that the real clusters, as they "exist" as diffracting entities (some icosahedral shells have an occupation rate less than unity) are drastically truncated at a suitable distance from the origin and the body - centered site. Though a Patterson analysis of this approximant crystal takes into consideration the full information in reciprocal space from the IPIC diffraction properties, we retain in the reconstructions of the structure in terms of icosahedral units only a small part of the icosahedral clusters, truncating them, trying to avoid interpenetration. This attitude seems somewhat contradictory to us and, of course, this remark will take its full meaning in the context of quasiperiodic packings of atoms constituted with IPIC, - quasicrystals -, where we do not know a priori the exact extent of the icosahedral local order.

3. IPIC MODEL

For convenience, many models of i - phase propose atomic configurations for which the icosahedral symmetry is locally exactly satisfied within the icosahedral clusters [8, 16]. However, in approximant crystals, we have already observed and analyzed some deviations (angular deviations, radial displacements and absence of atoms at some sites on some icosahedral shells in clusters (Verger - Gaugry, Tamura and Barbier, [12, 17]) with respect to the perfect icosahedral symmetry in the atomic configurations: so-called pseudo - icosahedral clusters. The angular and radial corrections are small for "good" approximants and may vary within a larger extent for other approximant crystals [6]. Similarly, it is totally reasonnable to think that small deformations of the perfect icosahedral clusters also occur in the i - phase but these deformations are mainly different and quasiperiodically defined by comparison with the case of the approximant crystal. For instance, some icosahedral phases such as in the Ti-TM system should generate pseudo - icosahedral clusters with fairly large deformations: indeed, in c - Ti_2Ni, the first icosahedral shell around Ni(e) atoms is constituted with two types of atoms (9 Ti and 3 Ni) conferring a "chemical and topological (size effects) disorder" for the action of the icosahedral group, and similarly for the subsequent shells.

In this framework, we have tried to understand how the quasiperiodic distribution of pseudo - icosahedral clusters and their associated perfectly icosahedral clusters [17] intervenes in the diffraction process. For this, we define a "$m\overline{3}5$ - cluster" to be a finite set of points in $\mathbf{R}^3$ which is $m\overline{3}5$ - invariant around a fixed center, center which belongs to this set or not. Let us denote it by $X(n)$, with the notations of [11], where n is the number of icosahedral layers in the cluster. When $m\overline{3}5$ acts transitively on one shell, the number of sites in this shell is: with zero degree of freedom, 12 (ico), 20 (dodeca), 30 (icosi) , then, with one degree of freedom in each case, 60 (3 possibilities: pentakis, hexe, trisico), then, 120 (general case), with two degrees of freedom. The color of a point of the cluster is the type of atom placed at this point and we consider as usual an atomic site as a couple (point, color). Let us denote by $\delta(X(n))$ a slight deformation of the colored perfectly icosahedral cluster, given by a collection of displacements, one for each site. A "formal" quasicrystal fqc, to name it, is merely a state of matter with no hole and no overlap of atoms and for which a compatibility exists between such deformed clusters, where atoms may belong simultaneously to different layers of several clusters.

We define a formal fqc as an infinite assembly of atoms in $\mathbf{R}^3$, by the following local conditions: 1) (homogeneity condition) there exists an enumerable infinite collection of centers of clusters, empty or occupied, called CC = $\{s_1, s_2, ... \}$, such that $0 < r_0 < \| s_i - s_j \| < R_0$, $i \neq j$, for some real numbers r_0 and R_0, 2) (orientational compatibility) there exists a common N ≥ 1 such that $\cap_i [X_i(n_i) - s_i]$ is a non - empty $m\overline{3}5$ - cluster $X(N)$ (coloration is not imposed for $X(N)$), 3) (covering) each atom in the assembly fqc belongs at least to one deformed colored cluster $\delta_i(X_i(n_i))$, centered at s_i, for a certain i. Let us denote by m the rank of the $\mathbf{Z}$ - module generated by CC.

To obtain interesting expressions of the structure factor, as a series, of the fqc assembly, we define and study the following scattering function [18]: for any $q \in E^*$ (= reciprocal space of $E = \mathbf{R}^3$),

$$H(q, \gamma_1, \gamma_2, \gamma_3) =$$
$$\Sigma_k \, \Sigma_j \; (K + \gamma_1 z_j) \; (\Lambda + \gamma_2 \lambda_j)^{-1} \; \exp[\, 2i\pi \, q \, . \, (x_j + \gamma_3 [\, \delta_k(x_j) \, - \, x_j \,]) \,] \tag{1}$$

with γ_1, γ_2, γ_3 in the interval [0 ; 1], where the first sum is taken over the infinite collection of $m\bar{3}\bar{5}$ - clusters, counted by their centers, where j runs over all the atomic sites of the k-th deformed $m\bar{3}\bar{5}$ - cluster $\delta_k(X_k(n_k))$; where K is the average atomic scattering factor, z_j the atomic scattering factor excess for the jth site, Λ is the average site multiplicity (average over all the atomic sites), λ_j the site multiplicity excess for the jth site, $S(q)$ is the structure factor at q of the perfect icosahedral unit $X(N)$ (optimally positioned $m\bar{3}\bar{5}$ - invariant cluster) which is the perfect icosahedral unit the closest to the distribution of slightly deformed $m\bar{3}\bar{5}$ - atomic clusters (in the sense of an appropriate least squares method [17]). We obtain the following $m\bar{3}\bar{5}$ - invariant expression for the intensities, at a scattering vector q in reciprocal space:

$$I(q) \;\; = \;\; (K\Lambda^{-1} T \, S(q) \,)^2 \;\; +$$
$$\text{linear and quadratic terms} + \; ... \tag{2}$$

The function T is a transparency function associated to the collection of centers CC, and its periodization in m - dimensional Euclidian space. T may take infinite values, but since we are mainly interested in models for which the intensities all form convergent series, whatever q is in E^*, T takes special forms associated to some integrable functions in $\mathbf{R}^{m-3}$. It is the analogue of the real - internal spaces matricial correspondances in the theory of phasons [19, 20]. Due to the lack of space, it is out of reach to justify such expressions here and we refer the reader to [11, 17, 18]. The other linear and quadratic terms give the couplings between deformations (topological disorder), icosahedral symmetry, chemical order (atomic number excess), and geometrical factors linked to interpenetration.

Since no particular hypothesis is made on the geometry of $X(N)$, this study shows that, at the first - order with the general above assumptions, the diffracting properties of an average perfectly icosahedral cluster, whatever it is, but finite, control the diffracting properties of the fqc. This cluster $X(N)$ remains of fundamental importance, with the possible successions of arithmetic and geometric sequences of icosahedral shells it may contain, which have their own specific diffracting properties [9]. The question of asking why such assemblies of atoms lead to intense Bragg peaks of small widths can be answered by saying that the width is roughly a decreasing function of the size of the object X(N) in real space, by well - known properties of the Fourier Transform. Physically, this means that the peak width is, in this approach, a measure of the spatial extent of the icosahedral order in the quasicrystal. Similarly, pseudo - inflation rules in diffraction patterns then arise from those given by X(N), as well as the asymmetry of intense Bragg peaks which seems to be an intrinsic character of a limited icosahedral local order [18]. The function T in eq. (2) acts on the diffraction pattern of X(N) as a modulating function able to annihilate some peaks and provide "m - dimensional lattice" icosahedral extinctions.

This model is easily generalized to formal quasicrystals for which the clusters are not all finite, but such that the average cluster X(N) has a convergent structure factor. This formalism is also well - adapted to describe approximant crystals [10 - 12] : there T has to be replaced by the quantity t, which is the number of clusters per unit cell of the Bravais lattice of the crystal; and the local conditions 1) to 4) of icosahedral interpenetrating clustering have slightly to be changed to have the crystallographic description of the atomic sites of the crystal true.

We have not investigated the problems of packings of icosahedral units as such, as in the icosahedral glass model [15], but the way the diffraction process can be treated when matter is obeying some local rules in a periodic case or a quasiperiodic one. This model takes into account the diffracting properties of the icosahedral glass model.

4. REFERENCES

1. Henley, C.L. (1990) 'Progress on the Atomic Structure of Quasicrystals', in T. Fujiwara and T. Ogawa (eds.), Quasicrystals, Springer Series in Solid - State Sciences 93, Springer - Verlag Berlin, Heidelberg, 38 - 47.
2. Poon, S.J., Drehman, A.J. and Lawless, K.R. (1991) 'Glassy to Icosahedral Phase Transformation in Pd-U-Si Alloys', Phys. Rev. Lett. 55, 2324 - 2327.
3. Zhang, Z., Ye, H.Q. and Kuo, K.H. (1985) 'A New Icosahedral Phase with $m\overline{3}\overline{5}$ Symmetry', Philos. Mag. Lett. B 52, L49; Zhang, Z. and Kuo, K.H. (1986) 'Orientation Relationship Between The Icosahedral and Crystalline Phases in $(Ti_{1-x}V_x)_2Ni$ Alloys', Philos. Mag. Lett. B 54, L83.
4. Tamura, N. and Verger - Gaugry, J.-L. (1991) 'Research of Quasicrystalline State in TiNiV System', Phase Transitions 32, 125 - 129.
5. Verger - Gaugry, J.-L. and Tamura, N. (1991) 'Construction of 6D Icosahedral Crystals From Approximate Icosahedral 3D Crystalline Structures and Phase Transitions', Phase Transitions 32, 89 - 101.
6. Tamura, N., François, A., Loiseau, A. and Verger - Gaugry, J.-L. (1991) 'Local Order in c - Ti_2Ni Approximant Phase and Icosahedral Symmetry', submitted to Phil. Mag. A.
7. Elser, V. and Henley, C.L. (1985) 'Crystal and Quasicrystal Structures in AlMnSi Alloys', Phys. Rev. Lett. 55, 2883 - 2886.
8. Audier, M. and Guyot, P. (1989) 'Quasi-Crystal Structure Models Related to Crystalline Structures', in M. Jaric and D. Gratias (eds), Aperiodicity and Order 3, Academic Press, Boston, 1 - 36.

9. Tamura, N. and Verger - Gaugry, J.-L. (1991) 'Systematic Study of Icosahedral Order in Crystalline Phases With Chemical Composition Close to Those of The Icosahedral Phases i-AlMn and i-AlMnSi', submitted to Philos. Mag. B.

10. Verger - Gaugry, J.-L. (1991) 'Quasicrystals and The Concept of Interpenetration in $m\bar{3}5$ - Approximant Crystals with Long - Range Icosahedral Atomic Clustering', in R. Yavari (ed.), European Workshop on Ordering and Disordering in Alloys, Grenoble, Elsevier Science Publishers, in press.

11. Verger - Gaugry, J.-L. (1991) 'Theory of G - Approximant Crystals With G, a Non - Crystallographic Point Group. I', J. Phys. I. France 1, 1303 - 1320.

12. Verger - Gaugry, J.-L., Tamura, N. and Barbier, J.-N. (1991) 'Spatial Extent of $m\bar{3}5$ - Atomic Clusters in 3D Icosahedral Quasicrystals and Diffraction Properties : The Interpenetration Concept' in J.M. Perez - Mato, F.J. Zuniga, G. Madariaga and A. Lopez - Echarri (eds.), Methods of Structural Analysis of Modulated Structures and Quasicrystals, World Scientific, Singapore, in press.

13. Yang, Q.B. (1988) Philos. Mag B 58, 47; Yang, Q.B.(1988) 'Structures of Ti2(Ni,V) in Crystalline and Quasicrystalline Phases', Philos. Mag. Lett. 57, 171 - 176.

14. Kramer, P., Papadopolos, Z. and Zeidler, D. (1991) 'The Root Lattice D6 and Icosahedral Quasicrystals' in A. Franck and K.B. Wolf (eds.), Proc. Symp. Symmetries in Physics, Cocoyoc, Mexico, Lecture Notes in Physics, Springer - Verlag, Berlin, Heidelberg, in press.

15. Stephens, P. W. (1989) 'The Icosahedral Glass Model', in M.V. Jaric and D. Gratias (eds.), Aperiodicity and Order 3, Academic Press, Boston, 37 - 104.

16. Yamamoto, A. (1990) 'Ideal Structures of The Icosahedral Al-Mn and Al-Cu-Li Quasicrystals', in T. Fujiwara and T. Ogawa (eds.), Quasicrystals, Springer Series in Solid - State Sciences 93, Springer - Verlag Berlin, 57 - 67.

17. Verger - Gaugry, J.-L., Tamura, N. and Barbier, J.-N. (1991) 'Theory of G - Approximant Crystals With G, a Non - Crystallographic Point Group. II: Optimal Action of The Icosahedral Group in $m\bar{3}5$ - Approximant Crystals Based on Local Order', submitted to J. Phys. I. France.

18. Verger - Gaugry, J.-L., Tamura, N. and Barbier, J.-N. (1991) 'Theory of G - Approximant Crystals With G, a Non - Crystallographic Point Group. III: IPIC Model Extended to Quasicrystals', to be submitted to J. Phys. I. France.

19. Jaric, M.V. and Qiu, Shi - Yue (1990) 'From Crystal Approximants to Quasicrystals', in T. Fujiwara and T. Ogawa (eds.), Quasicrystals, Springer Series in Solid - State Sciences 93, Springer - Verlag, Berlin, 48 - 56.

20. Socolar, J.E.S. (1986) 'Phason Strain in Quasicrystals', in D. Gratias and L. Michel (eds.), International Workshop on Quasicrystals, Les Houches, J. Physique France, Colloque C3 - 47, Supplément n°7, 217 - 227.

THE GEOMETRIC SHELL STRUCTURE OF METAL CLUSTERS

T.P. MARTIN, U. NÄHER, H. GÖHLICH and T. LANGE
Max-Planck-Institut für Festkörperforschung
Heisenbergstr. 1, 7000 Stuttgart 80, FRG

ABSTRACT. The mass spectrum of hot calcium clusters can be characterized as a recurring set of twenty peaks. Each occurence of the set corresponds to the formation of one complete icosahedral shell of atoms. Each of the twenty peaks (subshells) within a set (shell) can be correlated to the coverage of one triangular face. Groupings of five faces arranged around a common vertex play a special role in shell formation.

1. Introduction

One might think that the definition of a shell of atoms is straightforward. One layer of atoms arranged on the surface of a core such that the newly formed, larger unit has the same (overall) outer symmetry as the core itself. However, as one begins to construct examples, it becomes quickly clear that this definition might lead to confusion. For example, consider a cluster composed of atoms placed at the sites of a simple cubic lattice and having the overall outer shape of a cube. The first such cube that can be formed around a central atom contains 27 atoms, 3 atoms on a side; the next, 125 atoms, i.e., 5 atoms on a side, etc. But what happened to the 64 atom cube with 4 atoms on a side? It has no central atom. That is, this simple example might be considered to describe two distinct shell sequences, one set of shells possessing a central atom; the other set has a central 8 atom cube. One way of getting around this difficulty is to combine the two sets into a single set. Each successive member of this combined set is obtained by adding atoms to only three of the six faces of the preceeding member. We will see that it is useful to designate a set of such shells as irregular shells in order to distinguish them from, for example, the regular shells of an icosahedron where atoms must be added to <u>all</u> faces in order to complete the next shell.

1.1 SHELLS OBTAINED FROM CLOSE-PACKED SPHERES

A limited number of symmetric clusters can be constructed from the close-packing of hard spheres; e.g. tetrahedra, octahedra, and their truncated forms. The truncated forms can have triangular, square or hexagonal faces. For example, the cuboctahedron, an octahedron

P. Jena et al. (eds.), Physics and Chemistry of Finite Systems: From Clusters to Crystals, Vol. I, 147–156.
© 1992 *Kluwer Academic Publishers.*

148

truncated by a cube, has 6 square faces and 8 triangular faces. Although this figure is constructed from close-packed layers, the cut forming a square face reveals a surface which is not close-packed. Such a surface is relatively unfavorable energetically and is a good candidate to accept the first atoms of a new shell.

The hexagonal close-packed crystal is usually distinguished from a face-centered crystal by the ordering of the close-packed layers. The atoms in every other layer lie exactly above one another and one speaks of an ababa ... layer sequence. For the fcc close-packed crystal this sequence is abcabca... . It is convenient to use the same notion for clusters. The cuboctahedron obeys the sequence abca ... In this case the cuboctahedron can be said to have fcc structure, i.e., it can be cut out of an fcc crystal. However, if the bottom layers are rotated 60°, the sequence changes from abcabca to abcacba, for example. The result is a highly symmetric cluster, closely related to the cuboctahedron, however, it does not have fcc structure.

1.2 SHELL STRUCTURE RELATED TO THE bcc LATTICE

In the first paragraph we used the example of a cube-shaped cluster cut out of a simple cubic lattice. This was convenient to illustrate the concept of irregular shells, but is unrealistic in that elemental matter does not usually condense into a simple cubic structure. However, if such a cube is squeezed along a body diagonal, the cube deforms into a rhombohedron which can be cut out of a bcc lattice. The bcc rhombohedron represents a set of irregular shells containing, of course, the same number of atoms as simple cubic shells. The bcc lattice contains also a set of regular shells. The first member in this set is shown in Fig. 1a. The atoms of such clusters are contained within 12 rhombic faces.

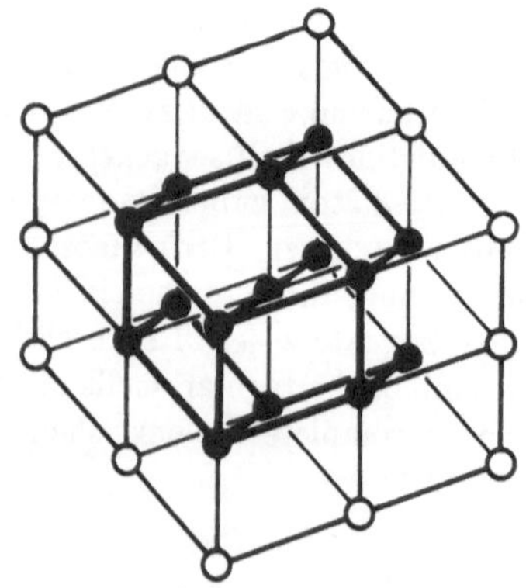

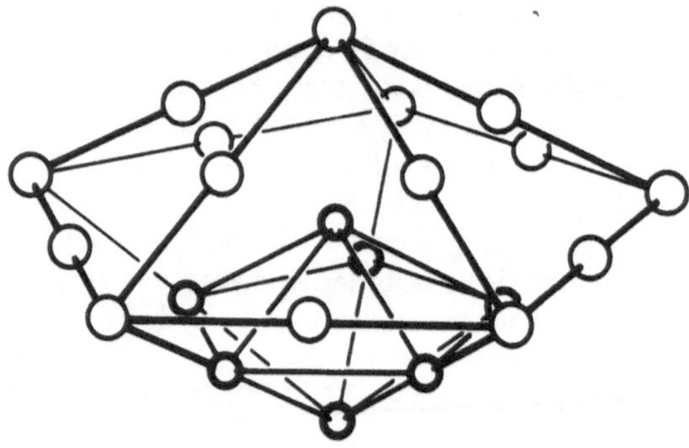

Fig. 1a: A geometric figure to be found in a bcc lattice is outlined with heavy lines. It forms the basis of a shell sequence.

Fig. 1b: Decahedra form a set of irregular shells. Successively large shells are formed by adding an umbrella-shaped partial layer.

1.3 SHELL STRUCTURES WITH FIVE-FOLD SYMMETRY

Until now we have discussed shell structure in clusters of close-packed atoms or of atoms on crystal lattice sites. Clusters in the form of icosahedra or decahedra are neither close-packed nor are they small pieces cut out of a crystal. A five-fold symmetry axis is not consistent with the crystalline requirement of translational symmetry. Icosahedra form a set of regular shells around a central atom. Nature has played a strange trick on us here. The number of atoms needed to complete icosahedral shells is <u>exactly</u> that needed to complete cuboctahedral shells. For this reason, the experimental observation of magic numbers corresponding to shell closings is not sufficient to allow us to distinguish between noncrystalline icosahedra and fcc cuboctahedra.

Decahedra represent a set of irregular shells. The shells possess alternately a central atom and a central 7-atom decahedron and are formed by placing a large overlapping "umbrella" on top of the previous member of the set, Fig. 1b.

2. Observation of Shells and Subshells of Atoms

Both calculations [1] and experiments [2–8] indicate that inert gas clusters containing from 13 to 923 atoms have icosahedral symmetry. These might be referred to as precrystalline structures since the inert gases are known to condense into fcc crystals. Precrystalline structures have also been observed for metallic materials in condensed units large enough to yield sharp electron diffraction patterns [9–11]. These quasicrystals present a fascinating challenge to scientists to develop methods for describing a regular but nonperiodic state of bulk matter. Smaller icosahedral metal particles have been observed directly using the newly developed technique of high-resolution electron microscopy [12].

Additional evidence exists for icosahedral symmetry in metal clusters [13]. Calculations predict that very small alkaline earth clusters prefer noncrystalline structures [14–17]. The pattern of NH_3 and H_2O binding energies with Co and Ni clusters has been interpreted as indicating icosahedral symmetry in metal clusters containing from 50 to 150 atoms [18,19]. Mass spectra of Ba and Ba-O clusters seem to indicate an icosahedral growth sequence in the size range from 13 to 35 atoms [20–22].

Recently, we observed a slow modulation in mass spectra of Na clusters which we interpreted as evidence for the existence of shell structures, i.e. a highly symmetric, onion-like cluster structure [23]. The modulation appeared only if the energy of the ionizing photons was chosen to coincide with the ionization potential of the clusters and was found to be almost periodic when plotten on a cube root of mass scale. The cusp-like minima of the mass spectra pointed to characteristic masses or numbers of atoms. Within the accuracy of reading the minima, these magic numbers correspond to the number of atoms in complete Mackay icosahedra [24]. However, on the basis of such observations, it is not possible to conclude that the clusters have icosahedral symmetry because icosahedral shells and fcc cuboctahedral shells contain exactly the same number of atoms. We have to look elsewhere for decisive experimental data. We believe these data are contained in the weaker mass peaks between shell closings.

Figure 2 shows a mass spectrum of calcium clusters containing up to 5000 atoms. The mass resolution is poor for several reasons. The clusters have been heated to encourage evaporation. Since loss of mass occurs in the acceleration region, the resolution suffers.

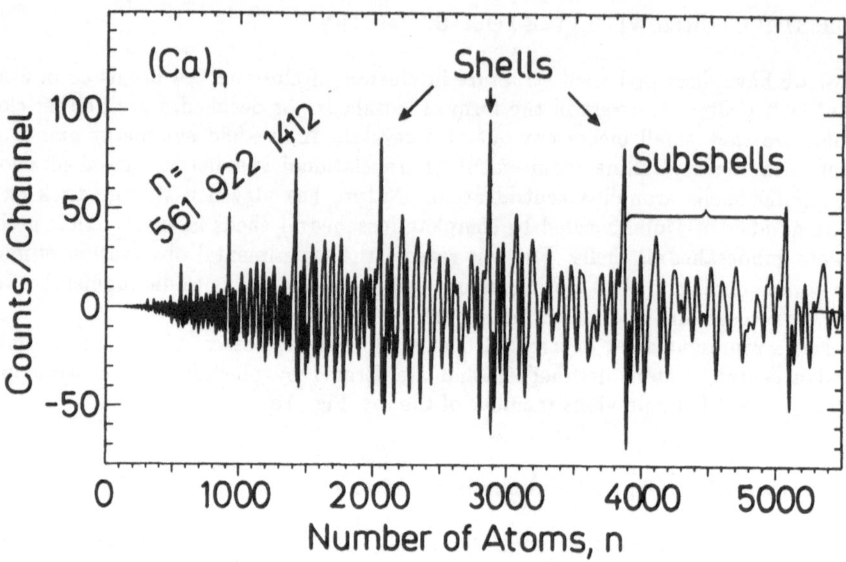

Fig. 2: Difference spectrum (highly smoothed envelope mass spectrum subtracted from a slightly smoothed spectrum) for Ca clusters. The strong peaks correspond to completely filled shells of atoms, the weaker peaks to so-called subshells.

Another factor contributing to our inability to separate cluster peaks is the fact that calcium posseses several relatively abundant natural isotopes including 97% ^{40}Ca and 2.1% ^{44}Ca. Since no additional information can be gained by using narrow time channels, the spectrum was averaged over 100, 64 ns time channels. An envelope function has been obtained by averaging over 3000 time channels. A convenient form for display is the difference of these averaged spectra shown in Fig. 2.

The difference spectrum for $(Ca)_n$ is chacterized by 7 strong peaks. Although these peaks are not equally spaced on a scale linear in mass, they do occur at equal intervals on an $n^{\frac{1}{3}}$ scale. This signals the presence of shell structure, in this case geometric shells of atoms. Clusters having closed shells with icosahedral or cuboctahedral structure would be expected to contain 561, 923, 1415, 2057, 2869, 2869, 3869 and 5083 atoms. In general, the observed peaks in the mass spectra occur about 0.2% lower in mass. The reason for this small discrepancy is not fully understood. This main sequence of mass peaks gives strong evidence that as atoms evaporate from clusters, perfect geometric shells are successively revealed. Cluster having perfect symmetry are rather immune to further loss of mass. The number of atoms in cluster belonging to the main sequence indicate the geometry is either icosahedral or cuboctahedral.

Notice the weaker subshell structure occuring between shell closings. In order to better compare subshell structure, the mass spectrum in Fig. 2 has been cut into segments, each segment corresponding to the formation of one shell. Since the number of atoms in a shell is approximately proportional to the square of the shell index, the mass scale of each segment is adjusted so that complete shells lie above one another, Fig. 3a. The same subshell structure is found in the mass spectra of strontium clusters, Fig. 3b. When compared in this way, the similarity of the subshell structure from shell to shell is quite apparent. But what are these subshells?

The clusters most probably grow by adding shells of atoms to a rigid core. The number of atoms contained in a growth shell is dependent on the preferred coordination and local symmetry of the atoms and on the overall symmetry of the shell. Clusters constructed of complete shells can be expected to be highly stable. For inert gas clusters both experiments and calculations indicate that partial icosahedral shells of atoms also show enhanced stability [5–7]. For example, one might expect that completely covered facets of a cluster surface represent intermediate structures of high stability. Since the facet structure of the icosahedron (20 triangular faces) and the cuboctahedron (8 triangular and 6 square faces) are quite different, a determination of partial shell sizes should make it possible to distinguish between the two structures.

The square faces of the cuboctahedron would be likely candidates to accept the first atoms of a newly deposited layer because the atoms in these faces are not close-packed. However, no arrangement of atoms on these faces alone or in combination with other cuboctahedral faces could be found which matched the observed subshell magic numbers. Next, we turned to the icosahedron for which subshell structure had already been studied [5–7]. The first atoms to form a new shell on an inert gas icosahedron apparently do not immediately take their final positions. This would force atoms on the border between two triangular faces to have contact with only two substrate atoms. Instead, the triangular faces are first filled with a close-packed layer. Only after the shell is more complete do the atoms rearrange into their final icosahedral positions. This shell filling sequence, observed in inert gas clusters, although close, seems to deviate significantly from the observed magic numbers for Ca and Sr clusters. Therefore, we would like to suggest an alternative sequence supported by a simple calculation.

First, we will try to explain the six strong, broad peaks shown in Fig. 3b for calcium and strontium clusters. The first method to be applied will be CNE (Complete Neglect of Everything). This is, we assume, the atoms in the new shell take immediately their final positions. We merely count the numbers of atoms needed to cover combinations of triangular faces. In order to assist this counting, in Fig. 4 the positions of the atoms in the sixth shell have been projected onto a plane in the manner of Northby [4]. We suggest that umbrella-shaped intermediate groups have enhanced stability. Each of the umbrellas contains 76 atoms and each has the same shape (although they appear distorted in the projection shown in Fig. 4). Only 51 additional atoms are necessary to complete the second umbrella because it shares atoms with the first. The third and fourth umbrellas overlap two others. Therefore, they require only 36 additional atoms for completion. This umbrella model is consistent with the observed subshell structure.

In order to explain the fine subshell structure occuring between completed umbrellas it is necessary to go one small step beyond CNE. The procedure is as follows. The outermost shell of a cluster is successively filled up with atoms. The already filled inner shells are considered to be rigid. Only icosahedral sites are allowed, in particular no fcc-substructure on the faces of the icosahedron is admitted. Although the atoms are added one by one, all atoms of the partly filled shell may change their sites when one new atom is added to the ensemble. Under these assumptions we tried to find the energy minimum of the cluster as a function of the total number of atoms.

The interactions of the atoms have been described by a Lennard-Jones potential, its minimum being the bonding length between neighboring shells. This is certainly not a very

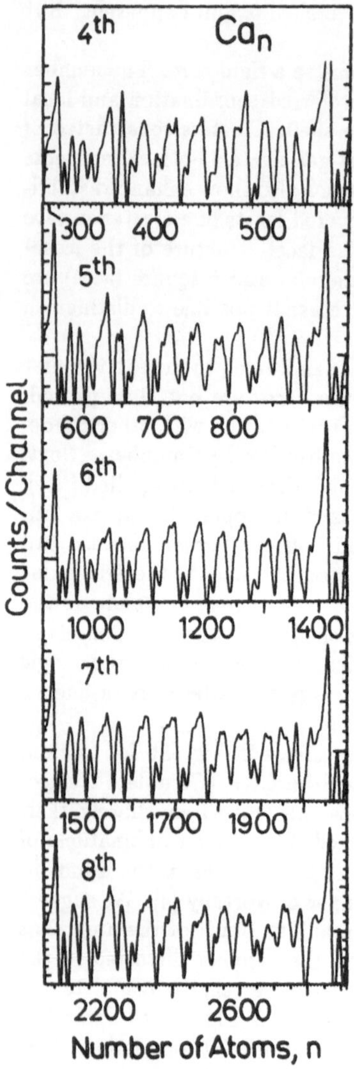

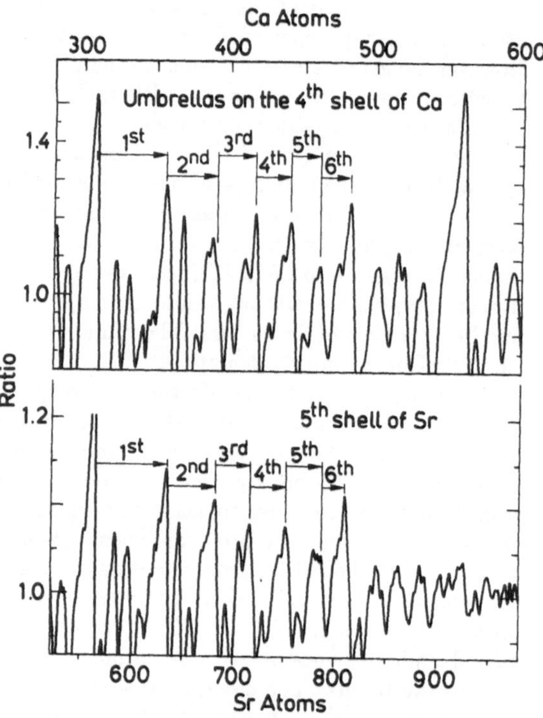

Fig. 3a: The same spectrum shown in Fig. 2 but cut into segments and each segment rescaled so they have the same length. Notice that the subshell structure repeats itself from shell to shell.

Fig. 3b: Mass spectra showing subshells in the 4th shell of Ca and the 5th shell of Sr.

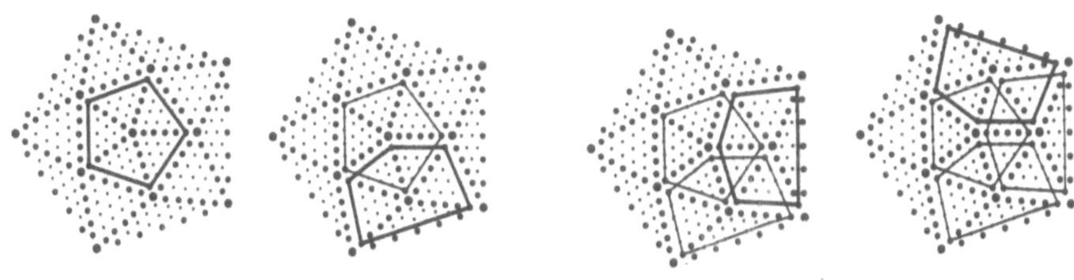

Fig. 4: The dots represent the atoms of the 6th shell of an icosahedron projected onto a plane. The bottom 76 atoms are not shown. The umbrella shaped structures are identical, each containing 76 atoms. We suggest that the umbrellas represent highly stable partial shells of calcium atoms.

realistic model for the binding energy of a metal cluster, but as we will show, the agreement with the experiment is worthy of note. Using such a potential, nearest neighbor interactions dominate the total energy because of the long range term in the order of r^{-6}. Therefore, the binding energy of one atom could be (although we have not done this) roughly estimated by simply counting the occupied neighbor sites and adding a nearly constant energy term as a substitute for the interactions with the rest of the cluster atoms. The most important factor determining the binding energy is the number of occupied nearest neighbor sites. This number can be as large as 9 for face atoms (6 nearest neighbors within the shell, 3 in the shell below), 8 for edge atoms (6+2) and 6 for corner atoms (5+1).

How will the atoms arrange as the shell is filled? The positions of the atoms will be determined by two, sometimes conflicting, tendencies. On the one hand, icosahedral corner and edge sites should be avoided because they provide less binding energy. On the other hand, the occupied surface area of the partial shell should have a minimum circumference without occupying corner or edge sites. A compromise has to be found.

The results of the calculation are summarized in Fig. 5. The 5th shell of an icosahedron is complete with 561 atoms and the 6th shell with 923 atoms. The mass interval shown in Fig. 5 represents, therefore, the growth of the 6th shell. The top curve shows the calculated maximum energy gained by the successive addition of one atom. The units of energy are left intentionally arbitrary in order to emphasize that the assumptions made in the calculation allow only a qualitative comparison with experiment. The 362 calculated energies (one for each atom added) take on one of 15 values. Each value can be associated with a specific nearest neighbor environment.

The calculations indicate that subshells are not characterized so much by particulary stable structures, as by particularly unstable islands. Notice that an atom at an edge site with only 4 nearest neighbors (two in the new shell and two in the substrate) is weakly bound. Such unstable conformations would be expected to be associated with minima in the mass spectra. In fact, the correlation with the minima in the measured spectrum, also shown in Fig. 5 is very good. The minima are associated with the atomic arrangements shown at the bottom of Fig. 5. The dots in the pentagonal figures represent the positions of atoms in the 6th shell of an icosahedron projected onto a plane. For clarity, the atoms on the bottom of the icosahedron are not shown in this projection. The shaded areas cover those sites which are occupied by atoms. In all cases the structures consist of compact

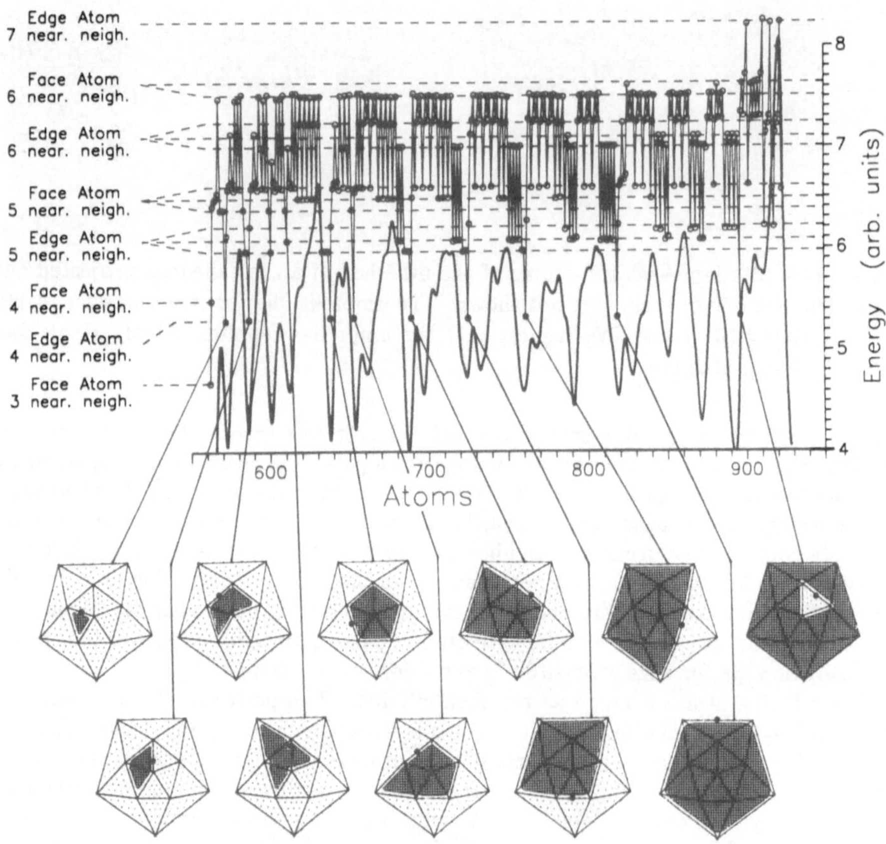

Fig. 5: Subshells (stable islands) observed during the formation of the 6th shell (561 atoms to 923 atoms) of a $(Ca)_n$ clusters. The experimental mass spectrum (smooth curve) is compared with the calculated energy gained with each new atom. The addition of one atom after completion of subshells leads to characteristic minima in the energy and the mass spectrum. The arrangements of surface atoms at the minima are shown by the shaded areas in the figures at the bottom. The dots in the pentagons represent the atoms of the 6th shell of an icosahedron projected onto a plane.

islands plus a single atom at an adjacent edge site (large black dot).

The growth of a shell seems to start by the successive coverage of 1,2,3 and 4 adjacent triangular faces. Then a reorganization of the atoms on the surface occurs prior to the completion of 5 triangular faces around a central vertex. This "umbrella" is a recurring motif. Four of the following minima can be described as overlapping umbrellas plus, of course, one atom.

3. Concluding Remarks

The mass spectra of Mg_n, Ca_n and Sr_n clusters show a characteristic series of peaks which are equally spaced on an $n^{\frac{1}{3}}$ scale, indicating the existence of a shell structure. In general, a knowledge of the number of atoms required to complete successive shells should allow the geometric form of the shells to be identified. However, nature has presented us with a strange coincidence in the present case. The numbers of atoms needed to complete icosahedral and cuboctahedral shells are identical. Fortunately, we could use the subshell structure to decide between the two geometries. After a lengthy discussion we came to the conclusion the clusters are icosahedral. In retrospect, however, perhaps the discussion need not have been so lengthy. The icosahedron has 20 faces, the cuboctahedron only 14. If we merely count the number of subshell peaks we arrive at the number 20, one peak for each face of the icosahedron. On the other hand, the discussion is necessary to explain why the peaks are not equally spaced and are not of equal intensity.

Subshell structure develops as material evaporates from well-defined geometric structures, i.e. it arises through the cluster equivalent of the sublimation process. Subshell structure would not be expected in those cases for which an appreciable evaporation rate can be achieved only above liquid, structureless clusters. The alkaline earth metals and the inert gases are among the few elemental solids that sublime. They are also the only elements to show, until now, geometric subshell structure in cluster fragmentation spectra.

References

1. Hoare, M.R. (1979) Adv. Chem. Phys. **40**, 49.

2. Echt, O. Sattler, K., Recknagel, E., (1981) Phys. Rev. Lett., **47**, 1121.

3. Farges, J., de Feraudy, M.F., Raoult, B., Torchet, G., (1986) J. Chem. Phys. **84**, 3491.

4. Northby, J.A. (1987) J. Chem. Phys. **86**, 6166; Harris, I.A., Kidwell, R.S., and Northby, J.A. (1984), **53**, 2390.

5. Lethbridge, P.G., Stace, A.J., J. Chem. Phys. (1989) **91**, 7685.

6. Miehle, W., Kandler, O., Leisner, T., Echt, O., J. Chem. Phys. (1989) **91**, 5940.

7. The subject of icosahedral shells of atoms in inert gas clusters has a long and interesting history. For a recent starting point into this extensive literature, see: 'Proceedings of Faraday Symposium on Large Gas Phase Clusters', J. Chem. Soc. Faraday Trans., (1990) 86.

8. Schriver, K.E., Hahn, M.Y., Persson, J.L., LaVilla, M.E., Whetten, R.L., (1989) J. Phys. Chem., **93**, 2869.

9. Schechtman, D., Blech, I., Gratias, D., Chan, J.W., Phys. Rev. Lett. (1984), **53** 1951.

10. Janot, C., Dubois, J.-M., J. Non-Crystalline Solids (1988), **106** 193.

11. Hall, B.D., Flüeli, M., Monot, R., Borel, J.-P. (1989), Z. Phys. D. **12**, 97.

12. See Proceedings: 4th Intl. Meeting on Small Particles and Inorganic Clusters, 1989, Chapon, C., Gillet, M.F., Henry, C.R. (eds.) Berlin-Heidelberg-New York: Springer.

13. Cleveland, C.L., Landman, U. (1991) J. Chem. Phys. **94**, 7376.

14. Blaisten-Barojas, E., Khanna, S.N. (1988) Phys. Rev. Lett., **61**, 1477.

15. Rao, B.K., Khanna, S.N., Meng, J., Jena, P. (1991) Z. Phys. D.

16. Pacchioni, G., Pewestorf, Koutecky, W. (1984) J. Chem. Phys., **83**, 201.

17. Kumar, V., Car, R. (1991) Z. Phys. D **19**, 177.

18. Klots, T.D., Winter, B.J., Parks, E.K., Riley, S.J. (1990), J. Chem. Phys. **92**, 2110.

19. Winter, B.J., Klots, T.D., Parks, E.K., Riley, S.J. (1991), Z. Phys. D. **19**, 391.

20. Rayane, D., Melinon, P., Cabaud, B., Hoareau, A., Tribollet, B., Broyer, M. (1989), Phys. Rev. **A39**, 6056.

21. Whetten, R.L, private communications.

22. Martin, T.P., Bergmann, T. (1990), J. Chem. Phys. **90**, 6664.

23. Martin, T.P., Bergmann, T., Göhlich, Lange, T. (1990), Chem. Phys. Lett. **172**, 209.

24. Mackay, A.L. (1962), Acta Cryst. **15**, 916.

25. Martin, T.P., Bergmann, T., Göhlich, H., Lange, T. (1991), Chem. Phys. Lett. **176**, 343.

ELECTRONIC SHELL STRUCTURE IN ICOSAHEDRAL METAL CLUSTERS

J. MANSIKKA-AHO, J. SUHONEN, S. VALKEALAHTI,
E. HAMMARÉN and M. MANNINEN
Department of Physics, University of Jyväskylä,
P.O. Box 35, SF-40351 Jyväskylä, Finland

ABSTRACT. The shell structure of valence electrons in icosahedral and cuboctahedral simple metal clusters is studied using the free electron model and the Hückel model. The shell structure in a 1415 atom icosahedral cluster has still similarities with that of a spherical cluster. The effect of the finite temperature on the shell structure in liquid clusters is discussed.

1. Introduction

The magic numbers of small alkali metal clusters are a consequence of a shell structure of the valence electrons [1]. The electronic shell structure has been observed up to the cluster size of about 3000 atoms [2]. The shell and supershell [3] structures correspond to electron energy levels in a spherical effective potential. It has been shown that the crystal field splitting due to the discrete ion pseudopotentials is negligible still in clusters containing thousands of atoms [4]. However, the overall cluster shape can produce a much stronger crystal field which will eventually wash out the shell structure [5] and geometrical arrangement of atoms will become dominating in determining the magic numbers. Experimentally it has been observed that for clusters containing thousands of atoms the magic numbers correspond to clusters with the geometry of a complete icosahedron [6]. The temperature of the clusters showing the large magic numbers of the electronic shell structure is close to the melting point of the metal whereas the clusters showing the icosahedral structures were colder.

In this work we have studied the electronic shell structure of icosahedral and cuboctahedral clusters in two different models. The Hückel model is used to compute numerically the shell structure in a potential box of desired symmetry [5] and a basis function expansion method has been used for calculating the shell structure in a finite potential well of icosahedral or cuboctahedral symmetry.

At high temperatures the clusters could be in a liquid state. Structural excitations will then change the shape of the droplet from a perfect sphere. We have used a simple model to estimate the effect of the droplet shape on the shell structure.

P. Jena et al. (eds.), Physics and Chemistry of Finite Systems: From Clusters to Crystals, Vol. I, 157–164.
© 1992 *Kluwer Academic Publishers.*

158

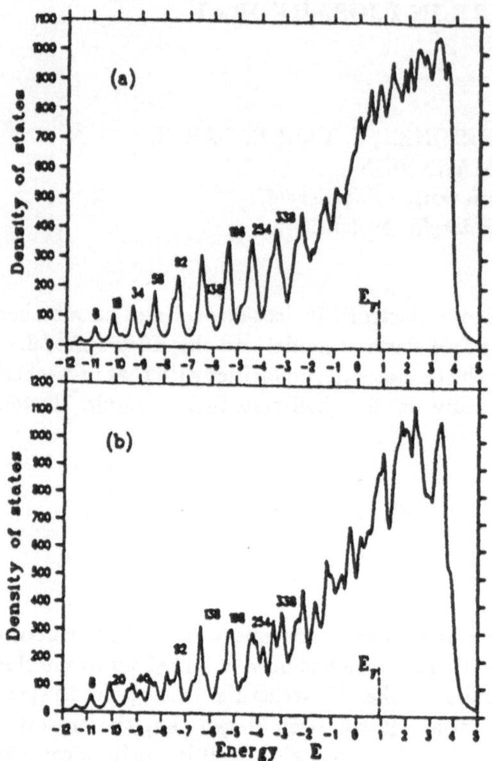

Figure 1. Density of states of Hückel clusters with (a) a spherical shape and with (b) an icosahedral shape. The discrete levels have been convoluted with a Lorenzian. The numbers show the shell fillings. E_F denotes the Fermi energy. The energy is in Hückel units ($\alpha = 0$, $\beta = 1$).

2. Hückel clusters

Although the electronic shell structure is most easily understood by studying the free electron model it is interesting to note that the geometries of the smallest clusters and the shell structure can be obtained also using the tight binding Hückel model with one state per atom [7]. The Hamiltonian to be diagonalized is then

$$H_{ij} = \begin{cases} \alpha, & \text{if } i = j; \\ -\beta, & \text{if } i \text{ and } j \text{ nearest neighbours}, \\ 0, & \text{otherwise} \end{cases} \tag{1}$$

where α and β are parameters. It is easy to see [8] that for clusters with a cubic lattice the Hückel model gives just a numerical solution for the free electron Schrödinger equation. Thus for spherical clusters the Hückel model reproduces the electronic shell structure. However, this is only true for energy levels close to the bottom of the band where the wave length of the wave function is large compared to the lattice constant. Figure 1 shows the density of states of two Hückel clusters with 923 atoms each. One of them is a sphere cut off from an fcc lattice and the other is an icosahedral cluster (for details see Ref. [5]). The spherical cluster shows the same shell structure as is obtained with an infinite spherical potential well. Close to the Fermi energy the density of states increases and the shell structure effects disappear in the Hückel model. The shell structure of the icosahedral cluster is very similar up to the shell closing with 338 electrons. This is an indication that the faceted surface of the icosahedron does not destroy the shell structure in so small clusters. The Hückel calculations for other geometries (like cuboctahedron or Wulff polyhedron) showed the disappearance of the shell structure in much smaller clusters [5].

3. Free electron clusters

The spherical sodium clusters can be described with the jellium model where the valence electrons move in a spherically symmetric effective potential which can be calculated self-consistently by using the density functional method [9]. A good approximation for this potential is a Wood-Saxon potential [3] which is a potential well with a rounded edge. In the present study we are mainly interested in the effect of the overall geometrical shape of the potential well on the electronic structure. To keep the calculation simple we approximate the effective potential with a finite square-well. The depth of the well is a sum of the work function and the Fermi energy. For a spherical cluster its radius is chosen to be $R = r_s N^{1/3}$, where r_s is the Wigner-Seitz radius and N the number of atoms in the cluster. For nonspherical clusters the well is assumed to have the same depth and volume as in the spherical case.

The electronic structure in a nonspherical potential is obtained by expanding the wave functions in terms of the solutions for the spherical potential well (for details see Ref. [10]). Figure 2 shows the results obtained for 309 atom icosahedral and cuboctahedral clusters. The shell structure of the icosahedral cluster is similar to that of the sphere whereas in the cuboctahedral cluster the shell structure is already strongly disturbed by the faceted surface. For larger clusters the electronic structure is calculated only close to the Fermi energy. Figure 3 shows the shell structure close to the Fermi energy for a 1415 atom icosahedral cluster. The main features of the shell structure of the spherical potential (also shown) remain when the potential is replaced with an icosahedral well.

The question now arises what is determining the observed magic numbers: is it the geometrical arrangement of the atoms or the shell structure effects. The observed intensities can be related to the second derivative of the total energy [1]

$$\Delta_2 E = E(N+1) - 2E(N) + E(N-1). \tag{2}$$

The shell structure energy is the sum of the individual eigenvalues. For a spherical cluster the maximum is obtained when N corresponds to a full shell and $\Delta_2 E$ will be the energy gap between the shells. The energy gap is proportional to $N^{-1/3}$ and vanishes for large clusters. The energy corresponding to the geometrical arrangement of atoms can be estimated for large clusters using a proper many-atom potential [11] and computer simulation. For large icosahedral clusters $\Delta_2 E$ will then be independent of the cluster size. This means that for large clusters the magic numbers will be determined by the cluster geometry whereas for small clusters they are determined by the electronic shell structure.

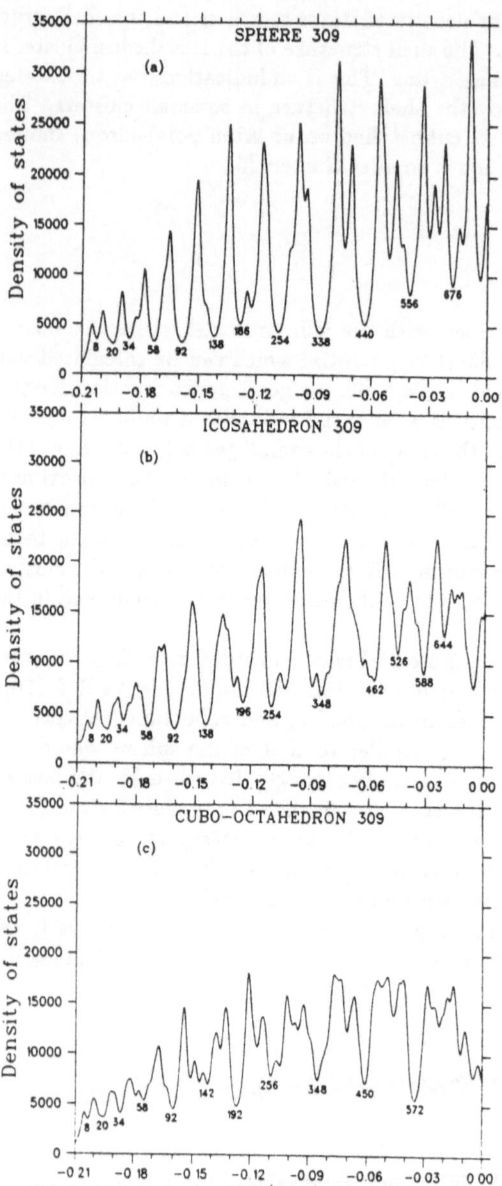

Figure 2. Density of states in a 309 atom sodium clusters with different shapes. In each case the potential is a square well with the same volume. The discrete levels have been convoluted with a Lorenzian.

Unfortunately, for sodium no reliable many-atom potentials have been derived. We have estimated the cross-over point for copper by using the effective medium theory [12] and molecular dynamics for calculating the structural part of $\Delta_2 E$ and estimated the electronic part from a spherical potential box. This estimation gives $N \approx 1300$. This could also be a typical value for other simple metals since both the structural and electronic part of the total energy depend similarly on the density of the metal (for nontransition metals). This result suggests that, if the cluster has less than about a thousand atoms, the electronic shell structure dominates the mass spectra even if the clusters would have icosahedral structures.

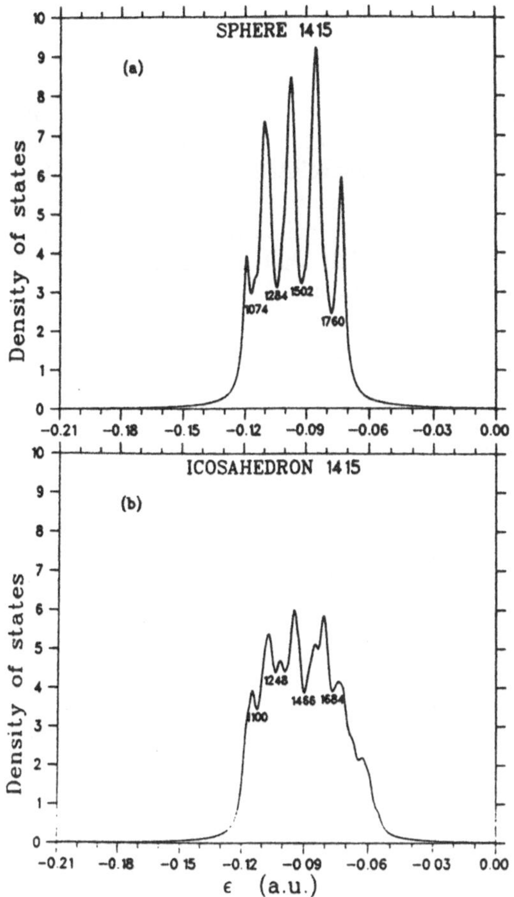

Figure 3. Density of states around the Fermi level in a spherical (a) and icosahedral (b) sodium cluster of 1415 atoms. The discrete levels have been convoluted with a Lorenzian.

4. Liquid clusters

In experiments the temperature of the clusters can be fairly high leaving the possibility that the clusters are not in the solid phase but are either liquid or have liquid-like surface. The shape of the cluster will then be different from that of a crystalline cluster. In order to minimize the surface tension the cluster would have a spherical shape. A finite temperature would mean excitations, e. g. surface waves, which could change the instantaneous shape of the cluster. It is then important to know the effect of the shape oscillations to the electronic level structure. In general, the energy associated to the surface waves can be estimated from the liquid drop model where the cluster energy is determined from the surface and curvature

162

energies. This model gives good average estimate for the size dependence of the total energy of jellium clusters [13]. The shape dependence of the energy levels has been studied using the Nilsson model, well established in nuclear physics, for quadrupole [14] and octupole [15] deformations.

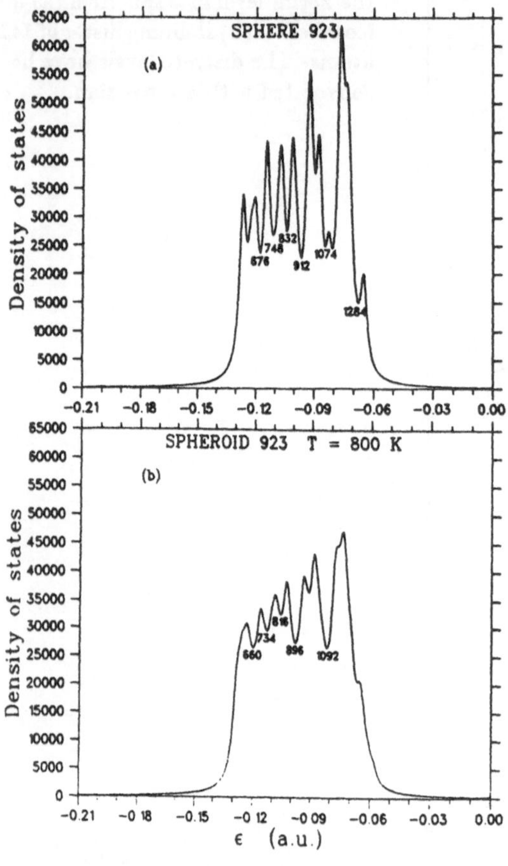

Figure 4. Density of states around the Fermi level in a spherical (a) and spheroidal (b) sodium cluster with 923 atoms. The discrete levels have been convoluted with a Lorenzian. The deformation energy of the spheroid correspond to a temperature of 800 K. Note that due to the supershell structure the shell structure in a 923 atom cluster is not as clear as in the 1415 atom cluster (Fig. 3 (a)).

We have studied the effect of the quadrupole deformation in our model of the equal-volume square-well potential. In these first calculations the potential well was taken to be a prolate spheroid. The energy associated to the deformation was estimated solely from the increase of the surface energy (in small clusters the change in the one-electron spectrum would also be an important part of the total energy). Assuming that at a given temperature the typical excitation energy would be of the order of kT we can relate the effect of the shape deformation to the cluster temperature. Figure 4 shows the density of states of a 923 atom spheroidal sodium cluster as compared to the density of states of a spherical cluster. The energy of the quadrupole deformation corresponds to the temperature of 800 K. It is seen that the deformation is still so small that the density of states has all the same features as that of the spherical potential. This result seems to indicate that the liquid clusters, well above the melting temperature, could have the shell structure of a spherical potential. A similar result for small copper clusters (tens of atoms) has been obtained by Christensen et al [16].

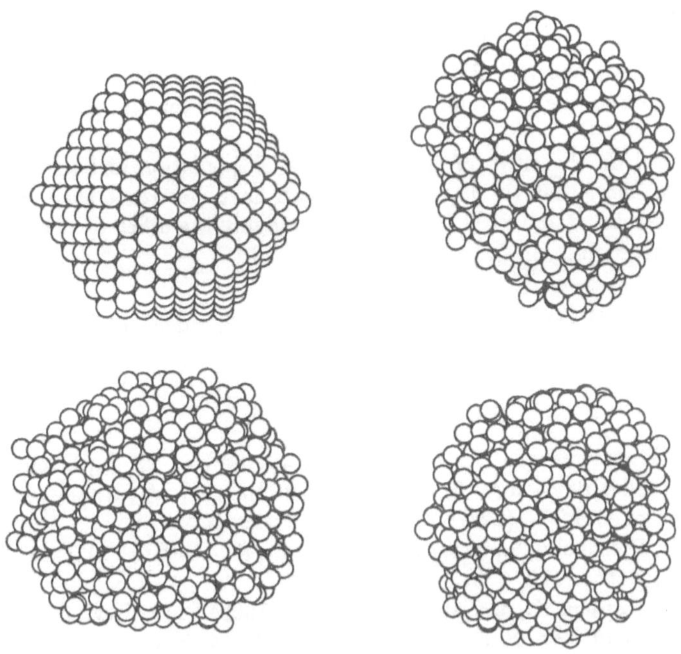

Figure 5. An example of the instantaneous cluster geometry in liquid copper clusters of 923 atoms at a temperature of 1400 K. Three projections of the cluster are shown. For comparison the structure of a perfect icosahedron of the same size is also shown.

Computer simulations can be used to study the fluctuations in the liquid cluster shape at finite temperatures. We have studied the 923 atom copper cluster at 1400 K, i.e. slightly above its melting point using the effective medium theory. Typical instantaneous structures of the liquid cluster are shown in Fig. 5. On the average the liquid clusters are as spherical as the icosahedral cluster (also shown in Fig. 5). This supports the result of Fig. 4 that in liquid clusters the electronic shell structure can still be seen. In the liquid clusters the structural part of the total free energy depends smoothly on the cluster size and $\Delta_2 E$ is always small. Then the electronic shell structure effects could be more easily seen in the liquid clusters than in the crystalline clusters.

5. Conclusions

The shape of the cluster surface becomes important in determining the electronic shell structure in a large cluster either due to the faceting of the surface of a solid cluster or due to surface waves of a liquid cluster. In faceted fcc clusters, cuboctahedron or Wulff construction, the shell structure (of a spherical potential) will be strongly disturbed already when the cluster has about 100 atoms. The icosahedral cluster shows the shell structure still when there is more than 1000 atoms in the cluster. However, even in the case of the icosahedral clusters the variation of the total energy with the number of atoms will be clearly dominated by the electronic shell structure only when the cluster has less than about 1000 atoms.

Estimates based on the effect of the quadrupole deformation in liquid clusters with about 1000 atoms show that the liquid clusters can have electronic shell structure well above the melting point. More work is needed for studying the different deformations as a function of the cluster size.

Acknowledgments: The authors would like to thank the Academy of Finland for financial support.

References

[1] Knight, W. D., Clemenger, K., de Heer, W. A., Saunders, W. A., Chou, M. Y., and Cohen, M. (1984) Phys. Rev. Lett. **52**, 2141; de Heer, W. A., Knight, W. D., Chou, M. Y., and Cohen, M. L. (1987) Solid State Physics **40**, 93.

[2] Pedersen, J., Bjørnholm, S., Borggren, J., Hansen, K., Martin, T. P., and Rasmussen, H. D. (1991) Nature **353**, 733.

[3] Nishioka, H., Hansen, K., and Mottelson, B. R. (1990) Phys. Rev. B **42**, 9377.

[4] Manninen, M. and Jena, P. (1991) Europhys. Lett. **14**, 643.

[5] Mansikka-aho, J., Manninen, M., and Hammarén, E. (1991) Z. Phys. D **21**, 271.

[6] Kappes, M. M., Schär, M., Radi, P., and Schumacher, E. (1986) J. Chem. Phys. **84**, 1863.

[7] Wang, Y., George, T. F., and Lindsay, D. M. (1987) J. Chem. Phys. **86**, 3493; Lindsay, D. M., Wang, Y., and George, T. F. (1987) J. Chem. Phys. **86**, 3500; Lindsay, D. M., Wang, Y., and George, T. F. (1990) J. Cluster Sci. **1**, 107.

[8] Manninen, M., Mansikka-aho, J., and Hammarén, E. (1991) Europhys. Lett. **15**, 423.

[9] Ekard, W. (1984) Phys. Rev. B **29**, 1558.

[10] Mansikka-aho, J., Hammarén, E., and Manninen, M. (1991) submitted to Phys. Rev. B.

[11] Nieminen, R. M., Puska, M. J., and Manninen, M. (1990) *Many-Atom Interactions in Solids* (Springer, Heidelberg).

[12] Jacobsen, K. W., Nørskov, J. K., and Puska, M. J. (1987) Phys. Rev. B **35**, 7423.

[13] Engel, E. and Perdew, J. P. (1991) Phys. Rev. B **43**, 1331.

[14] Clemenger, K. (1985) Phys. Rev. B **32**, 1359.

[15] Hamamoto, I., Mottelson, B., Xie, H., and Zhang, X. Z. (1991) Z. Phys. D **21**, 163.

[16] Christensen, O. B., Jacobsen, K. W., Nørskov, J. K., and Manninen, M. (1991) Phys. Rev. Lett. **66**, 2219.

Uzi Landman, R. N. Barnett, C. L. Cleveland
and G. Rajagopal
School of Physics
Georgia Institute of Technology
Atlanta, Georgia   30332

ABSTRACT. Investigations of variations of physical and chemical properties of clusters as a function of size allow systematic explorations of evolutionary patterns of materials properties from the molecular scale to the condensed phase regimes. Using classical and quantum molecular dynamics simulations we discuss size-evolutionary patterns of the geometrical structure of nickel clusters, fission energetics and dynamics of small doubly charged sodium clusters, and metallization of sodium-fluoride clusters, exhibiting an insulator-to-metal transition.

I. INTRODUCTION

Among the principal motivations for studies of clusters are a large number of observations that finite materials aggregates exhibit size-dependent physical and chemical properties, and the expectation that exploration of the systematics of such size effects and of their physical origins (both in the classical and quantum mechanical domains), would open avenues for understanding and elucidation of the size evolutionary patterns (SEPs) of materials properties, from the atomic and molecular scale to the condensed phase regime [1]. Size evolutionary patterns associated with a cluster sequence (A_n, n > 1, where A denotes an elementary unit such as an atom or a molecule) and their approach to the bulk limit (i.e., large n) often depend on the chemical constituents in the cluster, on the state (e.g., thermodynamic or charge state), and on the property whose size-dependence we interrogate; i.e., for a given material, different properties, such as cluster geometries, thermo-dynamic stability, abundance distributions, dissociation and ionization energies, electronic and vibrational spectra, and chemical reactivity, may exhibit different SEPs as the size of the system is varied.

The variation with size of certain properties of materials aggre-gates are commonly found to scale with the surface to volume ratio of the cluster (i.e., $S/V \sim R^{-1} \sim N^{-1/3}$, where S, V, R and N are the sur-face, volume, radius and number of particles, respectively). However,

P. Jena et al. (eds.), Physics and Chemistry of Finite Systems: From Clusters to Crystals, Vol. I, 165–176.
© 1992 *Kluwer Academic Publishers.*

even when applicable, the physical origins of scaling with S/V may vary for different properties. For example, the reduced melting point of clusters compared to bulk, can be related to weaker binding and consequently enhanced vibrational amplitudes of surface atoms (alternatively expressed as a reduced surface Debye temperature) whose relative contribution scales as $N^{-1/3}$ [2], while the linear variation with $N^{-1/3}$ of the vertical (and adiabatic) ionization energies of excess electrons localized in polar molecular cluster (such as in $(H_2O)_n^-$ or $(NH_3)_m^-$, for sufficiently large n and m) is a consequence of the long-range nature of the interaction between the electron and the cluster molecules, as shown via simulations [3a] and a cluster dielectric model [3b], and observed experimentally [4]. Furthermore, characterization of SEPs in clusters in terms of S/V is non-universal, in the sense of properties whose size-evolutions are of different physical origins; e.g., magic number patterns characteristic to simple metal clusters (portrayed in abundance distribution spectra and ionization potentials) whose origins are related to electronic shell structure [5] rather than surface-to-volume considerations.

In this paper we discuss investigations of SEPs for several cluster systems, employing classical and quantum molecular dynamics simulations. First we discuss investigations of geometrical structural motifs in nickel clusters [6], predicting a size-dependent evolution from icosahedral and decahedral structures to the onset of single crystal forms at 17000 atoms, and relating atomistic and macroscopic results. Next, patterns in the energetics and dynamics of fragmentation of doubly charged sodium clusters (Na_n^{+2}, $n \leq 12$) are discussed, demonstrating shell-closing effects on fission barriers and fragmentation channels [7]. Finally, we discuss SEPs of energetics and structures in metallization sequences of ionic clusters [8] (i.e., Na_nF_n, Na_nF_{n-1}, Na_nF_{n-2}, ..., Na_n).

II. SIZE EVOLUTIONARY PATTERNS: CASE STUDIES

A. Structures of Nickel Clusters

Determination of the optimal geometrical structures of clusters and their SEPs are particularly challenging problems due to the finite-size of the system, lack of translational symmetry, and plethora of possible structures (and isomeric forms). One of the main issues in this area relates to determination of the size boundaries for energetic stability of various structural motifs, and the size-evolution of structures toward the bulk crystallographic form [7,9].

While for small clusters (typically $N \lesssim 50$) structural optimizations employing quantum mechanical calculations (of various degrees of accuracy) are possible [10], in atomistic structural investigations of large clusters one must resort to approximate descriptions of the interatomic interactions. Until recently atomistic structural studies

of clusters, in the medium and large size domains, were limited to rare
gas (RG) systems, employing Lennard-Jones (LJ) pair interactions (see
Ref. 9 and citations in Ref. 6).

As is well known, the nature of cohesion in metals requires
description of the energetics of the material beyond pair-wise interatomic interactions. In our studies of nickel clusters we have employed the many-body potentials obtained via the embedded atom method
(EAM), where the dominant contribution to the energy of the metal is
viewed as the energy to embed an atom in the local electron density
provided by the other atoms of the system, supplemented by short-range
effective two-body interactions due to interionic core-core repulsion
[11,6].

Another approach to the investigation of the structure of metal
particles is founded on macroscopic concepts rather than atomistic
structural optimization. This approach is based on the Wulff construction [12] which determines the equilibrium macroscopic morphology of
the crystalite, and on the theory of elasticity which allows estimates
of the elastic strain energy for a given structure [13]. Using this
approach the total energy in a cluster of N particles can be expressed
as [6]

$$E_\alpha(N) = (\varepsilon_B + \varepsilon_\sigma{}^\alpha)N + C_\alpha N^{2/3} , \qquad (1)$$

where α denotes the structural form (i.e., icosahedral, decahedral,
cuboctahedral, etc.), ε_B and $\varepsilon_\sigma{}^\alpha$ are the per-particle bulk cohesive and
strain energies, respectively, and C_α in the surface term is a constant
determined by the geometry of the cluster and the surface energies,
γ_{hkl}, of the facets which make up the surface of the cluster (ε_B, $\varepsilon_\sigma{}^\alpha$
and γ_{hkl} can be obtained either from experimental data when available,
or calculated from model potentials as we do in our study [6]). In this
context we note an important generalization of the Wulff construction
introduced by Marks [14] to include twin boundaries. Motivated by
experimental observations, this modified construction allows for
reentrant faces at the twin boundaries of the decahedron (see Fig. 4
in Ref. 6) thus decreasing its surface energy.

Since one of the aims of our study is to test and assess the
adequacy of macroscopic concepts and variants thereof, we show in Fig. 1
results for a set of selected structures obtained via atomistic structural optimizations (in Fig. (1a)) and macroscopic formulae [6] (in Fig.
(1b)), and employing the EAM potentials (for definitions of model C and
HKI, which are single crystal motifs related to truncated octahedra,
see Refs. 6,9). (Note that the energies are plotted with reference to
those of the corresponding cuboctahedral structures which were found to
be of higher energy, i.e., less stable, compared to any of the structures considered by us; a similar result was also found for the LJ
structures [9]). From the favorable agreement between the results
shown in Figs. 1a and 1b, we conclude that such macroscopic formulae,
in conjunction with accurately calculated or experimentally measured

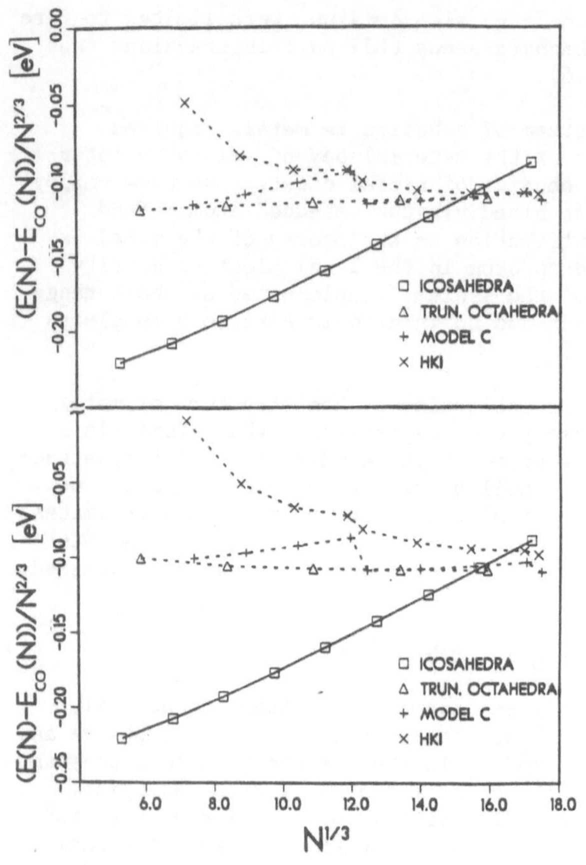

Figure 1: Top: energies obtained via atomistic structural optimization, plotted relative to the corresponding interpolated cuboctahedral energies.
Bottom: energies obtained by macroscopic formula (except for the icosahedra). N is the number of atoms in the cluster.

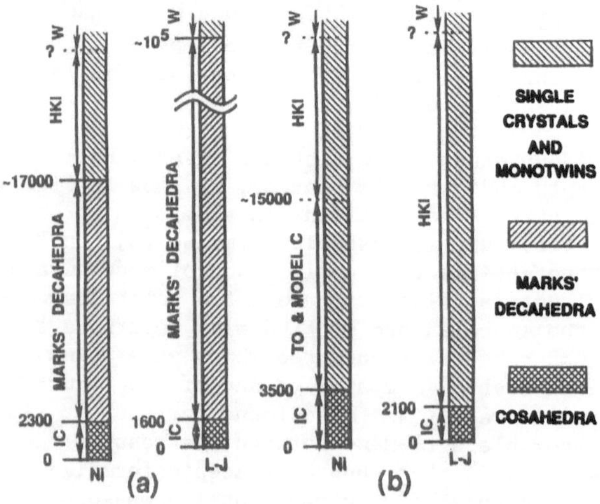

Figure 2: SEPs of optimal structures in nickel and LJ clusters. In (a) results including Marks' decahedra are shown, and in (b) without them. Solid lines divide structures belonging to different internal structural forms (icosahedra, IC; Marks' decahedra; and the single crystals and their monotwins, denoted by TO for truncated octahedra, model C, HKI or W for other Wulff-type structures). Dashed lines divide structures having the same internal form, but differing in their surface configuration (these are all single crystals and their monotwins).

cohesive, surface, and strain energies, provide a rather reliable framework for structural analysis at least in the size range N > 100.

The SEPs of optimal structures and the size-boundaries of energetic stability of various structural motifs for nickel, as well as Lennard-Jonseium (after Ref. 9), obtained via atomistic structural optimizations using the EAM potentials, are shown in Fig. 2 (where SEPs including Marks' decahedra (in a) and without them (in b) are shown). We predict that the optimal structures for clusters containing $\sim$ 2300 atoms belong to the icosahedral sequence. Larger clusters are predicted to assume Marks' decahedral geometries, while for clusters with $\sim$ 17000 atoms single crystal structures and their monotwins are predicted to be energetically favored. Note in particular, that the transition from Marks' decahedra, which are based on fivefold symmetrical non-single crystal structures, to the fcc single crystal form, is predicted to occur for the metal at a much smaller cluster size than that estimated [9] for rare gas (Lennard-Jonesium) clusters. Further details, including discussion about the energetics of low-symmetry clusters (i.e., incomplete atomic shells), curvature of the exposed facets, and thermal effects, may be found in Ref. 6.

B. Patterns and Barriers for Fission of Charged Small Metal Clsuters

Recent studies of metallic clusters [5] (particularly of simple metals) unveiled systematic energetic, stability, spectral [10,15] and fragmentation [16,17] trends. Moreover, several properties of atomic clusters (e.g., electronic shell structure [5], and most recently supershells [18], portrayed by the occurrence of magic numbers in the abundance spectra and ionization potentials; the influence of shape fluctuations analyzed within the jellium model [5]; giant spectral resonances interpreted as evidence for collective plasma oscillations [5,19]; and fragmentation, fission, patterns of ionized clusters [16,17]) bear close analogies to corresponding phenomena exhibited by atomic nuclei, suggesting an intriguing universality of the physical behavior of finite size aggregates, though governed by interactions of differing spatial and energy scales [20].

To investigate energetic patterns and dynamics of fission of small Na_n^{+2} ($n \leq 12$) clusters we have used our newly developed simulation method which combines classical molecular dynamics, or energy minimization, on the Born-Oppenheimer (BO) ground-state potential surface, with electronic structure calculations via the Kohn-Sham (KS) formulation of the local spin-density (LSD) functional method (for details of the BO-LSD method see Refs. 7,8). In dynamical simulations the ionic (classical) degrees of freedom evolve on the electronic BO surface which is calculated after each classical step. Minimum energy structures were obtained by a steepest-descent-like method, starting from configurations selected from finite temperature simulations. (Note that the existence of a (local) minimum-energy-configuration implies a barrier for fission.) Barrier heights and shapes were obtained by constrained energy minimization, with the center-of-mass distance, $R_{cm-cm'}$ between the fragments specified.

From the energetics of the clusters and of the various fragementation channels, given in Table I, we observe first that in all cases the energetically favored channel (see Δ) is $Na_n^{+2} \rightarrow Na_{n-3}^+ + Na_3^+$ ($n \leq 12$), i.e., asymmetric fission (except for $n = 6$), in contrast to results obtained from spherical jellium calculations where fragmentation via ejection of Na^+ is favored [21] (note the shell closing in the product Na_3^+ cluster). Secondly, the first vertical and adiabatic ionization potentials (vIP and aIP) exhibit an odd-even oscillation [5,22] in the number of particles, as well as shell closing effects for systems containing 8 electrons. Similar effects are seen for higher ionization energies though they are sometimes complicated by structural changes upon ionization [7]. Finally, our calculations show that for $n > 6$ fission involves energy barriers, in contrast to recent results [17b] obtained from adaptation of the liquid droplet model [23] to atomic clusters. The barriers for $n = 8$, 10 and 12 have been determined via constrained minimization to be: 0.16 eV, 0.71 eV and 0.29 eV for the energetically favored channel and larger barriers were found for the ejection of Na^+ from these clusters (0.43 eV and 1.03 eV for $n = 8$ and 10, respectively). The barriers for Na_{10}^{+2} are higher because of the closed-shell structure of this parent cluster.

Table I

Potential energies, E_p, for Na_n^{+2} ($4 \leq n \leq 12$), in the minimum energy configurations. Energies for systems fragmenting with no barrier, calculated for the minimum energy configuration of Na_n^+, are indicated by *. Dissociation energies, $\Delta_m = E(Na_{n-m}^+) + E(Na_m^+) - E(Na_n^{+2})$, for $4 \leq n \leq 12$. vIP and aIP, are vertical and adiabatic ionization energies for $Na_n \rightarrow Na_n^+$. Energies in eV.

/n	4	5	6	7	8	9	10	12
E_p	-10.50^*	-19.00^*	-23.46^*	-30.29	-36.95	-43.47	-50.36	-62.83
Δ_1	-2.48	0.10	-1.78	-1.07	-1.08	-0.74	-0.43	-0.26
Δ_2	-2.03	-0.24	-1.70	-1.22	-0.68	-0.82	-0.10	-0.23
Δ_3			-2.50	-1.59	-1.28	-0.87	-0.65	-0.94
Δ_4					-0.85	-0.67	0.10	-0.27
Δ_5							-0.13	-0.44
Δ_6								$+0.00$
vIP	4.41	4.33	4.67	4.24	4.62	3.83	4.11	3.93
aIP	4.38	4.11	4.40	4.07	4.35	3.70	4.04	

The potential energies along the reaction coordinates for the energetically favored channel and for Na^+ ejection, in the case of Na_{10}^{+2} are shown in Fig. 3. The most interesting feature seen from the

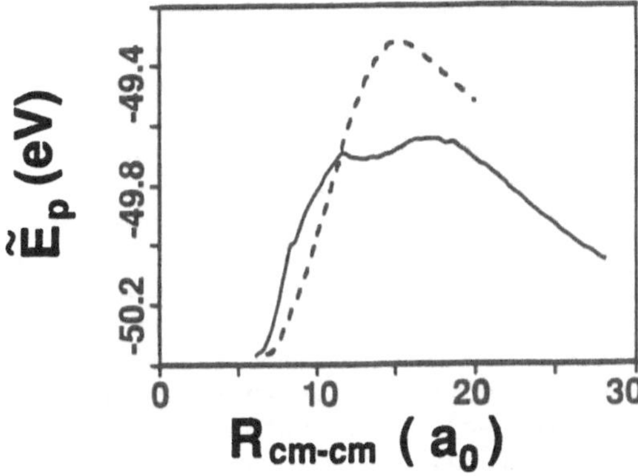

Figure 3: Potential energy vs distance between the center of mass for the fission $Na_{10}^{+2} \rightarrow Na_7^+ + Na_3^+$ (solid) and $Na_{10}^{+2} \rightarrow Na_9^+ + Na^+$ (dashed), obtained via constrained minimization of the LSD ground-state energy of the system.

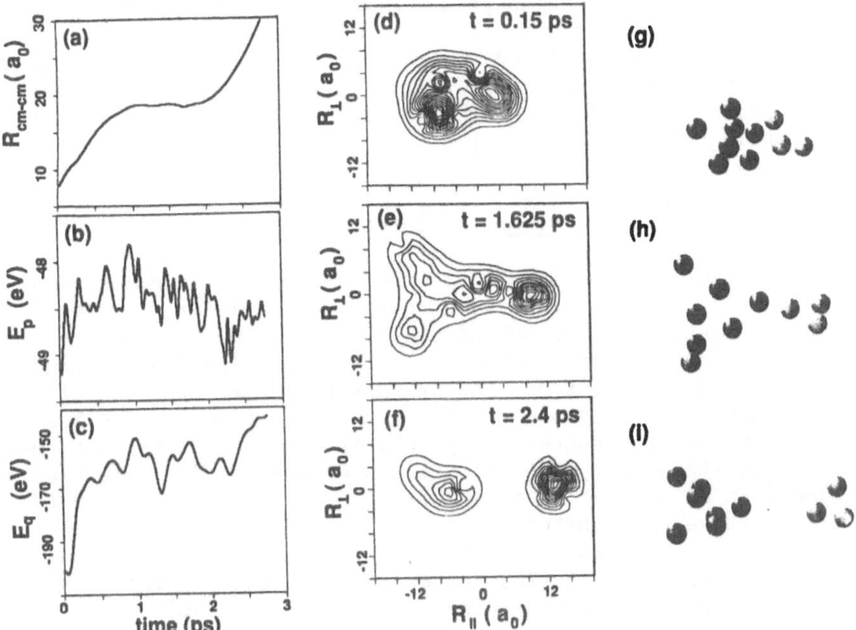

Figure 4: Fragmentation dynamics of Na_{10}^{+2}. (a-c) R_{cm-cm} between the fission products, total potential energy (E_p), and the electronic contribution (E_q), versus time. (d-f) Contours of the total electronic charge distribution, calculated in the plane containing the two centers of masses. The R axis is parallel to $\vec{R}_{cm-cm}$. (g-i) Configurations for the times given in (d-f). Dark and light spheres represent ions in the large and small fragments, respectively. Energy, distance, and time in units of eV, Bohr (a_0), and ps, respectively.

figure is the rather unusual shape of the barrier for the favored fission channel (also found for the asymmetric fragmentation of Na_{12}^{+2}). While double-hump barriers have been long discussed in the theory of nuclear fission [24], to our knowledge this is the first time that they have been calculated in the context of asymmetric fission of charged atomic clusters. The existence of the double-humped barrier is reflected in the dynamics of the fission process of Na_{10}^{+2} displayed in Fig. 4. This simulation started from a 600K Na_{10} cluster from which two electrons were removed (requiring 11.23 eV) and 0.77 eV was added to the classical ionic kinetic energy. The variation of the center of mass distance with time (Fig. 4a) exhibits a plateau for 750 fs $\leq$ t $\leq$ 2000 fs (see also the behavior of the electronic contribution to the potential energy of the system versus time in Fig. 4c). The contours of the electronic charge density of the system (Fig. 4(d-f)), at selected times, and the corresponding cluster configurations (Fig. 4(g-i)), reveal that the fission process involves a precursor state which undergoes a structural isomerization prior to the eventual separation of the Na_7^+ and Na_3^+ fission products. In this context we remark that examination of the contributions of individual Kohn-Sham orbitals to the total density for the intermediate stage (Fig. 4e) reveals that the lowest-energy orbital (s-like) is localized on the Na_7^+ fragment, the third is a p-like bonding orbital distributed over the two fragments, and the highest orbital is localized on the larger fragment.

C. Metallization Patterns of Ionic Clusters

Investigations of localizations modes, structure, dynamics and spectra of alkali-halide clusters containing multiple excess electrons (i.e., those electrons which substitute F^- anions in Na_nF_m, m < n) open new dimensions for studies of excess-electron localization and bonding in ionic clusters [8,25], since the process of "metallization" of an initially stoichiometric ionic cluster (i.e., starting from an ionic cluster with n = m, and successively substituting anions by electrons, ultimately resulting in a neutral metallic cluster Na_n) may portray a transition from an insulating to a metallic state in a finite system. In this context we note that while transitions from F-center to metallic behavior have been observed in bulk molten alkali halides at high excess-metal concentrations [26], experimental data on metal-rich alkali-halide clusters are preliminary in nature [25(f-h)].

Our investigations reveal systematic trends of the metallization process in small clusters, exhibiting a structural and energetic substitutional nature of excess electrons in halogen-anion-deficient alkali-halide clusters, structural transitions (which can involve a change in dimensionality) upon reaching the metallization limit, electronic structure odd-even and shell-closing effects portrayed in formation energies and ionization potentials, layer metallization (i.e., segregation of the excess metallic component on a face of the cluster) in an intermediate-size cluster ($Na_{14}F_9$), and indications of a transition to a metallic state for $Na_{14}F_m$ (for m $\leq$ 4).

In our BO-LSD simulations the interionic interactions are described by the Born and Huang parametrized potentials which we have tested in previous studies [25(c)], the interaction between the electrons and Na^+ is given by a norm-conserving pseudopotential, and a Coulomb repulsive potential describes the interaction between the electrons and F^- [8].

We begin with a metallization sequence (MS) of a small cluster, Na_4F_m ($0 \leq m \leq 4$), which for $m = 4$ is a stable ionic cuboid and at the other extreme ($m = 0$) is a stable planar rhombohedral Na_4 cluster [19] [see Figs. 5(a) - 5(e)]. The structures exhibited by this MS demonstrate a systematic trend: structures of neutral small clusters belonging to a MS are of the same dimensionality and symmetry of the corresponding parent cluster, converting to the structure of the corresponding metal cluster upon complete metallization ($m = 0$). Furthermore, the excess electrons in these clusters, substituting for halide anions, maintain the cohesion and structural integrity of the clusters and their distributions exhibit a delocalized, metallic character (with the regions occupied by the remaining anions, i.e., for $0 < m < n$, excluded).

The MS for a larger cluster $Na_{14}F_m$ ($0 \leq m \leq 14$), whose $Na_{14}F_{13}$ member is a particularly stable cluster [$Na_{14}F_{13}^+$ is a "magic number" ionic cluster [25a,c,d], i.e., a 3x3x3 filled cuboid structure, with three ions of alternating charges on an edge] exhibits another novel result: face (or atomic layer) metallization (segregation), as seen in Fig. 5(g). The optimal metallization sequence of this cluster proceeds via successive removal of neighboring halogens from one face of the cluster, resulting (for $Na_{14}F_9$) in a segregated metal layer.

Examination of the SEPs of metallization, and in particular the MS for $Na_{14}F_m$ (partly displayed in Fig. 6), reveals several trends:
(i) The F-center formation energies (E_f) (i.e., the difference between the total energies of Na_nF_{m-1} and Na_nF_m) are between 2.5 to 3.1 eV. It is of interest that within a MS the formation energies, as well as the adiabatic (and vertical) ionization potentials, exhibit odd-even oscillations in $n - m$ indicating that perhaps these excess-metal systems may be regarded [25(f)] as composed of a "metallic" component and a molecular-ionic one, symbolically represented as $Na_nF_m \equiv Na_{n-m}(NaF)_m$.
(ii) The first ionization potentials (vertical and adiabatic, vIP and aIP, respectively) of the clusters in each metallization sequence, $Na_{n-m}(NaF)_m$, are lower than the corresponding bare-metal (Na_{n-m}) values. For each of the MS's the IP's exhibit odd-even oscillations in $n-m$ and shell-closing effects (which are somewhat complicated for the larger cluster by structural rearrangement at various stages of metallization). Our results (structural and energetic) for the bare metal and mixed clusters are in good agreement with experimental data and previous calculations, when available [10].
A particularly interesting trend is exhibited by $\Delta vIP = vIP[Na_{n-m}] - vIP[Na_{n-m}(NaF)_m]$, plotted in Fig. 7 versus $n-m$ ($n = 14$ and $m=1,2,3,...$), which expresses the difference between the vertical ionization potential

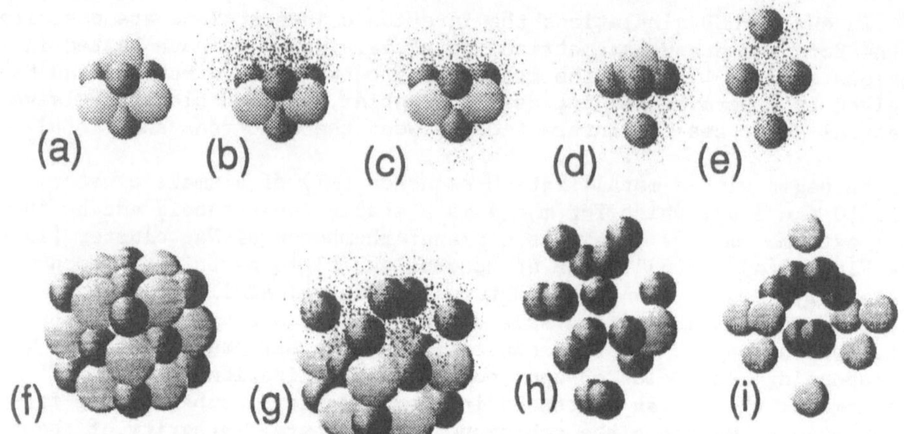

Figure 5: Optimal structures of Na_nF_m clusters. (a-e) MS for Na_4F_m $(0 \leq m \leq 4)$. (f-i) Structures of $Na_{14}F_m$, for m = 13, 9, 1 and 0, respectively. Large and small spheres denote F^- and Na^+. Small dots represent the total excess electronic distribution.

Figure 6: Energetics in the MS for $Na_{14}F_m$ vs. 14 - m, $(13 \geq m \geq 0)$. Formation, adiabatic, and ion-ion interaction energies in (a-c), respectively.

Figure 7: ΔvIP vs. 14 - m, $(13 \geq m \geq 0)$, for the MS of $Na_{14}F_m$. Note local minimum corresponding to $Na_{14}F_9$ and small values for m $\leq$ 4, indicating enhanced metallic character.

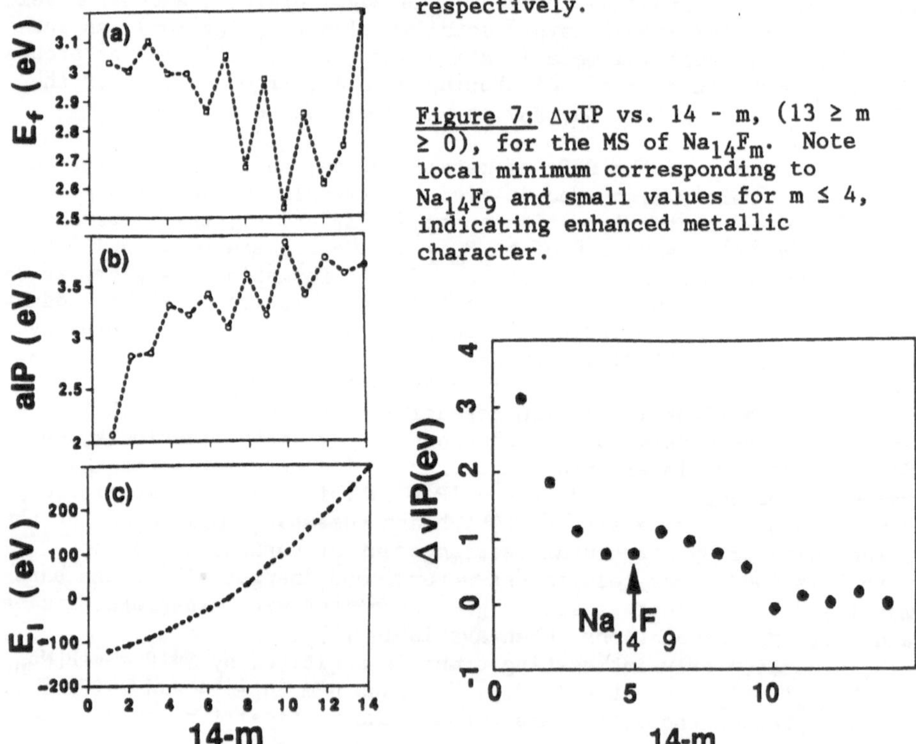

of a bare sodium cluster containing n-m atoms and that of the mixed $Na_nF_m \equiv Na_{n-m}(NaF)_m$ cluster. ΔvIP may be regarded as a measure of metallic behavior in the mixed cluster. As seen, for n-m = 5 (i.e., corresponding to $Na_{14}F_9$) ΔvIP attains a local minimum, suggesting enhanced metallic behavior for the face-segregated cluster. Moreover, for n-m $\geq$ 10 (i.e., m $\leq$ 4), the perturbing effect of the ionic component of the cluster is small, and the electronic properties of the clusters approach those of the corresponding bare metal ones.

(iii) Finally, we note that for the MS of $Na_{14}F_m$, a number of isomers are found, differing by the locations of the removed F^- anions. The results given in Figs. 5-7 correspond to the most stable ones, starting with $Na_{14}F_{13}$ where the excess electron forms a diffuse surface state localized on one of the faces of the cube [25(c)], proceeding by removing successively neighboring F^- anions from one face, yielding for m = 5 a metallized (segregated) layer [see the structure of $Na_{14}F_9$ in Fig. 5(g)], and ultimately resulting in an Na_{14} cluster.

ACKNOWLEDGEMENT: Research supported by U. S. Department of Energy, Grant No. FG05-86ER-45234. Calculations performed at the Florida State Computer Center through a grant by DOE.

REFERENCES

[1]   See articles in <u>Elemental and Molecular Clusters,</u> edited by G. Benedek and M. Pacchioni (Springer, Berlin, 1988); <u>Physics and Chemistry of Small Clusters</u>, edited by P. Jena, B. K. Rao and S. N. Khana (Plenum, New York, 1987).

[2]   P. R. Couchman and W. A. Jesser, Nature <u>269</u>, 481 (1977), Phil. Mag. <u>35</u>, 787 (1978); P. R. Couchman and C. L. Ryan, Phil. Mag. A<u>37</u>, 369 (1978), ibid. <u>40</u>, 637 (1979); P. R. Couchman and F. E. Karasz, Phys. Lett. A <u>62</u>, 59 (1977); R. Balian and C. Bloch, Ann. Phys. (N.Y.) <u>60</u>, 401 (1970).

[3]   (a) R. N. Barnett, U. Landman, C. L. Cleveland, and J. Jortner, J. Chem. Phys. <u>88</u>, 4429 (1988); (b) Chem. Phys. Lett. <u>145</u>, 382 (1988).

[4]   J. V. Coe et al., J. Chem. Phys. <u>92</u>, 3980 (1990).

[5]   See review by W. A. de Heer et al., Solid State Phys. <u>40</u>, 93 (1987), and references therein.

[6]   C. L. Cleveland and U. Landman, J. Chem. Phys. <u>94</u>, 7376 (1991).

[7]   R. N. Barnett, U. Landman and G. Rajagopal, Phys. Rev. Lett., in press (November, 1991).

[8]   G. Rajagopal, R. N. Barnett and U. Landman, Phys. Rev. Lett. <u>67</u>, 767 (1991), and Phys. Rev. (to be published).

[9]   B. Raoult, J. Farges, M. F. de Feraudy, and G. Torchet, Phil. Mag. B <u>60</u>, 881 (1989), and references therein.

[10] V. Bonacic-Koutecky, P. Fantucci, and J. Koutecky, Chem. Rev. (in press, 1991).

[11] S. M. Foils, M. I. Baskes, and M. S. Daw, Phys. Rev. B <u>33</u>, 7983 (1986); we use the EAM parametrization after J. B. Adams, S. M. Foiles, and W. G. Wolfer, J. Mater. Res. <u>4</u>, 102 (1989).

[12] G. Wulff, Z. Krist. 34, 449 (1901); C. Herring, Phys. Rev. 82 87 (1951).

[13] S. Ino, J. Phys. Soc. Jpn. 21, 346 (1966); 27, 941 (1969).

[14] L. D. Marks, J. Crys. Growth 61, 556 (1983); Surf. Sci. 150, 358 (1985); Phil. Mag. A 49, 81 (1984).

[15] V. Bonacic-Koutecky et al., Chem. Phys. Lett. 170, 26 (1990); and references therein.

[16] C. Brechignac et al., Phys. Rev. Lett. 63, 1368 (1989); ibid. 64, 2893 (1990).

[17] (a) W. A. Saunders, Phys. Rev. Lett. 64, 3046 (1990); (b) ibid. 66, 840 (1991).

[18] H. Golich et al., Phys. Rev. Lett. 65, 748 (1990).

[19] See refs. (6-13) in R. N. Barnett et al., J. Chem. Phys. 94, 608 (1991).

[20] See U. Landman et al., in Nuclear and Atomic Clusters 1991, Proceedings of a European Phys. Soc. Conf. Turku, Finland, edited by M. Brenner (Springer, Berlin, 1991).

[21] M. P. Iniguez et al., Phys. Rev. B 34, 2152 (1986).

[22] E. C. Honea et al., Chem. Phys. Lett. 171, 147 (1990).

[23] For studies of the liquid droplet model adapted to atomic clusters, see S. Sugano in Microclusters, edited by S. Sugano et al. (Springer, Berlin, 1987), p. 226; S. Sugano, A. Tawara, and Y. Ishii, Z. Phys. D 12, 213 (1989).

[24] For a discussion of shell corrections leading to double-hump nuclear fission barriers see M. A. Preston and R. K. Bhaduri, Structure of the Nucleus (Addison-Wesley, Reading, MA, 1975), p. 589. Shell effects on symmetric fragmentations of alkali-metal clusters have been studied using an approximate expression for the energy by M. Nakamura et al., Phys. Rev. A 42, 2267 (1990). We remark that examination of the variations in the KS orbital energies, calculated along the dynamical fission trajectory of the cluster, indicates a correlation between these variations (in particular level crossings) and the shape of the barrier which is thus related to the energetics of the deformation of the cluster (R. N. Barnett and U. Landman, unpublished).

[25] (a) U. Landman, D. Scharf, and J. Jortner, Phys. Rev. Lett. 54, 1860 (1995), (b) J. Chem. Phys. 87, 2716 (1987); (c) G. Rajagopal et al., Phys. Rev. Lett. 64, 2933 (1990); (d) E. C. Honea et al., Phys. Rev. Lett. 63, 394 (1989); (e) S. Pollack, C. R. C. Wang and M. M. Kappes, Chem. Phys. Lett. 175, 209 (1990); (f) S. Pollack et al., Z. Phys. D 12, 241 (1989); (g) T. Bergmann, H. Limberger, and T. P. Martin, Phys. Rev. Lett. 60, 1767 (1988); (h) E. C. Honea, Ph.D. thesis, University of California, Los Angeles, 1990 (unpublished).

[26] See review by W. W. Warren, Jr. in The Metallic and Non-Metallic States of Matter, edited by P. P. Edwards and C. N. Rao (Taylor and Francis, London, 1985); A. Selloni et al., Phys. Scr. T25, 261 (1989).

CHEMISTRY   OF   STABLE   QUASI-CRYSTALS

A. P. TSAI, Y.YOKOYAMA, A. INOUE and T. MASUMOTO
Institute for Materials Research,
Tohoku University, Sendai 980, Japan

ABSTRACT. Stable icosahedral(i-) and decagonal(D-) quasicrystals form in Al-(Cu,Ni,Pd)-transition metal(TM) ternary
and quarternary alloys.   Formation of the stable quasicrystals is closely related to the group of the TM elements
in the periodic table.   It stresses that the stability of
quasicrystal is mainly dominated by the electronic structure.   An empirical rule of the formation for the stable i-
quasicrystal is derived, that is, the composition of a
single stable i-phase lies in the valence concentration
(e/a) near 1.75. The universality of the empirical rule is
evidenced by the discovery of new stable i-phase.
A very strong composition dependence of phason strains
and chemical order is seen in rapidly quenched i-Al-Pd-Mn
alloys.   No difference in linewidths and dependence of
linewidths on phason momentum was observed in both as-
quenched and annealed states in a stable i-$Al_{70}Pd_{20}Mn_{10}$
alloy, indicating chemical composition plays an important
role in the phason strains in a quasi-lattice.   A part of
phase diagram including i-phase in Al-Pd-Mn system is determined.   The i-phase forms peritectically in a temperature
range of ~20 K which is much narrower than the other stable
i-phase.

1. INTRODUCTION

The discoveries of stable quasicrystal of Al-Cu-Fe
group[1,2] have emerged many exciting pictures in structure, stability and physical properties of quasicrystal. In
this group, icosahedral (i-) alloys have a highly ordered
structure described as a new class of i-lattice,i.e. a so
called face-centered icosahedral(FCI)[3] lattice.   In addition, they are stable up to peritectic temperature. These
alloys soon led a controversy in proposed structure among
the icosahedral-glass[4], the random-tiling[5] quasilattice
and the highly ordered[6] quasilattice.

P. Jena et al. (eds.), Physics and Chemistry of Finite Systems: From Clusters to Crystals, Vol. I, 177–187.
© 1992 *Kluwer Academic Publishers.*

The stable i-quasicrystal only appears at a critical composition range in some finite alloy systems. In a formation study, an empirical rule that the stable quasicrystals have a valency(e/a) of 1.75, was derived[2]. Consequently, there was realized experimentally and theoretically that a stable quasicrystal could be regard as a new class of Hume-Rothery phase stabilized by its electronic structure. The universality of the empirical rule was then identified by the discoveries of new stable quasicrystals in Al-Pd-TM(TM=transition metal)[7,8].

In this paper, we discuss the chemical effects on the formation of stable quasicrystals in Al-Cu-TM and Al-Pd-TM systems. Then, we focus on the Al-Pd-Mn alloy. A very strong chemical order in a face-centered icosahedral(FCI) $Al_{70}Pd_{20}Mn_{10}$ was found[9], which provided the contributions on stability and perfection of i-lattice. Finally, a part of phase diagram consisting of i-phases will be presented.

2. EXPERIMENTAL

Ternary and quarternary Al-Pd-TM alloy compositions of interest were melted in an argon atmosphere using an arc furnace. The slowly cooled samples were annealed in a sealed quartz tube. Rapidly solidified samples were prepared by using a single roller melt spinning apparatus. The samples were carefully powdered for X-ray diffraction measurement. The instrumental resolution was determined to be about 2.5×10^{-3} A^{-1} from Si powder. The microstructures of the quasicrystals were observed by using transmission electron microscopic(TEM) technique. Thermal stability was examined by differential thermal analyzer(DTA).

3. EMPIRICAL RULES FOR STABLE QUASICRYSTALS

We have previously reported stable i-$Al_{65}Cu_{20}TM_{15}$ (TM=Fe,Ru,Os) and stable decagonal(D-) $Al_{65}Cu_{15}TM_{20}$ (TM=Co,Rh,Ir) phase in Al-Cu-TM[2] system. By using a phenomenological model for transition metals developed by Pauling and assuming that the charge transfer from the conduction band into the d-band occurs compensate for unpaired spin of TM[10,11]. It was theoretically predicted [12] that quasicrystal can be regarded as a Hume-Rothery. With this viewpoint, we have derived the empirical rules; (1) the valency(e/a) of these stable quasicrystals were estimated to approximate to 1.75, (2) Al content was always kept in the vicinity of 65~70 at%[13]. Based on the rules, we succeed in finding new quasicrystals in Al-Pd-TM system. Figure 1 summaries the formation of Al based ternary stable quasicrystal with TM in periodic table. It is noticed some similar features in the periodic table for stable quasicrystals in Al-Cu-TM and Al-Pd-TM systems and

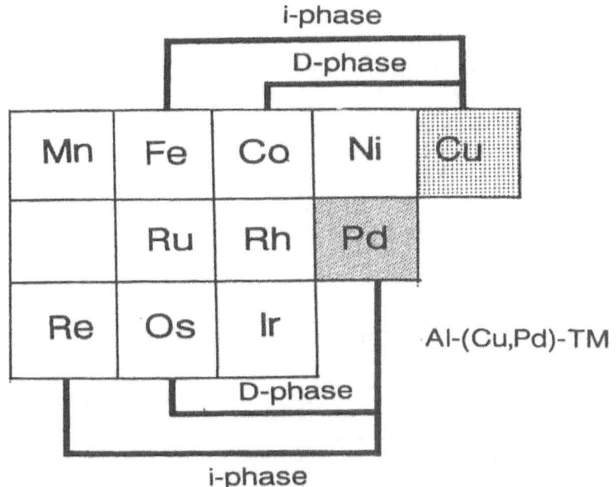

Fig.1   Formation of Al-based ternary stable quasicrystals
with transition metals(TM) in periodic table.

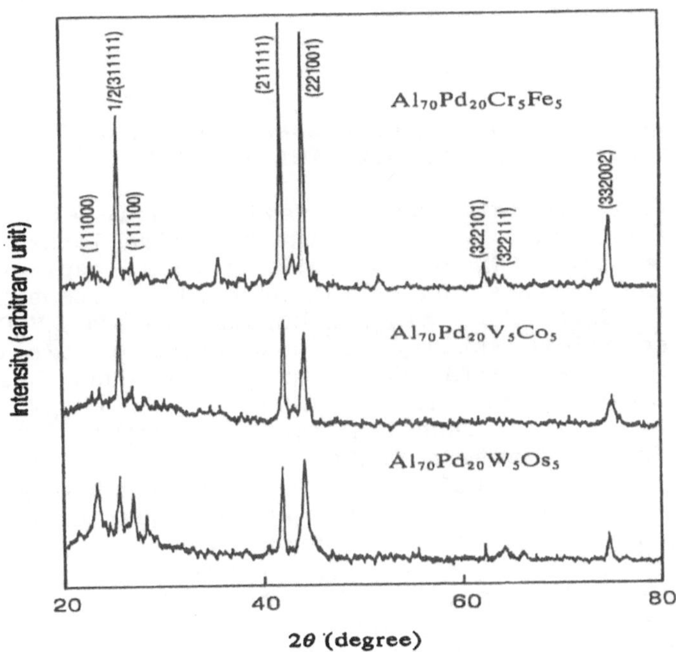

Fig.2   X-ray diffraction patterns of Al70Pd20Cr5Fe5,
$Al_{70}Pd_{20}V_5Co_5$ and $Al_{70}Pd_{20}W_5Os_5$ alloys annealed for
1173Kx24h after slow cooling.

described in the following. (1) TM elements locate at 3 column to the left of either Cu or Pd. (2) Pd locates at the left and one low down with respect to Cu. (3) TM in icosahedral phase locate one column left to that in decagonal phase. The systematic change in TM just satisfies the two criteria in the empirical rule i.e. e/a=1.75 and Al=65~70 at%. We should note that the stable quasicrystals commonly consist of three kinds of atoms; Al with 3p-valency, Pd or Cu with d-band-filled transition metal and transition metals like Mn, Fe or Co with d-valency.

Al-Pd-TM1-TM2 quarternary

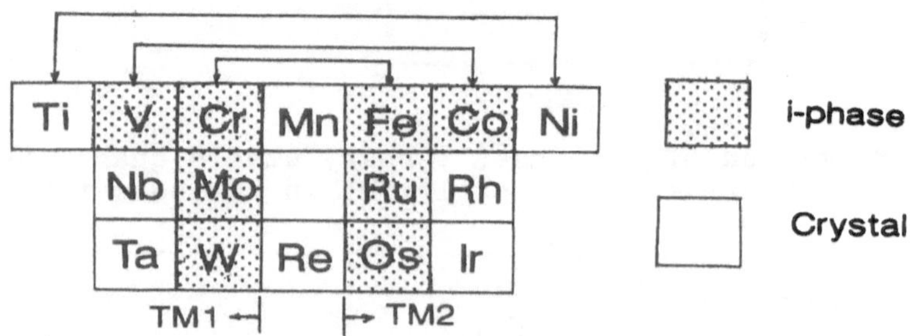

Fig 3   Change in the annealed phases in $Al_{70}Pd_{20}TM1_5TM2_5$ quarternary alloys with TM1 and TM2.

The empirical rule was also evidenced in Al-Pd based quarternary system.   we select the TM pairs located at the two opposite side of Mn or Re, both of them   replace the Mn or Re in half.   Hence, we obtain several quartenary alloys such as $Al_{70}Pd_{20}Cr_5Fe_5$, $Al_{70}Pd_{20}Mo_5Ru_5$, $Al_{70}Pd_{20}V_5Co_5$ etc. All of these alloys satisfy the conditions of e/a=1.75 and atomic ratio of Al:Pd:TM=7:2:1.   Figure 2 shows X-ray diffraction patterns of $Al_{70}Pd_{20}Cr_5Fe_5$, $Al_{70}Pd_{20}V_5Co_5$ and $Al_{70}Pd_{20}W_5Os_5$ alloys annealed for 1173Kx24hrs after slow cooling. The diffraction peaks   exhibit a FCI lattice as well as $Al_{70}Pd_{20}Mn_{10}$ but   visible peak broadening for the three alloys.   As expected, new stable quarternary stable i-quasicrystals were found and summarized in Fig.3. If the e/a deviates from the 1.75, the crystalline phases will appear as shown in Fig. 4.   Note that although the e/a of $Al_{70}Pd_{20}Mo_3Ru_7$(1.78) and $Al_{70}Pd_{20}Mo_7Ru_3$ (1.69) have little deviation from 1.75, a large amount of stable i-phase still can be identified in the alloys. This stresses that an i-phase is a really stable one. Although the e/a deviates slightly from the ideal one, it is still expected to obtain a local atomic arrangement satisfied the above-mentioned two criteria for constructing an i-lattice.

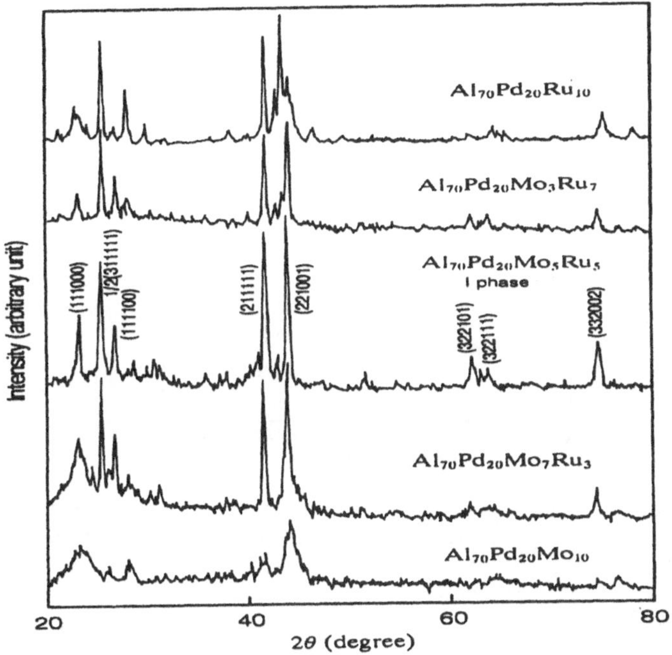

Fig.4 Change in X-ray diffraction patterns of $Al_{70}Pd_{20}Mo_{10-x}$ Ru_x (x=0,3,5,7 and 10 at%) alloys with Ru content.

4. CHEMICAL ORDER IN ICOSAHEDRAL Al-Pd-Mn ALLOYS

We have observed a very strong composition dependence of line widths[half width at half maximum(HWHM)] and chemical order in i-Al-Pd-Mn alloys. Figure 5 summarizes the HWHM's of the two alloy series of $Al_{90-x}Pd_xMn_{10}$ and $Al_{70}Pd_xMn_{30-x}$ in the as quenched state. The $Al_{70}Pd_{20}Mn_{10}$ sample has very small HWMH of 2.5×10^{-3} A^{-1}, comparable to that of well-annealed Al-Cu-Fe,Ru. In the first alloy series, the value of the HWMH increase by an order of magnitude and reaches ~2.5×10^{-2} A^{-1} for $Al_{80}Pd_{10}Mn_{10}$, while it increases only by a factor of 2-3 in the second alloy series. We note that first alloy series that the superlattice peak is much broader than the fundamental peak. The extremely strong dependence of the structure and peak width on Al content may suggest that the occurrence of superlattice ordering of SI lattice and suppression of phason disorder mainly result from the Al and Pd or Mn atomic order[14]. In Figs 6(a) and 6(b), we show for i-$Al_{70}Pd_{20}Mn_{10}$ [as-quenched(open) and annealed (solid)] the dependence of the widths on $G_\parallel$ and $G_\perp$, respectively. The figures clearly show that the peak width remains at a nearly constant value of 2.5×10^{-3} A^{-1} and is independent of both $G_\parallel$ and $G_\perp$. We do not see the linear increase of peak width reported in as quenched Al-Cu-Fe.

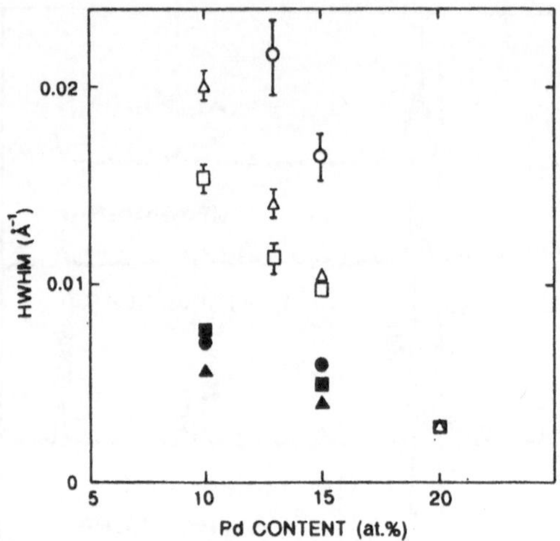

Fig.5 Pd content dependence of linewidths(HWHM) of as-quenched i-Al$_{90-x}$Pd$_x$Mn$_{10}$(open symbols) and i-Al$_{70}$Pd$_x$Mn$_{30-x}$(solid symbols) alloys. Circles, triangles and squares represent 1/2(311111),(111000) and (111100) peaks, respectively.

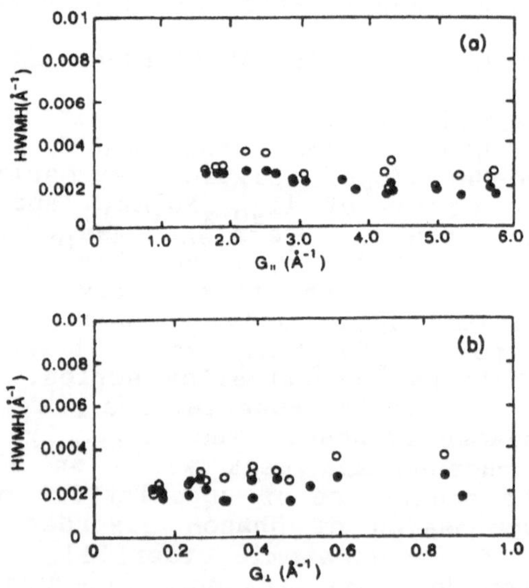

Fig.6 Dependence of linewidths(HWHM) on G$_{\parallel}$ and G$_{\perp}$ for as-quenched (open circles) and annealed(solid circles) i-Al$_{70}$Pd$_{20}$Mn$_{10}$.

We observed that the fully annealed alloy appeared to have a smaller peak width than the quenched samples. The small but finite peak width may be due to the finite domain size of ~2000 A, as reported by Calvayrac et al.[15] for an annealed i-$Al_{65}Cu_{20}Fe_{15}$ sample. Two-fold diffraction patterns of i-$Al_{70}Pd_{20}Mn_{10}$ prepared by melt-quenching(a), arc melting(b) and Bridgmann process(c) are shown in Fig. 7. No significant shift and broadening in diffraction peaks were seen, in fair agreement with X-ray data. These results, that the peak width of the i-$Al_{70}Pd_{20}Mn_{10}$ occurs only when Al replaces Pd and that there is no peak width with the replacement of Pd by Mn, suggest that long-range chemical ordering occurs between Al and Pd,Mn. In an ideal FCI i-$Al_{70}Pd_{20}Mn_{10}$, constituent atoms occupy finite sites, which

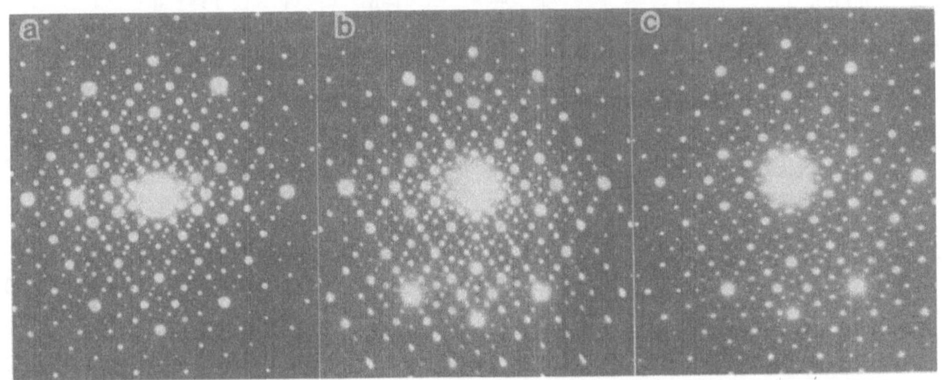

Fig.7   Electron diffraction patterns along two-fold
        direction of i-$Al_{70}Pd_{20}Mn_{10}$ prepared by melt
        quenching(a), arc melting(b) and Bridgmann(c)
        processes.

leads to a finite atomic ratio of Al,Pd and Mn and to an ordered i-structure. If Pd sites occupied by a small amount of Al atoms, the local bonding as well as icosahedral symmetry would be disturbed. In order to keep the whole system in icosahedral symmetry, phason strains would be introduced. Hence, we realize that the i-phase is stabilized by finite atomic decoration in a cluster, e.g., a Mackay icosahedron or space-filling tiles.

5. PERITECTIC GROWTH AND MORPHOLOGY OF STABLE ICOSAHEDRAL ALLOYS

The phase diagram including i-phase was firstly examined in Al-Li-Cu system by Chen et al.[16]   The i-phase grows

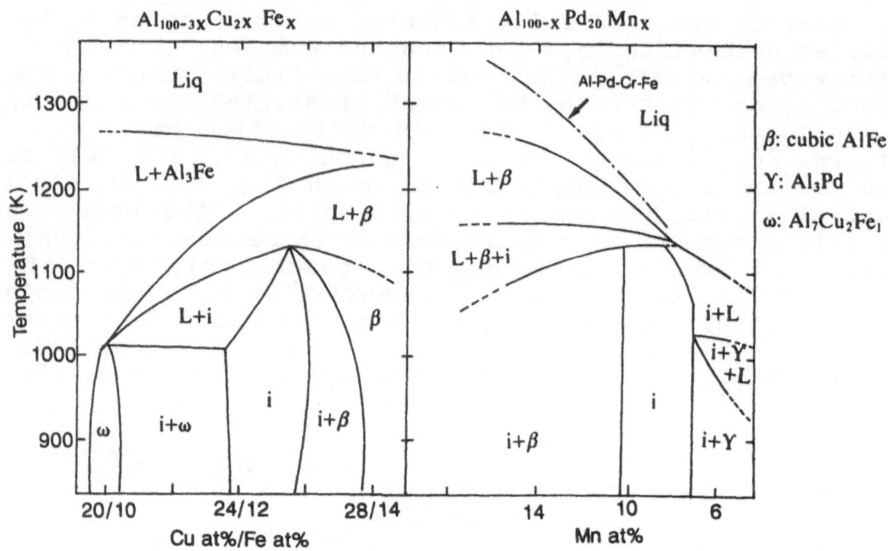

Fig. 8 Partial phase diagrams of Al$_{100-3x}$Cu$_{2x}$Fe$_x$(a)[ref 18] and Al$_{100-x}$Pd$_{20}$Mn$_x$(b). Broken line in (b) is the liquidus line of Al$_{70}$Pd$_{20}$Cr$_5$Fe$_5$.

through a peritectic reaction. In a thermal equilibrium reaction, an i-Al$_{5.1}$Li$_3$Cu forms peritectically from liquid and cubic R-phase. Later, the phase diagram of AlCuFe system was examined and indicated the i-AlCuFe formed through a peritectic solidification[17,18]. Figure 8 illustrates the partial composition-temperature section of i-AlCuFe and i-AlPdMn close to the ideal composition, determined from the DTA curves measured at a cooling rate of 5 K/min and X-ray diffraction technique. Both alloys exhibits the complicated peritectic reaction. The compositional range of stable i-phase determined from fully annealed state in Al-Cu-Fe and Al-Pd-Mn alloys are shown in Fig. 9. The temperature range($\Delta T=T_1-T_i$) between liquidus(T_1) and the formation of the i-phase(T_1) is larger for AlCuFe than for AlPdMn[19] in Fig.8, consequently, a wider formation region of stable i-phase was observed in AlPdMn in Fig. 9. We note in both cases the β phase acts as a nucleation site in the formation of stable i-phase. Up to date, there is no exception for stable i-phase which solidified through a peritectic reaction.

The morphologies in a SEM microgragh of slowly cooled i-Al$_{70}$Pd$_{20}$Mn$_{10}$ and i-Al$_{70}$Pd$_{20}$Fe$_5$Co$_5$ are shown in Fig. 10. As shown in Fig. 8, ΔT is only about ~20 K and a supercooled into i-phase region during solidification is very easy to be achieved. While the liquidus of iAlPdCrFe(the broken line in Fig.8) is higher than that of iAlPdMn by about ~50K,

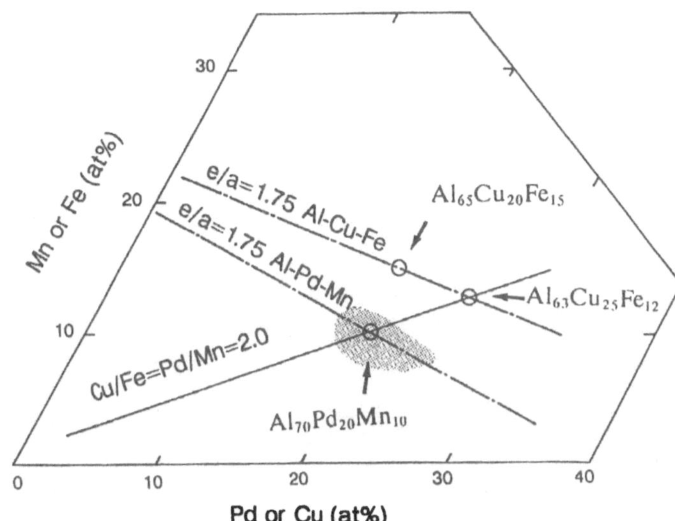

Fig. 9   Compositional range of stable i-phase in Al-Pd-Mn
          and Al-Cu-Fe alloys.

consequently, the slope of the liquidus is much larger for
iAlPdCrFe that for iAlPdMn. In a classical solidification
theory, the solidification morphology of crystalline grain
is determined by the stability of the solid-melt interface.
When the interface changes from stable to unstable, the
morphology of crystalline grains changes from equiaxial
grain with a planar interface to dendrite grains on the
basis of the constitutional undercooling criterion[20]. In
the criterion, it was described that the larger slope the
more unstable of the interface. Thus, we see equiaxial-
like grains characterized the planar growth with a size of
about 200 μm in iAlPdMn and dendritic grain sized in few μm
owing to interface instability in iAlPdCrFe. The iAlPdCrFe
exhibits the longitude arms along three fold axis, indicating that the i-grains grow dedritically along three fold
direction; While a flat five-fold plane is observed very
often in iAlPdMn suggests i-grains creating through a planar
growth. This proposed that a 5-fold plane is equilibrium in
i-phase, that is, the i-phase originally grows planarly
along 5-fold direction and changes to dendritic growth along
3-fold direction when the interface instability occurs. The
same tendency was observed in melt-quenched iAlPdCr[21].

ACKNOWLEDGEMENTS

We thank Dr. H.S.Chen for stimulating discussions. A
part of this work was supported by Grant-in-Aid for Scientific Research on Priority Areas "Crystal Growth Mechanism

in Atomic Scale".

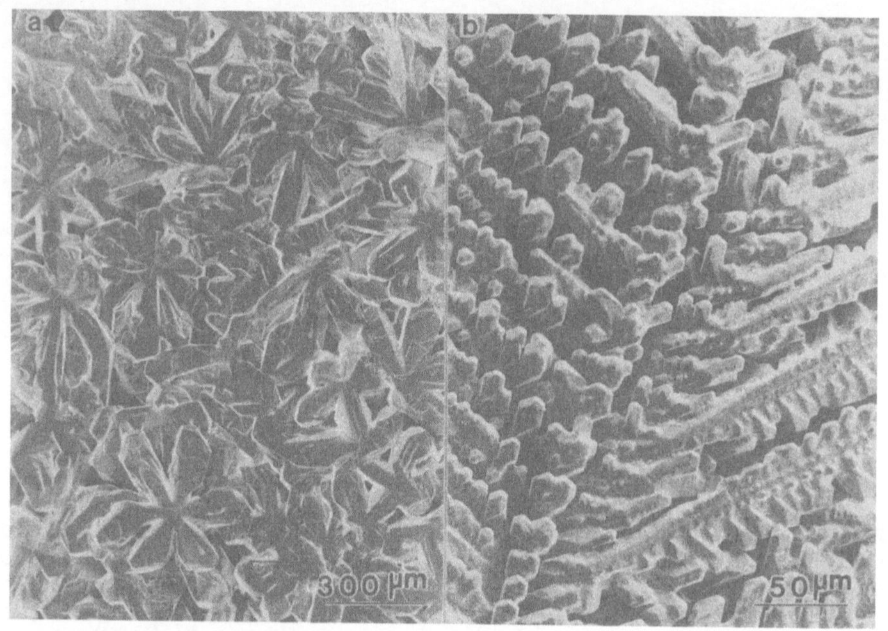

Fig. 10   SEM micrograph of the morphology for the slowly
cooled i-$Al_{70}Pd_{20}Mn_{10}$(a) and i-$Al_{70}Pd_{20}Fe_5Cr_5$(b)
alloys.

REFERENCES

1. A.P. Tsai, A. Inoue and T. Masumoto, Jpn.J.Appl.Phys.26,
   L1505(1987).
2. A.P. Tsai, A. Inoue and T. Masumoto, Mater.Trans.JIM 30,
   666(1989).
3. S.Ebalard and F.Spaepen, J.Mater.Res.4,39(1989).
4. P.W. Stephens and A.I.Goldman, Phys.Rev.Lett. 56,1168
   (1986).
5. M. Widom, K. Strandburg and R.H. Swendsen, Phys.Rev.Lett.
   58,706(1987).
6. D. Levine and P.J. Steinhardt, Phys.Rev.Lett. 53,2477
   (1984).
7. A.P. Tsai, Y. Yokoyama, A.Inoue and T. Masumoto,
   Jpn.J.Appl.Phys.29,L11(1990).
8. A.P. Tsai, A.Inoue, Y.Yokoyama and T. Masumoto, Mater.
   Trans.JIM 31,98(1990).
9. A.P. Tsai, A. Inoue and T. Masumoto, Philo.Mag.Lett.62,

95(1990).
10. L. Pauling, Phys.Rev.54,899(1938).
11. G.V. Raynor, Progress in Metal Physics 1531(1949).
12. J.Friedel, Helv. Phys. Acta.61,538(1988).
13. A.P. Tsai, A. Inoue and T. Masumoto,Sci.Rep..Res.Inst. Tohoku Univ.A36, 99(1991).
14. A.P. Tsai, H.S. Chen, A. Inoue and T. Masumoto, Phys. Rev.B.43, 8782(1991).
15. Y. Calvayrac, A. Quivy, M. Bessiere, S.Lalebrre, M. Cornier-Quiquandon and D. Gratias, J. Phys(paris)51,417 (1990).
16. H.S. Chen, A.R. Kortan and J.M. Parsey,Jr, Phys.Rev. B.36, 7681(1987).
17. C. Dong, M.De Boissieu and J.M. Dubois, J.Mater.Sci. Lett. 8.827(1989).
18. F. Faudot, A.Quivy, Y. Calvayrac, D.Gratias and Havme lin, Mater.Sci.Eng. A133,383(1991).
19. Y.Yokoyama, A.P.Tsai. A.Inoue and T.Masumoto, Mater. Trans. JIM to be published (1991).
20. M.C. Flemings, Solidification Process, McGraw-Hill, New York(1974)p273.
21. A.P. Tsai, H.S.Chen, A. Inoue and T Masumoto, Jpn.J. Appl.Phys.30,L1132(1991).

STRUCTURE AND EVOLUTION OF NANO- AND MICROCRYSTALLINE GRAIN AGGREGATES

N. RIVIER
Blackett Laboratory
Imperial College
London SW7 2BZ
Great Britain

ABSTRACT. Polycrystalline grain aggregates are cellular networks filling space at random, like soap froths. Their structure (in statistical equilibrium) and evolution (steady state coarsening) are universal, due to local elementary topological transformations (ETT), which are the "collisions" responsible for statistical equilibrium. The structure and its evolution can be represented as a many-body problem with short-ranged interactions. The bodies are paraboloids attached to the grains, with one additional degree of freedom beside their position in space. ETT are caused by simple and orthogonal motions of the bodies. The model describes quantitatively the sintering of polycrystalline mosaics, scaling, steady state distributions of grain sizes and shapes, with mean-field growth exponent. Individual grains obey von Neumann's law. The difference in coarsening rates between nano- and microcrystalline mosaics is related to the grain size distribution. The structure (Laguerre froth) has remarkable symmetries, stereology (it is identical to its own cut, and can be cut or lifted to higher dimension while preserving its geometry and statistics), and conformal invariance in any dimension and at all times.

1. Introduction

Materials grown from the melt, made by deposition, or by compaction and consolidation of individual clusters, form a polycrystalline mosaic which looks, after sintering, like an ideal geometrical **froth**: a random cellular network with cells (C), interfaces (edges E in 2D) and vertices (V) as constituents. Soap froths have the same structure, but very different grain (whether micro- or nanocrystalline), cell or bubble sizes [1].

Evolution of metallurgical and soap froths are also identical[1] [1,2]: Large grains (bubbles) grow at the expense of smaller ones, and the structure coarsens, asymptotically in a steady-state, scaling fashion: The distribution in linear size L of the grains,

$$P(L/\langle L \rangle) \, d(L/\langle L \rangle) \text{ is invariant in time t,}$$
$$\text{with } \langle L \rangle \sim \sqrt{t}. \tag{1}$$

---

[1] This is a conjecture by C.S. Smith [2]. The evolution has been studied experimentally in two-dimensional soap froths by Smith and Aboav, and by Glazier and Stavans [3], and in various computer simulations by Srolovitz *et al.* [4], Telley [5], Udler and Kawasaki. See Telley's thesis [5] for a review of simulations, and Glazier's [3] for experiments, up to 1989. The structure and evolution of 3D metallurgical mosaics is only inferred from observations on 2D cuts, or indirectly, from mechanical and thermodynamic properties.

P. Jena et al. (eds.), Physics and Chemistry of Finite Systems: From Clusters to Crystals, Vol. I, 189–198.
© 1992 *Kluwer Academic Publishers.*

(Ostwald ripening, with mean field exponent 1/2 [6]). Typically, P rises as a power of L, to decay as $\exp(-kL^D)$ at large L (D is the dimensionality of the structure). Nanocrystals, at low annealing temperature, coarsen much more slowly than ordinary, microcrystalline grain mosaics [7], if at all.[2] The distribution of grain shapes is also invariant.

The physical reason for this similar evolution is that the driving energy is concentrated on the interfaces between grains or bubbles (surface tension of the (double) liquid film in soap froths, grain boundary energy (independent of orientation for general grain boundaries) in sintered metallurgical aggregates). The driving force is the pressure difference between two neighboring bubbles given by Laplace's law in soap, and the chemical potential (Gibbs free energy per unit volume) difference between two neighboring grains given by the identical law of Gibbs-Thomson's in grain aggregates:

$$\Delta p = 4\sigma/R \qquad \text{(soap)}$$
$$\Delta\mu = 2\sigma/R \qquad \text{(metallurgy)} \qquad (2)$$

Here σ is the interface tension, and R, the radius of curvature of the interface between the cells. There is therefore considerable energy in the froth, carried by interfaces. The evolution is curvature-driven (by (2)), but it is a very slow process involving diffusion, because vertices and interfaces are locally in equilibrium at all times.[3]

These are the results which we shall explain and model in this paper.

The structure as a whole, and its evolution are related to the microscopic dynamics (2), but only insofar as the structure and dynamics of a many-body system reflect the pair potential. Equation (2) only explains the similarity between soap and grain mosaics. It implies neither their structure, nor their evolution.[4]

2. Elementary Topological Transformations

It is clear to anybody who has washed dishes that the evolution of the froth will be regulated by topological transformations: an interface shrinking to a point changes the topological character of the neighboring cells (Fig.1).

The key actors are **elementary topological transformations** (ETT), i.e. **local** topological fluctuations. (We now concentrate on 2D froths. The model introduced in §3 is valid in any dimension, where all the main concepts can easily be translated (*mutatis mutandis*)).

There are only two types of elementary topological transformations in 2D [1]: neighbor switching (T1) and cell disappearance (T2) (Fig.1).[5]

---

[2] We shall explain (in §3.5) the stability of nanograins by the fact that the grain sizes are more uniform in nano- than in microcrystals. Recent experiments can now rule out impurity or void pinning the (relatively) thicker grain boundaries in nanocrystals as the reason for their slow coarsening [7,8].

[3] The curvature driving the evolution is that of the interface (in (2)), or the difference in curvatures (=1/size) of neighboring bubbles [9]. Vertices are in equilibrium if interfaces meet at 120°. This implies that a hexagon will be in equilibrium, a pentagon shrinks, a heptagon grows, and an octagon grows faster, etc., as described by von Neumann's law (3) [10].

[4] For example, eq.(2) is only one of the ingredients necessary to prove von Neumann's law (3), which describes the rate of evolution of a given grain. The other is equilibrium at vertices of the froth.

[5] Mitosis (cell division, the topological agent of growth in biological tissues) is the inverse $(T2)^{-1}$, possibly composed with a few T1's.

Fig.1. Elementary topological transformations in 2D: T1 (neighbor switching (left) and T2 (cell disappearance, right)

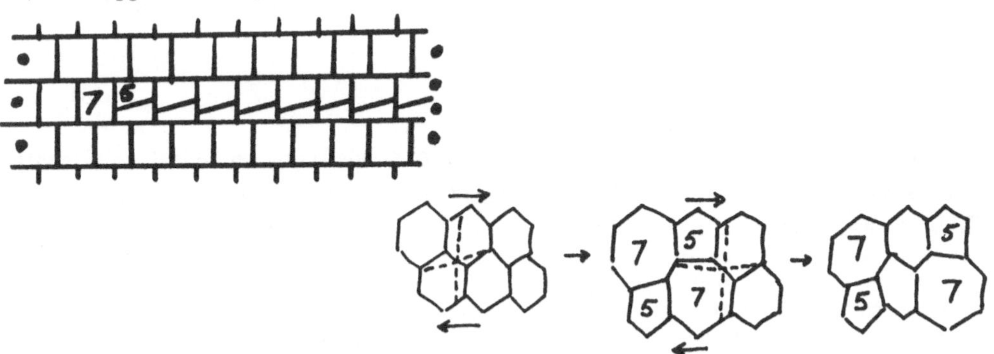

Fig.2. Topological dislocation (pair of 5- and 7-sided cell).
Fig.3. Creation and glide of a pair of topological dislocations under shear.

Note that T1 has a critical point with vertex coordination $z=4$, which is neither topologically nor dynamically stable, and occurs with negligible probability in a random froth which has $z=3$. Any fluctuation splits the critical vertex into two $z=3$ vertices, which are not in equilibrium and pull apart rapidly (stability requires $120°$ between interfaces since $\cos 60° =1/2$). Because its critical point is unstable, T1 is a rapid transformation. It is settled at large times. By contrast, <u>T2 is the main agent of slow coarsening</u>.

Elementary topological transformations act on the structure of the froth,

1) They change the topological character (number of sides n) of the cells involved, thereby creating local topological defects.

2) They make defects move (climb, glide) in response to external forces (applied shear, evolution in biological tissues, etc.).

3) They act as "collisions" to establish randomness, statistical equilibrium, and make the froth as the most probable (maximum entropy) archetype of a statistical ensemble (characterized by "random avoidance of the niceties of adjustment", i.e. of critical points) [1,12].

4) As we have stressed, T2 is the agent of slow coarsening.

with the following consequences: (i) Cells are hexagonal on average, $<n>=6$, for a large, $z=3$ froth ($C>>1$). $(n-6)$ measure the strength of the disclination carried by a n-sided cell.[6]

---

[6] A hexagonal froth will be in (indifferent) equilibrium. When fluctuations are large enough to shrink an edge and cause a neighbour switch, a dipole of dislocations (pentagon-heptagon) is formed (Fig.3). The two dislocations in the dipole can glide apart through the application of a local shear, but it is unlikely that a systematic separation of dislocations would occur, because shear stresses are screened by the randomness of the structure [11] (Examples of this screening are the thin amorphous layer protecting mechanically films of

192

(ii) Valencies relations, $<n>C = 2E = 3V$, imply that cells are the least numerous topological elements, suitable as independent degrees of freedom.

(iii) A dipole pentagon-heptagon is a (topological) dislocation (Fig.2). Topological dislocations respond to external forces like conventional dislocations in metals [12], but much more slowly [8], and locally [12]. For example, Fig.3 shows how shear creates a pair of topological dislocations, and make them glide apart. Note that a topological dislocation involves several grains ("grains talk to each other" [8]). In nanocrystals, it is the only mobile defect, smaller than material dislocations which traverse the nanograins and are pinned on the grain boundaries. It is seen in the strain rate sensitivity exponent discussed by R. Siegel [8].

(iv) The structure evolves slowly while remaining in statistical equilibrium. In 2D, it coarsens as

$$dA_n/dt = (2\pi\delta\sigma/3) (n-6) \tag{3}$$

(Von Neumann's law). Here, A_n is the average area of n-sided cells, and δ is the diffusivity of the atoms responding to Δp or $\Delta\mu$ in (2). This law can be derived [10] directly from the Gibbs-Thomson equation and equilibrium at vertices (120°). It is also a consequence of the structural equation of state (the structure remains in statistical equilibrium at all times). We will see it as a result of many-body dynamics.

(v) ETT establish the froth as the archetype of a statistical ensemble containing soap, grains, biological tissues, cracked mud, etc. [1], at all stages of evolution (if any).

3   The froth as a many-body problem

We aim to simulate and compute the structure of the froth and its evolution by some "molecular dynamics" of discrete, identical bodies. (If the froth is a territorial partition between foxes, then our aim is to study the population dynamics of the gas of foxes). A discrete description of the froth with local forces is indeed possible. It relies on a simple representation by H. Telley in 1989, of a geometrical partition of space (Laguerre or radical froth) invented early in the 19th. century [13], but forgotten until the beginning of the 1980, when it was uncovered by crystallographers [14,15] and computer programmers [16,5], apparently independently.

Instead of "how many bodies to have a problem?" [17], we want to know the number of degrees of freedom ($(D+1)C$, for C cells covering D-dimensional space), what are the bodies (the cells), and what are the potentials (defined by neighborhood rather than distance since the energy is carried by interfaces). We shall proceed in three steps: (i) represent a topological froth (§3.1-2), (ii) capable of supporting ETT of both types §3.3). And only then, (iii) compute the energy, the forces, and the evolution of the froth.

---

silicon on sapphire (D.L. Smith *et al.*, Inst. Phys. Conf. Ser. **67** (1983) 83), or the technique of vitrification of biological specimen to overcome water freezing damage in electron microscopy (J. Dubochet *et al.*, Quart. Rev. Biophys. **21** (1988) 129)). The pentagon tends to shrink, the heptagon to grow, through the 120° rule. The equilibrium becomes unstable and the froth coarsen, albeit slowly. Steady state is achieved because a disappearing 3-sided cell near a n>6 cell reduces n by 1, and slow down its growth rate. This **unstable equilibrium** is described by von Neumann's law (3) (exact for 2D froths with curved interfaces (2)).

This transition in a hexagonal froth from indifferent to unstable equilibrium can be studied by the Telley model of §3 in 1D, because in 1D there is no systematic force driving the coarsening, only fluctuations.

3.1. *Voronoi Froth*. The simplest way to generate a space-filling, random froth, is by Voronoi construction [1]. One begins with a Poisson distribution of points, which serve as seeds for the cells. A Voronoi cell contains all points in space nearest to its seed, and interfaces are perpendicular bisectors between seeds. One obtains a froth (because the perpendicular bisectors of a triangle formed by three seeds are concurrent) very simply, but it has an unrealistic structure (the cells are very anisotropic). Worse for our purpose is the fact that the number of seeds, hence of cells, is fixed from the beginning, so there can be no T2, no coarsening, and no evolution (only relaxation). The essential physics has been lost.

3.2. *Laguerre Froth*. A simple generalization of perpendicular bisector produces a froth which is both realistic and capable of evolution [5]. Consider circles instead of points as seeds, and define the <u>distance</u> $d(X,\Gamma)$ of a point X to a circle $\Gamma(r,x_0)$ (hypersphere if $D>2$) as the length of the tangent to Γ through X (d^2 is the power of X with respect to circle Γ),

$$d^2(X,\Gamma) = (X-x_0)^2 - r^2 . \tag{4}$$

The locus of points at equal distance between two circles is a straight perpendicular line, the radical axis (hyperplane if $D>2$). Like the perpendicular bisectors in a triangle, the three radical axes between three circles are concurrent (because the radical axis is an equivalence relation between two circles). They are the interfaces of a (radical or Laguerre) froth. The larger the circle seed, the larger its corresponding cell in the froth. Laguerre froth reduces to Voronoi's if all the seed circles are equal. Cell shapes are realistic even with little fluctuations in circle radii. Cell disappearance or division is easily accommodated, as we shall see. Generalization to 3 or more dimension is straightforward, as is the analysis of a section of the froth (a Laguerre froth itself [12]), important in the analysis of 3D polycrystalline aggregates in metallurgy.

3.3. *The Horizon of Paraboloids*. The radical axis between two circles of radii r_1 and r_2 depends on $r_1^2 - r_2^2$ only. This remark gave Telley the crucial idea of representing seeds as identical paraboloids (umbrellas) in one extra dimension, intersecting physical space as the Laguerre circles [5] (Fig.4). The altitude of the physical (hyper)plane is fixed but arbitrary. The umbrellas are the bodies of our problem, each specified by the 3 coordinates $q=(x,z)$ of their apex. x is the coordinate of the circle's centre in the 2-dimensional physical space, and z, the height above physical space, measures the circle radius, $z\sim r^2$, or, roughly, the cell size. The Laguerre froth has 3C coordinates (x,z) in 2 dimensions. (Note that the froth has $V=2C$ vertices, which would have required 4C coordinates if chosen independently).

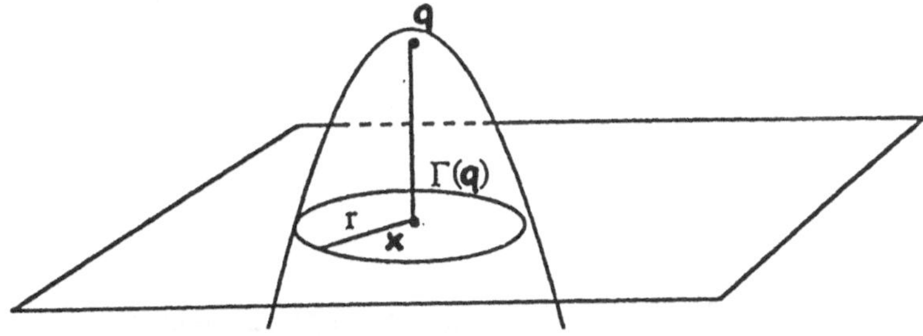

Fig.4. The seed of a cell (a circle) as a paraboloid (Telley [5]). The physical space is horizontal. The Laguerre froth generated is independent of the level of the physical "floor".

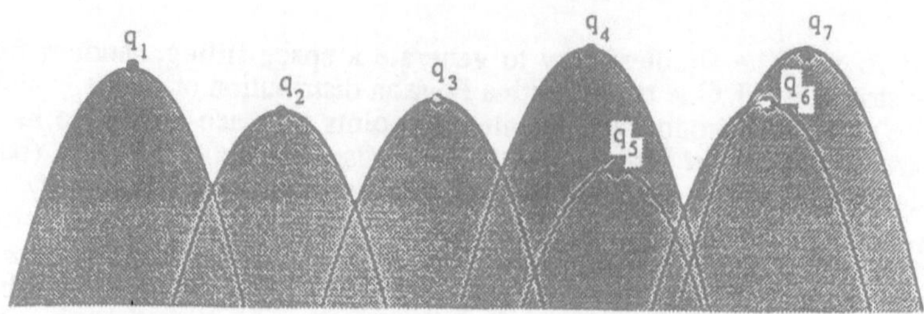

Fig.5. The "Yosemite" horizon of paraboloids. The horizontal axis represents the physical plane. q_5 is below the horizon and does not generate a cell in the froth. From [5].

Now for the ETT. The set of seeds looks like a Yosemite or Dolomite mountain profile (Fig.5). A paraboloid below the horizon (seed too small) does not generate any cell in the froth. Conversely, a paraboloid which is too high (seed too large) obscures smaller paraboloids nearby and gobbles up their representative cell. So, a T2 topological transformation occurs when a paraboloid is pushed below the horizon whereas, when a new paraboloid rises above the horizon, it divides the cell containing its apex. T1 transformations are produced by moving nearby seeds horizontally rather than vertically (The central pass switches from East-West to North-South direction). ETT are therefore naturally induced by orthogonal motions of the bodies, T1 by horizontal (x), fast motion, T2 or cell division by vertical (z), slow motion.

3.4. *Energy. Von Neumann dynamics*. We can now compute the forces on the cell paraboloids, given that the energy U is proportional to the total interfacial length of the froth, as in soap froths (surface tension) and in metallurgical grain aggregates (grain boundary's). When a seed paraboloid **q** moves, it displaces the interfaces and vertices of its own cell, hence changes the interfacial length, thus the energy of the froth. This defines a force

$$\mathbf{F} = -\mathbf{grad}\ U = \lambda_{x/z}\mathbf{v}, \tag{5}$$

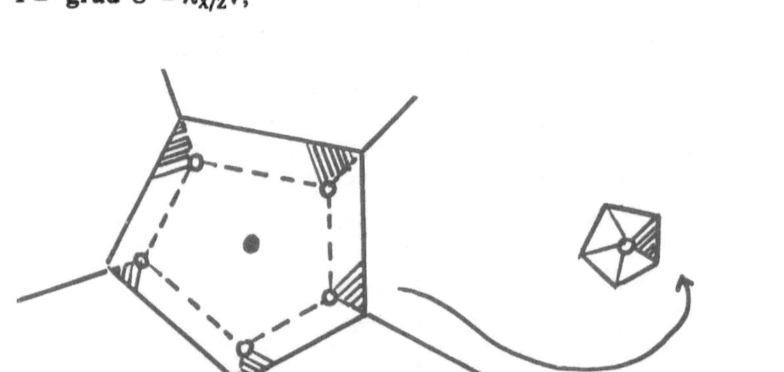

Fig.6. Vertical motion of the central paraboloid induces parallel translation of the interfaces of its cell (left). The energy balance is the difference between the perimeter and spokes of an infinitesimal polygon (right).

on $\mathbf{q}$ and generates its dynamics. Physically, the time constants are very different $\lambda_z \gg \lambda_x$: The seed relaxes rapidly to the center of gravity of its cell in physical space $\{\mathbf{x}\}$, whereas coarsening (motion in z) occurs at the much slower rate λ_z^{-1}. We now assume the system equilibrated (relaxed) in physical space.

The energy difference dU associated with the vertical motion dz_0 of seed 0 alone, keeping x_0 and all the other seeds (x_i, z_i) fixed, is proportional to the change in interfacial length when cell 0 is blown up by parallel displacement of its interfaces (Fig.6). dU is represented geometrically by an infinitesimal polygon (fairly regular if x_0 is already equilibrated) with n sides and n spokes, as

$$dU = \gamma \, [\text{perimeter} - \Sigma^n \, (\text{spokes})] \tag{6}$$

so that if perimeter $> \Sigma$(spokes) (as is the case if n<6), the cell shrinks, whereas it grows if n>6. One recovers, to a good approximation,[7] von Neumann's law $dU \approx \gamma \, [(6\text{-n}) \, dr]$.

Note that eq.(6) (and its derivation) can be read as a statement of the Gauss-Bonnet theorem, expressing the local integral of the Gaussian curvature as a sum of contributions of infinitesimal, equilateral triangles. Energy density is proportional to local curvature (rather than interfacial length), and the cell shape (6-n) is the intensity of a topological disclination. Von Neumann's law is therefore expected to be exact for a Laguerre froth on a locally curved substrate, and the evolution of the froth is curvature-driven (as is clear from Laplace's -in froths- or Gibbs-Thomson's law -in polycrystalline aggregates). Von Neumann's law is exact for circular interfaces [10].[8]

The great advantage of the Laguerre-Telley model is that it holds regardless of the space dimension, and that the cut of a Laguerre froth remains a Laguerre froth. The statistical analysis through cuts (stereology) is a standard method in metallurgy and in geology (fractured rocks). This implies that the 120° rule does not hold any longer[9], and von Neumann's law is only approximatively obeyed, but in any dimension (except 1D where there is no driving force). Simulations [4,5] yield steady-state coarsening, scaling as $\langle L \rangle \sim t^{1/2}$.

3.5.*Steady Coarsening.* We can now calculate the vertical dynamics of the froth (coarsening process). Let $\rho(z,t) = \int dx \, \rho(z,x,t)$ be the density of paraboloids making up the horizon, at height z and time t. It obeys a conservation law,

$$\partial \rho(z,t)/\partial t + (\partial/\partial z)[(\partial z/\partial t)\rho(z,t)] + \zeta(t)\rho(z,t) = 0 \tag{7}$$

The vertical current of paraboloids is convective, and they dip below the horizon at a rate $\zeta(t)$. Now, von Neumann's law $\partial z/\partial t = \gamma(n\text{-}6)$ is the "Langevin equation" for the random process n(t) caused by ETT. n(t) is certainly Markovian and Gaussian. Unlike the traditional Langevin equation in Brownian motion, the random noise is not white, and both systematic and random contributions to the velocity are lumped in the same term. Thus $\langle \partial z/\partial t \rangle = 0$

---

[7] The perimeter is very close to 6 dr, bounded by $6(\cos \pi/6)dr \approx 5.2$ dr (for n=3) and by 2π dr (for n=∞).

[8] A single, additional "vertical" parameter z can also be defined for every cell in this case. This is a consequence of the geometrical relations found by Plateau, for example, the fact that the centers of curvature of 3 interfaces incident at 120° at a vertex lie on a straight line [9].

[9] A cut soap froth is no longer a soap froth (the 120° law does not hold in the cut). Note also that, in a random structure, any cut is random.

196

(steady horizon), and the density obeys $\partial\rho(z,t)/\partial t = \Omega\,\rho(z,t)$, where $\Omega = -\gamma(n-6)\partial/\partial z - \zeta$ is Gaussian.[10,11] This yields a Fokker-Planck equation for the (non-normalized) distribution $P(z,t) = \langle\rho(z,t)\rangle = e^{-\int\zeta dt}\langle\exp\int^t du\,\Delta\Omega(u)\rangle\,P(z,0)$,[12]

$$\partial P(z,t)/\partial t = D_z\,\partial^2 P(z,t)/\partial z^2 - \zeta\,P(z,t) \tag{8}$$

which admits a <u>steady state</u> solution $P(z,t) = g(t)\,f(z)$, with $g(t) = G\exp(-\int dt[\zeta(t)-\delta])$ and a horizon profile which decays exponentially,[13]

$$f(z) = F\exp(-\eta z) = P(z) , \qquad\qquad \eta = \sqrt{[\delta/D_z]}. \tag{9}$$

δ relates coarsening rate to profile steepness, or to the width of grain size distribution. In nanocrystals, the distribution of grain sizes is very narrow, at least initially (since the grains are very small). The horizon profile is very steep and δ is as large as possible (it must be $<$ $\zeta(t)$). Hence, the rate of coarsening $\int^t du[\zeta(u)-\delta]/t$ is very slow in nanocrystals, as observed [7,8]. When the asymptotic, scaling regime is reached, coarsening proceeds as $1/t$, independently of the size distribution.

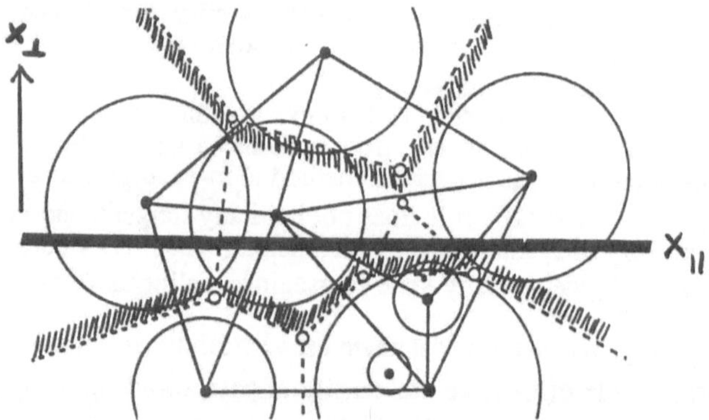

Fig.7. Stereology: Laguerre froth (dotted lines), generated by paraboloid seeds (circles). One seed here does not generate a cell (it lies below the horizon in Fig.5). Cut (thick line) of a Laguerre froth is a Laguerre froth (symmetry-preserving cut-and-projection). The strip (shaded), made of cut cells, has random width.

---

[10] $\langle\Delta\Omega(t_1)\,\Delta\Omega(t_2)\rangle = K^{-1}(t_1-t_2)\partial^2/\partial z^2$. $\langle\exp\int dt\,x(t)\,[n(t)-6]\rangle = \exp[(1/2)\int dt_1\int dt_2\,x(t_1)\,K^{-1}(t_1-t_2)\,x(t_2)]$.

[11] with a possible dependence on z, $n(z) = n_0+\alpha z$, absorbed in ζ.

[12] $D_z=\gamma^2\tau_c$ for $t>\tau_c$, where $\tau_c=\int^\infty du\,K^{-1}(u)$ is the correlation time between successive ETT affecting neighboring cells.

[13] Although the random process takes place here along the unphysical direction z, the structure is real and has a Hamiltonian function in $\{x\}$, the energy U, proportional to the total interfacial length of the resulting froth. Note that this Hamiltonian includes contribution from Gaussian curvature fluctuations.

The equilibrium profile is therefore expected to follow a Boltzmann distribution, and it does: (12) is the barometric law, and $\sqrt{[\delta/D_z]} = mg/(kT)$ would be the Einstein relation for our process n(t). Gravitational force along z and temperature are measured in arbitrary units. Also $\gamma^2 \propto D_z = \gamma^2\tau_c$.

If the steady state is scaling, $<L>\sim t^\nu$, thus $g(t)\sim<L>^{-2}\sim t^{-2\nu}$, δ is fixed by the rate of disappearance $\zeta(t) = \delta+2\nu/t$, or $\delta=\zeta(\infty)$. Simulations [4,5] yield $\nu=1/2$ (mean-field exponent [6], see also argument below), so that

$$P(z,t) = (C/t) \exp\{-\sqrt{[\zeta(\infty)/\tau_c]}\,[z/\gamma]\} \qquad t\rightarrow\infty \qquad (10)$$

in the asymptotic, scaling regime.

3.6. *Dimensional Reduction. Stereology.* The exponential profile can also be obtained from statics, simply by dimensional reduction. A (D-1)-dimensional cut of a D-dimensional Laguerre froth is also a Laguerre froth [12], retaining its statistics, geometry and symmetries (invariance by stereology) (Fig.7). Indeed, any vertical (z) cut of a paraboloid is a parabola, with now $z' = z - kx_\perp^2$, so that the parabola may be below the horizon in $x_\parallel$ if the apex of its original paraboloid is very far from the cut (Mount McKinley, even though very high, is not felt on Franklin St., Richmond). The D-dimensional physical space spanned by coordinate x is split, by the cut, into parallel, $x_\parallel$, and perpendicular, $x_\perp$, components.[14,15]

If statistics of the froth are invariant under cut, then $P'(z') \propto P(z)$ are proportional, hence $P(z)$ must be exponential, as obtained above (eq.(9)).

In addition to the stereological (cut) invariance, the froth has another remarkable symmetry: Steady-state coarsening of the froth (while remaining in statistical equilibrium) is a local dilatation symmetry, an example of conformal invariance.

3.7. *Growth Exponent.* We have shown that the froth has a steady state solution. To obtain the growth exponent ν in the scaling regime, assume that the seeds of the Laguerre froth are nucleated at random times t_i, and grow according to

$$r_i(t) = \alpha\,(t-t_i)^\nu \qquad (11)$$

Steady state imposes the same exponent ν throughout the froth and through its evolution. Recall that the Laguerre froth geometry is determined by $\Delta z = k^2(r_i^2-r_j^2)$, and assume that we are in the late stage of coarsening, $\Delta t=t_i-t_j<<t-t_i$. Then, $\Delta z \approx (k\alpha)^2\,2\nu\,\Delta t\,(t-t_i)^{2\nu-1}$. The froth becomes coarser if $\nu<1/2$, finer if $\nu>1/2$, but remains in a steady state if $\nu=1/2$ (even though there is a steady loss of paraboloids below the horizon). In the steady state, the linear cell size L_i will be proportional to the radius r_i of its seed. The mean-field [6] exponent $\nu=1/2$ is observed in simulations [4,5].

---

[14] In a way reminiscent of the cut-and-projection, or strip method introduced recently to describe quasicrystals and incommensurate structures [18,19]. The froth is the "atomic surface". But here the acceptance region, or width of the strip, is <u>random</u>, defined by the horizon. Neighbourhood is topological rather than metric. And the full symmetry survives intact the cut and projection!

[15] Note that: (i) Any cut of a Voronoi froth is a Laguerre froth (a result obtained by Imai et al. [16]).
(ii) A random (irrational slope) cut of a Laguerre crystal is a Laguerre froth.
(iii) In a random froth, all cuts are random (irrational).

4. Conclusion

Laguerre's froth, with its paraboloid seeds as "bodies" of the "gas", constitutes an excellent model for understanding and simulating the dynamics of a coarsening metallurgical aggregate. It generates the froth and its dynamics simply and accurately, in arbitrary space dimensions, and the fundamental elementary topological transformations appear naturally through local displacement of the seeds. The potential between two seeds is a function of neighborhood, itself defined by the horizon.

Coarsening is described as motion in the vertical direction: A cell is driven slowly by a systematic force proportional to the number of its neighbors minus 6 (von Neumann), and by curvature fluctuations (by the latter only in 1D). The aggregate is in an unstable equilibrium, and undergoes steady coarsening, asymptotically scaling. In the steady state, there is a correlation between the (pre-scaling) coarsening rate and the width of the grain size distribution. As a consequence, nanocrystals are very stable.

Topological froths, with energy concentrated on interfaces, abound in nature [1]. They have characteristic, universal structure and evolution, regardless of the cell size and of the microscopic diffusion mechanism.

References

[1]   D. Weaire and N. Rivier, Contemp. Physics **25** (1984) 59.
[2]   C.S. Smith, Scient. Amer. **190** (1954) 58.
      H.V. Atkinson, Acta Met. **36** (1988) 469.
[3]   J.A. Glazier, S.P. Gross and J. Stavans, 1987, Phys. Rev. A **36** (1987) 306.
      J.A. Glazier, *Dynamics of Cellular Patterns*, Ph.D Thesis, U. Chicago (1989).
[4]   D.J. Srolovitz, M.P. Anderson, P.S. Sahni and G.S. Grest, Acta Metall. **32** (1984) 793; Phil. Mag.B (1991).
[5]   H. Telley, 1989, *Modélisation et Simulation Bidimensionnelle de la Croissance des Polycristaux*, PhD Thesis, EPFL, Lausanne.
[6]   W.W. Mullins, J. Appl. Phys. **27** (1956) 900; **59** (1986) 1341.
      B. Castaing, in Appendix A of Glazier *et al.* [3].
[7]   R.W. Siegel, MRS Bull., Oct.1990, 60.
      H. Hahn, J.L. Logas and R.S. Averback, J. Mater. Res. **5** (1990) 609.
[8]   R.W. Siegel, this volume.
[9]   N. Rivier, "Geometry and evolution of soap froth", preprint.
      See also T. Herdtle and H. Aref, Proc. Roy. Soc. A (1991), to appear.
[10] J. von Neumann, in *Metal Interfaces*, Amer. Soc. Metals, Cleveland (1952) 108.
[11] D.M. Duffy, Thesis, University of London, 1982; N. Rivier, in *Amorphous Materials, Modeling of Structure and Properties*, V. Vitek, ed., AIME (1983), 81.
[12] N. Rivier, J. Physique (Coll) **51** (1990) C7-309.
[13] L. Gaultier, J. Ecole Polytechn. **16** (1813) 147.
[14] W. Fischer and E. Koch, Zeits. Kristallogr. **150** (1979) 248.
[15] B.J. Gellatly and J.L. Finney, J. Non-cryst. Solids **50** (1982) 313.
[16] H. Imai, M. Iri and K. Murota, SIAM J. Comput. **14**, (1985) 93.
      F. Aurenhammer, SIAM J. Comput. **16** (1987) 76. The power of a point with respect to a circle is called Laguerre's "distance", hence the name for the partition [5].
[17] G.E. Brown, in R.D. Mattuck, *A Guide to Feynman Diagrams*, McGraw Hill, NY (1967), §1.1.
[18] M. Duneau and A. Katz, Phys. Rev. Lett. **54** (1985) 2688; discovered independently by P.A. Kalugin, A.Yu. Kitayev and L.S. Levitov, by V. Elser, and by R.K.P. Zia and W.J. Dallas.
[19] P. Bak, Phys. Rev. B**32** (1985) 5764. H. Bohr, Acta Math. **45** (1925) 29.

M. KLEMAN
Laboratoire de Physique des Solides, Université Paris–Sud, 91405 Orsay
Cédex (France)

ABSTRACT. In spite of the absence of translational symmetries, defects can still be defined in quasicrystals ; the method consists in considering the quasicrystal as the boundary of a high dimensional crystal (d = 6 for the icosahedral case). There are therefore two types of defects, those of the high dimensional crystal which intersect the boundary (essentially : dislocations) and those specific of the boundary itself (essentially : phason defects, also called disvections). Those defects will be presented in their topological context and their relation to plastic deformation evoked.

1. The classical theory of plastic deformation.

In a crystal which contains no impurities of any sort, the plastic deformation processes are, as it is well known, related to the presence of dislocations ; they move, multiply, annihilate, etc.. under applied forces, and are responsible of the fact that low applied stresses suffice to provoke irreversible deformations [1].

Dislocations present in a crystal break locally its symmetries of translation ; each line of dislocation is therefore defined by a Burgers vector precisely equal to the lattice translational symmetry which is broken. Simple geometrical models, which have been developed since the pionneering work of Taylor, Polanyi and Orowan, tell us that the motion of a dislocation can be decomposed in a component of glide (in the surface defined by the dislocation line and its Burgers vector), which is conservative, and a component of pure climb, perpendicular to the first component, which is non-conservative. Finally the interaction between dislocations has a very simple geometrical feature : there is no topological obstruction to the crossing of line defects, which interact essentially through their long distance stresses. This property originates in the fact that the group of lattice translations is abelian.

This quite idyllic geometrical picture is of course somewhat altered by the question of the core of the dislocation, which is a region where linear elasticity does not apply. When the stacking fault energy is large on any crystallographic plane, one

P. Jena et al. (eds.), Physics and Chemistry of Finite Systems: From Clusters to Crystals, Vol. I, 199–210.

can convince oneself that the core is of atomic dimensions. If this is not the case, it is most often possible to describe the core as splitted into imperfect dislocations, also called partials, separated by stacking faults ; the sum of their Burgers vectors is of course a perfect Burgers vector. These stacking faults introduce a special type of atomic disorder, which can be the precursor of a phase transformation (most well known is the case of the order-disorder transformation in many intermetallic compounds, where stacking faults are usually called antiphase boundaries).

Finally let us recall that, due to the absence of translational symmetries, dislocations in amorphous materials are necessarily imperfect (the term partial would there be improper). There is however little doubt that they exist. It is all the more important to mention here their existence that it has been argued that there is local icosahedral symmetry in amorphous metals (and simple liquids) and that the crystallographic analogies between the amorphous state, Frank and Kasper phases, and quasicrystals, have been stressed quite often [2].

In principle, dislocations of rotation (disclinations) are not forbidden in usual crystals. However their elastic energy, which scales like the square of the distance, is very large compared to the elastic energy of dislocations. The plastic deformation they would give rise to would be very special since, because of the non-commutativity of the group of rotations, there would be strong phenomena of obstruction to their crossing.

2. Dislocations of translation in quasicrystals.

A quasicrystal is, in essence, an atomic arrangement modulated by a certain number $d_\perp$ of incommensurate periods, such chosen that symmetries which are forbidden in classical crystallography do appear (for example : decagonal symmetries in bidimensional quasicrystals, $d_\parallel = 2\, d_\perp = 2$). Therefore, as for more usual incommensurate phases, the "crystallography" of a quasicrystal makes use of a high dimensional crystal (HC) of dimension $d = d_\parallel + d_\perp$; the atoms of the quasicrystal QC are the intersections of the so-called atomic surfaces (the $d_\perp$-dimensional decoration of HC) with a $d_\parallel$-dimensional irrational planar cut $P_\parallel$ (fig. 1). In order to simplify the picture, we shall hereunder adopt the usual convention that the hypercell decoration is congruent to the projection $A_\perp$ of the hypercell in $P_\perp$ (the space complementary to $P_\parallel$ in $R^d = P_\parallel * P_\perp$), and is attached to each of its vertices. The QC then appears as made of very simple tiles, viz. two types of rhombi in the decagonal case, two types of rhombohedra in the icosahedral case, and can be as well obtained by the well-known cut-and-strip method [3].

A d = 2 representation of the crystallography of a quasicrystal.
The orientation of $P_\parallel$ yields the quasisymmetries of the QC ; the global atomic arrangement in the QC depends on a 'phase' variable, the position of $P_\parallel$ which is parameterized by its intersection $M_\perp$ with $P_\perp$.

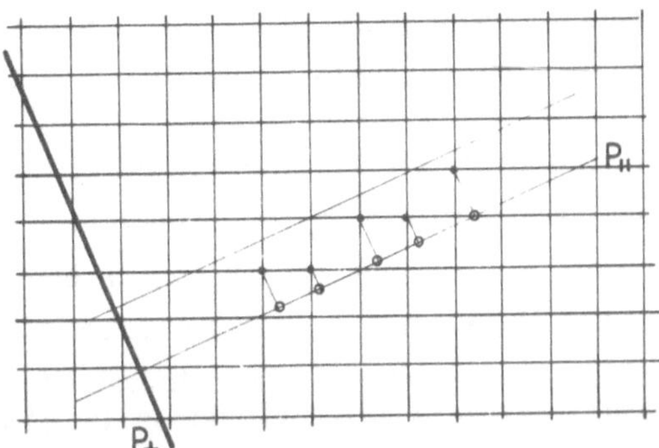

Figure 1. This high-dimensional crystallographic image of the QC can be used to define and classify its topological defects, which are of two sorts :

- those which are the intersections of the defects of HC with $P_\parallel$;

- those which are specific of $P_\parallel$ and correspond to the singularities of the phase variable hereabove defined.

In this paragraph, we shall discuss the first type of defects, viz. the intersections of the dislocations in HC with the QC. The intersection defects can reasonably be defined also as dislocations ; they have been observed by a number of people by electron microscopy imaging diffraction methods (see for example [4] ; selection rules established [5], similar to those of normal dislocations, and even in some cases the d-dimensional Burgers vector measured !

A dislocation line L_d in a d-dimensional crystal is by definition a singularity such that when circumnavigating about it along any 1D loop, the sum of the elastic displacements is a Burgers vector constant $\vec{b}$. The singular region of the 'line' L_d is in fact a (d-2)-dimensional manifold, to which we refer as a 'hyperline', and whose intersection with a $d_\parallel$-dimensional space is generically $d - 2 - d_\parallel = d_\perp - 2$ dimensional. This is obviously a line (resp. a point) in the icosahedral case (resp. the decagonal case), as expected. In non-generic cases, the intersection can be of higher dimensions, yielding intriguing objects like walls or volumes (when $d_\parallel = 3$), as lines or areas (when $d_\parallel = 2$) and we shall not consider such possibilities. In principle, the calculation of the strain field goes in two steps : first get the strain field $\vec{u}$ of the dislocation in the high dimensional crystal, then take its value $\vec{u}$ on $P_\parallel$: the so-called 'phonon' part (the displacement field) is the projection $\vec{u}_\parallel$ of $\vec{u}$ on $\vec{P}_\parallel$; the 'phason' part is $\vec{u}_\perp$ ($\vec{u} = \vec{u}_\parallel + \vec{u}_\perp$) since it is precisely this quantity which measures the perpendicular displacement of the lattice, which is equivalent to a displacement of the cut $P_{\parallel}$ parallel to itself by a quantity - $\vec{u}_{\perp}$ [6].

The new position of the atoms in $P_{\parallel}$ are the intersections of the displaced (and possibly deformed) atomic surfaces with a $P_{\parallel}$ supposedly fixed. Consider, as a guide, the pedagogical case $d = 3$, $d_{\parallel} = 2$, $d_{\perp} = 1$ (fig. 2). L_3 here is a line, which can in principle have any shape. If this is so, the singular region in $P_{\parallel}$ has the shape of a furrow, which is considerably 'phonon' strained. In order to reduce the core energy, it appears physical to take L_3 perpendicular to $P_{\parallel}$ ($L_3 = P_{\perp}$) ; more generally the same argument would yield :

$$L_d = L_{\parallel} * P_{\perp} \tag{1}$$

L_d is the direct product of the (dislocation) line $L_{\parallel}$ by the complementary $d_{\perp}$-dimensional space $P_{\perp}$. This type of dislocation hyperline, apart of reducing the core energy, has also the advantage of yielding in $P_{\parallel}$ a strain which is very little anisotropic, which is precisely what one would expect in a medium which has such a large point symmetry than a quasicrystal (an icosahedral phase is more symmetric than a cubic phase).

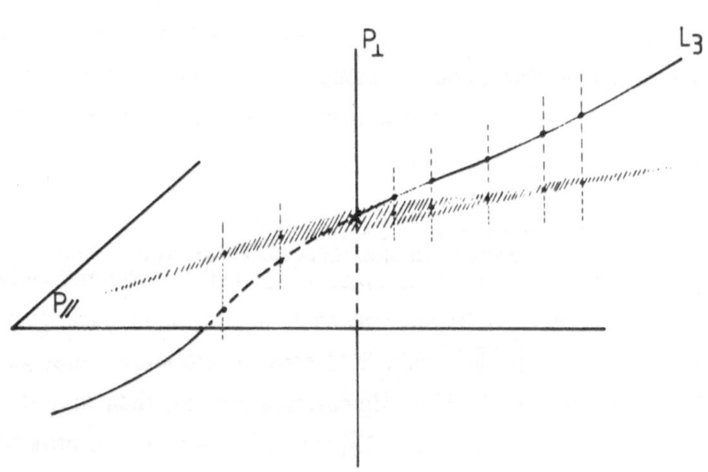

Figure 2. Construction of a dislocation in a quasicrystal as the intersection of a dislocation in a high dimensional crystal (here $d = 3$; $d_{\parallel} = 2$; $d_{\perp} = 1$).

With these restrictions, the analytical calculation of a dislocation field in $P_{\shortparallel}$ does not present too many difficulties, at least for the representation of its principal values (fig. 3). The $\vec{b}_{\shortparallel}$ component introduces a global phase shift from one side of the cut surface to another, much akin to a stacking fault which can be dispersed in the form of elementary stacking faults (fig. 3) usually called 'phason defects'. As we shall comment later these phason defects are true topological defects.

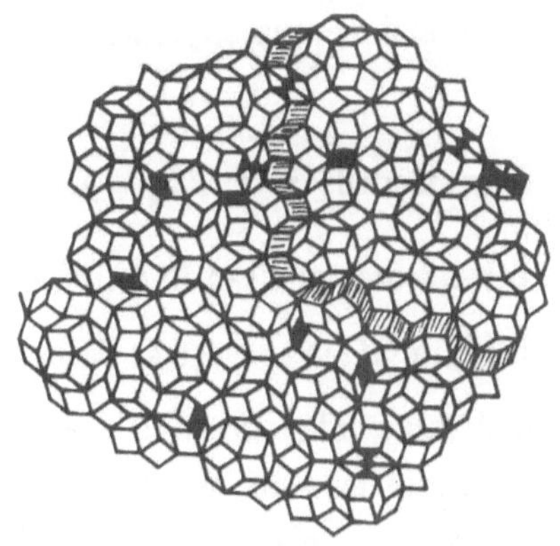

Figure 3. Simulation of a dislocation of Burgers vector $\vec{b} = (1,\bar{1},0,0,0)$. Extra-matter is shadowed. The edges which are balckened are 'phason defects' (mismatches).

Before discussing phase defects, a number of remarks are in order concerning dislocations (see [6] for details) :

- the cut surface Σ_d of the dislocation in $\mathbb{R}^d$ is $(d-1)$-dimensional and can be described as the locus of a half-line $\lambda(\vec{x})$ attached to the point $\vec{x} \in L_d = L_{\shortparallel} * P_{\perp}$. Since $\lambda(\vec{x})$ does not belong, by definition of the cut surface Σ_d, to L_d, and since L_d contains $P_{\perp}, \lambda(\vec{x})$ belongs to a $d_{\shortparallel}$-plane $P_{\shortparallel}(\vec{x})$ parallel to $P_{\shortparallel}$ at location $\vec{x}$. Hence the intersection of the cut surface Σ_d with $P_{\shortparallel}$ is $\Sigma_{d_{\shortparallel}} = U\{L_{\shortparallel} + \underset{\vec{x}\,\in\,L_{\shortparallel}}{U} \lambda(\vec{x})\}$, which is

a true cut surface in $P_{\shortparallel}$ for the dislocation of Burgers' vector $\vec{b}_{\shortparallel}$.

204

- a 'glide manifold' $G = L_d * \beta$ can be defined in $\mathbb{R}^d$. β is a direction parallel to the HC Burgers vector $\vec{b} = \vec{b}_\| + \vec{b}_\perp$. But since $\vec{b}_\perp$ belongs to $P_\perp$, i.e. to L_d, we also have $G = L_d * \beta_\|$, where $\beta_\|$ is a direction parallel to the QC Burgers vector $\vec{b}_\|$. Again, since $\beta_\|$ belongs to $P_\|$, the intersection of G with $P_\|$ is $G_\| = L_\| * \beta_\|$ and is a true glide surface in the usual sense for the QC Burgers vector $\vec{b}_\|$. Any motion of $L_\|$ along $G_\|$ is conservative (glide), while any motion of $L_\|$ which brings it out of $G_\|$ is non-conservative (climb). In conclusion, the categories of glide and climb are equally valid in usual crystals and in quasicrystals, provided L_d is chosen as in eq. 1.

- since $\vec{b} = \Sigma \, n_i \, \vec{e}_i$ projects irrationally in $P_\|$, $\vec{b}_\| = \Sigma \, n_i \, \vec{e}_{i_\|}$ takes a set of values which are dense in $P_\|$. In other words one can find (infinitely) many vectors $\vec{b}^\alpha$ in HC whose projection length $b_\|^\alpha$ in QC is as close as desired to any value $b_\|$, and in particular to any infinitesimal value. Therefore one expects that any dislocation $\vec{b}_\|$ would tend to split into dislocations of small $b_\|$ components (see [6] for illustrations). However this reduction in $b_\|$ always happens at the expense of an increase in the $b_\perp$ component, which measures the content in 'phason' defects, since the total Burgers vector $\vec{b} = \vec{b}_\| + \vec{b}_\perp$ is rational. For example, taking $\vec{b} = (- f_{n+1},\ f_{n+1},\ 0,\ f_n,\ f_n,\ 0)$ in an icosahedral crystal $(d = 6)$, one gets $b_\| = \sqrt{\dfrac{2}{\tau+2}}\, \tau^{-n}$, $b_\perp = \sqrt{\dfrac{2}{\tau+2}}\, \tau^{n+1}$; the f_n's are the Fibonacci numbers ($f_0 = 1$, $f_1 = 1$, $f_2 = 2$, $f_3 = 3$, ... $f_{n+1} = f_n + f_{n-1}$) and τ is the golden mean : $\tau = (1 + \sqrt{5})/2$. Therefore the splitting of a finite dislocation into small ones depends on a crucial way on the relative energy of the 'phonon' and the 'phason' parts. In a continuous description, introducing K_u and K_w as typical phonon and phason elastic constants, and assuming no interaction between the two fields, the energy per unit length of dislocation is :

$$W = K_u \, b_\|^2 \, \ell n \frac{D}{a} + K_w \, b_\perp^2 \, \ell n \frac{D}{a} + w_c \qquad (2)$$

where D is a typical distance between defects, a the core size (which is of the order of max $(b_\perp, b_\|)$), and w_c a core energy [6].

3. Phase defects.

In the above, in order to construct a dislocation in the QC, we have deformed HC and observed how its intersection with a fixed $P_{||}$ is modified. For simplicity, we have decorated each cell with an atomic surface equal in size and orientation to the projection of the hypercubic cell on $P_{\perp}$. Another way of introducing a deformation of the quasicrystal is to deform $P_{||}$, letting the hyperlattice fixed.

The two ways are of course equivalent. The second one makes easier the perception of what a stacking fault is : by displacing a piece P of $P_{||}$ parallel to itself by a quantity $\vec{w}$ ($\vec{w}$ belonging to $P_{\perp}$), one obtains an abrupt phase change along the boundary of P. Since $\vec{w}$ cannot be equal to a hyperlattice vector, this phase change manifests by a sequence of mismatches between the tiles on both part of the boundary. More precisely, some of the atomic surfaces which cut $P_{||}$ before the introduction of the fault no longer cut it, while new atomic surfaces cut it now. It can be shown that as many new vertices enter the deformed quasicrystal as old vertices have left it. Density is conserved.

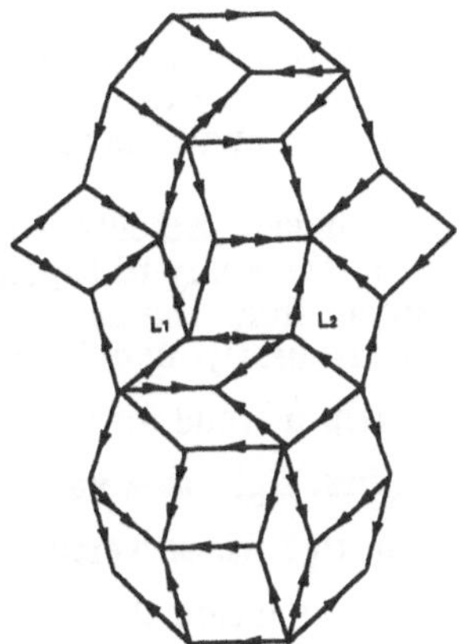

Figure 4. a) An elementary phase defect (mismatch) L_1L_2.

It is easy to show that the stacking fault can be dispersed into elementary objects (fig. 4), by a sequence of local shifts of the vertices equivalent to a small motion of parts of the boundary. It is these elementary mismatches one observes around the dislocations. They are point-like defects in the decagonal case, line-like defects in the icosahedral case. Clearly, they should play a role in the motion of any dislocation, since they attend this motion, to which they contribute by some friction, or by being emitted or absorbed on the core of the dislocation. Their existence has been clearly demonstrated experimentally, by the distortions they bring to the diffraction patterns [8].

These elementary mismatches are true stacking faults in the usual sense, since they can be constructed as ribbons separating two dislocations of opposite signs. This is illustrated on fig. 5.

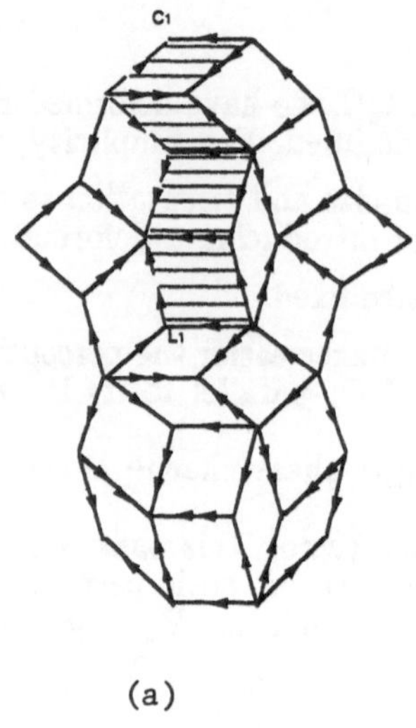

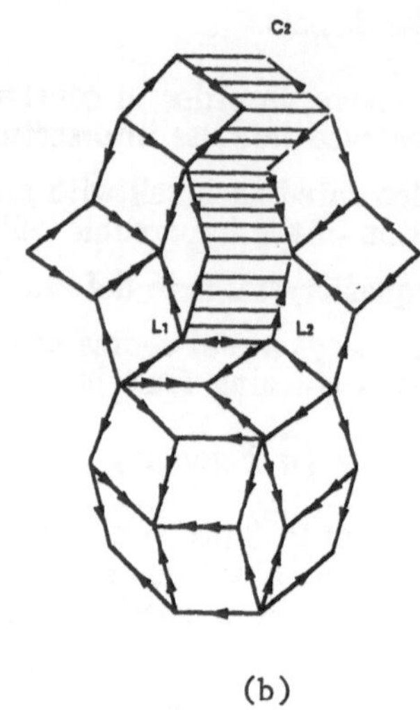

Figure 5

An elementary mismatch as the sum of dislocations of opposite signs.
a) Cut surface $L_1 C_1$. The construction of the dislocation in L_1 requires the removal of the shadowed tiles and the subsequent glueing of the lips of the cut ; b) Cut surface $L_2 C_2$. The shadowed tiles represent here the matter to be added in order to construct L_2. The result of a) and b) is as in fig. 4.

The topological nature of this type of defect presents a number of features which are of a rather unprecedented nature in the study of topological imperfections. The main difficulty, as we shall see, stems from the fact that the order parameter describing the phase of a <u>finite</u> QC is partly undetermined. Consider indeed a perfect QC, defined by the cut-and-strip method for simplicity : the value $\vec{w}$ of the intersection of $P_{\parallel}$ with $P_{\perp}$ appears at first sight as a valid order parameter ; $\vec{w}$ fixes the bottom of the strip in the hyperlattice whose vertices project on $P_{\parallel}$ along the QC, and on $P_{\perp}$ along a dense set of points inside the so-called acceptance domain.

But $\vec{w}$ defines an infinite perfect crystal, since then and only then is the acceptance domain uniformly filled. For a still perfect but finite quasi-crystal, a finite part of the strip is filled, and its $P_{\parallel}$ boundary is not defined, since it can be moved with the strip, by some quantity without letting the projected vertices escape out of the strip. This remark makes all the more difficult the definition of the order parameter at each point of a deformed QC, even if it is infinite.

This remarkable property, which has no equivalent in usual crystals, can be given a precise geometrical description, in the Bravais lattice picture of the QC, as follows :

Consider, for example, an icosahedral crystal tiled with rhombohedra, pick one of them, say R_0, and lift it to the hypercubic lattice, where it belongs to some hypercube C_0 (R_0 is the projection of one of its 3-faces). Some arbitrariness exists in the choice of C_0, since in 6-space a 3-face of a cube in an hypercubic lattice belongs to 8 hypercubes. At the moment, we make no restriction on the choice of C_0. C_0 projects on $P_\perp$ along a rhombic triacontahedron TR_0 which spans a strip S_0. Using this strip to construct a QC_0 (by the cut-and-strip method), a certain number of vertices of the QC already belong to this QC_0, and, at least for this finite part of the Q, we can use the projection $\vec{w}_0$ in $P_\perp$ of the center of C_0 as the value of their order parameter, as a first approach to the value of the order parameter. Consider now a path in QC, which meets an increasing number of rhombohedra R_i. If all their vertices belong in projection on $P_\perp$ to S_0, we keep S_0. If some other vertices appear, it is always possible, in the case of a perfect QC, to find another C_i whose strip S_i contains all the previous vertices and the new ones. This process can be repeated a number of times and converges towards the strip S of the infinite, perfect quasi crystal. The sequence of $\vec{w}_i$'s converge towards $\vec{w}$. Note that, while the $\vec{w}_i$'s are all rational, $\vec{w}$ is not, so that there is some similarity between this process and the process of definition of an irrational number as the limit of a sequence of rational ones. Note also that the process does not converge if the quasicrystal, otherwise perfect, is finite.

If the quasicrystal is not perfect, the question is even more involved, and it might become difficult to find a S_i containing all the previous vertices of S_{i-1}, at some stage i. This is clearly true for the case of a phason defect, in a perfect quasicrystal, because there is at least one vertex outside the strip S. But it is possible to test the presence of the defect by considering a closed path surrounding it, along which we test successive R_i's and C_i's. But, for the purpose of finding a topological invariant which tests and measures the presence of the defect, it is no longer the sequence of corresponding strips S_i that we consider, but the sequence of vectors $\vec{\gamma}_i = \vec{w}_{i+1} - \vec{w}_i$ joining the centers of consecutive C_i's. Also, since it is no longer the global undeterminacy on the order parameter that we want to identify, but its undeterminacy due to the presence of the defect, we consider a special sequence of cubes C_i, the so-called unstable cubes first introduced in the context of the geometrical properties of quasicrystals by Frenkel et al [9].

Let Γ be a closed loop which passes through a sequence of rhombohedra R_i, traversing their 2-faces F_i in a phason (only)-strained QC. Each F_i is projected from some 2-face ϕ_i in the hypercubic lattice and belongs to 16 hypercubes, which are opposed two by two on F_i, this meaning that two by two their centers O_i, O_{i+1} are aligned along a perpendicular to F_i. But amongst these pairs, one of them is particular in the sense that its projection on $P_\perp$ is a common face to two polyhedra TR_i, TR_{i+1}. ϕ_i is a 'silhouetting' face for the hypercubes C_i and C_{i+1}, which are called unstable cubes for the face ϕ_i. Any rhombohedron R_i is the projection of some 3-face of a well defined unstable cube C_i with respect to a (selected) face F_i. Now, following the sequence of unstable cubes along Γ, one comes back after a complete circumnavigation, to the same unstable cube C_0 from which one started, with now at our disposal a set $\{\gamma\}$ of <u>ordered</u> $\vec{\gamma}_i$ vectors whose sum is zero. This set is not trivial if Γ surrounds a defect of the phase.

The $\vec{\gamma}_i$ vectors are the generators of the displacement under which a special, non-simply connected, atomic surface Σa is invariant. We shall limit ourselves here to only a few facts on the way the topology of the phase proceeds from Σ_a. More details can be found in [10].

- the Σ_a surface is obtained by lifting to each cube C its $d_\perp$-dimensional projection TR in $P_\perp$. One gets in this way (see [9,10]) a continuous $d_\perp$-surface embedded in the hyperlattice. This 'decoration' of the cubic hyperlattice makes it look very much, mutatis mutandis, to a surfactant cubic phase (where the surfactant layer plays the role of Σ_a). The $d_\shortparallel$ ($d_\perp + 2$) $\vec{\gamma}_i$'s generate a d-dimensional subgroup of Z^d, the group of translations of the hypercubic lattice.

- there is a one-to-one correspondance between the set of all TR's in $P_\perp$ (where TR's are overlapping) and Σ_a, which appears in this way as an unfolding of all these TR's, with conservation of their contacts along faces.

- Σ_a being not simply connected (Σ_a is full of handles and passages), a path of the type $\vec{\gamma}_i\vec{\gamma}_d\vec{\gamma}_i^{-1}\vec{\gamma}_j^{-1}$ is generally not homotopic to zero. But such paths on Σ_a are precisely those which have been considered above when following the variation of the phase along Γ. Therefore it is not the vectorial character of the $\vec{\gamma}_i$'s which is relevant, but the fact that, in projection on $P_\perp$, they act as operators of translation of the TR's, or equivalently, as operators of translations of the lifts of the TR's on Σ_a.

- Pursuing this analysis, it can then be shown that the first homotopy group $\pi_1(\Sigma_a)$ of Σ_a, which classifies indeed the singularities of the phase, is a subgroup of commutators of the first homotopy group $\pi_1(TR)$ of the TR with faces identified, i.e. a subgroup $N(\varphi_{ij})$ generated by elements of the type $\varphi_{ij} = \gamma_i\gamma_j\gamma_i^{-1}\gamma_j^{-1}$.

- It is not difficult to convince oneself that the operators γ_i themselves classify the dislocations of the quasicrystal. Therefore the defects classified by the elements of the group $N(\varphi_{ij})$ are topologically speaking, multipoles of dislocations. This remark justifies the construction of fig. 5. $N(\varphi_{ij})$ is a non-abelian group ; it reduces to the trivial group for the approximants of the quasicrystal. Because it can be shown that the symmetries broken by the phase defects are Cartan transvections in some space of negative Gaussian curvature, we have called these defects disvections.

All the discussion above, which has been led at a very general level, does apply perfectly to the octagonal quasicrystal ($d_\shortparallel = d_\perp = 2$) and with some minor precautions to the icosahedral quasicrystal. But the pentagonal case offers some difficulties, which can be overcome anyway, but of which we shall not discuss in the course of this article. Notice finally that our analysis, which is based uniquely on the symmetry properties of the quasicrystal, does predict the existence of singularities of the phase (disvections) even in structures which can accommodate continuous phasons [11].

4. Conclusion

The question of the nature of the defects in quasicrystals is probably one of the most attractive one has met in the physics of defects, and raises new problems concerning the topological nature of defects as well as the role they play in plastic deformation. And it is probably the first time that one has to deal with a system whose order parameter is not defined, while its singularities are. This remark should draw our attention to the possible misuse of a Landau-Ginzburg description in the case of quasicrystals.

5. References

[1] Friedel J. (1964). Dislocations, Pergamon Press Ltd, Oxford.

[2] Widom M. (1988). 'Short and long range icosahedral order in crystals, glass, and quasicrystals', in M.V. Jaric (ed.), Aperiodicity and Order, Acad. Press, San Diego, pp. 59-110.
Kléman M. (1989). 'Curved crystals, defects and disorder', Adv. in Physics 38, 605-667.

[3] Duneau M. and Katz A. (1985). 'Quasiperiodic patterns'. Phys. Rev. Lett. 54, 2688-2691.

[4] Devaud-Rzepski J., Cornier-Quiquandon M. and Gratias D. (1990). 'Analysis of Dislocations in Icosahedral Al Cu Fe alloy', in M. José Yacaman et al. (eds), Quasicrystals and Incommensurable structures in Condensed Matter, World scientific, Singapore, pp. 498-515.
Zhang Z. and Urban K. (1990). 'Dislocations in decagonal Al Cu Co alloy'. M.V. Jaric and S. Lundqvist (eds), Quasicrystals, world scientific, Singapore, pp. 269-276.

[5] Hatwalne Y. and Ramaswamy S. (1990). 'How to see the Burgers vector of a quasicrystal dislocation'. Phil. Mag. Lett. 61, 169-174.
Wollgarten M., Gratias D., Zhang Z. and Urban K. (1990). 'On the determination of the Burgers vector of quasicrystal dislocations by transmission electron microscopy', preprint.
Kléman M., Staiger W. and Yu Dapeng (1991). Electron microscopy diffraction contrast in quasicrystals : some rules', preprint.

[6] Kléman M. and Sommers C. (1991). 'Dislocations in a Penrose lattice'. Acta metall. Mater. 39, 287-293.

[7] Socolar J.E.S., Lubensky T.C. and Steinhardt P.J. (1986). 'Phasons, phonons and dislocations in quasicrystals', Phys. Rev. B34, 3345-3360.

[8] See, for instance, Lubensky T.C., Socolar J.E.S., Steinhardt P.J., Bancel P.A. and Heiney P.A. (1986). 'Distortion and Peak Broadening in Quasicrystal Diffraction Patterns'. Phys. Rev. Lett. 57, 1440-1443.

[9] Frenkel D.M., Henley C.L. and Siggia E.D. (1986). 'Topological constraints on quasicrystals transformations'. Phys. Rev. B34, 3649-3669.

[10] Kléman M. (1990). 'Topology of the phase in aperiodic crystals'. J. Phys.
France 51, 2431-2447. (1991). 'Dislocations and disvections in aperiodic crystals',
J. Phys. France. (1991) in press. 'Singularities of the phase in aperiodic crystals :
disvections', in preparation.

[11] Levitov L.S. (1989). 'Icosahedral crystals with continuous phasons', J. Phys.
France 50, 3181-3190.

Acknowledgements. I wish to thank Gilles Abramovici, Ovidiu Radulescu, and Hans-Rainer
Trebin for fruitful comments and discussions. This work is part of a 'PROCOPE' European
program involving the Institut fur theoretische und Angewandte Physik (Suttgart) and the
Laboratoire de Physique des Solides (Orsay).

A COMPUTATIONAL QUANTUM MECHANICAL STUDY ON WEAKLY BONDED CLUSTERS OF NH_3 WITH SO_2 AND HNC

SOUMITRA CHATTOPADHYAY[1] and PATRICIA L.M. PLUMMER[2]

1 Science Division, Columbia College, Columbia, MO 65216

2 Departments of Physics and Chemistry, University of Missouri-Columbia, Columbia, MO 65211

ABSTRACT. Electronic structure calculations employing *ab initio* methods have been performed for two dimers of ammonia, one with hydrogen isocyanide and the other with sulfur dioxide. The results are reported at HF and MP levels of calculation using large polarization basis sets for energies, optimized geometries, rotational constants, vibrational frequencies, charge distribution and dipole moments for the systems.

1. Introduction

In recent years, research on weakly bonded molecular clusters has received considerable attention in part because such clusters can be regarded as an intermediate phase between the gaseous and the condensed phases of matter. Molecular clusters can be formed because of different kinds of bonding, e.g. hydrogen bonding, anti-hydrogen bonding, electrostatic interactions, weak van der Waals' type interaction, charge transfer between the monomer fragments and/or a combination of all of these. The complexes under consideration in this study have the potential to show one or more of the above types of bonding. In addition, sulfur dioxide is a major pollutant of the atmosphere and is capable of forming complexes with a large variety of chemical compounds --- both organic and inorganic substances. Hydrogen isocyanide and ammonia can serve as important prototypes for forming hydrogen-bonded (h-bonded) complexes with sulfur dioxide. It is quite difficult to detect a Y-H$\cdots$X (where X and Y are two different elements) type bond. As a result, even though the study of h-bonded complexes have been a subject of interest to scientists for almost a half a century, only in the last decade or so have the theoretical and experimental tools have become sufficiently sensitive to provide qualitative characterization of the h-bonded systems. With the advancement of both the hardware and software aspects of computers, it is now possible to predict the electronic structures, stability and frequencies of different vibrational and rotational modes for these weakly bonded clusters with reasonable accuracy.

For some of the heterodimers (a dimer of two different monomers) it has been found that anti-hydrogen-bonded (a-h-bonded) dimers also exist. In the last few years, we have examined the properties of a series [1] of dimers using *ab initio* methods. This paper is one of the series of calculations we have done on various weakly bonded clusters. The question of identification of the most stable structures for the dimers and the type(s) of bonding present in each case motivated the quantum mechanical study of these systems.

2. Summary of Previous Work: Dimers of HCN with SO_2 and NH_3

Much research has been done on weakly bonded dimers in recent years. Of these studies, the studies done by Legon and Millen [2] and Buckingham and Fowler [3] are of particular interest to us. In [2], microwave spectroscopy was used to examine a number of such complexes. In [3], an electrostatic model was used to study the structures of weakly bonded dimers. The dimers under consideration there include $H_3N\cdots HCN$ and $SO_2\cdots HCN$. For both of these complexes, a h-bonded structure was predicted to be stable. In addition to

P. Jena et al. (eds.), Physics and Chemistry of Finite Systems: From Clusters to Crystals, Vol. I, 211–216.
© 1992 *Kluwer Academic Publishers.*

the h-bonded structure, an a-h-bonded structure was also predicted to be stable for the $SO_2\cdots HCN$ complex. Later, in an experimental study on the system $SO_2\cdots HCN$, Goodwin and Legon [4] found that there are four possible a-h-bonded structures for the dimer consistent with their data, but no h-bonded one. Subsequently, our calculations [1] have shown that only one of these a-h-bonded structures for the dimer is stable.

Another paper on the results obtained from the calculations on the dimer between HCN and NH_3 has been submitted for publication [5]. For the dimer between HCN and NH_3, we have found several interesting features. According to the previous studies on the system [6], which were done quite some time ago, only a C_{3v} structure was considered for the dimer $H_3N\cdots HCN$. From our studies [5] using 6-31G* basis set, we have identified several stationary states on the potential surface. In two of them, the monomer units were associated by means of a hydrogen bond with HCN serving as a hydrogen donor. One of the structures had a C_{3v} symmetry with a binding energy of 9.29 kcal/mol. The second had a C_1 symmetry and a h-bond stabilization energy of 4.82 kcal/mol. After calculating the frequencies associated with this dimer, it was concluded that the C_1 state was a transition state on the potential surface. There was a third stable structure with an association energy of 2.71 kcal/mol. It too was a h-bonded structure, but with the nitrogen of the HCN unit serving as a hydrogen acceptor. The association energies quoted above are obtained from the MP2 level of calculation which has been found to describe of the systems best.

3. Method of Calculation

Geometry optimizations were done for the monomer fragments and the dimers using the GAUSSIAN [7] series of programs. For some of the calculations, the GAMESS [8] series of programs were used.For some of the calculations, we have used large basis sets such as the 6-311G** or the D95* , but for the dimers discussed in this paper only the results of calculations with the 6-31G* [9] basis set are reported. With the optimized geometries obtained from above Hartree-Fock (HF) calculations, Møller-Plesset [10] (MPn; n=2,3,4) calculations were done both for frozen-core (FC) and full correlations (FULL). Energies at the configuration interaction with single and double excitations excluding the inner-shell excitations (CISD) were also found. Counterpoise calculations [11] at the optimized geometry for all of the dimers have been performed to estimate the respective stabilities. The second derivatives for the frequency calculations were found using the analytical method. The calculations were done using the following computers: IBM-4381 and IBM-3090 with FPS-164 attached processor at the University of Missouri-Columbia, Columbia, MO and IBM scientific centers at Los Angeles, CA and Cambridge, MA and the CRAY X-MP at Champaign, IL. For plotting the structures of the dimers, we have used the package program PLUTO [12].

4. Results and Discussion

4.1. THE DIMER OF NH_3 AND HNC

For the dimer between ammonia and hydrogen isocyanide, the calculations were carried out at both the HF and the MP2 levels using the 6-31G*[9] basis set. As in the case of $H_3N\cdots HCN$, it is found that the MP2 level of calculation gives a better description of this system. In addition to optimization of the monomer units and the dimer, counterpoise calculations [11] were done to check the stability of the dimer. The Hessian matrix was examined to ensure that each state is a true minimum on the potential surface. The primary stable structure for the dimer has a hydrogen bonding between the N of ammonia and the H of hydrogen isocyanide. Comparison of the charge distribution on the monomer fragments with that of the dimer showed us that there is considerable charge transfer between the N of the NH_3 and the C and H of the HNC subunits on dimer formation. Also it shows that there is a total of $0.0355e^-$ transfer from the NH_3 subunit to the HNC subunit. Thus the association between the two subunits appears to result from a combination of charge transfer and hydrogen bonding. Hence the dimer can also be regarded as a *charge transfer complex*. The bond lengths and bond angles for the monomer units and the dimer unit are presented in detail in Table 1. The energies are reported in Table 2, the dipole moments in Table 3, the rotational constants in Table 4, the charge distribution in Table 5 and the harmonic vibrational frequencies and the suggestion for contributions of the individual monomer units to the dimer spectrum in Table 6. The final optimized structure obtained for the dimer from the calculations at MP2 level is shown in Fig. 1.

TABLE 1. Bond lengths (Å) and bond angles (deg.)

Parameter	Reference	NH_3	HNC	$NH_3 \cdot HNC$
R(N1-H1)	MP2 calc.	1.0167		1.0178
	Expt. [a]	1.012		
R(N2-H4)	MP2 calc.		1.0022	1.0362
	Expt. [b]		0.994	
R(N1-H4)	MP2 calc.			1.8203
R(N2-C1)	MP2 calc.		1.1856	1.1855
	Expt. [b]		1.169	
∠H1N1H2	MP2 calc.	106.36		106.79
	Expt. [a]	106.7		
∠H1N1H4	MP2 calc.			111.99

a) Ref. 15, b) Ref. 16

TABLE 4: Rotational constants (in GHz.)

System	Ref.	A_0	B_0	C_0
NH_3	HF	306	306	193
	MP2	296	296	189
	Expt. [a]	---	283.3	185.9
HNC	HF	0.0	47	47
	MP2	0.0	44	44
	Expt.	---	---	---
$H_3N \cdot HNC$	HF	192	4	4
	MP2	188	4	4

a) Ref. 15

TABLE 6: Calculated vibrational constants for the dimer $H_3N.HNC$ and corresponding monomer frequencies (cm^{-1})

Harmonic Freq. #	ω at HF level	ω at MP2 level	Corresponding monomer freq. MP2	exptl.	Remarks
1	133	139			
2	134	139			
3	216	246			
4	326	345	514		ω_2 (HNC)
5	328	345	514		ω_2 (HNC)
6	969	1060			
7	970	1060			
8	1290	1236	1160	968	ω_2 (NH_3)
9	1843	1745	1755	1627	ω_4 (NH_3)
10	1843	1745	1755	1627	ω_4 (NH_3)
11	2308	2029	2047		ω_1 (HNC)
12	3671	3262	3853		ω_3 (HNC)
13	3695	3509	3505	3337	ω_1 (NH_3)
14	3822	3658	3661	3444	ω_3 (NH_3)
15	3822	3658	3661	3444	ω_3 (NH_3)

TABLE 2: Optimized energies and stabilities (in a.u.)

Reference	NH_3	HNC	$NH_3 \cdot HNC$
E(RHF)	-56.1852	-92.8553	-149.0570
E(MP2)	-56.3574	-93.1339	-149.5139

Stability (in kcal/mol) w.r.t. monomer fragments at optimized geometry: (i) at HF level: 11.19, (ii) at MP2: 14.18; at Counterpoise HF: 10.42

TABLE 3: Dipole moments of the systems (in D)

Reference	NH_3	HNC	$NH_3 \cdot HNC$
HF calc.	1.9218	2.8321	5.8605
MP2 calc.	1.9687	2.7352	6.0145
Exptl. [a]	2.3	---	---

a) Kollman *et al*, Ref. 6

TABLE 5: Charge distribution on the monomers and the dimer (in terms of e⁻)

System	H1	H2	H3	N1	H4	N2	C
NH_3/HF	.332	.332	.332	-.996			
NH_3/MP2	.330	.330	.330	-.989			
HNC/HF					.430	-.582	.152
HNC/MP2					.437	-.617	.180
$H_3N \cdot HNC$ /HF	.361	.361	.361	-1.046	.517	-.611	.059
$H_3N \cdot HNC$ /MP2	.364	.364	.364	-1.050	.536	-.657	.080

(Refer to Figure 1 for the labeling of atoms)

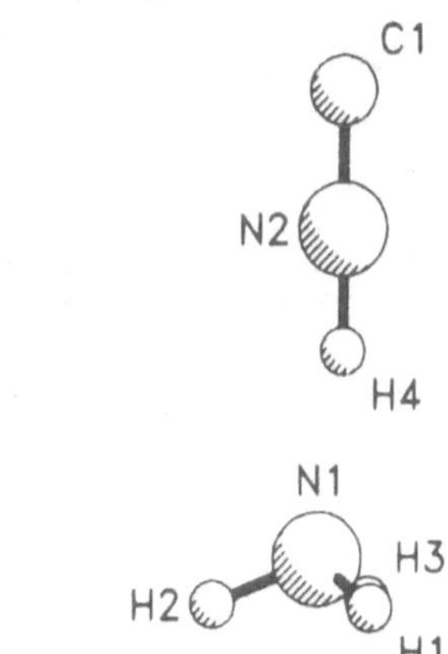

FIG. 1: The optimized geometry of $H_3N \cdot HNC$ dimer at MP2 level with 6-31G* basis set

4.2. THE DIMER OF NH_3 AND SO_2

Both ammonia and sulfur dioxide are present in the atmosphere. Knowledge of the details of their association and the possibility of their subsequent reaction are of concern to atmospheric scientists in addition to inorganic and physical chemists. An interesting observation is the large number of experiments that have been conducted on this system with the first one dating back to 1826 [13]. However, even with the numerous experiments, the structure of the complex is not agreed on. Furthermore, the system has been the subject of only one theoretical study carried out by Lucchese *et al* [14]. In their study, all the bond lengths and angles were held fixed at the experimental geometry and the N···S bond length was allowed to vary between 2.70 - 3.54 Å. The work was done in 1976, and the quantum chemical programs have evolved and matured in the interim. Also the hardware did not permit the use of polarization functions for this size system at that time. In addition, the Hessian matrix was seldom calculated. As a result of these deficiencies, one could not conclusively tell whether the optimized structures were true minima or whether they were transition states or other stationary states on the potential surface. We found the need for a more complete study of the system and investigation of the structures involving h-bonded and a-h-bonded geometries. The geometry obtained from [14] was the starting point for this study.

In the course of our calculations with varying orientations of the monomer units, an optimization run was started with the plane of the SO_2 moiety perpendicular to the C_3 plane of NH_3. Even though this configuration had a very high energy, the energy rapidly decreased and the calculations were allowed to proceed. The new structure obtained as a result of these calculations was no longer identifiable as a heterodimer of NH_3 and SO_2. In the course of optimization, two of the H's from the NH_3 group migrated to the two O's of the SO_2 group, thus resulting in the formation of a dihydroxide of sulfur. This reduced SO_2 was found to be weakly associated with the imide, NH, through an a-h-bonded interaction.

Thus, in addition to the geometries for NH_3 and SO_2, optimizations were done for HOSOH and NH. From the results of these HF calculations, limited MP calculations were carried out. Rotational and vibrational frequencies for $NH_3 \cdot SO_2$ as well as NH·HOSOH were found. Some tightening of the gradients

TABLE 7: Bond-lengths (in Å) and bond-angles (in degrees)

Parameter	Reference	SO_2	NH_3	$NH_3 \cdot SO_2$	NH·HOSOH
R(S-O)	HF/6-31G*	1.4151		1.4153	1.6615,1.6276
	Experimental[a]	1.4321			
R(N-H)	HF/6-31G*		1.0026	1.003	1.014
	Experimental[a]		1.012		
R(S-N)	HF/6-31G*			2.7649	2.7652
∠OSO	HF/6-31G*	118.8		100.48	117.71
	Experimental[a]	119.5			100.36
∠HNS	HF/6-31G*			101.76	103.96,106.93, 122.55
∠HOS	HF/6-31G*			109.8	110.2

TABLE 8: Optimized HF energies

System	RHF/6-31G*	CP/6-31G*
NH_3	-56.18385	-56.18531
SO_2	-547.16900	-547.16925
NH	-54.84988	---
HOSOH	-548.33631	548.33752
$NH_3 \cdot SO_2$	-603.36506	---
NH·HOSOH	-603.20090	---

a) in Table 7: Ref. 15

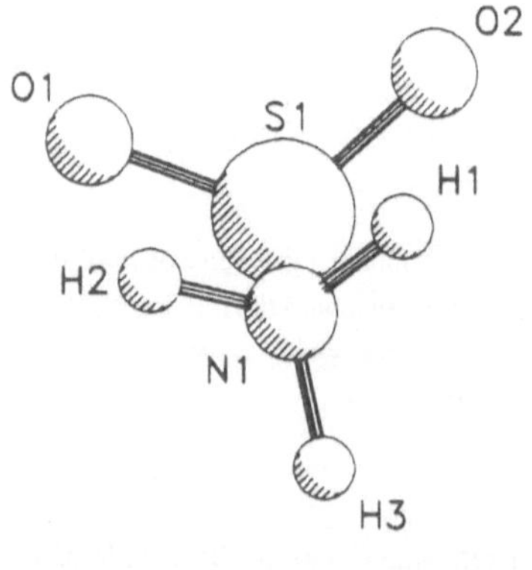

FIG. 2: The most stable state for $NH_3 \cdot SO_2$

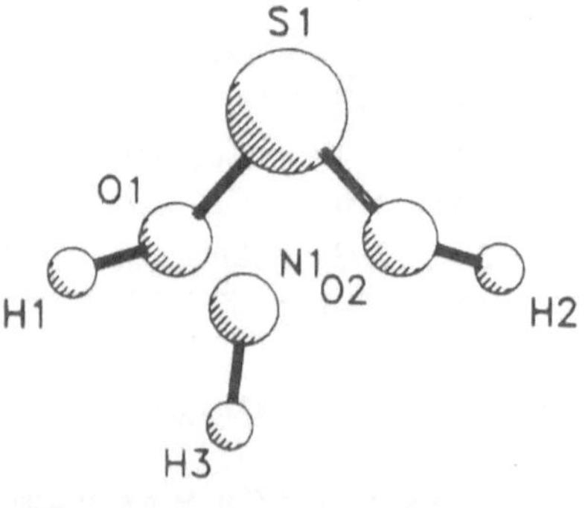

FIG. 3: The most stable state for NH·HOSOH

TABLE 9: Comparison of rotational frequencies
(All frequencies are in GHz)

System	Reference	A_o	B_o	C_o
Ammonia	HF/6-31G*	305.7	305.7	192.6
	experimental[a]	---	283.3	185.9
	% difference	---	7.9	3.6
	MP2/6-31G*	295.7	295.7	189.3
Sulfur	HF/6-31G*	60.908	10.648	9.064
Dioxide	experimental[a]	60.821	10.325	8.806
	% difference	0.14	3.13	2.93
	MP2/6-31G*	57.681	9.676	8.286
NH	HF/6-31G*	0.0	516.8	516.8
HOSOH	HF/6-31G*	25.95	8.79	6.82
NH·HOSOH	HF/6-31G*	21.45	3.25	2.91
$NH_3 \cdot SO_2$	HF/6-31G*	8.7	3.8	3.0

a: Ref 15

TABLE 10: Calculated and experimental dipole moments

Ref.	NH_3	SO_2	NH	HOSOH	$NH_3 \cdot SO_2$	NH.HOSOH
HF	1.92	2.23	1.80	0.48	3.66	2.37
MP2	1.97	2.35	---	---	---	---
Expt.[a]	1.47	1.57	---	---	---	---

a) Ref. 15

TABLE 11: Charge distribution on the atoms (in e-)

System	Ref.	N1	H1	H2	H3	S	O1	O2
NH_3	HF	-.996	.332	.332	.332	--	--	--
SO_2	HF					1.08	-.54	-.54
NH	HF	-.342	.342					
HOSOH	HF	---	.473	.473	---	.609	-.777	-.777
$NH_3.SO_2$	HF	-.300	.105	.105	.105	1.30	-.653	-.653
NH·HOSOH	HF	-.452	.504	.476	.343	.675	-.769	-.769

were needed for the complex in order to get a true minimum. Two minima along with two transition states were found. The binding energy at the HF level with the 6-31G* basis set was found to be 7.66 kcal/mol. For the NH·HOSOH structure, the binding energy was 9.23 kcal/mol with respect to the imide and sulfur dihydroxide fragments. We report the results obtained from the calculations for this system in Tables 7 through 11. The geometries for the structures predicted to be most stable are shown in Figures 2 and 3. By comparison of the charge transfer between the monomer units and the dimers, one can tell that the bonding for the $NH_3 \cdot SO_2$ dimer was due to charge transfer whereas that in case of NH·HOSOH was due to weak electrostatic interaction. We also conclude from our calculations that the structure reported in the earlier study [14] was a stationary state which was not a true minima on the potential surface.

TABLE 11: Calculated Harmonic Vibrational Frequencies for $NH_3 \cdot SO_2$ and NH·HOSOH (unscaled) and corresponding monomer frequencies calculated at MP2 (and HF) level with 6-31G* basis set and the experimental values (cm^{-1})

Freq.#	ω / $NH_3 \cdot SO_2$ (HF)	Monomer freq. at MP2 level	experimental	Remarks	ω/NH.HOSOH (HF)	Monomer freq at HF	Remarks
1	120.8	---		---	105.9	---	---
2	124.1	---		---	145.5	---	---
3	152.1	---		---	212.5	---	---
4	209.3	---		---	303.4	---	---
5	363.1	---		---	417.0	370.1	ω_1 of HOSOH
6	384.9	---		---	557.3	533.0	ω_2 of HOSOH
7	607.9	487.8	519.0	ω_2 of SO_2	710.6	547.1	ω_2 of HOSOH
8	1264.8	1160.0	968.0	ω_2 of NH_3	847.2	895.2	ω_2 of HOSOH
9	1425.3	1083.8	1151.0	ω_1 of SO_2	902.3	898.9	ω_2 of HOSOH
10	1502.3	1312.5	1361.0	ω_3 of SO_2	1120.8	---	---
11	1840.8	1755.0	1627.0	ω_4 of NH_3	1313.1	1337.0	ω_4 of HOSOH
12	1849.2	1755.0	1627.0	ω_4 of NH_3	1339.1	1337.0	ω_4 of HOSOH
13	3745.1	3505.0	3337.0	ω_1 of NH_3	3671.8	3606.9	ω of NH
14	3804.4	3661.0	3444.0	ω_3 of NH_3	4061.5	4064.3	ω_3 of HOSOH
15	3928.8	3661.0	3444.0	ω_3 of NH_3	4063.9	4066.5	ω_3 of HOSOH

5. Summary

Our previous studies on the dimers of HCN with SO_2 and NH_3 predicted an a-h-bonded structure for the dimer $SO_2\cdots HCN$ where the association was solely due to weak van der Waals' (including electrostatic) interaction, while for the dimer $H_3N\cdots HCN$, it was due to hydrogen bond formation. In this latter case, the NH_3 group can serve as either a proton donor or a proton acceptor. In contrast, for the complex of ammonia with hydrogen isocyanide, both charge transfer and hydrogen bond formation serve to stabilize the dimer.

It is interesting to note that for the two complexes we studied with SO_2 as one of the monomers, the most stable structures were both anti-hydrogen bonded. To determine if this trend persists for other heterodimers involving SO_2, investigation of additional systems is needed. In addition to a stable $NH_3 \cdot SO_2$, we found an isomerized structure (NH·HOSOH) as a result of optimization of all the geometric parameters starting from an initial conformation which had the hydrogens of NH_3 pointing towards the oxygens of SO_2. Here, too, an a-h-bonded structure was preferred. However, to fully characterize this system, calculations using more extensive basis sets and additional post Hartree-Fock studies should be carried out.

6. References

1. Chen, T.S. and P.L.M. Plummer (1985), J. Chem. Phys, **89**, 3689; P.L.M. Plummer (1985), J. Rech. Atmos., **19**, 165; S. Chattopadhyay and P.L.M. Plummer (1989), Bull. APS, **34**, Abs. M 25 10; S. Chattopadhyay and P.L.M. Plummer (1990), J. Chem. Phys., **93**, 4187; S. Chattopadhyay (1991), A Computational Study on Molecular Systems Using *Ab Initio*, Semi-Empirical and Molecular Dynamics Methods, Ph.D. dissertation, University of Missouri-Columbia, Columbia, MO, U.S.A.
2. Legon, A.C. and D.J. Millen (1982), Chem. Soc., **73**, 71; A.C. Legon and D.J. Millen (1986), Chem. Rev., **86**, 635.
3. Buckingham, A.D. and P.W. Fowler (1985), Can. J. Chem., **63**, 2018.
4. Goodwin, E.J. and A.C. Legon (1986), J. Chem. Phys., **85**, 6828.
5. Chattopadhyay S. and P.L.M. Plummer (1991), *Ab Initio* Studies on the Dimer of Ammonia and Hydrogen Cyanide, Submitted for publication to J. Chem. Phys.
6. Goel, A. and C.N.R. Rao (1971), Trans. Faraday Soc., **67**, 2828; D. Bonchev and P. Cremaschi (1974), Theo. Chim. Acta., **35**, 69; P. Kollman, J. McKelvey, A. Johansson and S. Rothenberg (1975), J. Am. Chem. Soc., **97**, 955; S. Vishveshwara (1978), Chem. Phys. Let., **59**, 26; M.J. Frisch, J.A. Pople and J.E. Delbane (1985), J. Phys. Chem., **89**, 3664; K. Somasundaram, R.D. Amos and N.C. Handy (1986), Theo. Chim. Acta., **69**, 491; J.G.R. Tostes, C.A. Taft and M.N. Ramos (1981), J. Phys. Chem., **85**, 2007; G.T. Fraser, K.R. Leopold, D.D. Nelson Jr., A. Tung and W. Klemperer (1984), J. Chem. Phys., **80**, 3073; R.B. Bohn and L. Andrews (1989), J. Phys. Chem., **93**, 3974.
7. Frisch M.J., J.S. Binkley H.B. Schlegel, K. Raghavachari, C.F. Melius, R.L. Martin, J.J.P. Stewart, F.W. Bobrowitz, C.M. Rohlfing, L.R. Kahn, D.J. DeFrees, R. Seeger, R.A. Whiteside, D.J. Fox, E.M. Flauder and J.A. Pople (1986), *GAUSSIAN 86*, Carnegie-Mellon Quantum Chemistry Publishing Unit, Pittsburgh, PA 15213, U.S.A.
8. Dupuis, M., D. Spangler, J.Wendoloski, M. Schmidt and S. Elbert (1987), *GAMESS*, Version 1.02, North Dakota State University and Iowa State University, U.S.A.
9. Hehre, W.J., R. Ditchfield and J.A. Pople (1972), J. Chem. Phys., **56**, 2257; P.C. Hariharan and J.A.Pople (1973), Theo. Chim. Acta, **28**, 213.
10. Møller, C. and M.S. Plesset (1934), Phys. Rev., **46**, 618.
11. Frisch, M.J., J.E. DelBane, J.S. Binkley and H.F. Schaefer III (1986), J. Chem. Phys., **84**, 2279; Schwenke, D.W. and D.G. Truhlar (1985), J. Chem. Phys., **82**, 2418.
12. Motherwell, W.D.S. (1976), *PLUTO*, Plotting program, Cambridge Univ. Press, England.
13. Döbereiner, J. (1826), J. Chem. Phys., **47**, 119; E. Divers and M. Ogawa (1900), J. Chem. Soc., **77**, 335; I. C. Hisatsune and J. Heicklen (1975), Can. J. Chem., 53, 2646. There are many more references on experimental work on the system available. For example, the reader may refer to B. Meyer (1977), *Sulfur, Energy and Environment*, Elsevier, New York, for additional references.
14. Lucchese, R.R., K. Haber and H.F. Schaefer III (1976), J. Phys. Chem., **98**, 7617.
15. Herzberg, G. (1966), *Electronic Spectra of Polyatomic Molecules*, von Nostrand, New York.
16. Pulay, P., G. Fogarasi, F. Pang and J.E. Boggs (1979), J. Am. Chem. Soc., **101**, 2550.

STRUCTURAL INSTABILITY BY PHASON SOFTENING IN ICOSAHEDRAL QUASICRYSTALS

Y. ISHII

Department of Material Science
Himeji Institute of Technology
Kamigouri-cho, Ako-gun, Hyogo 678-12, Japan

ABSTRACT. The structural instability driven by soft phason modes in icosahedral quasicrystals is studied. Hydrodynamical instability due to slowing down of relaxation of the phason fields is discussed in comparison with thermodynamical instability, which leads to a spontaneous onset of the uniform phason strains. Diffuse scattering line shapes due to the phason softening are investigated.

1. Introduction

As in conventional incommensurate systems, we have phason degrees of freedom besides phonon degrees of freedom in quasicrystals. In analogy with martensitic transformation, which is a spontaneous freezing of shear strains, we may consider a spontaneous freezing of the phason strains [1]. The resulting modulated quasicrystalline structures could be either periodic or non-periodic. Actually some of crystalline intermetallic compounds found in a similar stoichiometry to the quasicrystalline phases can be viewed as modulated quasicrystals with the uniform phason strains [2].

Recently it is found that heat treatment of rapidly quenched Al-Cu-Fe alloys produces distinct broadening of diffraction peak lines at about 600°C while the resolution limited sharp spots characterizing the icosahedral phase are obtained at higher temperatures [3,4]. Samples annealed at lower temperatures or slowly solidified exhibit a periodic microcrystalline structure [5]. Goldman, Shield, Guryan and Stephens [3] and Bancel [4] have speculated that the structural transition at about 600°C is driven by some soft phason modes. Ishii [6] has derived a special mode of the phason strains giving the rhombohedral crystalline phase of Al-Cu-Fe alloys and has shown that the splitting of the Bragg peaks due to the freezing of this special mode is consistent with the broad line shapes around the structural "transition" temperature.

Energetics of modulated phases and possible scenario of structural transformation are of special interest. Widom [7] and Henley [8] have discussed elastic instability due to the phason softening in a context of the random tiling model for quasicrystals. There the quasicrystalline order is stabilized by entropy and the phason stiffness constant should be proportional to temperature. Then the system becomes unstable against onset of the uniform phason strain at a sufficiently low temperature. They have found that singularity occurs at a temperature different from that associated with the onset of the uniform phason strains for two-dimensional pentagonal quasicrystals. Recently Hatwalne and Ramaswamy [9] have pointed out that the soft phason mode in icosahedral solids could induce characteristic deformation of the reciprocal lattice, which is different from that expected for the uniform phason strain configuration favored by the higher order terms in the generalized elastic free

P. Jena et al. (eds.), Physics and Chemistry of Finite Systems: From Clusters to Crystals, Vol. I, 217–222.
© 1992 *Kluwer Academic Publishers.*

218

energy.

In this paper, we shall discuss structural instability induced by the phason degrees of freedom in icosahedral quasicrystals within a phenomenological theory. Detailed analysis will be published elsewhere [10]. Recently Widom [11] has obtained similar results to those presented here independently.

2. Hydrodynamical vs. Thermodynamical Instabilities

Two types of instabilities associated with the phason degrees of freedom can be imagined: One is softening of the phason modes, at which a diffusion constant of a particular phason mode goes to zero. Then diffusion and relaxation of the phason fluctuation become extremely slow and eventually the soft mode fluctuation is frozen (hydrodynamical instability). The other possibility is violation of thermodynamical stability conditions associated with an onset of the uniform phason strains (thermodynamical instability). From the symmetry consideration, the onset of the uniform phason strains proceeds as a first order transition.

More precisely, let us consider icosahedral quasicrystals. For icosahedral quasicrystals, the elastic free energy is given as

$$\frac{f}{k_{\mathrm{B}}T} = \frac{1}{2}\sum_{\mathbf{q}}\left(\vec{u}_{\mathbf{q}}^{\dagger}\widehat{A}\vec{u}_{\mathbf{q}} + \vec{v}_{\mathbf{q}}^{\dagger}\widehat{B}\vec{v}_{\mathbf{q}} + 2\vec{v}_{\mathbf{q}}^{\dagger}\widehat{C}\vec{u}_{\mathbf{q}}\right), \tag{2.1}$$

where

$$\vec{u}_{\mathbf{q}} = \frac{1}{V}\int d\mathbf{r}\,\exp(-i\mathbf{q}\mathbf{r})\vec{u}(\mathbf{r}), \qquad \vec{v}_{\mathbf{q}} = \frac{1}{V}\int d\mathbf{r}\,\exp(-i\mathbf{q}\mathbf{r})\vec{v}(\mathbf{r}), \tag{2.2}$$

are the phonon and phason fields with wavevector $\mathbf{q} = (q_1, q_2, q_3)$. The stiffness tensors are given by

$$(\widehat{A})_{ij} = \mu q^2 \delta_{ij} + (\lambda + \mu)q_i q_j, \tag{2.3}$$

$$\widehat{B} = \frac{K_4 + 2K_5}{3}q^2\;\widehat{1} + \frac{K_4 - K_5}{3}\begin{pmatrix} \tau q_2^2 - \frac{1}{\tau}q_3^2 & 2q_1 q_2 & 2q_1 q_3 \\ 2q_2 q_1 & \tau q_3^2 - \frac{1}{\tau}q_1^2 & 2q_2 q_3 \\ 2q_3 q_1 & 2q_3 q_2 & \tau q_1^2 - \frac{1}{\tau}q_2^2 \end{pmatrix}, \tag{2.4}$$

$$\widehat{C} = K' \times \begin{pmatrix} -q_1^2 - \frac{1}{\tau}q_2^2 + \tau q_3^2 & -2/\tau q_1 q_2 & 2\tau q_1 q_3 \\ 2\tau q_2 q_1 & \tau q_1^2 - q_2^2 - \frac{1}{\tau}q_3^2 & -2/\tau q_2 q_3 \\ -2/\tau q_3 q_1 & 2\tau q_3 q_2 & -\frac{1}{\tau}q_1^2 + \tau q_2^2 - q_3^2 \end{pmatrix}, \tag{2.5}$$

where $q = |\mathbf{q}|$ and τ is the golden mean, $(\sqrt{5}+1)/2$. λ and μ are the conventional Lamè coefficients and K_4 and K_5 are phason stiffness constants associated with two normal modes of the phason strains. K' is a coupling constant of the phonon-phason mixing.

The stability conditions against uniform deformation, i.e., the onset of the uniform phonon (elastic) and/or phason strains, (*thermodynamical* stability), are

$$\lambda + \frac{2}{3}\mu > 0, \qquad K_4 > 0, \qquad \mu > 0, \qquad K_5 > 0, \qquad \text{and} \qquad \mu K_5 > 3K'^2. \tag{2.6}$$

From the symmetry consideration, the structural transition caused by the onset of the uniform strains proceeds as a first order transition. Then one should go beyond the quadratic couplings and examine the third order invariants of the phason strain fields to find the most favorable structure of the strained quasicrystal. For icosahedral quasicrystals, there are 20 minimal third order couplings in the elastic free energy; 4 for the elastic (phonon) strains, 5 for the phason strains and 11 for the phonon-phason coupling. The third order invariants for the phason strains may cause instability of the icosahedral phase, which is assumed to be locally stable. The analysis of the most favorable structure has been done by Biham, Mukamel and Strinkman [12] and more thoroughly by Ishii [13].

The equilibrium fluctuations of the phonon and phason fields are readily calculated as

$$< u_{\mathbf{q}}^{(i)} u_{-\mathbf{q}}^{(j)} >= \widehat{A}^{-1} + \widehat{A}^{-1}\widehat{C}^{\dagger}\left(\widehat{B} - \widehat{C}\widehat{A}^{-1}\widehat{C}^{\dagger}\right)^{-1}\widehat{C}\widehat{A}^{-1}, \qquad (2.7a)$$

$$< v_{\mathbf{q}}^{(i)} v_{-\mathbf{q}}^{(j)} >= \left(\widehat{B} - \widehat{C}\widehat{A}^{-1}\widehat{C}^{\dagger}\right)^{-1}, \qquad (2.7b)$$

$$< u_{\mathbf{q}}^{(i)} v_{-\mathbf{q}}^{(j)} >= -\widehat{A}^{-1}\widehat{C}^{\dagger}\left(\widehat{B} - \widehat{C}\widehat{A}^{-1}\widehat{C}^{\dagger}\right)^{-1}, \qquad (2.7c)$$

where $< \cdots >$ means thermal average with respect to the elastic free energy defined in Eq.(2.1). Because the response of the phonon fields is expected to be fast compared with that of the phason fields, the phason motion is effectively followed by a phonon cloud or a lattice distortion. Thus the phason fluctuation is given by an inverse of the effective phason stiffness tensor $\widehat{B} - \widehat{C}\widehat{A}^{-1}\widehat{C}^{\dagger}$ rather than that of the bare one. For the phonon fluctuation in Eq.(2.7a), the first term is a contribution from a fast fluctuation around the "equilibrium" displacement due to the phonon-phason coupling while the second term is a contribution from the slow fluctuation of the "equilibrium" displacement.

According to hydrodynamic theory of quasicrystals, the phason degrees of freedom is a diffusive mode whereas the phonon is a propagating one [14]. The diffusion constants of the phason degrees of freedom are proportional to eigenvalues of $\widehat{B} - \widehat{C}\widehat{A}^{-1}\widehat{C}^{\dagger}$. Let us assume that the smallest eigenvalue $\lambda_{*}^{\perp} \equiv D_{*}q^{2}$ of $\widehat{B} - \widehat{C}\widehat{A}^{-1}\widehat{C}^{\dagger}$ with a normalized eigenvector $\vec{e}_{*}^{\perp}$ becomes minimal when $\mathbf{q}$ points to a particular direction, $\mathbf{q}_{*}$. When D_{*} goes to zero, diffusion and relaxation of the phason field, which is polarized in $\vec{e}_{*}^{\perp}$ and varies along the $\mathbf{q}_{*}$ direction, become extremely slow and eventually the soft mode fluctuation is frozen. This is *hydrodynamical* instability of quasicrystals.

The stiffness tensors and their eigenvalues depend on the direction of $\mathbf{q}$. By examining the eigenvalues as a function of the direction of $\mathbf{q}$, we find that the minimal eigenvalue is obtained when $\mathbf{q}$ points to one of the special symmetric directions of an icosahedron except for a small region in the physical parameter space (Fig.1). The hydrodynamical stability conditions are then obtained as

$$K_{5} - \frac{4K'^{2}}{\lambda + 2\mu} > 0, \qquad (2.8)$$

for the minimal eigenvalue obtained when $\mathbf{q}$ points to the vertex direction of an icosahedron,

$$\frac{8K_{4} + K_{5}}{9} - \frac{4K'^{2}}{9(\lambda + 2\mu)} > 0, \qquad (2.9)$$

for the minimal eigenvalue obtained when $\mathbf{q}$ points to the face direction and

$$\frac{\tau^{-2}K_{4} + \tau^{2}K_{5}}{3} - \frac{\tau^{2}K'^{2}}{\mu} > 0 \quad \text{and} \quad \frac{K_{4} - K_{5}}{3} + \frac{K'^{2}}{\mu} > 0, \qquad (2.10a)$$

$$\frac{\tau^{2}K_{4} + \tau^{-2}K_{5}}{3} - \frac{\tau^{-2}K'^{2}}{\mu} > 0 \quad \text{and} \quad \frac{K_{4} - K_{5}}{3} + \frac{K'^{2}}{\mu} < 0, \qquad (2.10b)$$

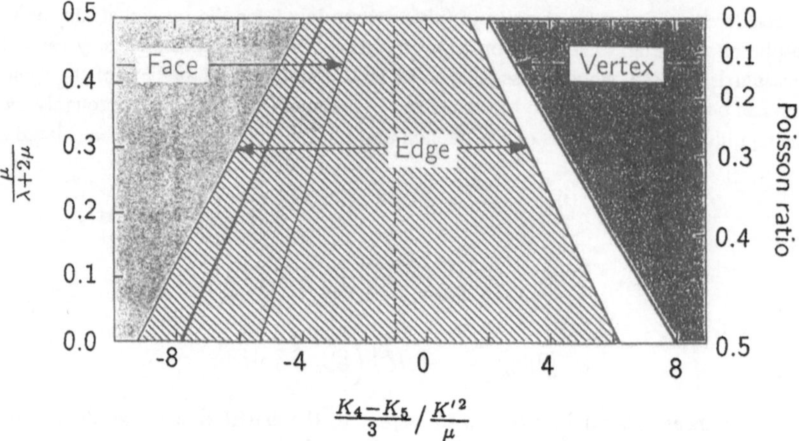

Fig.1: Phase diagram of the minimal directions. Thick lines denote phase boundaries. Dark-shaded, hatched and bright-shaded regions represent that the vertex, edge and face eigenvalues, respectively, can be minimum.

for the minimal eigenvalue obtained when $\mathbf{q}$ points to the edge direction.

The soft phason mode giving the minimal eigenvalue induces the special configuration of phason fluctuations, which is described by a matrix $\underline{\Phi} \equiv (\hat{q}_{*i} e^{\perp}_{*j})$ where $\hat{q}_{*i} = q_{*i}/|\mathbf{q}_*|$. Approaching to the instability (2.9), for example, we have the large amplitude of the phason modes described by

$$\underline{\Phi} = \frac{1}{3} \begin{pmatrix} 1 & 1 & 1 \\ 1 & 1 & 1 \\ 1 & 1 & 1 \end{pmatrix}, \tag{2.11}$$

where $\mathbf{q}_* \parallel (1,1,1)$. (Equivalent expressions are obtained for $\mathbf{q}_*$ in the other face directions.) The phason fluctuation described by Eq.(2.11) is regarded as a precursor to a uniform deformation associated with the symmetry change $I_h \rightarrow D_{3d}$ because $\underline{\Phi}$ in Eq.(2.11) belongs to the unit representation of D_{3d}. It should be noted that the soft mode fluctuation (2.11) is quite different from the phason fields frozen in the low-temperature phase of Al-Cu-Fe [6]. Similarly the hydrodynamical instabilities described by Eqs. (2.8) and (2.10) induce the precursor fluctuations to uniform deformations associated with the symmetry change $I_h \rightarrow D_{5d}$ and $I_h \rightarrow C_{2h}$, respectively.

Note that the hydrodynamical stability conditions (2.8)-(2.10) can be satisfied even if one or more thermodynamical stability conditions (2.6) are violated. This means that the thermodynamical instability and the onset of the uniform strains always interrupt into growing of the soft phason fluctuations. Then soft phason modes described by the phason matrix $\underline{\Phi}$ is not relevant to the actual structural transition. Physically the special configuration of the phason strains favored by the higher order terms in the structural energy, which is generally different from that determined by the soft mode, freezes via a first order transition [13].

3. Diffuse Scattering Line Shape

The soft phason mode fluctuation leading to hydrodynamical instability can be observed in scattering experiments. Theoretically the x-ray scattering intensity is proportional to the structure factor.

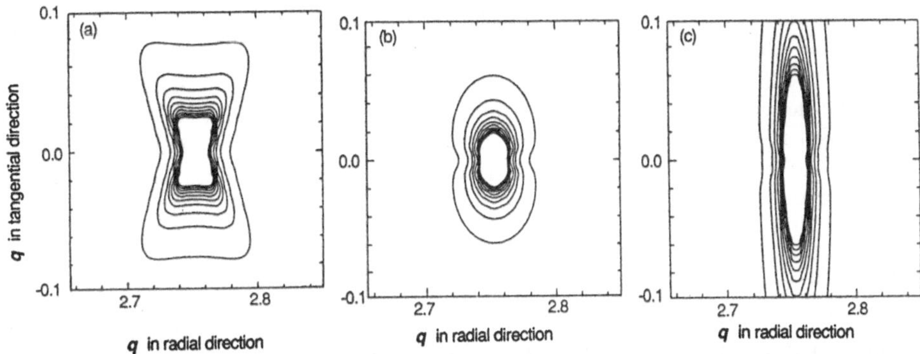

Fig. 2 : Diffuse scattering line shapes of the (222200) spot in the 2-fold symmetry plane. Contours represent $|F_{\mathbf{G}}(\mathbf{q})| = 20 \times n(n = 1, \cdots 10)$. Horizontal axis is the radial direction of $\mathbf{q}$. Lamè coefficients are taken as $\lambda = 750$ and $\mu = 150$. The other parameters are taken as (a)$K_4 = 150$, $K_5 = 5$ and $K' = 15$, (b)$K_4 = 0$, $K_5 = 75$ and $K' = 15$, and (c)$K_4 = 50$, $K_5 = 50$ and $K' = 50$ corresponding to the vertex, face and edge instabilities, respectively.

General formulas for diffuse scattering from quasicrystals are derived by Jarić and Nelson [15]. Here we adopt a simple expression as

$$S(\mathbf{q}) = \sum_{\mathbf{G}} |\rho_{\mathbf{G}}|^2 \exp(-2W) \left[\delta(\mathbf{q} - \mathbf{G}^{\parallel}) + F_{\mathbf{G}}(\mathbf{q} - \mathbf{G}^{\parallel}) \right] \tag{3.1}$$

where the amplitude of the mass density wave $|\rho_{\mathbf{G}}|$ is assumed to be independent of the spatial coordinates. The Debye-Waller (DW) factor and the diffuse scattering part are defined as

$$2W = \sum_{\mathbf{q}} F_{\mathbf{G}}(\mathbf{q}), \tag{3.2}$$

$$F_{\mathbf{G}}(\mathbf{q}) = \sum_{i,j} \left[G_i^{\parallel} G_j^{\parallel} < u_{\mathbf{q}}^{(i)} u_{-\mathbf{q}}^{(j)} > + G_i^{\perp} G_j^{\perp} < v_{\mathbf{q}}^{(i)} v_{-\mathbf{q}}^{(j)} > + 2 G_i^{\parallel} G_j^{\perp} < u_{\mathbf{q}}^{(i)} v_{-\mathbf{q}}^{(j)} > \right], \tag{3.3}$$

where $\mathbf{G}^{\parallel}$ and $\mathbf{G}^{\perp}$ are real and perpendicular components of the reciprocal lattice vector $\mathbf{G}$, respectively.

Figure 2 illustrates the diffuse scattering line shapes corresponding to the three hydrodynamical instability of the phason softening. Since the growing phason fluctuation may be interrupted by the onset of the uniform phason strains, which proceeds as a first order transition, it is unlikely to observe a critical behaviour so clearly but we still expect that the characteristic diffuse scatterings are observed before the first order transition takes place. Careful single crystal measurements are certainly desired to test the idea of the soft phason modes.

4. Conclusion

The structural instability driven by the soft phason mode in icosahedral quasicrystals are discussed. Two possible instabilities associated with the phason degrees of freedom, i.e., hydrodynamical and thermodynamical instabilities, are investigated. It is shown that three distinct hydrodynamical

222

instabilities are expected depending on the physical parameters. The three hydrodynamical instabilities induce the precursor fluctuations to uniform deformations associated with the symmetry change, $I_h \to D_{5d}$, $I_h \to D_{3d}$ and $I_h \to C_{2h}$. If the system stays in an equilibrium, however, the thermodynamical instability interrupts into growing of the soft phason fluctuations to lead to a first order transition into a different strained phase with the frozen phason strain configuration. Even so we expect that the characteristic diffuse scattering line shape is observed before the first order transition takes place.

References

[1] M. V. Jarić and U. Mohanty, Phys. Rev. Lett. **58**, 230 (1987); Phys. Rev. B**38**, 9434 (1988).

[2] Y. Ishii, J. Non-cryst. Solids, **117/118**, 840 (1990).

[3] A. I. Goldman, J. E. Shield, C. A. Guryan and P. W. Stephens, *Proc. of the Adriatico Research Conference on Quasicrystals*, edited by M. V. Jarić and S. Lundqvist, (World Scientific, Singapore, 1990), pp.60–73.

[4] P. Bancel, Phys. Rev. Lett. **63**, 2741 (1989), *ibid* **64**, 496 (1990).

[5] M. Audier and P Guyot, *Quasicrystals and Incommensurate Structures in Condensed Matter*, edited by M. J. Yacaman, D. Romeu, V. Castaño and A. Gómez, (World Scientific, Singapore, 1990); see also *Proc. of the Adriatico Research Conference on Quasicrystals*, edited by M. V. Jarić and S. Lundqvist, (World Scientific, Singapore, 1990), pp.74–91.

[6] Y. Ishii, Phil. Mag. Lett. **62**, 393 (1990).

[7] M. Widom, *Proc. of the Adriatico Research Conference on Quasicrystals*, edited by M. V. Jarić and S. Lundqvist, (World Scientific, Singapore, 1990), pp.337–355.

[8] C. L. Henley, in *Quasicrystals and Incommensurate Structures in Condensed Matter*, edited by M. J. Yacaman, D. Romeu, V. Castaño and A. Gómez, (World Scientific, Singapore, 1990).

[9] Y. Hatwalne and S. Ramaswamy, Phys. Rev. Lett. **65**, 68 (1990).

[10] Y. Ishii, To be published.

[11] M. Widom, To be published in Phil. Mag. Lett.

[12] O. Biham, D. Mukamel, and S. Strinkman, *Introduction to Quasicrystals*. edited by M. V. Jarić (Academic, Boston, 1988), Chap.5.

[13] Y. Ishii, *Quasicrystals*, edited by T. Fujiwara and and T. Ogawa, (Springer, Berlin, 1990) pp. 129–138.

[14] T. C. Lubensky, S. Ramaswamy and J. Toner, Phys. Rev. B**32**, 7444 (1985).

[15] M. V. Jarić and D. R. Nelson, Phys. Rev. B**37**, 4458 (1988).

Positive and Negative Helium Cluster Ions Formed From Supercritical Expansions

T. Jiang, S. Sun, and J. A. Northby
Physics Department, University of Rhode Island
Kingston, RI, 02881

ABSTRACT. When helium gas expands through a 5 μm nozzle into a vacuum under conditions for which the stagnation entropy is lower than the critical point entropy, cluster formation proceeds by liquid fragmentation rather than vapor condensation. When these clusters are bombarded by 35 eV electrons, predominantly large positive cluster ions are produced. Their size, determined by combining energy analysis with velocity measurements, ranges from less than 10^3 atoms to more than 10^6. The fall-off at large masses is approximately exponential. When the electron energy is reduced to 2 ± 2 eV, it is no longer possible to produce positive cluster ions, but when the stagnation entropy drops below a threshold of $S_0 \approx 5.5$ kJ/kg·K a strong negative cluster ion signal appears. In contrast to positive clusters, no small negative clusters are found: we set a lower bound on their size of about 5×10^5 atoms. We also find that their lifetime in a field of 10^3 V/cm exceeds 10^{-4} sec. While their minimum size is consistent with a model in which an electron is trapped on the surface by the polarization potential, the observed stability in a field argues against it. We suggest instead that the structure may be an electron metastably trapped in a bubble state inside the droplet.

1. Introduction.

In early studies of helium cluster beams by Gspann and co-workers[1], and by Stephens and King[2], neutral clusters were formed by condensation of supersaturated helium gas in a low temperature, low pressure free jet expansion. Detection took place by electron impact ionization followed by mass analysis of the resulting positively charged cluster ions. Gspann's operating conditions led to quite large cluster ions, typically in the range $N = 10^6 \rightarrow 10^8$ atoms, while Stephens and King utilized conditions in which only small cluster ions were produced. It is likely that significant evaporation accompanies the ionization process so, particularly for small clusters, information about the parent neutral cluster mass distribution is masked. In more recent experiments by Toennies and co-workers[3-6], neutral clusters beams have been formed with small nozzles and much higher stagnation pressures. At very high pressures, the entropy of the gas in the stagnation chamber may be reduced below the critical point entropy. In an isentropic expansion from such conditions the local thermodynamic state of the system passes continuously into the liquid state and enters the metastable region as a liquid. In such a "supercritical expansion" one expects that clusters form by fragmentation of the liquid jet and not by condensation. In these experiments clusters were detected by electron impact ionization also, but only relatively small cluster fragment ions were studied. If large cluster ions were present they could not have been observed. We have constructed an experiment to determine the size of the larger cluster ions formed in such supercritical expansions directly, by measuring their energy. This can be combined with knowledge of their velocity[3,7] to determine the mass distribution. We have found that by varying the energy of the ionizing electrons it is possible to produce either positively or negatively charged clusters. Our positive

P. Jena et al. (eds.), Physics and Chemistry of Finite Systems: From Clusters to Crystals, Vol. I, 223–228.
© 1992 *Kluwer Academic Publishers.*

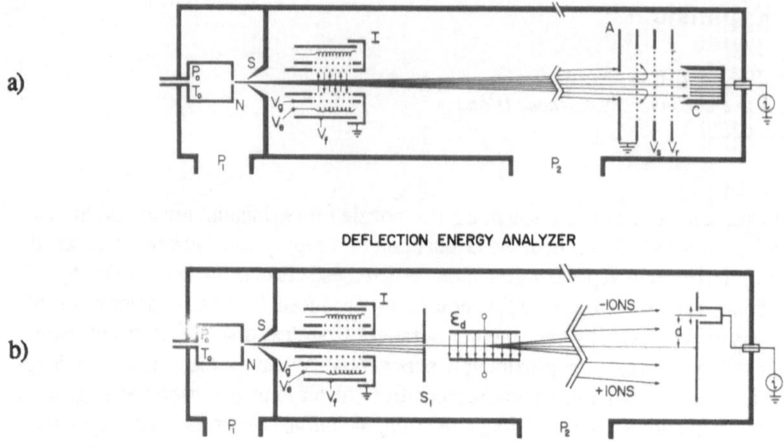

Figure 1. Schematic of apparatus showing the two different methods of energy analysis.

ion results will be described elsewhere[8]. This discussion will focus primarily on our studies of negatively charged clusters. We note that negative helium clusters have been recently observed by Gspann[9] also.

2. Experiments.

A schematic representation of our apparatus is shown in figure 1. Highly purified helium gas is expanded from a stagnation chamber at temperature (T_0) and pressure (p_0) through a 5 μm nozzle (N) into a diffusion pumped vacuum chamber ($p_1 < 10^{-3}$ torr). The cluster beam which forms from the expansion passes through a skimmer (S) into a second vacuum chamber (p_2: operating pressure $\approx 1 \times 10^{-6}$ torr; base pressure $\approx 5 \times 10^{-8}$ torr) where it is ionized by an electron impact ionizer (I), whose electron energy (eV_e) can be varied. Since the clusters of interest here are fairly large they should by largely undeflected by the ionization process. After this point we measure their energy by one of two methods. In the first (1a), a stopping potential energy analyzer (A), has an entrance aperture which collimates the beam to an angular spread of 1.4°, followed by three grids. The first is at ground potential (as is the entire flight path). A variable positive or negative potential (V_s) is applied to the second, and the third grid is normally operated at $V_r = -10$ V to prevent electrons emitted from the collector by photons or metastable atoms from contributing to the current. When V_s is positive, for example, the DC current ($i^+(V_s)$) reaching the Faraday collector measures the current of positively charged cluster ions whose energy exceeds eV_s, less any negative ion current which is present. The energy distribution is obtained by differentiating the collector current with respect to the stopping potential. One significant advantage of this method is that the collector current is measured by a DC ammeter and there is no uncertainty associated with the mass dependent sensitivity of charge amplifying detectors. The energy resolution of this method is quite good also ($\approx 2\%$). It has two drawbacks, however. First, when both positive and negative carriers are present it is complicated to distinguish between them. Second, as constructed the apparatus is limited to 1 kV repelling voltage and as a result we can't stop clusters whose energy exceeds 1 keV. These disadvantages are not present in the second method, however. As shown in 1b, in this

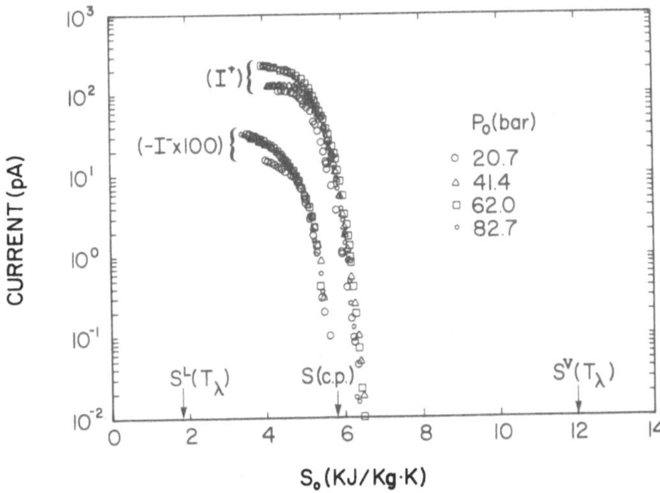

Figure 2. Positive (I^+) and negative (I^-) cluster ion current at four different source pressures as a function of the stagnation entropy. The critical point entropy, S(c.p.), and the liquid and vapor entropy at the lambda point ($S^L(T_\lambda)$ and $S^V(T_\lambda)$) are indicated for comparison.

method the beam passes through a collimating aperture S_1 and between two deflecting plates. The electric field ($\mathcal{E}_d$) necessary to deflect a cluster to a detector which is mounted a distance (d) off the axis is proportional to its energy. Thus a measurement of the detector current vs. deflecting field gives us the energy distribution directly. This allows us to separate positive and negative contributions, and to measure much more energetic clusters. We can also vary d to measure the beam profile at fixed $\mathcal{E}_d$, and to determine the energy resolution. The principle disadvantage of this method is that its energy resolution is not as high as that of the stopping potential method.

3. Results.

Our first measurements were made with the stopping potential energy analyzer, with an ionizing electron energy eV_e=35 eV. Under these conditions, mostly positively charged clusters are produced. With stopping potential V_s = 30 V, the current $I^+ \equiv i(V_s$=30 V:V_e=35 V) measures the number of cluster ions whose energy exceeds 30 eV ($N>10^4$). If we decrease T_0 at fixed p_0 there is a fairly well defined threshold below which a massive cluster ion signal appears and grows rapidly in intensity. This threshold occurs when the stagnation entropy approaches the entropy at the critical point, $S_0 \equiv S(p_0,T_0) \approx S(c.p.)$, and we interpret I^+ as the current of fragmentation clusters. In fact, I^+ proves to be a universal function of $S(p_0,T_0)$, as shown in figure 2. With the ionizer energy at 35 eV we found evidence of a small negatively charged component of the beam, but it was not possible to study it in the presence of the much larger positive ion signal. Thus we reduced the ionizer energy to $eV_e \approx 2 \pm 2$ eV. With this low energy it is not possible to produce positive ions of any kind. We found, however, that below a well defined threshold value of the stagnation entropy we were able to generate a strong beam of negatively charged cluster ions, $I^- \equiv i(V_s$=0,V_e=2 V). This threshold was significantly lower than that of the positive ion. Again, as shown in figure 2, I^- also proves to be a universal function of the stagnation entropy. Since our positive ion data indicate that the mean cluster size increases with decreasing S_0, this suggests that negative clusters are produced

226

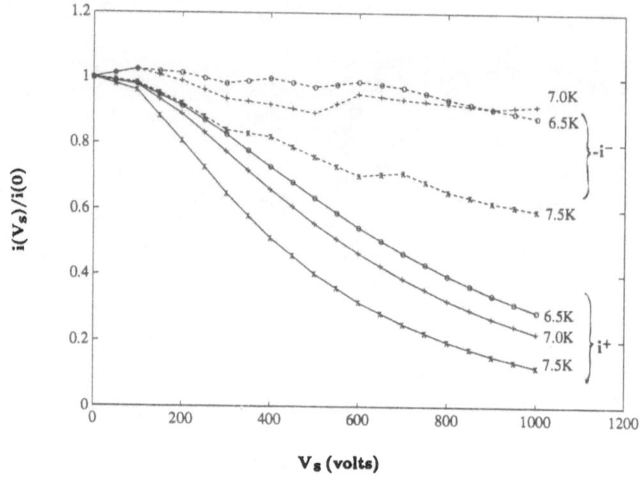

Figure 3. Normalized positive and negative ion currents, $i^+(V_s)$ and $i^-(V_s)$

only from the heavier fraction of the beam. More evidence on this point is provided in figure 3, where we compare the V_s dependence of the positive and negative cluster currents, $i^+(V_s)$ and $i^-(V_s)$, at p_0=20.7 bar for three different temperatures. We see, for example, that at 6.5 K only about 30% of the positive clusters have energies in excess of 1000 eV, while more than 90% of the negative clusters do. The situation at 7 K is similar. At 7.5 K it is possible to attenuate the negative signal with V_s < 1000 V, but not nearly as much as the positive signal. All this suggests that negative clusters are significantly more massive than positive clusters. The positive ion signal indicates that small clusters are present in the neutral beam, and thus the explanation for the lack of small negative clusters must lie in the stability and/or formation dynamics of the cluster ions themselves. There is another important but less obvious conclusion which we can draw from these data. The analyzer grid spacing is 1 cm and thus the analyzer repelling field was as large as 1000 V/cm. If electrons could be detached from clusters by this field during their 10^{-4} sec transit of this region they would not reach the collector. Since no significant current reduction was observed, we must conclude that the lifetime of the negative ion exceeds 10^{-4} sec in a field of 1000 V/cm.

Finally, in figure 4 we show the log of the energy distribution of the negative cluster current, as measured with the deflection energy analyzer, at several temperatures for p_0=20.7 bar. For fixed p_0 and T_0, mass is proportional to energy, but the proportionality constant $(2/v^2)$ varies from curve to curve[3]. For example, at 6.5 K, 1 keV corresponds to $N \approx 9 \times 10^5$ atoms while at 8.5 K, 1 keV corresponds to $N \approx 7 \times 10^5$ atoms. The energy resolution of this analyzer is not very sharp at present. The response to a monoenergetic beam is approximately gaussian with a FWHM of about 70% of the energy. In fact, the measured distribution at 8.5 K is only slightly wider than that we would expect for a monoenergetic beam of 700 eV clusters. Given this resolution we conclude that : (1) The smallest negative clusters detected have energies of about 700 eV at 8.5K, which corresponds to $N \approx 5 \times 10^5$ atoms. (2) There appears to be a rather sharp cutoff near this value under all source conditions. (3) Above this value the distributions pass through a temperature dependent peak before decaying exponentially at large energies (masses). (4) The decay constant is a strong func-

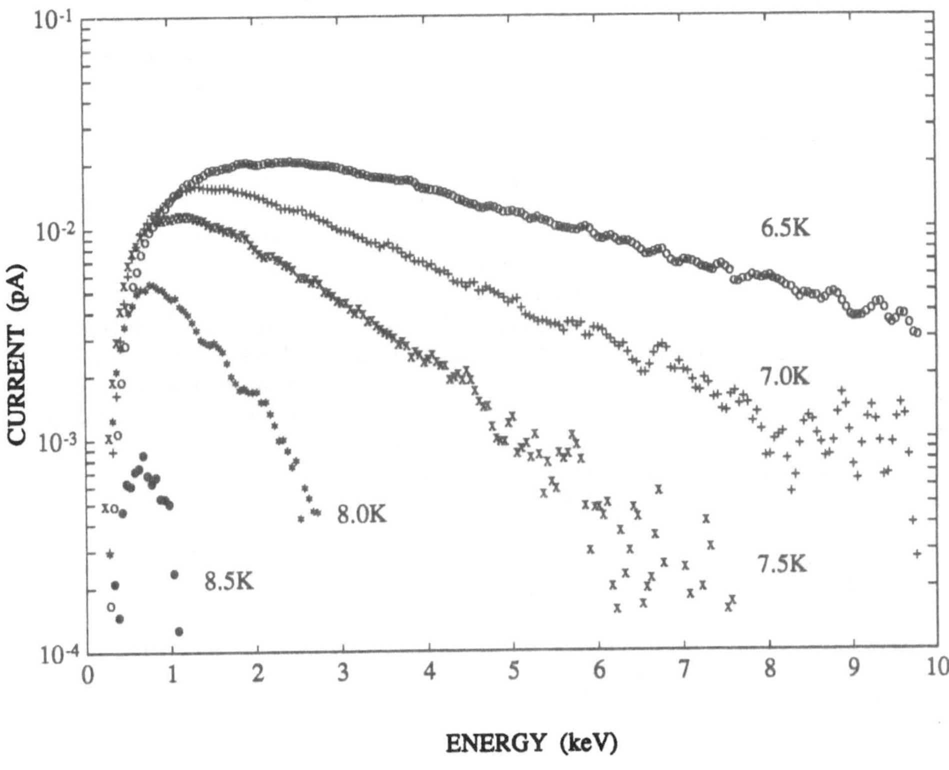

Figure 4. Energy distribution of negative cluster ion current measured with deflection energy analyzer for several source temperatures at $p_0 = 20.7$ bar.

tion of temperature. (5) At low temperatures there is a significant intensity of very energetic clusters (>10 keV corresponding to $N>10^7$ atoms).

4. Discussion.

We first ask: What we can learn about the fragmentation process from the negative ion data? The behavior of the energy distribution at low masses is clearly dominated by the ionization probability and not by the neutral beam mass distribution. It is likely, however, that the exponential decay at large energies (masses) results from a similar behavior of the precursor neutral cluster distribution. To relate the two distributions quantitatively would require an understanding of the size dependence of the ionization process. It is possible that, for these very large clusters, ionization probability is nearly unity. If this is so, and if multiply charged clusters are unstable and decay back to singly charged clusters then it is possible that the two distributions are in fact equal. We note that exponential decays at large masses are often characteristic of fragmentation processes[10].

Another interesting question is: What can we learn about the nature of the negative cluster ion itself from these data? The only model of a negative helium cluster which has been discussed to date[9,11-13] has the electron bound by the polarization potential in a surface state outside the cluster. It has been calculated[11] that the binding of such a surface electron would vanish for a cluster smaller than about 5×10^5 atoms, which is suggestively close to our lower limit on the cluster ion mass. However we have calculated[14] the lifetime of an electron in such a state on a plane surface in the presence of a normal electric field and find that in fields of >100 V/cm it is very short (< 10^{-9} sec.). Our observed stability in fields of 10^3 V/cm would appear to rule this model out. Surface states might play a role in the formation process, however. Two other possible structures are an electron in a bubble state (as in the bulk liquid[15]) metastably trapped by the polarization force <u>inside</u> the cluster, and a negatively charged impurity ion in or on the cluster. This could arise from capture of background gas atoms[6] before or after ionization of the cluster takes place. We have utilized experimental measurements of the tunneling from bulk bubble states into the vacuum[16] to estimate the field dependence of the lifetime of such a state in a cluster and we find that it should have the stability required by our experiments even for clusters smaller than our observed minimum size. We have no detailed explanation for the origin of a minimum size in this model. It would have to originate from the formation process, and not from stability considerations. We can't rule out an impurity state either, but the measurements appear to be insensitive to rather large changes in the base pressure of the system, and thus to the background impurity concentration. This makes it seem less likely that they are important. Whatever the nature of the negative ion, it is certain to be optically active. This offers both a most convenient method of determining its structure and hopefully, a means to learn about the dynamics of the cluster itself.

5.Acknowledgements.

This work has been supported by the NSF (DMR8816482)

6. References.

[1] J.Gspann,in Physics of Electronic and Atomic Collisions, S.Datz, ed.,
North Holland, Amsterdam, 1982, page 79f.

[2] P.W.Stephens and J.G.King, Phys.Rev.Lett.<u>51</u>,1538(1983)

[3] H.Buchenau, E.L.Knuth, J.A.Northby, J.P.Toennies, and C.Winkler,
J.Chem.Phys.<u>92</u>,6875,(1990)

[4] H.Buchenau, J.P.Toennies, and J.A.Northby, J .Chem.Phys. (to be published)

[5] K.Martini, J.P.Toennies, and C.Winkler, Chem.Phys.Lett.<u>178</u>,429 (1991), and private communication.

[6] A.Sheidemann, J.P.Toennies, and J.A.Northby, Phys.Rev.Lett.<u>64</u>, 1899(1990);
A.Sheidemann, B.Shilling,J.P.Toennies, and J.A.Northby, Physica<u>B165</u>, 135(1990)

[7] Typical cluster velocities are from 200 to 300m/s corresponding to energies of 0.8 $\rightarrow$ 1.3 meV/atom (ref.3). We have also measured the cluster velocity in our apparatus by synchronous gating of both V_g and V_s. If the beam has a unique velocity the result is a triangular functional dependence of the current on the gating frequency from which the time of flight can be obtained. Our results are in satisfactory agreement with those of Buchenau, et al., and we have used the latter to relate mass and energy in our experiments.

[8] T.Jiang and J.A.Northby, to be published

[9] J.Gspann, Physica <u>B169</u>,519(1991)

[10] R.Englman, J. Phys.: Condens. Matter <u>3</u>,1019(1991) ; and refs. therein.

[11] M.V.Rama Krishna and K.B.Whaley, Phys.Rev.<u>B38</u>,11839(1988)

[12] V.M.Nabutovskii and D.A.Romanov, Sov.J.Low Temp.Phys.<u>11</u>, 277(1985)

[13] J.L.Ballester and P.R.Antoniewicz, J.Chem.Phys.<u>85</u>,5204,(1986): J.L.Ballester,Thesis (Univ..of Texas at Austin, 1989, unpublished)

[14] P.Bianco and J.A.Northby, to be published.

[15] A.L.Fetter, in "The Physics of Liquid and Solid Helium", K.H.Bennemann and J.B.Ketterson, eds. (John Wiley and Sons, New York,1976), Part I, p.207f

[16] W.Schoepe and G.W.Rayfield, Phys.Rev.<u>A7</u>,2111(1973)

ELECTRONIC STRUCTURE AND REACTIVITY OF BIMETALLIC CLUSTERS CONTAINING SODIUM ATOMS

K. KAYA, A. NAKAJIMA, T. NAGANUMA, AND K. HOSHINO
Department of Chemistry, Faculty of Science and
Technology,Keio University, 3-14-1 Hiyoshi,Kohoku-Ku,
Yokohama 223, Japan

ABSTRACT.   Ionization potentials of $Al_n Na_m$, and $Co_n Na_m$ were
measured.   Appreciable decrease in the Ip value was observed as a
general tendency when Na atoms are doped to the Al or Co clusters.
In Al-Na systems, Ip increases when the total number of electrons
satisfies the electronic shell closing.   Reactivity change   toward
hydrogen adsorption was found to have nothing to do with the
behaviors of the change in the ionization potentials of $Co_n Na_m$.

1. INTRODUCTION

For the last few years, surface reactivity of metal clusters
has been studied extensively because they are suitable candidates
for active catalysts.   Dramatic cluster size dependence of the
reactivity was unveiled in the clusters composed of transition
metal elements[1].   Whether the reactivity comes from electronic
factor or geometric one has been the subject of controversy.
Anticorrelation between ionization potentials and dissociative
chemisorption rates of several transition metal clusters has been
reported. That is to say, higher the ioniztion potential, the lower
is the reaction rate.   To a first approximation, this
explanation seems to be plausible in view of the fact that the
dissociative chemisorption takes place through the electron
migration from metal to the antibonding orbital of the adsorbed
species such as hydrogen molecule[2].   Cluster size dependence of
the reactivities toward hydrogen of Co and Fe clusters has been
interpreted in terms of the variation in ionization potentials of
individual clusters.   Winker et al. have investigated the
relationship between ionization potentials, geometric structure,
and surface reactivity of copper clusters[3].   Water adsorption
equilibrium reaction on the copper cluster surface has higher
adsorption energy when cluster size is by one atom larger than
that of icosahedral geometrical closed shell packing.   That means
that one extra atom after the geometrical shell closing plays the
role of active reaction site. Their result suggests the important
role of geometrical structure.

P. Jena et al. (eds.), Physics and Chemistry of Finite Systems: From Clusters to Crystals, Vol. I, 229–234.
© 1992 Kluwer Academic Publishers.

we have developed the modified laser vaporization method for the binary cluster production using two different metal rods. This method has various advantages over hitherto examined methods for the binary cluster formation such as double crucible method and the single laser vaporization method using alloy rod. Most advantageous point is that one can choose arbitrarily two metal elements for the binary cluster formation. Various binary clusters have been produced by this method, and surface reactivity toward hydrogen of these binary clusters has been found to vary in an unexpected way[4-6]. These results were qualitatively explained either by electronic or geometric structure of the binary clusters. However, no informations have been obtained on the electronic and geometric structures of these clusters. As a basic physical constant to understand electronic property of a cluster, ionization potential seems to be essentially important.

When alkai metal atoms such as Na, K, and Cs are adsorbed on the bulk metal surface, the work function of that metal decreases appreciably and surface reactivity is enhanced enormously[7]. This phenomena were first discovered long time ago by Langmuir[8]. On the analogy of bulk metal-alkali metal atom system, binary system of alkali atom doped metal clusters seems to be a suitable model to examine the relationship between surface reactivity and ionization energy of the clusters.

In this work, ionization potentials of $Al_n Na_m$ (n>m) and $Co_n Na_m$ (n>m) were determined by the photoionization efficiency method. Surface reactivity toward hydrogen of $Co_n Na_m$ was compared with the measured Ip values.

2. EXPERIMENTAL

Details of the modified laser vaporization method was published elsewhere[5]. Two target rods of Al(or Co) and Na were vaporized by two pulsed Nd^{3+}:YAG lasers (532 nm). Generated hot metal atoms undergo cooling by He pulse and binary clusters are grown in a channel(8 cm length and 3 mm diameter). The skimmed cluster beam was irradiated by a frequency doubled pulsed dye laser or an ArF excimer laser to ionize the clusters. Mass analysis was made by a TOF mass spectrometer equipped with a reflectron. The ions were detected by a two stage multichannel plate and signal processing was done by a transient oscilloscope assisted by a microcomputer. In order to avoid multiphoton processes and fragmentation of photoionized cluster ions, the fluence of ionization laser photon was kept below 500 $\mu J/cm^2$. The photoionization efficiency spectra were measured at 0.05-0.1 eV interval. PIE curve of the binary clusters was determined after the normalization taking the mass spectra ionized by an ArF laser as standard ones. The ionization potentials were estimated from the linear extrapolation of the ion intensity as a function of ionization energy. Surface reaction rate (relative) of $Co_n Na_m$ with hydrogen was determined from the depletion of the $Co_n Na_m$ signal after the reaction with hydrogen gas in a fast flow tube.

3.RESULTS AND DISCUSSION

3-1. IONIZATION POTENTIALS OF $Al_n Na_m$

Figure 1 depicts the ionization potentials (vertical) of $Al_n Na_1$ as a function of cluster size n. Ionization potentials of bare clusters Al_n have been determined by Schriver et al.[9]. The Ip value goes up from n=1 to higher clusters and decreases suddenly from n=13 to 14. Then the value decreases gradually in line with the predicted one from spherical droplet model. As seen in the figure, Ip of $Al_n Na_1$ decreases appreciably in comparison with bare Al_n when n is less than 12, and small change in Ip is observed at the cluster size larger than 14. Inspection of the details reveals that $Al_6 Na$, $Al_9 Na$, $Al_{13} Na$, and $Al_{23} Na$ have particularly high

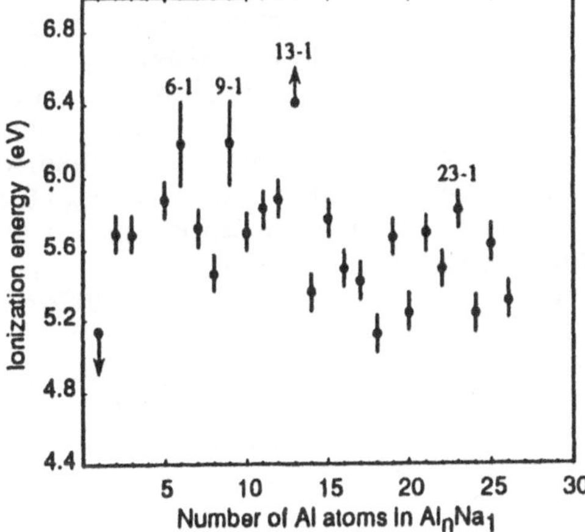

Figure 1. Ionization Potentials of $Al_n Na_1$ (n<27). The upward arrow indicates Ip higher than the solid point.

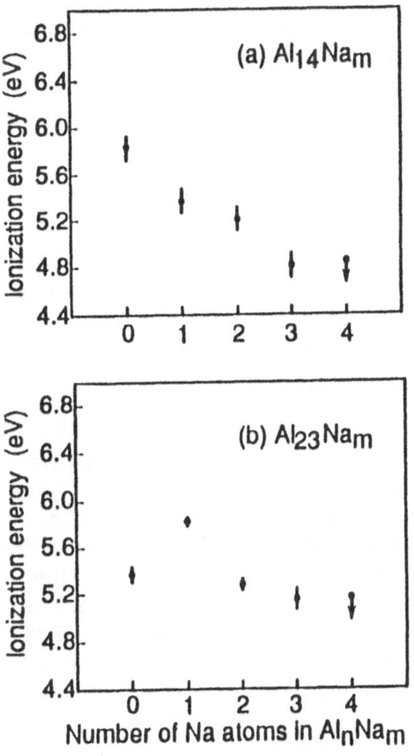

Figure 2. Ionization Potentials of $Al_n Na_m$ plotted against m for n=14(fig. a) and 23(fig. b).

Ip value. Among them, $Al_6 Na$, $Al_{13} Na$, and $Al_{23} Na$ have 19, 40 ,and 70 valence electrons, which correspond to nearly shell closing of 2s, and complete closings of 2p and 3s shell. So stability due to the electronic shell closing causes high Ip in the above three binary clusters. No explanation is given for the high Ip value of Al_9Na because the total number of valence electrons does not satisfy the electronic shell closing. In Fig. 2, ionization potential of Al_nNa_m plotted against the number of Na atoms m is shown taking n=14, and 23 as examples. $Al_{14}Na_m$ has no oppotunity to close its electronic shell. The Ip value is the monotonically decreasing function of m as seen in the figure. In contrast to this, the Ip's of $Al_{23}Na_m$ are, 5.36 eV(m=0), 5.82 eV(m=1), 5.26 eV(m=2), and 5.15 eV (m=3) with an experimental error of 0.1 eV. $Al_{13}Na_m$ behaves similarly to $Al_{23}Na_m$, i. e., Al_{13} has Ip value near 6.42 eV and $Al_{13}Na_1$ has the value far more than 6.4 eV, then the value drops abruptly to 4.7 eV in $Al_{13}Na_2$. Al_6Na_m and Al_9Na_m do not exhibit increase of the Ip when m goes from 0 to 1. These results confirm the fact that $Al_{13}Na_1$ and $Al_{23}Na_1$ are stabilized by the electronic shell closing corresponding to 2p and 3s shell, respectively.

As a conclusion, in Al_nNa_m, a valence electron of Na atom is delocalized in the Al cluster and contributes to the formation of the electron shell.

3-2. Co_nNa_m CLUSTERS; REACTIVITY AND IONIZATION POTENTIALS.

The first attempt to investigate the surface chemistry of gas phase metal clusters was reported by Smalley's group[10,11]. According to their work, dramatic cluster size dependence of the hydrogen adsorption reactivity of the pure cobalt clusters was revealed. the result was explained in terms of the anticorrelation between ionization potentials and surface reactivity of the clusters. Rosen et al. made theoretical calculation of Co clusters and found deep relationship between Fermi energy of respective clusters and the reactivity[12]. However, several examples on the surface reactivity of metal clusters suggest another factor to be more important, i.e., geometric structure. In the previous section, it was concluded that sodium atom doped on the metal cluster works as an electron donor and that the ionization potential of the Na doped cluster changes greatly as compared to the genuine metal cluster. In order to testify the influence of ionization energy on the surface reactivity, sodium doped Co clusters seem to be suitable model systems.

Cluster size dependences of the adsorption reaction rate and ionization potentials of bare cobalt clusters have been reported independently by Morse et al.[10,11], and Yang and Knickelbein (and recently by Menezes and Knickelbein)[13,14].
Combining two results altogether, anticorrelation between reactivity and ionization energy of clusters seems to hold approximately at least in bare clusters.

Figure 3 shows the ionization potentials of Co_nNa_m (m=0-3) plotted against the number of cobalt atom n. As seen in the figure, adsorption of Na atoms decreases appreciably the ionization potentials of Co clusters as a whole. Especially, Na doping affects enormously on the Co clusters with small n number (n<10). Largest variation in Ip value is at n=6 where the difference in Ip between Co_6 and Co_6Na_1 is more than 1 eV. In going from small n to the larger clusters, Ip variation becomes constant of the cluster size (about 0.4 eV). In small clusters, first atomic shell is not completed thus small Co clusters are easily deformed structurally and electronically, which results into the large decrease in Ip. Once, cluster size becomes large enough to form atomic shell, structural perturbation due to the sodium adsorption may be not important. Na atoms simply behave as an electron donor to the Co clusters and decrease the Ip value.

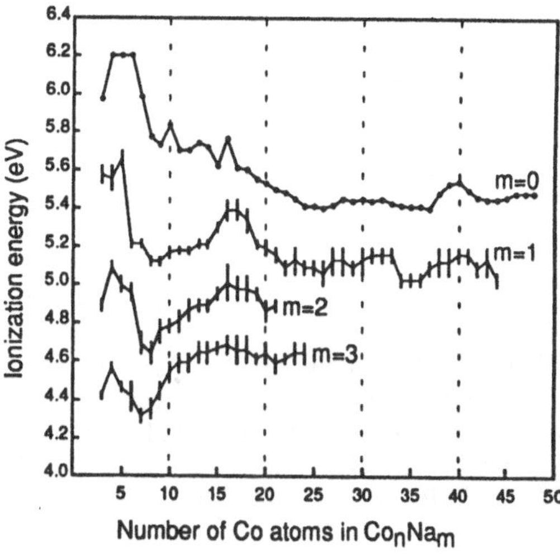

Figure 3. Ionization Potentials of Co_nNa_m plotted against n for m =0, 1, 2, and 3.

In Figure 4, adsorption reaction rate (relative) with hydrogen and Ip value of Co_nNa_m are plotted against sodium atom number m at n=8 and 15. Obviously, sodium atoms decreases the Ip value in both clusters in a similar manner. Still, reactivity changes completely differently from each other. This indicates that Ip-reactivity anti-correlation does not always hold and that geometric factor should play key role in the surface reaction.

In order to understand more about

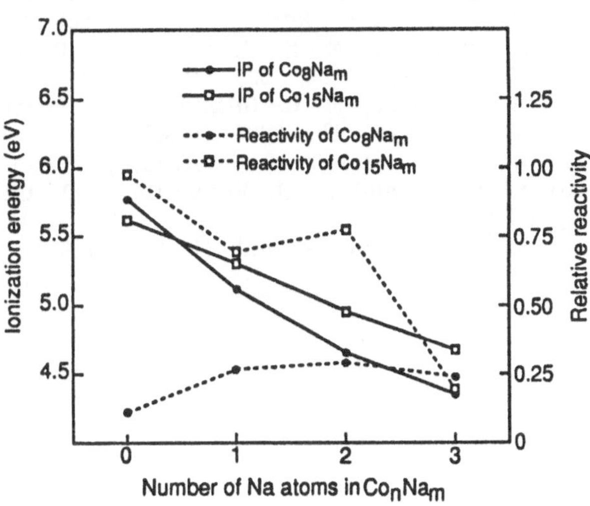

Figure 4. Ionization Potentials and Reactivity of Co_nNa_m plotted against m for n= 8 and 15.

the mechanism of the surface chemistry of clusters, information on the geometric structure seems to be essentially important.

4. ACKNOWLEGEMENT

This work is financially supported by the Grant-in-Aid for Priority Area from Ministry of Education, Science and Culture.

5. REFERENCES

[1] A. Kaldor, D. M. Cox, and M. Zakin, Adv. Chem. Phys.(1988) 211.

[2] R. L. Whetten, D. M. Cox, D. J. Trevor, and A. Kaldor, Phys. Rev. Lett. 54, 1494 (1985).

[3] B. J. Winter, E. K. Parks, and S. J. Riley, J. Chem. Phys. 94, 8618 (1991).

[4] S. Nonose, Y. Sone, K. Onodera, S. Sudo, and K. Kaya, Chem. Phys. Lett.164, 427 (1989).

[5] S. Nonose, Y. Sone, K. Onodera, S. Sudo, and K. Kaya, J. Phys. Chem. 94, 2744 (1990).

[6] A. Nakajima, T. Kishi, T. Sugioka, Y. Sone, and K. Kaya, J. Phys. Chem. 95, 6833 (1991).

[7] H. P. Bonzel, Surf. Sci. Rept., 8, 43 (1987).

[8] J. B. Taylor, and I. Langmuir, J. Am. Chem. Soc. 53, 486 (1931).

[9] K. E. Schriver, J. L. Person, E. C. Honea, and R. L. Whetten, Phys. Rev.Lett., 64, 2539 (1990).

[10] M. D. Morse, M. E. Geusic, J. R. Heath, and R. E. Smalley, J. Chem. Phys., 82, 590 (1985).

[11] M. D. Morse, M. E. Geusic, J. R. Heath, and R. E. Smalley, J. Chem. Phys. 83, 2293 (1985).

[12] A. Rosen, and T. T. Rantala, Z. Phys. D 3. 205 (1986).

[13] S. Yang, and M. B. Knickelbein, J. Chem. Phys., 93, 1533 (1990).

[14] W. J. M. Menezes, and M. B. Knickelbein, Chem. Phys. Lett. 183, 357 (1991).

METAL CLUSTERS: HOW MANY ATOMS ARE REQUIRED FOR THERMIONIC EMISSION ?

T. Leisner, K. Athanassenas, O. Echt[*], D. Kreisle, and E. Recknagel
Fakultät für Physik, Universität Konstanz, W-7750 Konstanz, Germany
[*]Dept. of Physics, University of New Hampshire, Durham, NH 03824, USA

ABSTRACT. We observe thermionic emission from free refractory metal clusters W_n, $n \geq 5$ and Ta_n, $n \geq 4$, following excitation by a Q-switched Nd:YAG laser at its 2nd, 3rd and 4th harmonic. The ion yield from this process, which extends up to 5 μs for large n, dominates over prompt multiphoton ionization if the laser fluence is chosen properly. The appearance of pronounced wavelength dependent intensity anomalies in the size dependence of the yield is traced to the quantization of the attainable excess energy in the clusters.

1. Introduction

One aim of cluster science is to pursue the quantitative change of a specific property as a function of size. For example, many experimental and theoretical studies have been devoted to determine the ionization potential, the bond length, the dissociation energy, and the melting temperature of clusters. Another crucial question pertains to the size beyond which a particular (collective) phenomenon occurs. Among these, structural (including solid-liquid) phase transitions, the transition to the metallic state, superconductivity, and ferromagnetism have been investigated. However, excepting an early report on transition metal oxides [1], the occurrence of thermionic emission from hot, isolated clusters has been noted only recently [2 - 7]. There is no a-priori reason why a very small cluster, or molecule, should not exhibit this phenomenon, and an analysis of "metastable decay" of clusters/cluster ions might be misinterpreted [8] if cluster dissociation is assumed to be the only decay channel.

Thermionic emission involves the non-adiabatic coupling of ionic and electronic motion, it is a ubiquitous, technically very important phenomenon in bulk matter. It is well described by the quantized free-electron model of metals, the Richardson-Dushman equation relates the emission rate to the temperature and work function [9]. In case of free metal clusters, excited by a short (10 ns wide) laser pulse, the obvious signature of thermionic emission is the appearance of delayed ions (and electrons). Our results to be presented below prove that delayed ionization from tungsten and tantalum clusters does occur over a time range of several microseconds after excitation, depending on cluster size.

2. Experimental

Metal clusters are formed by pulsed laser vaporization in a carrier gas. After passing two differential pumping stages, the relatively cold, neutral clusters reach the ionization zone of a time-of-flight mass spectrometer (Fig. 1). Here they are excited by the light of a Q-switched

P. Jena et al. (eds.), Physics and Chemistry of Finite Systems: From Clusters to Crystals, Vol. 1, 235–240.
© 1992 *Kluwer Academic Publishers.*

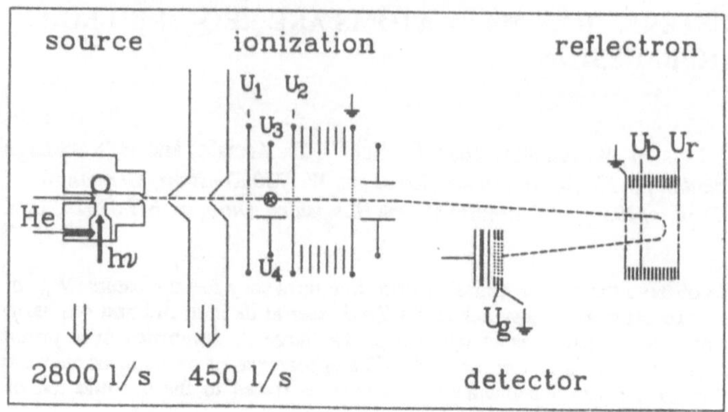

Fig. 1: Schematic diagram of the experimental setup. $\otimes$ = laser beam.

Nd:YAG laser operating on the 2nd, 3rd, or 4th harmonic (532 nm, 355 nm and 266 nm, respectively, corresponding to photon energies of 2.33 eV, 3.50 eV, and 4.66 eV). Direct one-photon ionization is energetically excluded for the species being discussed. Details of the mass spectrometer, which features collinear extraction of the ions into the drift tube and an electrostatic mirror, have been described earlier [10].

A new feature of the instrument is the special design of the ion acceleration region which enables us to unambiguously distinguish between prompt ions (generated during the laser pulse) and delayed ions. Two electrodes (kept at potential U_3 and U_4, respectively, see Fig. 1) are mounted perpendicular to the standard extraction electrodes (kept at U_1 and U_2, respectively). Slightly before and during the laser pulse, which occurs at t = 0, a strong electric field (3000 V/cm) is applied perpendicular to the neutral cluster beam and the laser beam, by setting U_3 « $U_1 = U_2 = U_4$. This blocking field is switched to zero by switching U_3 at t_1. If an ion is exposed to the blocking field for more than a few tens of nanoseconds, it will be deflected out of the beam and will not be able to reach the detector, cf. below. The remaining clusters continue to move with the characteristic velocity of the neutral cluster beam, which is approximately 1500 m/s. After 5 μs, they would enter the second acceleration stage of our ion lens. Ions born herein are suppressed by a repelling potential in front of the detector, while any positive ions formed in the time window [t_1, 5 μs] by thermionic emission (or some other process) are extracted into the drift tube towards the reflectron and the ion detector by simultaneously switching U_1 and U_2 to obtain $U_1 > U_3 = U_4 > U_2$ at $t_2 < 5$ μs.

The power of this new technique, which requires rapid switching of 3 different high voltages by as much as 3 kV, has been subjected to various tests. Most importantly, one needs to assess the minimum value of t_1 that will completely block any prompt ions. Test experiments were run with W_n^{2+} (which show up under high laser fluence) and with Ag_n^+. These species are not expected to be formed by delayed (thermionic) emission, because the (estimated) ionization potentials of their precursors, W_n^+ and Ag_n, are so much greater than their (estimated) dissociation energies. Thermionic emission is therefore quenched in favor of dissociation or, in other words, the dissociation energy limits the effective cluster temperature to a value which renders thermionic emission negligible [11]. We find that a value of $t_1 = 20$ ns is sufficient to block W_9^{2+} from the spectrum, while $t_1 = 40$ ns is required for W_{39}^{2+} (these values are

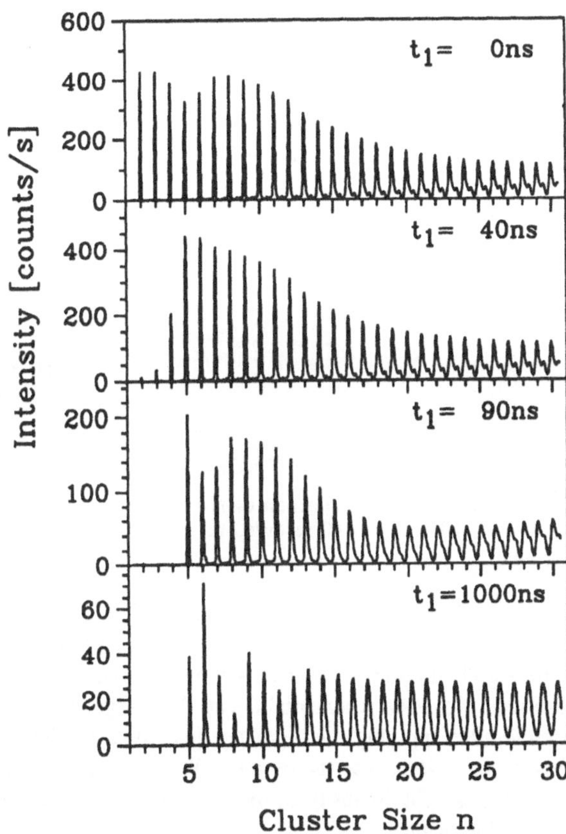

Fig. 2:
Mass spectra of tungsten clusters recorded at 355nm for different values of t_1

corrected for offsets by cable delay etc.). Quite generally, the results for W_n^{2+} ($9 < n < 39$) agree with the expected scaling law t_1 (min) = const $\sqrt{n}$.

Compared to another recent study of thermionic emission from refractory metal clusters [5], our experimental arrangement features the following advantage: The yield of delayed ionization is obtained by integrating over a specific mass peak; the time dependence of the yield is then obtained by scanning t_1. Therefore, a small signal of delayed ions can be detected in the presence of strong prompt ionization. Also, the time dependence of the yield is measured directly and need not be extracted from the peak shape of possibly contaminated mass peaks.

3. Results

So far, we have detected delayed ionization for clusters of tungsten and tantalum. As the results for both metals are quite similar, we will mainly display the results obtained for tungsten, and merely note if tantalum clusters behave differently.

The mass spectra of tungsten (Fig. 2) and tantalum clusters exhibit a large yield of ions for delay times t_1 as long as microseconds. If the fluence of the excitation laser I_L is chosen

238

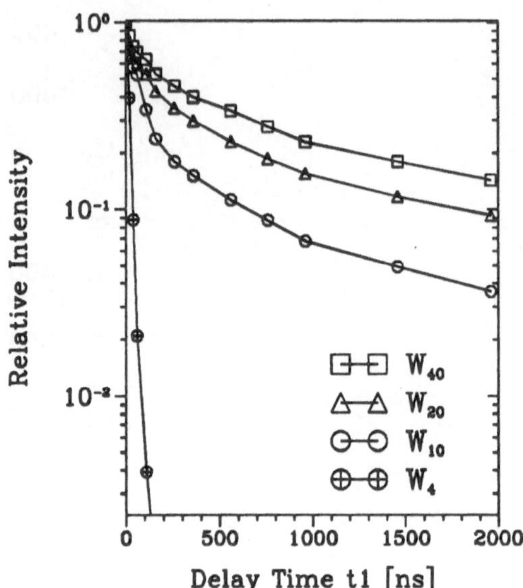

Fig. 3:
Yield of W_n^+ ions as a function of t_1
for n = 4, 10, 20, and 40 recorded at
355 nm

properly, all clusters with a size n at and above a certain minimum size n_c could be detected
with high intensity. The minimum size is found to be n_c = 4 for Ta and n_c = 5 for W.

In Fig. 3, the ion abundance is plotted versus the blocking time t_1 for some cluster sizes. It
features a slow nonexponential decay. For large clusters, more than 10% of the intensity is still
present after applying the blocking field for 2 μs. The overall decline flattens with increasing
cluster size. Note, that the tetramer W_4^+ is removed from the spectra as fast as expected for
prompt ions. The intensity of the delayed ions depends on the power of the exciting laser in a
remarkable manner:

The delayed ion signal becomes detectable at very low laser fluence ($I_L < 1$ mJ/cm^2) and
increases with increasing I_L until it reaches a pronounced maximum at moderate laser power
($I_L \approx 10$ mJ/cm^2). At higher fluence, the signal decreases and eventually vanishes (for
$I_L > 1000$ mJ/cm^2). As discussed later, this feature is well understood under the assumption
that the delayed ions are due to thermionic emission of the neutral metal clusters.

Furthermore, the mass spectra of delayed ions (Fig. 2 bottom) show pronounced intensity
modulations, which were not present in the conventional spectra recorded without blocking
pulse (Fig. 2 top). The amplitude and period of this modulation depends on the wavelength of
the excitation laser, as illustrated in Fig. 4. The 8-mer and 9-mer, e.g., represent either a local
maximum or minimum in abundance.

4. Discussion

The experimental findings presented above can easily be understood, if thermionic emission of
electrons from hot metal clusters is considered to give rise to delayed ions in our spectra.

We assume that the electronic excitations in transition metal clusters caused by single photon
absorption are strongly coupled to vibrations [5, 12]. In this case, the successive absorption of

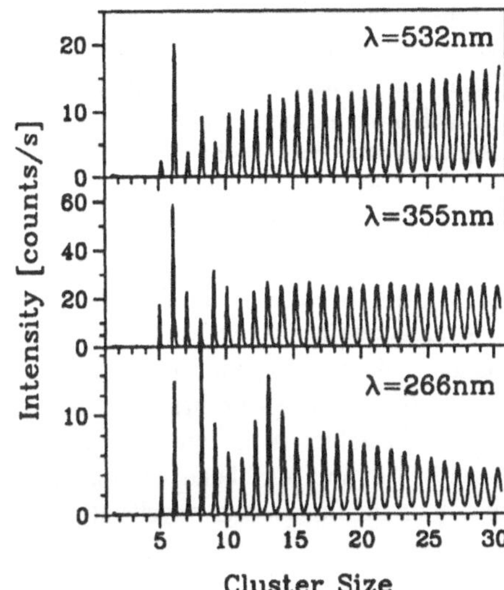

Fig. 4:
Mass spectra of W_n^+ clusters recorded at $t_1 = 1500$ ns for different laser wavelengths

photons leads to an effective heating of the clusters rather than to direct multiphoton ionization [1]. There are three possible decay channels for hot clusters:

a - dissociation (evaporation of atoms or molecules)

b - thermoluminescence ("evaporation" of photons)

c - thermionic emission ("evaporation" of electrons)

While (a) is the process usually observed for clusters prepared with high internal energy, very little is known about process (b). So far only the thermoluminescence of nanometer-sized particles has been investigated [13], and it was rationalized using Planck's law and Mie theory. A very crude estimate for W_{40} clusters suggests that above 3500 K the cooling efficiency via black-body radiation is small compared to both dissociation and thermionic emission [2]. Therefore we tentatively ignore it in the following analysis. Process (c), thermionic emission, might occur in clusters of refractory metals because their high cohesive energy allows them to become very hot without dissociating. This process may lead to the production of positive ions and is discussed in the next paragraph.

If thermionic emission is treated as a statistical process and the angular momentum of the cluster is neglected, the rate of electron evaporation from a hot cluster, k_e, is a function of the cluster's internal energy E and its ionization potential IP. Obviously, by energy conservation, k_e is zero if E is below IP. With E above the ionization potential, k_e is a very strongly increasing function of E (compare, e.g., the Richardson-Dushman eqation). Let us now consider the probability for the emission of one electron from a neutral cluster of size n within a time interval $[t_a, t_b]$ after excitation. Even if the exact dependence of k_e on E is unknown, it is easily rationalized that this probability is negligible if E lies outside a narrow interval centered at the optimum energy E_0. Clusters with $E \ll E_0$ are not hot enough to emit an electron before

---

[1] This holds true as long as the mean time between the absorption of two subsequent photons is large compared to the relaxation time of the electronic excitation in the cluster. If we use an excitation laser beam with a very high fluence, we observe many prompt ions due to direct multiphoton ionization.

t_b, while clusters with $E \gg E_o$ are very likely to emit the electron earlier than t_a [2]. This qualitatively explains our experimental observation: The laser fluence I_L needs to be chosen carefully for preparing clusters with $E \approx E_o$ in order to observe a large fraction of delayed ions born within our experimental time window $[t_1, 5\ \mu s]$.

But even if I_L is optimized, not all clusters can exactly reach $E_o(n)$, because their internal energy can only be raised in integer multiples of the photon energy. At a given photon energy, some cluster sizes can be heated to $E \approx E_o(n)$ and appear as maxima in our spectra, whereas others are either too cold (after absorption of m photons) or too hot (after absorption of $m+1$ photons), and therefore form abundance minima.

The ideas outlined above can be formulated quantitatively by using the Richardson-Dushman equation and replacing the bulk workfunction W_f by the size dependent cluster ionization potential IP(n). Some care has to be taken to use an appropriate expression for the effective cluster temperature as a function of E [7]. The resulting simple model accounts for the dependence of the ion abundance on the time t_1 as well as for the dependence of the ion intensity on the fluence and wavelength of the excitation laser.

ACKNOWLEDGEMENT

This work was financially supported by the Deutsche Forschungsgemeinschaft

REFERENCES

[1] G. C. Nieman, E. K. Parks, S. C. Richtsmeier, K. Liu, L. G. Pobo, and S. J. Riley, High Temp. Science 22 (1986) 115

[2] T. Leisner, Ph. D. Thesis, University of Konstanz, 1991

[3] T. Leisner, K. Athanassenas, O. Echt, O. Kandler, D. Kreisle, and E. Recknagel, Z. Phys. D20 (1991) 127; T. Leisner, K. Athanassenas, O. Kandler, D. Kreisle, E. Recknagel, and O. Echt, Mat. Res. Soc. Symp. Proc. Vol. 206 (1991) 259

[4] S. Maruyama, M. Y. Lee, R. E. Haufler, Y. Chai, and R. E. Smalley, Z. Phys. D19 (1991) 409

[5] A. Amrein, R. Simpson, and P. Hackett, J. Chem. Phys. 95 (1991) 1781; and this conference

[6] E. E. B. Campbell, G. Ulmer, and I. V. Hertel, submitted to Phys. Rev. Lett.

[7] C. E. Klots, Z. Phys. D20 (1991) 105; and preprint, submitted to Chem. Phys. Lett.

[8] S. K. Cole and K. Liu, J. Chem. Phys. 89 (1988) 780

[9] C. Herring and M.H. Nichols, Rev. Mod. Phys. 21 (1949) 185

[10] T. Leisner, O. Echt, O.Kandler, D. Kreisle, and E. Recknagel, Int. J. Mass Spectrom. Ion Proc. 87 (1989) R19

[11] C.E. Klots, J. Phys. Chem. 92 (1988) 5864

[12] P. J. Brucat, L. S. Zheng, C. L. Pettiette, S. Yang, and R. E. Smalley, J. Chem. Phys. 84 (1986) 3078

[13] E. A. Rohlfing, J. Chem. Phys. 89 (1988) 6103; R. Scholl and B. Weber, this conference

[14] C. Brechignac, Ph. Cahuzac, J. Leygnier, and J. Weiner, J. Chem. Phys. 90 (1989) 1492

---

[2] A similar argument is frequently used in the description of the unimolecular decay of cluster ions following electron impact or multiphoton ionization [14]. The energy $E_o(n)$ depends only slightly on t_a and t_b, if these are chosen realistically.

STRUCTURAL AND ELECTRONIC PROPERTIES OF NEUTRAL AND CHARGED Mg_n CLUSTERS ($n \leq 13$ AND $n = 57, 69$).

F. REUSE, M.J. LOPEZ*, S.N. KHANNA[+],V. DE COULON[†] and J. BUTTET
Institut de physique expérimentale, Ecole Polytechnique fédérale de Lausanne, PHB-Ecublens, CH 1015 Lausanne, Switzerland
** Departemento de Fisica Teorica y Fisica Atomica, Molecular y Nuclear, Universidad de Vallodalid, Spain*
[+]Virginia Commenwealth University, Richmond, Virginia 23284, U.S.A.
[†] I.R.R.M.A., Ecole Polytechnique Fédérale de Lausanne, PHB - Ecublens, CH 1015 Lausanne, Switzerland

ABSTRACT We present electronic calculations on neutral, cationic, anionic and doubly ionized Mg_n clusters for $n \leq 13$ and $n = 57, 69$. Our studies employ Gaussian basis functions, treat exchange-correlation effects via the local spin density approximation (LDA), and use pseudopotentials. We investigate the size evolution of the bonding for neutral and charged clusters, and show that already around $n=10$ there are clear signs of a jellium type behavior, suggesting that within LDA neutral and charged clusters from $n=10$ have acquired several of the characteristic features of the metallic state. The transition from metastability to stability for doubly charged cations is also studied.

Introduction

The elements of groups IIA and IIB of the periodic table are interesting in relation with cluster physics from several points of view. First there is a large increase in cohesion energy in going from the dimer to the solid. The dimer is a very weakly bound molecule, while the bulk phase is metallic and has a relatively large cohesion energy. Attention has recently been focused on mercury clusters, where a van der Waals to metallic transition has been suggested [1]. However it is difficult to carry out ab-initio calculations on mercury, since the presence of d electrons and relativistic effects have to be taken into account. Therefore ab-initio theoretical studies have mostly studied the size evolution of the structural and electronic properties of the light elements of column II. In particular LDA computations have been reported for Be_n [2] and Mg_n [3] , several calculations at the Hartree-Fock or Configuration interaction levels have also been published [4].

A second interest of group II elements is related to the strong modifications in the nature and strength of the bonding as a function of the charge of the clusters. This is easy to understand for dimers, where the removal of an electron increases strongly the cohesive energy, since an electron is taken out of an antibonding orbital, which is doubly occupied in the neutral dimer. The removal of an additional electron to create a doubly charged dimer further increases the binding energy. However doubly charged dimers are metastable, with a rather large potential barrier (0.48 eV for Mg_2^{++}), since there is a strong Coulomb repulsion between the two holes. In a recent paper [3] we have studied singly charged anionic Mg_n

P. Jena et al. (eds.), Physics and Chemistry of Finite Systems: From Clusters to Crystals, Vol. I, 241–247.
© 1992 *Kluwer Academic Publishers.*

clusters, as well as singly and doubly ionized Mg_n ($n{\leq}7$) clusters and investigated their structural and electronic properties. In the present paper we extend these calculations to larger clusters ($n{\leq}13$, n=57, 69) in order to follow the evolution of their properties towards the metallic behaviour. We shall in particular report results on doubly charged clusters, which become stable around n=10.

It is known that the main features of clusters formed from elements of the IA and IB columns of the periodic table are well explained within a shell model [5]. Recently it has been shown [6] that the dissociation behaviour of doubly charged clusters is also influenced by shell effects. Although very small Mg clusters are different from alkali clusters, bulk Mg is a nearly free electron metal, a shell model picture should thus appear in Mg aggregates when their size increases, it is therefore of interest to explore the occurence of shell closing effects in column II elements.

Method of calculation.

The general framework for these computations is provided by the Local Spin Density Approximation (LSDA), core effects are incorporated through the use of non-local pseudo-potentials [7]. The exchange and correlation contributions use the potential form proposed by Ceperley and Alder [8]. The molecular program is based on the expansion of the Kohn-Sham orbitals in localized gaussian basis functions, we have used for all calculations a basis formed of 4s and 2p gaussians. More details about the method and basis set are given in reference [3].

For $n{\leq}13$ the ground state geometries were obtained by systematic exploration of subsets of the Born-Oppenheimer surface, characterized by symmetry constraints. These subsets were chosen taking into account symmetry arguments, the possibility of Jahn-Teller distortions, and for the neutrals the ground state geometries obtained by ab-initio molecular dynamic calculations [9]. In the case of Mg_{57} and Mg_{69}, which are formed from bulk subunits, we have employed a newly developped program, that uses explicitly their symmetry group in order to facilitate calculations. Information about the representations carried by the eigenspaces are also obtained. This is useful in the prediction of possible Jahn-Teller distortions.

Neutral and singly charged clusters.

The geometries of neutral clusters up to n=7 have been reported [3]. They are characterized by three dimensional compact geometries. The structures of lowest energy for Mg_8 - Mg_{13} have been briefly reported in reference [9], they show strong similarities, but also diffe-rences, with the ground state geometries of Be clusters [2]. Mg_8 - Mg_{11} are characterized by a central body of six atoms at the vertices of a triangular prism. Mg_8 is a distorted triangular prism with two capping atoms on the side faces. There is however a geometry which can be visualized as a pentagonal bipyramid where an atom of the pentagonal basis is replaced by two atoms distant of 5.56 a.u., which is only 21 meV higher in energy. Mg_9 is a rectangular triangular prism (C_{3h}) with 3 capping atoms on the side faces. The structures of Mg_{10} and Mg_{11} are obtained from Mg_9 by adding respectively one and two capping atoms on the triangular bases of the prism. Mg_{12} can be viewed as a strongly distorted cubooctaedron, Mg_{13} is obtained from Mg_{12} by adding a capping atom on one face of the distorted cubo-octaedron. All the ground state structures with $n{\leq}13$ are singlet states, but in most cases we have found isomers of zero or non-zero spin very close in energy. Notice that up to n=13 all atoms are surface atoms, the first cluster with a central atom that we have found is n=16. A visual inspection of the ground state geometries shows that the 60° angles are dominant and that for several clusters it is possible to recognize one (distorted) plane with the hexagonal structure.

Mg_{57} is a subunit of the bulk structure. This has been obtained in adding to a central atom the first layer of 12 nearest neighbours, and to that layer a second shell of nearest neighbour atoms. Mg_{69} has been obtained in adding to Mg_{57} part of the third shell of nearest neighbours. The two free parameters in the structure and the spin state have been optimized for Mg_{57} in order to get the minimum energy for this particular symmetrical configuration (symmetry D_{3h}), neutral Mg_{57} has spin 3 and Mg_{69} spin 1.

The singly charged anions and cations (n = 8-13) have been obtained by keeping the symmetry of the neutral clusters, and letting the interatomic distances relax. In the case of Mg_9^+ and Mg_{11}^- we have diminished the symmetry constraints in order to accomodate a Jahn-Teller distortion, since in these two cases there are non-fully occupied degenerate levels. In all cases the bond length changes by at most 0.5 a.u., and the ground state structures have spin 1/2. In the case of Mg_{57} and Mg_{69} we have kept the geometry of the neutral cluster.

In Fig. 1 we give the atomization energies per atom of neutral, singly and doubly charged clusters. These data show that around n =10 the atomization energies per atom of charged clusters are similar, while the values for the neutral clusters are still lower by about 0.2 eV (notice that the LSD approximation tends to enhance the binding energy, this is especially true for the small Mg clusters). This behaviour is compatible with the differences of atomization energies deduced from Born-Haber cycles, and hence are related to the ionization potentials and electron affinities. Notice also that for small n there is a dramatic increase of atomization energies between neutral and singly charged cations. It is due to the removal of an electron from an antibonding orbital, as already mentioned.

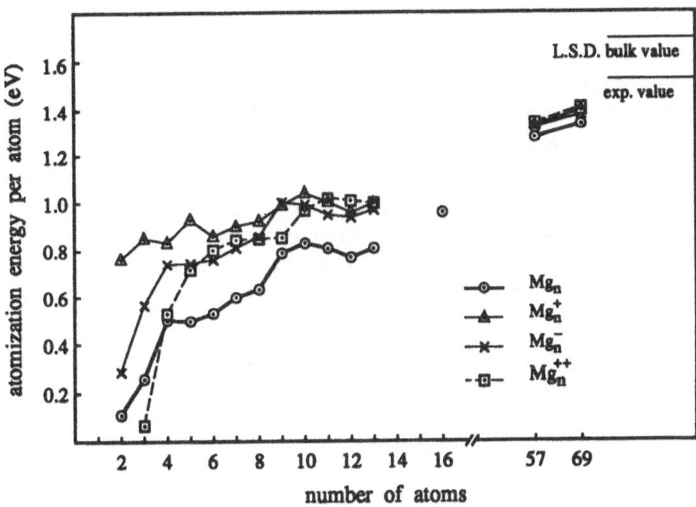

Figure 1   Atomization energies per atom of neutral, anionic, cationic and doubly ionized Mg_n clusters. The LSD bulk value is taken from ref. [10].

The analysis of the extra electron orbital for singly charged anions (see also ref. [3]) indicates that it has a bonding character between some atoms and an antibonding character between other's, but globally its effect is bonding and increases the atomization energy per atom compared to the neutral cluster. For Mg_{57} and Mg_{69} neutral and charged clusters have approximately the same atomization energies, which correspond to about 80% of the LSD bulk value, although for Mg_{57} still 75% of the atoms are surface atoms.

A closer look at the atomization energies indicates that the neutral clusters having 4, 9 and 10 atoms are particularly stable, they correspond to shell closing numbers in a spherical jellium model, having respectively 8, 18 and 20 electrons. The analysis of the Kohn-Sham energy levels shows that there is a tendency of the energy levels for Mg_9 and Mg_{10} to be grouped by subsets having 1, 3, and 5+1 molecular orbitals. Furthermore if we expand the Kohn-Sham orbitals in spherical harmonics around their center of mass, we find that the orbitals are delocalized and that their symmetries are of 1s, 1p, 1d and 2s type, in agreement with the spherical jellium model. For example we obtain for Mg_{10}, with increasing energies, orbitals having 98% s character, 93, 95 and 95% p character, 87, 93, 93, 95 and 95% d character, 79% s character. This grouping of levels is not as clear for other clusters, although it is sill present. It is also interesting to notice that these remarks are still valid for charged clusters, which also show a pronounced s, p, d, s character. The maximum atomization energy per atom appears for the anions at n=9 corresponding to 19 electrons and for singly charged cations at n=10 corresponding also to 19 electrons. The validity of a shell behaviour is still more clear for the doubly charged clusters, whose maximum atomization energy appears at n=11, corresponding to the shell closing number 20.

From this point of view one could conclude that the neutral and charged clusters of Mg have already reached the metallic behaviour, implicit in a shell model description, around n=10.

Doubly charged cationic clusters.

We have analysed in ref. [3] the structural and electronic properties of doubly charged Mg_n^{++} clusters (n ≤ 7) and shown that, although the dissociation of Mg_n^{++} in two singly charged fragments leads to a lower energy, it is still possible to calculate the equilibrium geometries which correspond to a metastable state protected from dissociation by potential barriers. For very small clusters (n ≤ 6) the equilibrium geometrical structure corresponds to a linear chain, at n=7 the stabilizing effect of the bonds already appear, and Mg_7^{++} is a pentagonal bipyramid with one atom added on each summit. However Mg_7^{++} is still metastable (0.27 eV) towards dissociation in singly charged fragments. When the size of the clusters increases we expect to obtain more compact geometries, since the stabilizing effect of the bonds prevails over the repulsion of the holes left behind by the doubly ionizing process.

In order to study these effects, and to find in particular the critical number n_c above which the doubly charged Mg clusters are stable, we have pursued our studies for larger clusters. It is however difficult to find the equilibrium geometries which, contrary to the case of singly charged clusters, are often different of the neutral ground state structures. The geometries that we propose here should thus be taken with caution. Although they correspond to the lowest energy geometrical structure obtained after having tried several other geometries with different symmetry constraints.

We have found that Mg_8 is an hexagonal bipyramid, the linear chain is only 40 meV higher in energy and a distorted cube only 120 meV higher. The lowest energy for Mg_9 keeps the symmetry of the neutral, it is still a regular triangular prism with three capped atoms, but the side of the equilateral triangle has increased from 5.8 to 7.1 a.u., while the distance of the capped atoms to the triangular prism atoms has not changed. This increasing length reflects the repulsive effect of the missing charges.This is still more clear for Mg_{10}^{++},wich keeps the symmetry of the neutral cluster, but whose length of the equilateral triangle situated near the atom capped on the triangular face increases from 5.9 a.u. to 9.25 a.u. Mg_{11} is a triangular bipyramid with one atom capped on each face of the bipyramid, Mg_{12} has a C_{3h} symmetry with 6 atoms in a middle plane and 3 atoms on each side of this plane. The lowest energy geometrical structure that we have found for n=13 is probably not final, it is a distorted cubooctaedron with spin S = 1.

$Mg_{n+m}^{++} \rightarrow Mg_n^+ + Mg_m^+$	$E_n^+ + E_m^+ - E_{n+m}^{++}$	$E_{n-1}^{++} + E_1 - E_n^{++}$
$Mg_2^{++} \rightarrow Mg^+ + Mg^+$	- 2.54	4.88
$Mg_3^{++} \rightarrow Mg_2^+ + Mg^+$	- 1.30	2.76
$Mg_4^{++} \rightarrow Mg_2^+ + Mg_2^+$	- 0.92	1.90
$Mg_5^{++} \rightarrow Mg_3^+ + Mg_2^+$	- 0.48	1.46
$Mg_6^{++} \rightarrow Mg_3^+ + Mg_3^+$	- 0.29	1.20
$Mg_7^{++} \rightarrow Mg_5^+ + Mg_2^+$	- 0.27	1.10
$Mg_8^{++} \rightarrow Mg_5^+ + Mg_3^+$	- 0.40	0.88
$Mg_9^{++} \rightarrow Mg_5^+ + Mg_4^+$	- 0.33	0.87
$Mg_{10}^{++} \rightarrow Mg_5^+ + Mg_5^+$	+ 0.39	2.03
$Mg_{11}^{++} \rightarrow Mg_9^+ + Mg_2^+$	+ 0.81	1.58
$Mg_{12}^{++} \rightarrow Mg_{10}^+ + Mg_2^+$	+ 0.21	0.84
$Mg_{13}^{++} \rightarrow Mg_{10}^+ + Mg_3^+$	- 0.02	0.79

Table 1. Lowest energy fragmentation channel and evaporation energy of doubly charged Mg_n clusters. E_n , E_n^+ and E_n^{++} correspond respectively to the total energy of neutral, singly and doubly charged Mg_n clusters. All values are given in eV.

We give in Table 1 the fission energy of Mg_n^{++} clusters, a negative sign indicating that the energy of the final products is lower than that of the doubly charged cation. As Table 1 indicates the transition from metastability to stability occurs at n=10. The slightly metastable state for Mg_{13}^{++} may be due to the fact that we have not found the real equilibrium geometry, however the maximum in stability at Mg_{11}^{++} is certainly real and reflects the shell structure of the doubly charged Mg clusters. Notice also that the dissociation products for Mg_{11} to Mg_{13} correspond to clusters near shell closing. This is in agreement with recent experiments done on noble metal clusters [6].

Several papers have recently studied the competition between fission and neutral atom evaporation in doubly charged alkali [11] and noble metal clusters [12]. This effect is governed by the relative values of the height of the potential barrier for fission and the evaporation energy neccessary to take a neutral atom out of the doubly charged cation. We give in Table 1 the evaporation energy for n=2-13, it is clearly shown that around the shell closing number there is a large increase of the evaporation energy. The behavior as a function of size of the potential barrier for fission is more difficult to determine. When the size increases the Coulomb repulsion term decreases, which tends to diminish the height of the barrier. On the other hand the fission energy tends to increase, which contributes to increase the height of the barrier. From these considerations we thus expect the relative importance of the fission and evaporation energies to be anomalous around the shell closing numbers.

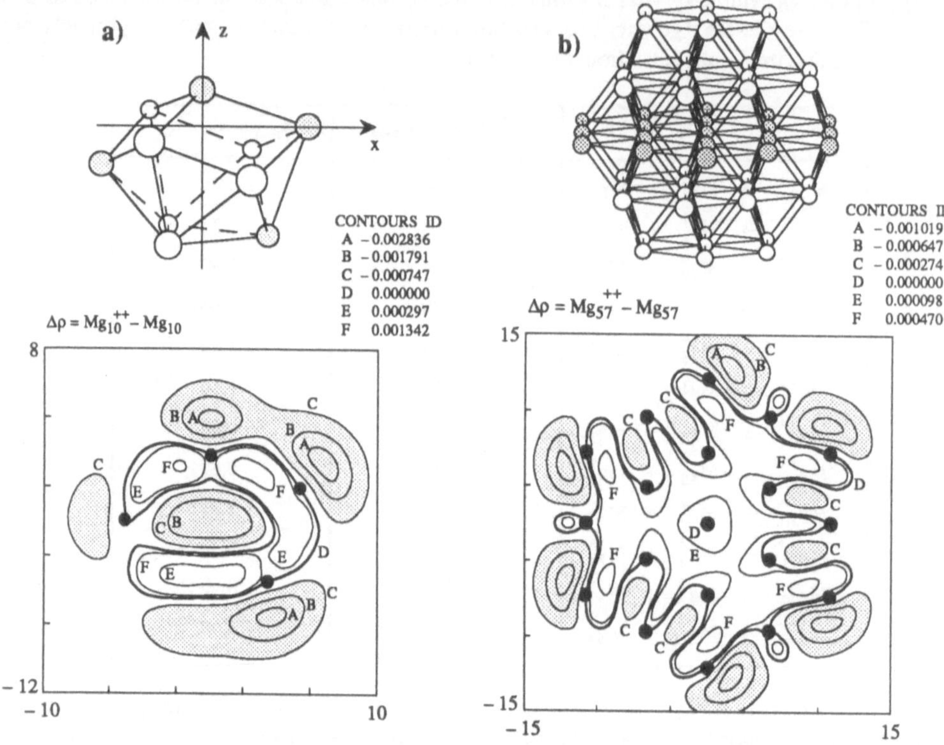

Figure 2   Contour plot of the difference between the electronic density of Mg_n^{++} and that of Mg_n with the same geometrical structure as Mg_n^{++}. The length scale in given in atomic units, the electronic density (table on the right hand side) is given in units of electron per cubic atomic unit. The shaded areas show the region of missing charges. Fig. 2a corresponds to n = 10 and Fig. 2b to n = 57.

To conclude this section we give in Fig. 2 the difference between the electronic density of Mg_n^{++} and that of a neutral cluster with the same geometrical structure. For n=57 (Fig. 2b) all the charge deficit is located at the surface of the cluster. This is strongly reminiscent of a metallic behavior. For n=10 most of the charge deficit is situated at the surface of the cluster, there are however also missing charges in the center of the cluster, which creates dipoles with a strong stabilizing effect. From that point of view Mg_{10}^{++} behaves differently of a metal, since it does not screen efficiently the interior of the cluster.

Conclusions.

We have studied neutral and charged Mg clusters and shown that already around n=10 there are clear signs of a jellium type behaviour, suggesting that neutral and charged Mg_{10} clusters have already reached the metallic state implicit in a jellium description. There are however no signs of a sharp transition between a van der Waals type of bonding and a metallic bonding, this could be related to the LDA approximation which overestimates the metallic character of condensed systems. On the other hand the screening properties of Mg_{10}^{++} cations indicate that a full metallic behavior has not yet been reached. In contrast Mg_{57} can certainly be considered as metallic. We have also studied the transition of doubly charged cations from metastability to stability and shown that it occurs between n=9 and n=10.

Acknowledgements

This work has been supported by the Swiss National Foundation under Grant No 20-26535.89.

References

1. Rademann K., (1989) Ber. Bunsenges. Phys. Chem. 93, 653. Haberland H., Kornmeier H., Langosch H., Oschwald M., and Tanner G., (1990) J. Chem. Soc. Faraday Trans. 86, 1.
2. Kawai H., and Weare J. H., (1990) Phys. Rev. Lett. 65, 80.
3. Reuse F., Khanna S. N., de Coulon V., and Buttet J., (1990) Phys. Rev. B 41, 11743.
4. Lee T. J., and Rendell A. P., (1990) J. Chem. Phys. 92, 489 and references therein.
5. de Heer W. A., Knight W. D., Chou M. Y., Chou M. L., (1987) in Seitz F., and Turnbull D. (eds.), *Solid State Physics 40*, Academic Press, New York.
6. Katakuse I., Itoh H., Ichihara T., (1990) Int. J. Mass. Spectrom. Ion Proc. 97, 47. Rabin I., Jackschath C., and Schulze W., (1991) Z. Phys. D 19, 153.
7. Bachelet G. B., Hammann D. R., and Schlüter M., (1982) Phys. Rev. B 26, 4199.
8. Perdew J. P., and Zunger A., (1981) Phys. Rev. B 23, 5048.
9. de Coulon V., Delaly P., Ballone P., Buttet J., and Reuse F., (1991) Z. Phys. D 19, 173. Kumar V., and Car. R., (1991) Z. Phys. D 19, 177
10. Moruzzi V. L., Janak J. F., and Williams A. R., (1978) Calculated electronic properties of metals, Pergamon, New York.
11. Bréchignac C., Cahuzac Ph., Carlier F., and de Frutos M., (1990) Phys. Rev. Lett. 64, 2893.
12. Saunders W. A., (1990) Phys. Rev. Lett. 64, 3046.

INTRINSIC STABILITY OF QUASICRYSTALS UNDER THE GENERATION OF FRENKEL DEFECTS

J. ROTH
Institut für Theoretische und Angewandte Physik
der Universität Stuttgart
Pfaffenwaldring 57/6
D-W-7000 Stuttgart 80
Germany

ABSTRACT. For a threedimensional diatomic icosahedral quasicrystal of the Mg(AlZn) type we have investigated the behaviour under a load of Frenkel defects for a wide range of defect concentration. We have applied relaxation methods, which give us the behaviour for zero temperature. To study the influence of a finite temperature we have randomly displace the atoms. We find a continuous transition to the amorphous phase, intensified at higher temperatures. The results are discussed with respect to experimental results.

1 Introduction

The stability of quasicrystals is an important, in general still unresolved problem. Recently we have investigated the question using perfect particularly decorated threedimensional Penrose patterns with atoms interacting via a Lennard-Jones-potential [1]. Similar work was performed later by Lançon and Billard [2].

Defects are important for the physical properties both of crystals as well as of quasicrystals. In particular, they influence the stability and can induce reversible or irreversible phase transitions.

Concerning the type of metastability of quasicrystals there are experimental and theoretical results for the relevance of point defects: In 1985 Urban et al. [3,4] studied quasicrystals with high voltage electron microscopy and showed that quasicrystals of the AlMn and AlV type can be transformed in situ into an amorphous state when exposed to 500 keV electron radiation. The transition is reversible as annealing of the irradiated sample reconstructs the quasicrystalline state. Bauer [5] extended the investigation to representatives of all known quasicrystalline classes and found, that they all perform the transition.

A possible mechanism for the transformation has been studied by Kronmüller et al. [6,7] by a rate equation theory. Assuming that as in ordinary crystals the radiation defects are Frenkel pairs of vacancies and interstitials the shape of the phase boundary between the quasicrystalline and amorphous phase was deduced in an irradiation-temperature diagram. The phase transition is driven by generation and recombination of radiation defects.

P. Jena et al. (eds.), Physics and Chemistry of Finite Systems: From Clusters to Crystals, Vol. I, 249–258.

In a previous paper[8] we have classified the interstitial sites of a monatomic three-dimensional quasicrystal and investigated the properties of Frenkel defects, particularly the transformation to an amorphous state with increasing defect concentration.

In this article we study the behaviour of Frenkel defects in a diatomic quasicrystal using two schemes. In the first one we explicitly built in the defects, in the second scheme they are generated by random displacements of all atoms. In both cases we find a continuous transition from the icosahedral to an amorphous structure, although the behaviour in the intermediate region between low and high concentration of the defects is slightly different. For the simulation *method* we refer to our previous article [1]. The method to find the interstitial sites and to generate the defects has been described in detail for the monatomic case[8]. Contrary to the monatomic case we use only stable relaxed interstitial sites chosen at random in the present simulation.

In section 2 we give a short introduction on the basis of our investigations, and in section 3 we describe the generated defects. In section 4 we investigate the quasicrystal stability as a function of the defect concentration per atom n_d. Summary and conclusions are drawn in the final section.

2 Generation, Numerical Relaxation, and Analysis of the Samples

Details of the model used, its generation, the simulation method and the tools for analysis have been described in a previous paper [1], which we review briefly. The initial configuration is a three-dimensional Ammann-Kramer-Penrose lattice. The quasilattice was decorated with two types of atoms, which were endowed with truncated Lennard-Jones-potentials. The atoms of type A are placed at the vertices and edge centers of the rhombohedra. Two atoms of type B lie on the long diagonal of the prolate rhombohedron dividing it into three parts with ratio 1:τ:1. The model(Fig. 2) is a simplified structure for (MgZn)Al[1]. Boundary effects are avoided by using rational approximants of the golden mean $\tau = \frac{1}{2}(\sqrt{5} + 1)$ and periodic boundary conditions. The initial configuration was relaxed with the help of the steepest descent method.

3 Frenkel Defects in Quasicrystals

One part of a Frenkel defect is a vacancy which as in periodic crystals is created by removing an atom of the ideal structure. The other part is a nearby interstitial. The generalization of Frenkel defects is described in detail for the monatomic case in [8]. To find the interstitial sites we use the Voronoi decomposition of the structure. The corners of the Voronoi cells are the interstitial sites, their neighbouring atoms form cells which for the present structure are slightly distorted tetrahedra. We find in total 11 different types by considering the rhombohedral cells and their possible connections. The properties of the tetrahedra are summarized in Tab. 1. By studying the potential surface in the tetrahedra it is possible to show that there exists a local minimum near the center of all cells. Only for one type (No. 10 in Tab. 1) there is no minimum. Therefore all interstitial sites are stable exept No. 10. The latter will not be used for the generation of Frenkel defects.

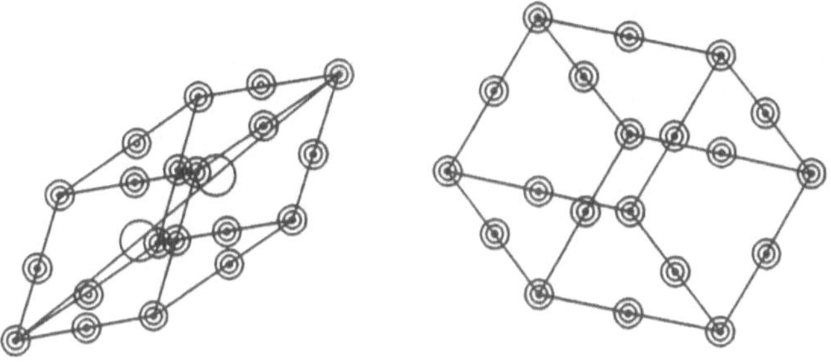

Figure 1: The two rhombohedra. The filled and open circles are A and B atoms, respectively.

number	quasilattice cell or face	radius	number
1	prolate rhombohedron	0.6292	712
2		0.6928	712
3		0.6910	2136
4		0.7102	2136
5	oblate rhombohedron	0.7292	1320
6	prolate-prolate face	0.7332	1082
7		0.7397	2164
3	polate-oblate face	0.6910	2108
8		0.7112	2108
9		0.7449	2108
1	oblate-oblate face	0.6292	266
10	unstable	0.6924	266
11		0.7484	266
total			17384
stable			17118

Table 1: The interstitial sites in the Mg(AlZn) model structure. The quasilattice contains 3016 lattice points with 2304 A-atoms and 712 are B-Atoms. The second column indicates the type of cell or face where the interstital site occurs and the third column the radius of the interstitial sites that is the unique distance to the nearest lattice atom in units of the nearest neighbour distance. The last column displays the number of these interstital sites.

For the simulation, a certain number of vacancies was generated randomly and their corresponding atoms were distributed randomly over the stable holes sites.

Apart from the systematic investigation of one realization of interstitial sites for a large number of concentrations we have studied a number of additional different realizations for a few defect concentrations to improve the statistical significance of our results. The results for all concentrations yield only small deviations from the general trend which lie within the range of the statistical error bars.

4 Results

The defect configurations generated as described in section 3 were analysed by several tools:

1. By determining E_b, the binding energy per atom, which is the negative of the potential energy per atom and is a measure of stability.

2. By the radial distribution function $G(r)$ characterizing the translational order.

3. By the bond order parameters Q_l of Ref. [1,9] displaying the orientational order. Icosahedral order can be present if Q_6, Q_{10}, Q_{12}, ... have a nonzero value and Q_4, Q_8, and Q_{14}, ... are zero.

4.1 RESULTS FOR BUILT-IN DEFECTS

For the computer simulation of the Frenkel defects we used a rational approximant with 3016 atoms, 2304 of type A and 712 of type B. We have studied a number of defects up to 1200 which yields a concentration of approximately 40%. We find a very simple behaviour. As the concentration increases, the binding energy(Fig. 2) decreases exponentially, described by the law

$$E(n_D) = a \exp(bn_D) + E_a$$

where $E(n_D)$ is the binding energy per atom, n_D is the defect concentration, E_a is the energy of an amorphous structure with the same number of atoms and a an b are two coefficients. We find $a = 0.33$, $b = -2 \cdot 10^{-3}$, $E(0) = 2.40$ and $E_a = 2.07$. To estimate the length of the error bars we have study five different realisations for 250, 500, 750 and 1000 defects and found that the deviation of the energies from the exponential law caused by different vacancies and the occupation of different types of interstitials can be explained within the error range. A simulation for a number of defect concentrations for the next higher approximant with 12776 atoms yields no different behaviour. We conclude that the decreasing binding energy is independent of the number of atoms for these sizes of the model.

The structure of the final state has been analysed by a number of structure functions. If we use the radial distribution function(Fig. 3) we find that the sharp peaks of the perfect structure are more and more washed out with increasing defect concentration and that the radial distribution function (rdf) finally becomes typically amorphous. If we consider the bond order parameter (bop) we find that the icosahedral ordering decreases monotonically with the increase of the defect concentration(Fig. 4). If we compare the rdf for a certain

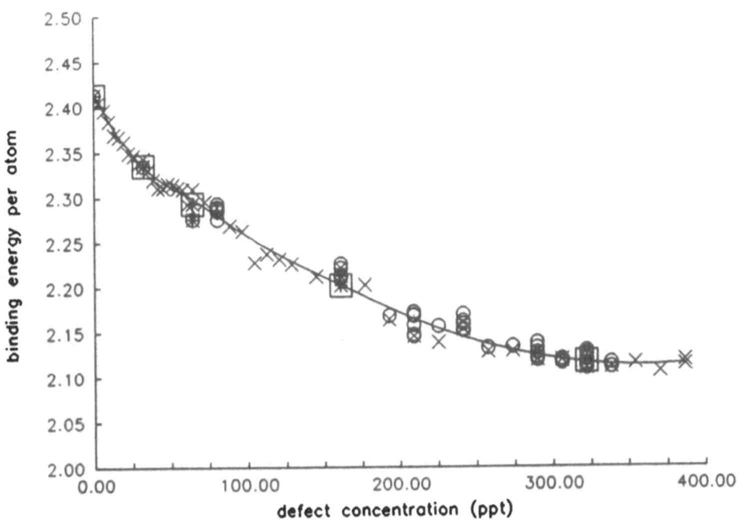

Figure 2: The binding energy for the built-in defects. The × indicate one series of simulations, o additional series with different realizations of defects. The □ stand for the simulation with 12776 atoms, ppt means defects per thousand atoms

concentration (for example at 20%) with the corresponding bop we find that the translational order, which is experimentally observed in the powder diffraction pattern, decays more rapidly than the bond orientational order. This important fact will be discussed later with respect to the recent experimental results.

4.2   RANDOM DISPLACEMENTS VS. FRENKEL DEFECTS

We use the same approximant as for the other scheme. All atoms are shifted from their ideal position by a small amount obeying an isotropic gaussian distribution. The amount of the displacement is measured by the average displacement for all atoms. We find that the binding energy(Fig. 5) of the final structure is nearly constant as long as the average displacement is smaller than about 20%. The structure is not changed in this range, all atoms remain at there place. If the displacement is larger, the energy decreases rapidly and reaches E_a at about 50%. In this range we find a tetrahedrally close packed structure without defects as for our perfect model, but some of the atoms have exchanged there position. The translational(Fig. 6) as well as the orientational order(Fig. 7) decrease monotonously. Thus we find a sharp transition in contrast to the built-in defects.

We will now look for a connection between the average displacement length and the number of Frenkel defects generated. Compared with the first scheme we assume that defects are generated when the average displacement is about 20% of the nearest neighbour distance, which yields a maximal displacement of about 80%. This must be compared to the average size of a tetrahedron which is about 70%, and thus clearly supports our assumption, because the size of the tetrahedra is the same as the distance from an atom to the lowest energy barrier on the path to an adjacent cell. We have calculated the number of atoms

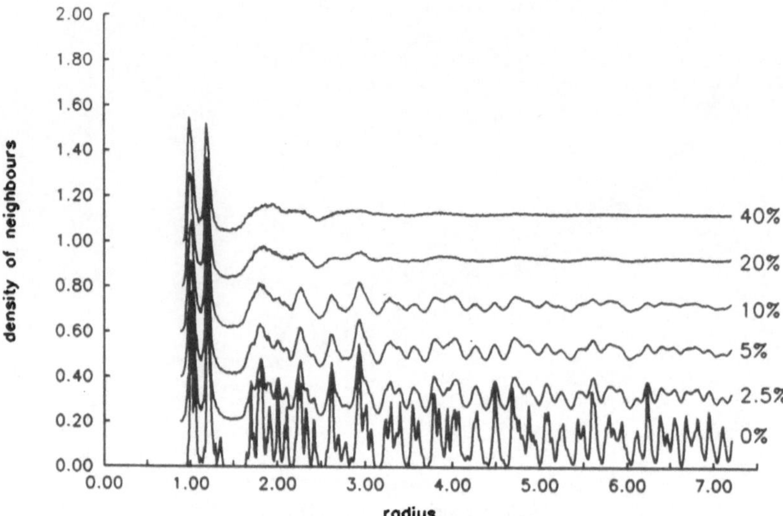

Figure 3: The radial distribution function for the built-in defects for several concentrations as indicated at the right side.

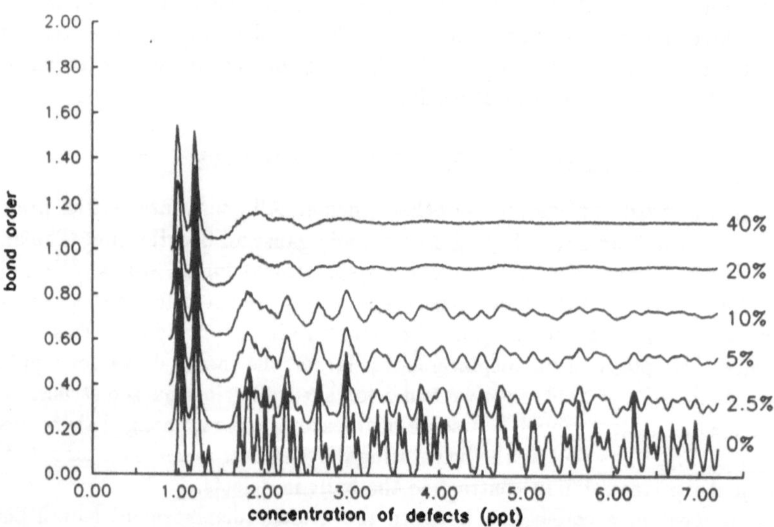

Figure 4: The bond orientational order parameters for the built-in defects for several concentrations. The signs are the following functions: $\triangle$ is Q_6, $\square$ is Q_{10}, and ∇ is Q_{12}, ppt means defects per thousand atoms.

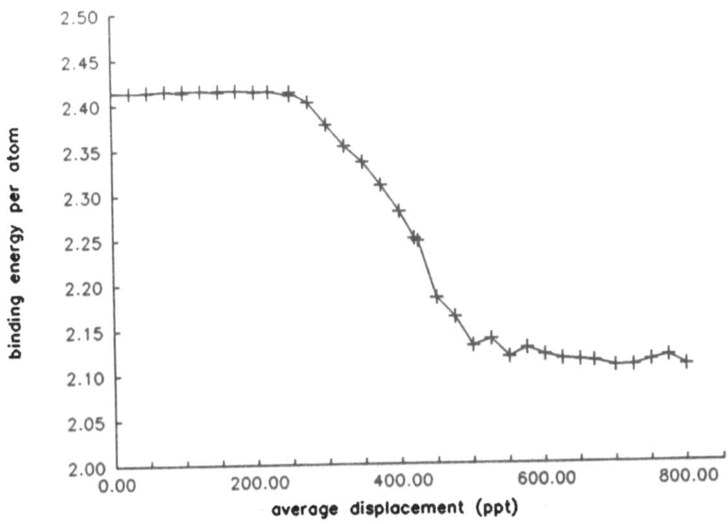

Figure 5: The binding energy for the random displacements, ppt means the thousandth part of the nearest neighbour distance.

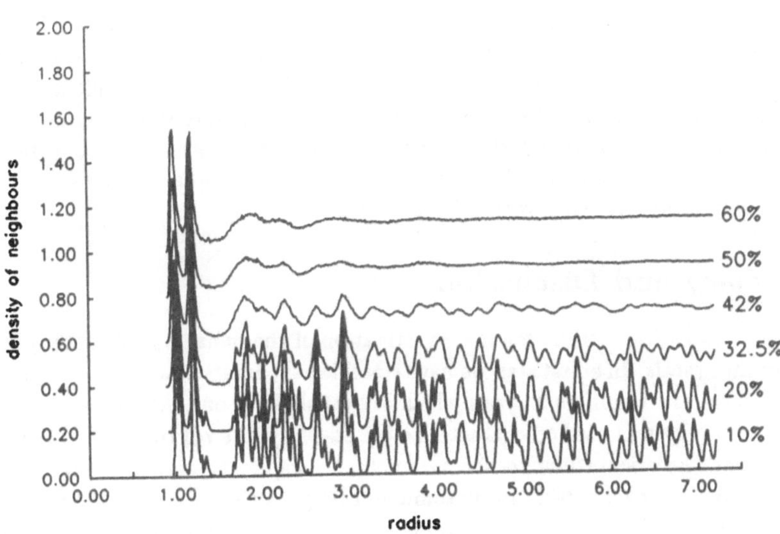

Figure 6: The radial distribution function for the random displacements for several values of the average shift, as indicated at the right side.

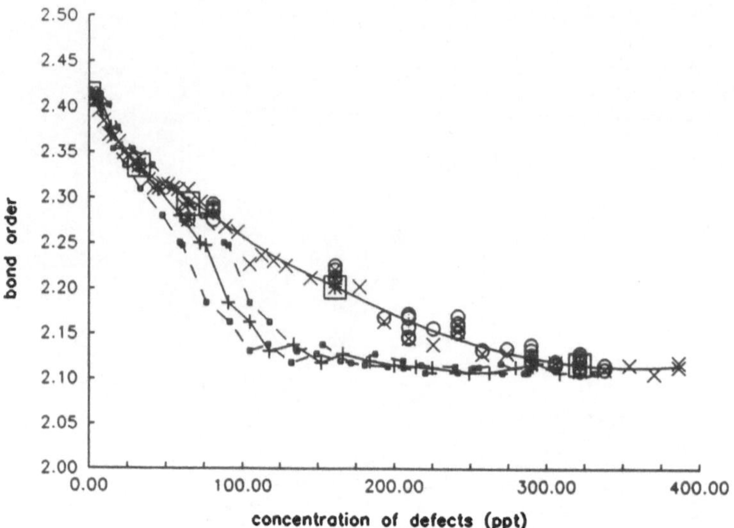

Figure 7: The bond orientational order parameters for the random displacements for several values of the average shift. The signs are the following functions: $\triangle$ is Q_6, $\square$ is Q_{10}, and ∇ is Q_{12}, ppt means the thousandth part of the nearest neighbour distance.

shifted more than a certain limit distance l, and we assume these atoms generate Frenkel defects. Thus we can translate the average displacement into a concentration of defects. We have used different l and find that the data of the first and the second scheme can be fitted best for a low defect concentration for $l = 19\%$(Fig. 8). If the average displacement and the number of defects are larger, the energy decreases more rapidly in the second scheme because all atoms are shifted. Interpreting the random displacement of the atoms in terms of a finite temperature we find that the decay of the icosahedral order is intensified due to the distortion caused by the temperature fluctuations.

5   Summary and Discussion

We have used a relaxation method for the investigation of the behaviour of Frenkel defects in diatomic quasicrystals. In a first step we have classified the interstitial sites for a diatomic quasicrystal. There are 11 different types which are stable with one exception.

The binding energy was used to characterize the stability, the rdf to describe the translational order, and the orientational order parameter to display the orientational order.

We have studied Frenkel defects in diatomic quasicrystals using two different schemes of generation. We have found that if the atoms are shifted randomly by a certain amount there exists a quit sharply defined limit $l = 19\%$ of the nearest neighbour distance above which the structure becomes amorphous. If we construct the defects explicitly, and leave the structure outside the defects undisturbed, we find a continuous transition to the amorphous state.

Within this model it is not possible to study the influence of a temperature high enough

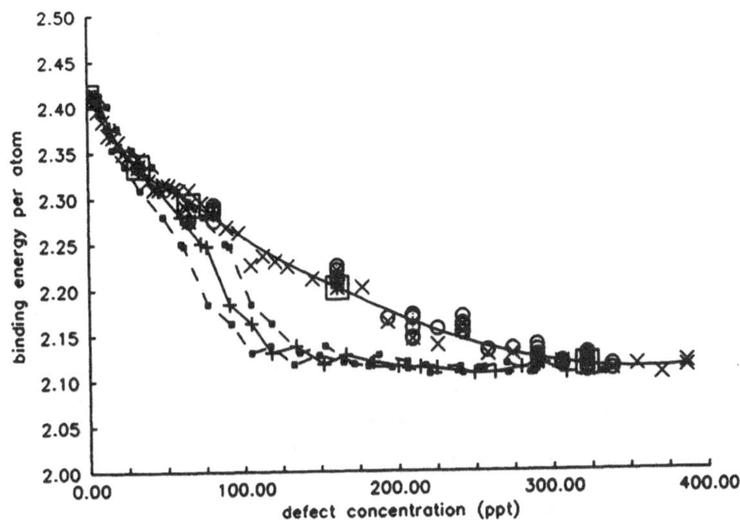

Figure 8: The binding energy for both schemes compared. The upper solid line are the built-in defects, the lower solid line the random displacements for $l = 19\%$. The dashed lines display the result for a treshold of 18 and 20% respectively. The signs are the same as for Fig. 2 and 5.

to allows the interstitials as well as the vacancies to diffuse and to recombine. For a further comparison of our simulations with the experimental results we plan to anneal the quasicrystal and look for the growth of the icosahedral seeds. This simulation seems to be possible, because the icosahedral structure is more stable than the amorphous phase and a crystalline phase has not been observed up to now.

In contrast to the theory of Kronmüller *et al.* [6], which presents critical defect concentrations of 0.2% [6] and 2.0% [7] we find a continous transition from the icosahedral state to an amorphous structure at zero temperature. But if we simulate a temperature influence the decay is intensified and at about 12% defects the structure is completely amorphous.

Recent investigations[10,11] have thrown new light on the question of the nature of the amorphous phase. It this not sure, if the amorphous phase is truely amorphous or if the icosahedral symmetry is only hidden. It is possible that there are still icosahedral seeds which grow on annealing. This fact would fit well to our results: we find that the translational order decreases more rapidly than the orientational order. We find a range which seems to be amorphous but is not truely amorphous.

The author gratefully acknowledges the collaboration with Prof. H.-R. Trebin and Prof. R. Schilling.

References

[1] J. Roth, R. Schilling, H.-R. Trebin, Phys.Rev. B **41**, 2774 (1990)

[2] F. Lançon and L. Billard, J.Non-Cryst.Sol. **117/118** 836 (1990)

[3] K. Urban, N. Moser, H. Kronmüller, phys.stat.sol. (a) **91**, 411 (1985)

[4] K. Urban, J. Mayer, M. Rapp, M. Wilkens, A. Csanady,J. Fidler, J.Phys. (Paris) **47**, C3-465 (1986)

[5] M. Bauer, Diploma Thesis, Stuttgart (1987)

[6] H. Kronmüller, Cryst.Latt.Def. & Amorph.Mat. **14**, 137 (1987)

[7] H. Kronmüller, A. Hoffmann, M. Bauer in **Quasicrystalline Materials**, Proc. of the I.L.L./CODEST Workshop in Grenoble, eds. Ch. Janot, J.M. Dubois, World Scientific, Singapore, p. 29 (1988)

[8] J. Roth, R. Schilling, H.-R. Trebin, to be published in Z.Physik Cond.Matt.

[9] P.J. Steinhardt, D.R. Nelson, M. Ronchetti, Phys.Rev. **B28**, 784 (1983)

[10] J.L. Robertson, S.C. Moss, K.G. Kreider, Phys.Rev.Lett. **60**, 2062 (1988)

[11] L.C. Chen, F. Spaepen, J.L. Robertson, S.C. Moss, K. Hiraga, J.Mater.Res. **9**, 1871 (1990)

DETERMINATION OF THE SPIN STATE FOR CARBON CLUSTERS:
FOCUSING EFFECT BY HEXAPOLE MAGNETIC FIELD

H. SHIROMARU, M. MIZUMACHI and Y. ACHIBA
Department of Chemistry, Tokyo Metropolitan University
Hachioji, Tokyo 192-03
Japan

N. YANASE, N. KOBAYASHI and Y. KANEKO
Department of Physics, Tokyo Metropolitan University
Hachioji, Tokyo 192-03
Japan

ABSTRACT. The existence of carbon clusters C_4 and C_6 with triplet
electronic states has been verified from the observation of a focusing
effect in a hexapole magnetic field. The present results directly show
the existence of even-numbered carbon clusters with a linear chain
structure.

1. Introduction

Carbon clusters have been extensively studied on account of their
importance in processes such as interstellar opacity and combustion [1].
The recent success in the macroscopic production of C_{60} has brought new
aspects in the research of carbon clusters, leading them to the field of
material science [2]. Even for rather smaller carbon clusters, however,
there still remain many interesting and important problems to be
understood. For example, understanding relatively small carbon clusters
provides an important clue to help clarify molecular evolutions in
universe. Indeed, C_3 and C_5 clusters have actually been detected in the
circumstellar shell of IRC+10216 [3,4].
    The stability of linear form for the carbon clusters C_n, with n-odd
have been confirmed on a basis of experimental and theoretical works.
For n-even carbon clusters, however, their geometrical structures are
still controversial. Two distinct types of equilibrium structure, linear
chain and ring, have been proposed to characterize their stability. Most
of the theoretical reports have shown that the chain form is almost
isoenergetic compared with the ring form for C_4 [5] and larger clusters
[6,7]. According to the results of ab initio calculations, the ground
electronic states of even numbered clusters are triplet ($^3\Sigma$) if they
exist as a linear form, whereas those of rhombic or monocyclic ring form
are singlet. Thus, the determination of the spin states of the carbon

P. Jena et al. (eds.), Physics and Chemistry of Finite Systems: From Clusters to Crystals, Vol. I, 259–264.
© 1992 *Kluwer Academic Publishers.*

260

usters directly leads to understanding of their geometrical structures.

For n-even clusters, so far two different conclusions have been deduced from the gas phase experiments. The high values of the electron affinities for n-even clusters implied that these clusters up to n=8 are linear possessing triplet character [8], whereas the preferential rhombic structure of C_4 was suggested by the results of a Coulomb explosion experiment [9]. Recently, the presence of the ring form C_6 (and also C_5) molecules was also deduced from the results of Coulomb explosion work [10]. By means of high resolution spectroscopy, Heath and Saykally [11] have shown the existence of a linear form of C_4.

On the other hand, in rare gas matrices, evidence for a linear or slightly bent chain form for n-even carbon clusters has been demonstrated by observing the triplet states in ESR spectra [12,13]. However, as pointed out by the authors of ref. 13, some ambiguities arising from the matrix effect can not be eliminated. In the present study, the spin states of carbon clusters up to n=6 were successfully investigated in the gas phase by measuring the degree of deflection of the carbon cluster beam in an inhomogeneous magnetic field.

The inhomogeneous magnetic field provided by a hexapole focusing lens [14] was used to identify the magnetic characters of carbon neutral clusters. The existence of the clusters possessing triplet in character was distinguished by measuring the enhancement of ion signals for clusters of a specific size. The magnetic moments of the clusters of specific size can be evaluated from the magnetic field strength (that is, the focal length of the lens) required to obtain the best focusing conditions.

2. Experimental

Figure 1 shows a schematic drawing of the apparatus employed in this work. Carbon clusters were produced by a laser vaporization method [15]. The 2nd harmonic from Nd-YAG laser was focused onto a rotating

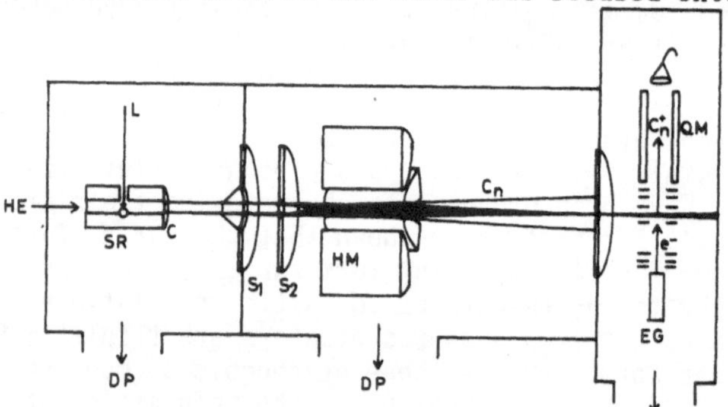

Figure 1. Schematic drawing of the apparatus used in the present study. SR:sample rod, L:laser for vaporization, C:channel, S1:skimmer, S2:collimator, M:hexapole magnet, S3:slit, EG:electron gun, QM:Q-pole mass filter, EM:electron multiplier, DP:diffusion pump, TP:turbo-molecular pump.

graphite rod. A pulsed flow of He gas was used for cooling and clustering.

The neutral carbon clusters seeded in He gas were introduced into the hexapole magnetic field through a pin hole of a skimmer. The hexapole magnet (16cm long) was actuated by a water cooled coil of copper tube. After passing through the magnetic field and a field free drift tube, clusters were introduced into the detection chamber. Two apertures of 3mm in diameter were placed at the entrance of the magnet and the exit of the drift tube so that only the beam which passed through the exit slit could be ionized in the detection chamber. Here only small part of the singlet species in the beam can enter into the detection region, because the singlet components diverge, being independent of the magnetic field. The neutral clusters in the detection chamber were ionized by an electron impact and analyzed by a quadrupole mass filter. The raw signals were measured using a storage oscilloscope and signals synchronized with the pulsed flow were averaged and recorded. The intensity of magnetic field was determined by a Hall generator.

3. Results and Discussion

The averaged time profile of ion signals without hexapole magnetic field (magnetic lens 'off') is shown in fig. 2a. The curve triggered by the

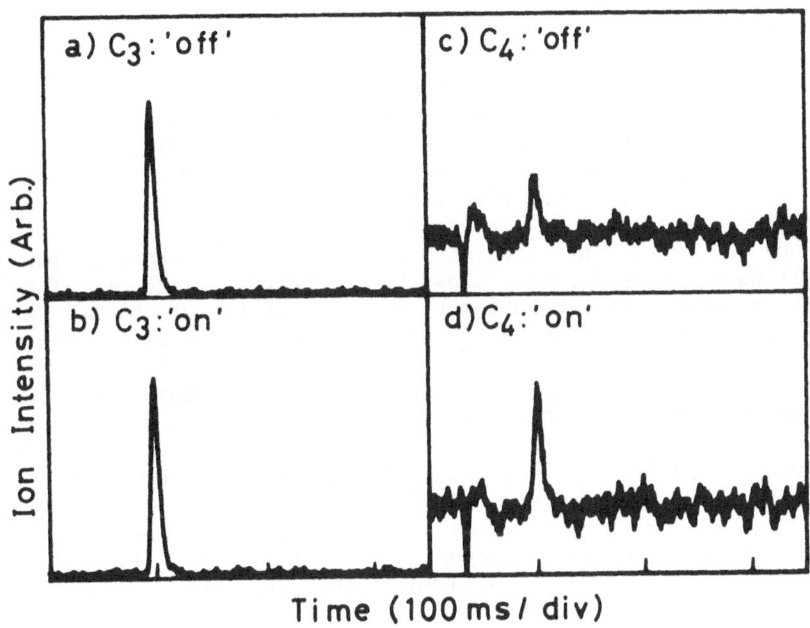

Figure 2. Ion signals recorded in the storage oscilloscope. The scale of the ordinate for each cluster is fixed for the lens 'on' and 'off'.
(a) C_3, magnetic lens 'off' (b) C_3, magnetic lens 'on'
(c) C_4, magnetic lens 'off' (d) C_4, magnetic lens 'on'

gas injector opening pulse was obtained by tuning the ion lens system and quadrupole mass filter with optimized for m/e = 36. The peak shown in this figure, that is, ion signal synchronized with the timing of laser irradiation is considered to be produced by the electron impact ionization of neutral C_3. The contribution of larger neutral cluster to this peak will be mentioned later. In the 'bunch' of clusters, there exist cluster ions produced by laser vaporization. The contribution of these ions can be excluded because the lens system was set to reject the ions coming from outside. It should be noted in this figure that the time profile of the peak corresponds to not only the flight time of neutral clusters but also the time profile of the action of pulsed injector of He gas. The flight time after ionization is negligible.

The effect of magnetic deflection was measured by changing magnetic field intensity while other parameters were kept constant. Fig. 2b shows the ion signal obtained with magnetic lens 'on' optimized for triplet molecules of 36 amu. As seen in this figure, no signal enhancement was observed in the case of C_3. In contrast with this, the signal intensity of C_4, which is shown in figs. 2c (lens 'off') and 2d (lens 'on') clearly depended on the magnetic field intensity.

The plots of the beam intensities as a function of the magnetic field strength are shown in fig. 3a-f for C_1 - C_6 carbon clusters, respectively. The ground electronic state of the carbon atom is $^3P_0$ with no magnetic moment. However, as seen in fig. 3a, an enhancement was observed. This fact can be explained as follows: Under the present beam conditions, considerable amounts of $^3P_1$ carbon atom are significantly populated during vaporization, since it lies only $16cm^{-1}$ above $^3P_0$ state. On the basis of the present data, the velocity of the beam was estimated by using the value of magnetic moment for $^3P_1$ state as well as the magnetic field, giving an optimum focusing condition for the C_1 beam. As a result, the velocity of 1800 m/s was deduced and its value can be accepted as reasonable as a helium seeded beam.

The ground electronic states of C_2, C_3 and C_5 have so far been well explained by a linear singlet character. As expected, the beam intensities of these clusters shows no dependence on the magnetic field. However a very small enhancement is seen for C_2 at higher magnetic fields. The origin of this slight magnetic dependence of C_2 beam will be discussed later.

For the clusters of C_4 and C_6, on the other hand, a large enhancement of the beam intensity was clearly observed as shown in figs. 3 (b) and (c). This fact undoubtedly indicates that at least a considerable portion of C_4 and C_6 are of linear form with triplet character in the ground state.

Recent theoretical studies of the electronic states of C_4 have suggested that the singlet rhombus form and the triplet linear form are almost isoenergetic [5]. In principle, the percentage of paramagnetic portion can be evaluated from the present experiment, by comparing the degree of enhancement for the clusters of a particular size with that of standard sample. In this sense, O_2 seems to be suitable as a standard sample because the electronic states of O_2 and n-even C_n of the chain form are both $^3\Sigma$. That is, if the degree of enhancement is the same for C_n (n is specific size) and O_2, for example, all the portion of C_n in the beam would be considered to exist as triplet species. On the other hand, if the degree of enhancement for C_n is reduced, singlet species would be formed in the C_n beam. From present results, we should point out that the existence of the singlet rhombic C_4 can not be easily ruled out. Actually, the amount of magnetic field enhancement on C_4 was found to be smaller than that shown for O_2 [16]. This evidence rather seems to imply that the singlet rhombic form partly exists in the beam. In order to clarify this point, a more quantitative experiment is now under way.

Finally, we would like to point out the influence of fragmentation on the present data. The C_2^+ signal intensity is slightly enhanced in the relatively strong magnetic field as mentioned already. However, the singlet character of C_2 has well been established, and the energy level of the lowest triplet state of C_2 (a$^3\Pi$, 716 cm^{-1} above X$^1\Sigma$) is too high to be thermally populated under the present experimental conditions. The enhancement of C_2^+ signal seems to be inconsistent with these facts. The most plausible interpretation for this discrepancy at present is that fragmentation of higher clusters with triplet character, induced by electron impact ionization, contributes to the observed C_2^+ signal.

4. Acknowledgments

This work was partially supported by Grant-in Aid from the Ministry of Education, Science and Culture.

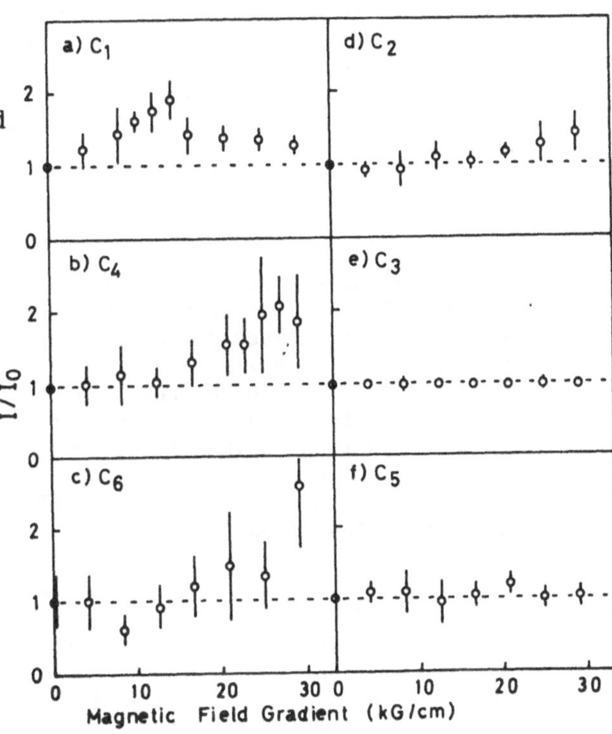

Figure 3. The plots of relative intensity of cluster beam versus magnetic field gradient. The results for C_1 to C_6 are arranged in following order; the plots showing signal enhancement are drawn in right side (a,b,c). Each plot is normalized by the signal intensity obtained at zero magnetic field.

5. References

[1] Weltner Jr. W., and Van Zee R. J. (1989) 'Carbon molecules, ions, and clusters', Chem. Rev. 89, 1713-1747.

[2] Kraetschmer W., Fostiropoulos K. and Huffman D. R. (1990) 'The infrared and ultraviolet absorption spectra of laboratory-produced carbon dust: evidence for the presence of the C_{60} molecule', Chem. Phys. Lett. 170, 167-170.

[3] Hinkle K. H., Keady J. J. and Bernath P. F. (1988) 'Detection of C_3 in the circumstellar shell of IRC+10216', Science 241, 1319-1322.

[4] Bernath P. F., Hinkle K. H. and Keady J. J. (1989) 'Detection of C_5 in the circumstellar shell of IRC+10216', Science 244, 562-564.

[5] Slanina Z. (1990) 'Ab estimation of the energy difference between the linear and rhombic structure of $C_4(g)$', Chem. Phys. Lett. 173, 164-168, and references therein.

[6] Raghavachari K. and Binkley J. S. (1987) 'Structure, stability, and fragmentation of small carbon clusters', J. Chem. Phys 87, 2191-2197.

[7] Martin J. M. L., Francois J. P. and Gijbels R. (1990) 'Ab initio study of the infrared spectra of linear C_n clusters (n=6-9)', J. Chem. Phys. 93, 8850-8861.

[8] Yang S., Taylor K. S., Craycraft M. J., Conceicao J., Pettiette C. L., Cheshnovsky O. and Smalley R. E. (1988) 'UPS of 2-30-atom carbon clusters: Chain and rings', Chem. Phys. Lett. 144, 431-436.

[9] Algranati M., Feldman H., Kella D., Malkin E., Miklazky E., Naaman R., Vager Z., and Zajfman Z. (1989) 'The structure of C_4 as studied by the Coulomb explosion method', J. Chem. Phys. 90, 4617-4618.

[10] Vager Z., Feldman H., Kella D., Malkin E., Miklazky E., Zajfman E. and Naaman R. (1991) 'The structure of small carbon clusters', Z. Phys. D - Atoms, Molecules and Clusters 19, 413-418.

[11] Heath J. R. and Saykally R. J. (1991) 'The structure of the C_4 cluster radical', J. Chem. Phys. 94, 3271-3273.

[12] Van Zee R. J., Ferrante R. F., Zeringue K. J. and Weltner Jr. W. (1988) 'Electron spin resonance of the C_6, C_8, and C_{10} molecules', J. Chem. Phys. 88, 3465-3474.

[13] Cheung H. M. and Graham W. R. M. (1989) 'Electron-spin-resonance characterization of nonlinear C_4', J. Chem. Phys. 91, 6664-6670.

[14] Christensen R. L. and Hamilton D. R. (1959) 'Permanent magnet for atomic beam focusing', Rev. Sci. Instrum. 30, 356-358.

[15] Diets T. G., Duncan M. A., Powers D. E. and Smalley R. E. (1981) 'Laser production of supersonic metal cluster beams', J. Chem. Phys. 74, 6511-6512.

[16] Yanase N., Miyake K., Kobayashi N. and Kaneko Y. (1991) 'Measurement of magnetic moments of $(O_2)_2$ and $(O_2)_3$', Proc. 17th International Conference of the Physics on Electronic and Atomic Collision, Brisbane, pp. 595.

CLUSTER RELATIVE-STABILITY INTERCHANGES

Zdeněk SLANINA*
Max-Planck-Institut für Chemie (Otto-Hahn-Institut)
W-6500 Mainz, FRG

ABSTRACT. Relative-stability interchanges in isomeric cluster systems are studied and illustrated with the C_4 two-component and 2HF-HCN and HF-2HCN three-component sets.

It has been proved that clusters can frequently exhibit isomerism, i.e. that more than one minimum energy structure can be present. Recently, structural or spectroscopic features and energetics of several gas-phase molecular clusters exhibiting isomerism have been investigated; thermodynamics of some of them were even evaluated [1].

C_4 cluster system represents an interesting demonstrating example. On the basis of theoretical calculations (for a review, see Ref. 2) there exists a consensus that there are two energetically low-lying structures: the rhombic (D_{2h}) and the linear triplet ($D_{\infty h}$). The rhombic structure appears to lie mostly lower in terms of potential energy, but there are quite different estimates of the energy separation of the two structures [2] (ranging, for example, from -5 to 65 kJ/mol). The critical evaluation [3] recommends the value of $\Delta H_0^o = 20.92$ kJ/mol for the ground-state energy difference between the linear triplet and the rhombic structures. It corresponds to the potential energy difference $\Delta E = 20.63$ kJ/mol (the MP2/6-31G* vibrational frequencies [4] were used for the zero-point vibrational correction). However, the energy difference itself is not sufficient for stability evaluations at the considerably high temperatures of carbon cluster observations [5] - entropy is to be included into a relevant reasoning.

An n-membered isomeric mixture is characterized by the equilibrium mole fractions w_i of the individual isomers. The populations can be evaluated [6] using the standard enthalpy terms at the absolute zero temperature $\Delta H_{0,i}^o$ and the isomeric partition functions q_i:

265

P. Jena et al. (eds.), Physics and Chemistry of Finite Systems: From Clusters to Crystals, Vol. I, 265–270.
© 1992 *Kluwer Academic Publishers.*

$$w_i = \frac{q_i exp[-\Delta H^o_{0,i}/(RT)]}{\sum_{j=1}^{n} q_j exp[-\Delta H^o_{0,j}/(RT)]},\tag{1}$$

where R denotes the gas constant and T temperature. If rotational-vibrational motions are neglected in Eq. (1) simple Boltzmann factors w'_i follow [6]. For isomeric systems it has become customary to distinguish [6] two categories of quantities. The standard partial terms ΔX^o_i belonging to the individual isomers and the standard overall terms ΔX^o_T belonging to the total process with a pseudospecies represented by the equilibrium isomeric mixture Finally, so called isomerism contributions to thermodynamic terms δX_1 were introduced:

$$\delta X_1 = \Delta X^o_T - \Delta X^o_1.\tag{2}$$

The values of δX_1 generally depend on the choice of the reference isomer labelled by $i = 1$. It is convenient the most stable species (in low temperature region) be chosen as the reference structure.

Fig. 1 presents temperature evolution of the relative populations w_i of the C_4 isomers evaluated for the recommended energetics [3] and MP2/6-31G* structural and vibrational data. Although at low temperatures the ground-state structure is prevailing, already at a temperature of 1320 K there is relative stability crossing. Consequently, beyond this point the linear triplet structure becomes relatively more populated species. This finding is particularly important at the temperatures of the observation [5] (2400 K). It turns out that the species higher in potential energy is still more important for the interpretation of the observed equilibrium constants of C_4 formation. It is also clear from Fig. 1 that the simple Boltzmann factors do not work well in this case.

Fig. 1 also serves an illustration of Eq. (2) for heat capacity at constant pressure term ($X = C_p$). The isomerism contribution to heat capacity has a characteristic dependency with a sharp temperature maximum. The maximum position is correlated with the stability-interchange temperature (it is shifted to somewhat lower temperature). The maximum can be conserved in the overall heat capacity term and thus, it could serve as an experimental proof of the isomerism (for the first direct experimental evidence of the carbon cluster isomerism, see Ref. 7). It has been argued that the temperature maximum is in fact related to fast temperature changes of the isomeric relative stabilities (i.e., to the dw_i/dT terms). The dashed line in Fig. 1 represents so called [1] isofractional term, i.e. term derived for an artificial condition $dw_i/dT = 0$.

The second illustrative example in this report deals with hydrogen-bonded molecular complexes. Quite recently, two trimeric systems were subjected to calculations, viz. 2HF-HCN and HF-2HCN. Three different local energy minima were found [8] on energy hypersurface of each system: HCN.HF.HF [shortly, HCN.(HF)$_2$],

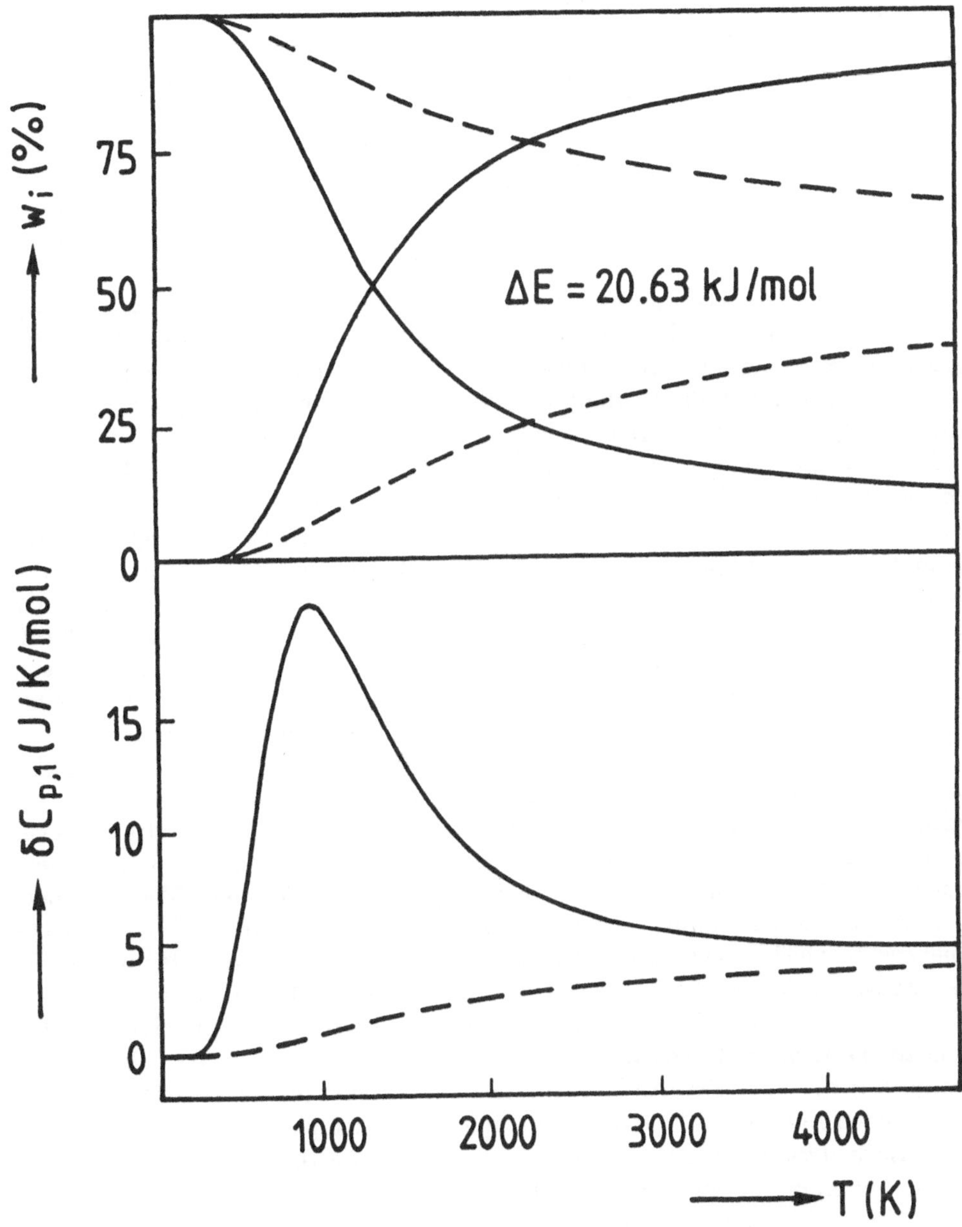

Figure 1. Temperature dependences of the mole fractions w_i in the C_4 isomeric system (top, - - simple Boltzmann factors) and of the isomerism contributions to heat capacity $\delta C_{p,1}$ (bottom, - - isofractional term).

HF.HCN.HF, and HF.HF.HCN [shortly, $(HF)_2$.HCN] structures, only the second being linear, and HCN.HCN.HF [shortly, $(HCN)_2$.HF], HCN.HF.HCN, and HF.HCN.HCN [shortly, HF.$(HCN)_2$] species, all of them possessing linear structure.

Fig. 2 presents temperature evolution of the w_i terms in both systems. The ground-state structure in the 2HF-HCN system, HCN.$(HF)_2$, is the most stable one only till temperature of 590 K; after this temperature threshold the isomer highest in potential energy, $(HF)_2$.HCN, becomes the most populated of the three structures. In fact, another crossing is not well seen in Fig. 2 because of its scale - already at a temperature of about 350 K the two structures higher in energy (i.e., HF.HCN.HF and $(HF)_2$HCN) interchange their relative stabilities though the ground-state structure is still prevailing. Finally, a third crossing appears at a temperature of about 830 K, the third possible pair of isomers being involved (i.e., HCN.$(HF)_2$ and HF.HCN.HF).

The lower part of Fig. 2 deals with the HF-2HCN system. Again, three interchanges of relative stability are present in the system (though the third crossing temperature is too high from practical point of view). A striking feature of the isomeric interplay in the system is the particularly low position of the first relative-stability interchange. The two structures lower in potential energy (i.e., $(HCN)_2$.HF and HCN.HF.HCN) interchange their relative stabilities already at about 130 K. The ground-state structure becomes the least stable at about 660 K, being surpassed by the HF.$(HCN)_2$ isomer in the point. Finally, the two structures higher in energy interchange their stabilities mutually at about 1020 K.

The reported results clearly have some more general consequences. Comparable stabilities of some structures or even interchanges of their relative stabilities are well expectable in an isomeric cluster system, at least in some temperature regions. Potential-energy difference between isomers even of several dozens of kJ/mol need not be a sufficient separation to prevent an isomeric coexistence. The treatment describes conditions of inter-isomeric thermodynamic equilibrium. Thus, it is to be analyzed to which extent the conditions could be reached in a particular experimental situation before applying the calculated w_i terms. It should also be recognized that in a real gas-phase mixture (for example in the HF and HCN case) in addition to the trimers studied here, monomers and other clusters (e.g., dimers and tetramers) will act. Nevertheless, the relative isomeric populations found within our isolated sets should be still transferable to the complex mixture. In fact, the isomeric analysis represents a necessary pre-requisite for the complex-mixture comprehensive study. Finally, quality of the mole fractions (1) can further be improved by an inclusion of corrections to anharmonicity and non-rigidity effects. Such corrections would certainly be desirable in spite of the related computational demands. Till now, however, there are only some qualitative arguments [6] indicating that such correction terms may substantially cancel out between the numerator and denominator in Eq. (1).

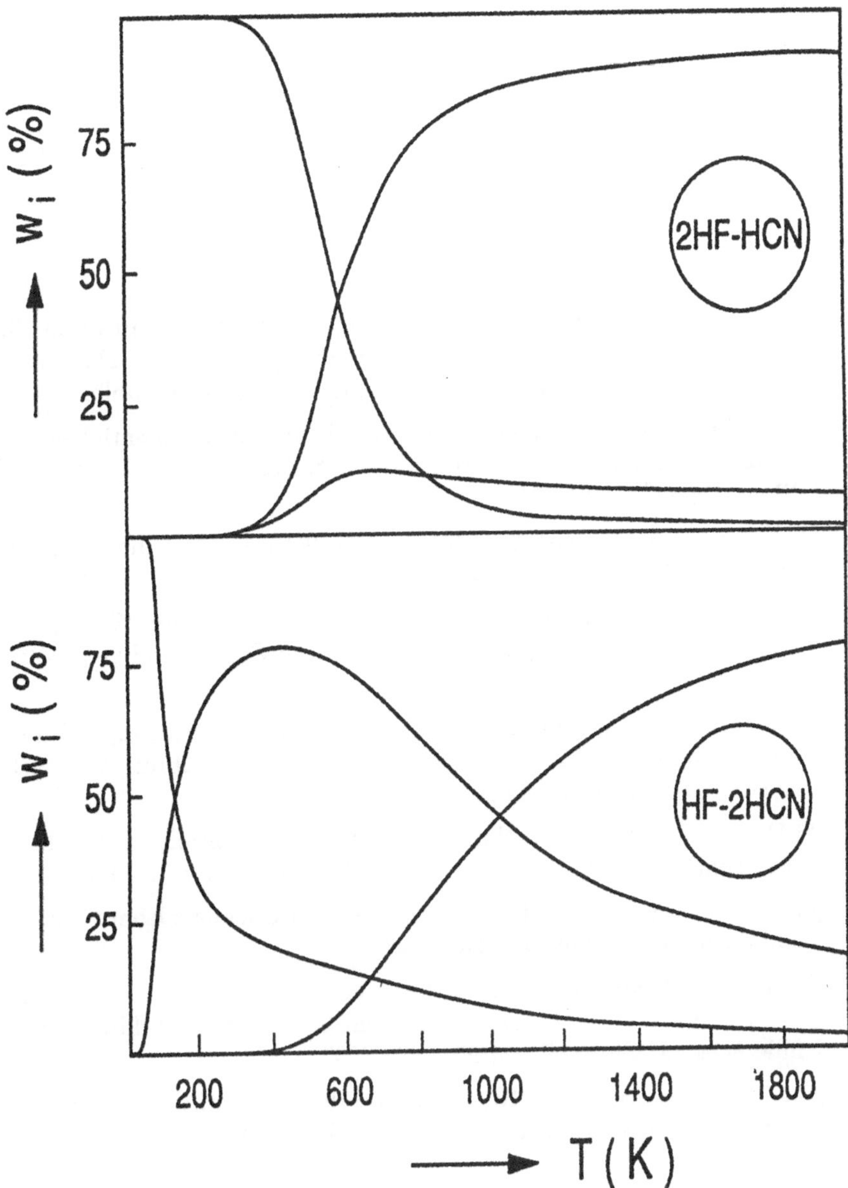

Figure 2. Temperature dependences of the mole fractions w_i for 2HF-HCN [top, high temperature stability order from top: $(HF)_2.HCN$, $HF.HCN.HF$, and $HCN.(HF)_2$], and for HF-2HCN [bottom, the order: $HF.(HCN)_2$, $HCN.HF.HCN$, and $(HCN)_2.HF$] isomers.

Acknowledgements

The report was prepared during a research stay at the MPI für Chemie (Otto-Hahn-Institut) supported by the Alexander von Humboldt-Stiftung. The support as well as the valuable discussions with Prof. K. Heinzinger and the kind hospitality of him, of his group and of the the MPI für Chemie are gratefully acknowledged.

References

* Correspondence address: The J. Heyrovský Institute of Physical Chemistry and Electrochemistry, Czechoslovak Academy of Sciences, Dolejškova 3, CS-18223 Prague 8-Kobylisy, Czech and Slovak Federal Republic.

[1] Slanina, Z. (1987) 'Equilibrium isomeric mixtures: Potential energy hypersurfaces as the origin of the overall thermodynamics and kinetics', Int. Rev. Phys. Chem. 6, 251-267.

[2] Weltner, W. Jr. and Zee, R. J. van (1989) 'Carbon molecules, ions, and clusters', Chem. Rev. 89, 1713-1747.

[3] Zee, R. J. van, Ferrante, R. F., Zeringue, K. J., Weltner, W. Jr., and Ewing, D. W. (1988) 'Electron spin resonance of the C_6, C_8, and C_{10} molecules', J. Chem. Phys. 88, 3465-3474.

[4] Michalska, D., Chojnacki, H., Hess, B. A. Jr., and Schaad, L. J. (1987) 'Ab initio second order Møller-Plesset calculation of the vibrational spectra of C_4 clusters', Chem. Phys. Lett. 141, 376-379.

[5] Drowart, J., Burns, R. P., DeMaria, G., and Inghram, M. G. (1959) 'Mass spectrometric study of carbon vapor', J. Chem. Phys. 31, 1131-1132.

[6] Slanina, Z. (1986) Contemporary Theory of Chemical Isomerism, Academia and D. Reidel, Prague and Dordrecht.

[7] Helden, G. von, Hsu, M.-T, Kemper, P. R., and Bowers, M. T. (1991) 'Structures of carbon cluster ions from 3 to 60 atoms: Linears to rings to fullerenes', J. Chem. Phys. 95, 3835-3837.

[8] Almeida, W. B. de and Hinchliffe, A. (1990) '*Ab initio* vibrational spectrum of the H-bonded trimers $(HCN)_2 HF$ and $HCN(HF)_2$', Mol. Phys. 69, 305-318.

FORMATION OF CLUSTERS IN CHARGE SEEDED EXPANSIONS OF RARE GAS MIXTURES

S. Sun and J. A. Northby

Physics Department, University of Rhode Island

Kingston, Rhode Island, 02881

ABSTRACT. We have measured the argon cluster ion mass spectrum (Ar_N^+, $N = 2 \rightarrow 60$) produced in a corona discharge charge seeded expansion (CSE) source of 10% argon in helium at 87K for a range of pressures. We have found a set of "magic" cluster sizes in this mass range identical to those found previously in a CSE of pure argon, and again in clear distinction to those produced by ionization of neutral clusters. With the 10% mixture we have been able to study smaller cluster ions than in the pure CSE case, and in this range we have seen especially strong peaks at $N = 3$ and 7. We also have carried out studies of how the operation of the CSE source depends on conditions in the stagnation region other than pressure and temperature. In particular, we have studied how the mass spectrum is altered by changing the placement of the corona discharge relative to the nozzle, and we will discuss these results as well.

1. Introduction.

A well known feature of the mass distributions of cluster ions which are formed from free jet expansions of rare gases is the presence of particular cluster sizes which are significantly more intense than their neighbors. These "magic numbers" are thought to represent cluster sizes which are especially stable against evaporative decay[1]. The most common method of producing a charged cluster beam is by electron impact ionization of neutral clusters produced in a high pressure expansion[2]. In our laboratory we have utilized a second method[3-5], the charge seeded expansion (CSE) method introduced by Searcy and Fenn[6], and first applied to rare gases by Beuhler and Friedman[7]. In the CSE method positive ions are produced in the stagnation chamber by a corona discharge. The ions are then co-expanded with the gas and, in the supersaturated conditions in the jet, act as condensation nuclei upon which clusters grow. One somewhat surprising result is that for the smaller clusters ($N \leq 120$) the magic numbers for argon depend on which method is used to produce them. For larger argon clusters they are the same, however[3-5].This presents a difficulty in

P. Jena et al. (eds.), Physics and Chemistry of Finite Systems: From Clusters to Crystals, Vol. I, 271–276.

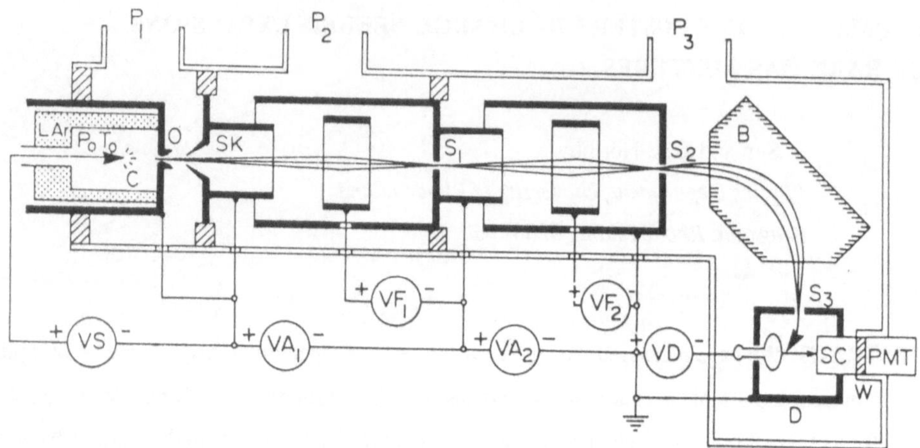

Figure 1. Apparatus schematic.

any attempt to interpret magic numbers simply in terms of ground state energetics, since those can't depend on the production method.

The objective of the experiments we will describe below was to try to confirm our previous results obtained by the CSE method, to extend them to smaller cluster sizes, and to attempt to understand a bit more about how a CSE cluster ion source works. In order to produce small clusters by this method it is necessary to reduce the stagnation pressure, but this can only be carried so far before the corona discharge becomes unstable. Thus we have chosen to study the expansion of a mixture of 10% Ar in helium at 87K. This allows us to work at a high enough pressure to operate the discharge stably. We can still produce significant intensities of small pure Ar_N^+ cluster ions, however, since cluster formation is determined mostly by the Ar partial pressure and the He gas is too weakly interacting to be incorporated into the clusters.

2. Experiments

A schematic diagram of the experimental apparatus is shown in figure 1. It consists of a low temperature ionized cluster beam source, ion optics, and a mass spectrometer. In the present experiments we have used a smaller nozzle (d = 110 μm) than was used in the original pure argon studies[3-5], but otherwise the apparatus is very similar. A high purity rare gas mixture in a stagnation chamber (P_0= 90 $\rightarrow$ 210 torr) surrounded by a liquid argon bath (T_0 = 87K), expands through a nozzle (O) into a differentially pumped vacuum chamber ($P_1 < 2\times10^{-4}$ torr). A corona discharge (C) is maintained on a sharp tungsten tip located behind the nozzle. The discharge produces positive ions which are swept into the nozzle where they become condensation nuclei and clusters grow about them. After the pressure drops to a point where collisions effectively cease, the clusters cool

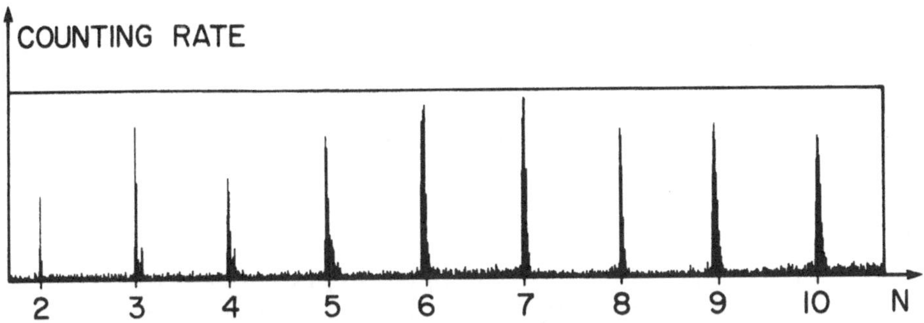

Figure 2. Experimental spectrum of Ar_N^+, N=2 → 10, at 87K.

rapidly by evaporation until they approach or reach a bound state. The resulting cluster ion beam passes through a skimmer (SK) into a second chamber ($P_2 = 3x10^{-6}$ torr), where it is accelerated (VA_1) and focussed (VF_1) through a 1 mm pinhole (S_1) into another chamber ($P_3 = 1x10^{-8}$ torr) . Here it is further accelerated (VA_2) and focussed on the entrance slit (S_2) of a 30 cm, 90° magnetic sector mass spectrometer. Finally, the mass selected clusters are focussed through the exit slit (S_3) into a Daly-type[8] detector (D).

3. Results and Discussion

Figure 2 shows a typical mass spectrum for Ar_N^+ clusters in the mass range N = 2 →10. Ar_3^+ and, to a lesser extent, Ar_7^+ appear to stand out relative to their neighbors. It is difficult to draw any definite conclusions about magic numbers in this low mass region, since peak intensities are really quite sensitive to the source pressure and temperature, to the discharge conditions, and to the presence of a weak magnetic field near the nozzle. That said, however, we do feel that Ar_3^+ is an especially strong peak under most conditions and that it is probably significantly more tightly bound than Ar_4^+.

In figure 3 we present three different mass spectra. Each was taken in a separate run with a different spacing between tip and nozzle, D_{tn}, and each was constructed from a series of scans taken at different pressures. Each scan was normalized by dividing by a gaussian envelope fit by eye, as in our previous measurements[3-5]. They were then linked together to produce the complete spectrum by overlaping one or more peaks from adjacent scans. In the first run, (a), with D_{tn} approximately equal to 2.0 mm, we were able to observe mass peaks in the range from 12 to 41 atoms for P_0 between 150 and 200 torr. The pumping speed of our vacuum system limited us to operating the source at pressures below 210 torr and in this case N= 41 was the largest cluster produced. In this spectrum we can see magic numbers 13, 19, 23, 26, and 29. In the second run, (b), with D_{tn} ≈

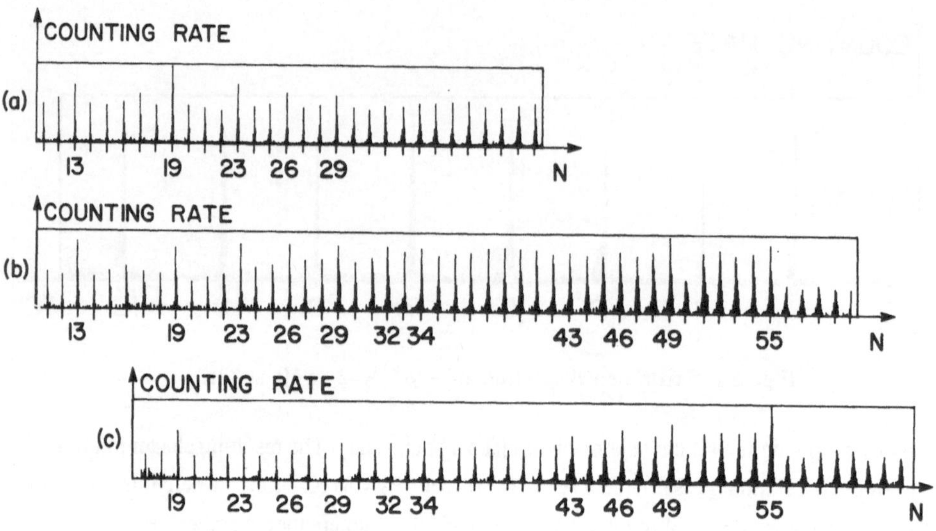

Figure 3. Experimental spectra of Ar_N^+, $N > 10$, for three different D_{tn}. (a) $D_{tn} \approx 2.0$mm. (b) $D_{tn} \approx 2.5$mm. (c) $D_{tn} \approx 3.0$mm.

2.5mm we obtained significant intensities in the mass range $N = 2 \rightarrow 60$ for P_0 between 170 and 200 torr. This covered the closure of the first and second icosahedral shells at $N = 13$ and 55. The magic numbers are 13, 19, 23, 26, 29, 32, 34, 43, 46, 49, and 55, which are the same as our previous results for pure argon[3-5], but the peaks 43 and 46 do not stand out so clearly in this spectrum. The third run, (c), with $D_{tn} \approx 3.0$ mm, produced mass peaks from $N = 17$ to 60 for P_0 between 100 and 180 torr The intensities of peaks 43 and 46 are stronger here than for the second run. In our experiment the magic numbers 13, 19, 23, 26, 29, 49, and 55 are always reproducible and stable, but 32, 34, 43, and 46 are not. The relative intensities of these four masses depend more sensitively on source conditions. They appear to stand out often enough in our data so that we are sure they are "special", but they are not sufficiently reproducible to be designated as "magic". Perhaps we could call them "weakly magic numbers".

In general, as the tip-nozzle distance increased, the mean cluster size for a given pressure also increased. Thus to obtain the same cluster size as for smaller D_{tn}, we had to operate at a lower stagnation pressure. It turned out that it was not possible to sustain a stable discharge for pressures below 100 torr, and at this pressure the smallest clusters produced with $D_{tn} \approx 3.0$ mm already contained 17 atoms. That's why spectrum (c) cuts off at this value. This suggests that 17 atoms may be the size of the cluster ions already present in the stagnation chamber at equilibrium under these

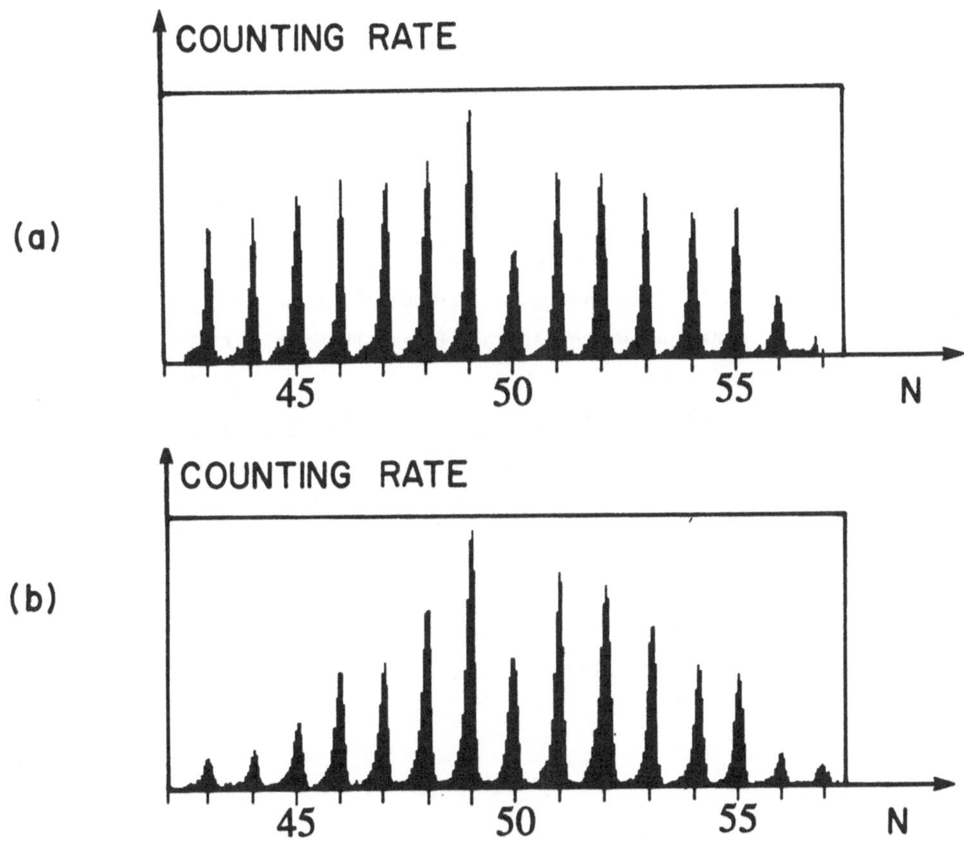

Figure 4. Two non-normalized mass scans near N = 50 taken at (a) $D_{tn}\approx2.5mm$, and (b) $D_{tn}\approx3.0mm$.

source conditions[9]. Another significant result of varying D_{tn} is illustrated in figure 4. The first non-normalized scan (figure 4a) was obtained during the second run with $D_{tn} \approx 2.5$ mm. It covers the mass range from N = 43 to 56 and is relatively broad. The second spectrum, (figure 4b) was obtained during the third run with the largest D_{tn} (3 mm). The source pressure was chosen to cover the same mass range but the distribution is considerably more sharply peaked. Our qualitative observation was that larger values of D_{tn} gave narrower distributions and larger clusters. This seems reasonable since the gas temperature at the nozzle should be lower if it is further removed from the discharge. Also, cluster ions have large time to grow to their equilibrium size in the stagnation chamber before entering the nozzle. Finally, lower temperatures would be expected to lead to larger clusters at equilibrium, and this in turn should produce a more deterministic growth process.

4. Acknowledgements

We would like to thank Isaac Harris for help in the initial stages of these measurements. This work was partially supported by the NSF under grants DMR8405190 and DMR8816482.

5. References

[1] Casero, R., and Soler, J. M., J. Chem. Phys. 95, 2927 (1991), and references therein.

[2] Miehle. W., Kandler, O., Leisner, T. and Echt, O., J. Chem. Phys. 91, 5940 (1989)

[3] Harris,I.A., Kidwell, R.S., and Northby, J.A., in Procedings of the Seventeenth International Conference on Low Temperature Physics, edited by Eckern, U., Schmidt, A., Weber, W., and Wuhl, H. (North-Holland, Amsterdam, 1984)

[4] Harris,I.A., Kidwell, R.S., and Northby, J.A., Phys. Rev. Lett. 53, 2390 (1984)

[5] Harris,I.A., Norman, K. A., Mulkern, R. V., and Northby, J.A., Chem. Phys. Lett. 130, 316 (1986) : S.Sun, M.S. thesis (Univ. of Rhode Island, 1991, unpublished)

[6] Searcy, J. Q., and Fenn, J. B., J. Chem. Phys. 61, 5282 (1974)

[7] Beuhler, R. J., and Friedman, L., J. Chem. Phys. 78, 4669 (1983)

[8] Daly, N. R., Rev. Sci. Inst. 31, 264 (1960) ; Rev. Sci. Inst. 34 1116 (1963).

[9] Akinci, G., and Northby, J. A., Phys. Rev. Lett. , 573 (1979)

PHASE CHANGES FOR CLUSTERS AND FOR BULK MATTER

R. STEPHEN BERRY AND HAI-PING CHENG
Department of Chemistry and The James Franck Institute
The University of Chicago
5735 South Ellis Avenue
Chicago, Illinois 60637
U. S. A.

ABSTRACT. Two kinds of phase changes are analyzed and compared, in the context of clusters and in terms of their implications for bulk matter. The homogeneous melting and freezing transition is reviewed briefly with regard especially to the simultaneous coexistence throughout a range of temperatures or energies of solid-like and liquid-like clusters. This picture is extended from the zero-pressure conditions applied in earlier analyses to a full range of pressures, so that a full phase diagram is obtained. Coexistence persists to cluster sizes of at least 147 for Lennard-Jones clusters. Surface melting is a different kind of transition, showing no coexistence, i.e. only one form that is *locally* stable at any temperature, and a very small but apparently nonzero latent heat. The nature of homogeneous melting of clusters implies metastability (undercooling in particular) but with a lower sharp temperature limit for bulk matter, and likewise for superheating *if there is no free surface or the melt does not wet the solid.* However if the bulk solid has a free surface and the melt wets the solid, then the behavior of cluster surfaces implies that such a solid cannot exhibit superheating.

1. Introduction

The modern theory of melting and freezing of clusters began with simulations that indicated first that small clusters could exhibit liquid-like behavior [1-3] and display a kind of phase equilibrium between solid-like and liquid-like forms of clusters of the same size and composition [4-7]. This background was reviewed in the context of the developments that followed[8].

Not all clusters exhibit clearly defined solid-like and liquid-like forms. Clusters of some kinds and sizes, such as clusters of 7, 13 or 19 atoms bound by Lennard-Jones pairwise interactions with icosahedral or other closed-shelll character, do display distinct phase-like forms, at least in simulations but clusters of 12, 14 or 17 atoms bound by the same forces pass through slush-like forms in the energy or temperature ranges in which one might hope thay would show solid-like and liquid-like forms [9, 10]. Whether the cluster shows distinct, phase-like forms or appears to pass through a slush-like form depends on whether or not the potential surface has a deep, solid-like well separated by a relatively high free energy barrier from a rolling plain with many readily accessible catchment basins [11].

In addition to homogeneous melting and freezing, in which the entire cluster is liquid-like or solid-like, clusters of enough particles may also exhibit surface melting. This was suggested by Briant and Burton[4] and demonstrated with Monte Carlo simulations by

P. Jena et al. (eds.), Physics and Chemistry of Finite Systems: From Clusters to Crystals, Vol. I, 277–286.
© 1992 *Kluwer Academic Publishers.*

Nauchitel and Pertsin[12]. While the generalities that determine which small cluster sizes may exhibit surface melting remain unknown, it is very likely that the capacity to exhibit surface melting does not set in suddenly and ambiguously at some critical cluster size. Clusters of 45, 50 and 54 argon atoms show surface melting but some of the clusters with sizes between these do not. Lennard-Jones clusters of 55 or more atoms do appear to exhibit surface melting.

The questions we address here are these. 1. What kind of equilibrium conditions characterize homogeneous solid-like and liquid-like clusters? 2. How are these conditions different from those of bulk matter and how do they merge into those of bulk matter? 3. What can we learn about phase stability and equilibrium for freezing and melting of bulk matter from the phase equibrium of clusters? 4. What are the equilibrium conditions characterizing thermal (canonical) or mechanical (microcanonical) equilibrium for a cluster which may exhibit a molten surface? 5. What are the implications of these conditions for surface melting of bulk matte and can we resolve issues such as determining the order of the surface melting transition? There are other questions of equal interest that we shall not be able to discuss within the scope of this presentation; for example, how, in precise terms, does the capability of a cluster to exhibit distinguishable phases depend on the shape of its multidimensional potential surface? What qualities of that surface are particularly relevant? How are rates of passage from one catchment basin to another related to the phase characteristics of a cluster? How does ergodic (or apparently ergodic) behavior evolve in time and how does it change when a cluster goes through a melting or freezing transition? We shall at most only touch on this second set of problems.

2. Homogeneous Melting

2.1. SUMMARY OF BEHAVIOR OF CLUSTERS

The freezing and melting of clusters has been studied extensively by simulations, many of which were reviewed fairly recently [8, 13] and by analytic theory [8, 14, 15] but the experimetal evidence is still very meagre [16-19]. In fact some of the experiments designed to show the existence and possibly the coexistence of solid and liquid clusters turned out to show only that solid clusters may hold foreign probe molecules in interior and in surface sites, and that the spectral transitions of molecules in these two sites differ in frequency and width [20, 21]. Nevertheless the theoretical and computational evidence is so persuasive that it seems worth pursuing the subject from both theoretical and experimental directions.

The theoretical results indicate that at very low energies, the density of states of solid-like clusters is higher than that of nonrigid, liquid-like clusters. Typically, there is some threshold energy required to introduce defects or some other structural change that enables mobility, so that at the lowest energies, the only states available to the cluster are solid-like. The density of states of the solid-like cluster grows with energy. However the density of states of the nonrigid, liquid-like cluster grows even faster with increasing energy. This means that a comparison of the partition functions for extremely rigid, solid-like clusters and extremely nonrigid, liquid-like clusters shows that the solid form has the larger partition function and therefore the lower free energy at low temperatures, where only the low-energy states are populated. The free energy $F(t,\gamma)$ can be considered a function of both temperature and the degree of nonrigidity γ; thus, at a fixed, low temperature, the free energy is a monotonically increasing function of nonrigidity. (This picture can be made quantitative with the density of defects serving as the parameter γ that measures nonrigidity [14, 15].) However aa the temperature of the cluster is increased and it is capable of reaching higher and higher energy levels, the rapid increase of the density of states of the nonrigid, liquid-like form makes the free energy curve—still a function of nonrigidity at successively higher fixed temperatures—flatten near its nonrigid limit and, at some temperature T_f, develop a point of zero slope. At fixed γ, and at temperatures just above T_f, $F(T,\gamma)$ has two minima, one near or possibly at the rigid limit and the other, near or at the nonrigid limit. Each of these minima corresponds to a locally stable state of the cluster.

As the temperature is raised, the density of states of the liquid-like, nonrigid form continues to outweigh, more and more, the density of states of the solid-like form, so $F(T,\gamma)$ tips more and more toward the nonrigid limit. The entropy contribution to the nonrigid form makes its free energy *relatively* lower and lower than the that of the solid-like form, so that a temperature T_m is reached at which the curve of $F(T,\gamma)$ has a point of zero slope at or near the rigid, solid-like limit. Above T_m, $F(T,\gamma)$ is a monotonic, decreasing function of γ, and only the liquid-like form is stable at all.

Between T_f and T_m, the locally stable solid-like and liquid-like forms differ in free energy but only by relatively small amounts, not by many orders of magnitude. This means that the equilibrium constant, expressed in terms of that free energy difference, $\Delta F = F_{liquid} - F_{solid}$, is

$$K_{eq} = [\text{amount of liquid}]/[\text{amount of solid}] \equiv [l] \,/\, [s]$$

$$= \exp(-\Delta F/k_b T). \tag{1}$$

A more useful quantity for our purposes is

$$D_{eq} \equiv (K_{eq} - 1)\,/\,(K_{eq} + 1)$$

$$= (\,[l] - [s]\,)\,/\,(\,[l] + [s]\,), \tag{2}$$

which has the advantage of varying between -1 and $+1$ instead of between 0 and ∞ as K_{eq} does. It is also useful to recall that $F_i = N\,\Delta\mu_i$, where μ_i is the mean chemical potential of the atoms in the cluster in phase i, solid or liquid.

The distribution of clusters of a given size in a canonical ensemble of solid and liquid clusters is now clear: D_{eq} is zero for temperatures below T_f, jumps discontinuously at T_f to some nonzero value above -1, probably below zero indicating that the solid dominates. This function then increases smoothly with temperature through 0, its value at the temperature T_{eq} at which $\mu_{solid} = \mu_{liquid}$, and continues increasing smoothly up to some value below $+1$ at T_m, where it again jumps discontinuously, here to $+1$, indicating all liquid in the equilibrium ensemble at temperatures above T_m. Smaller clusters show larger discontinuities and smaller slopes in the continuous region. That is, the thermodynamic discontinuities are likely to be apparent in equilibrium systems only in small clusters, because the steps at the discontinuities grow smaller with cluster size. This is because of the factor of N in the exponent of K_{eq}; if N is not large, then a change in sign of ΔF produces a relatively small swing in D_{eq} and K_{eq}. However if N is large, D_{eq} and K_{eq} swing over a very narrow range of temperature from very close to their all-solid limits to very close to their all-liquid limits. The results of molecular dynamics and Monte Carlo simulations are entirely in accord with these findings so long as one is careful to keep step sizes short enough to maintain the vibrational mechanics, the intervals for determining diffusion constants short enough to avoid showing only that the clustered particles stay together, and the total trajectory long enough to see approximate equilibrium distributions of times spent in the various accessible regions, e.g. solid-like and liquid-like [8].

Before going on to the behavior of bulk matter, we must raise several caveats. These are connected with the fact that not all clusters show distinct solid-like and liquid-like behavior. This can be due to at least two reasons. In some cases, the cluster can become quite nonrigid and its atoms or molecules may undergo large-amplitude motions without exploring more than a few minima around its initial, solid-like configuration. This is the case for the Lennard-Jones and similar clusters of six atoms, which we designate M_6: the geometry of lowest energy for these clusters is an octahedron. On the potential surface of this cluster, with its 12 spatial dimensions, each octahedral minimum is surrounded by 12 distorted octahedra. These are linked by low-energy saddles to their "home base" octahedron and by somewhat higher-energy saddles to other distorted octahedra, which surround permutations of the initial regular octahedron [11, 22, 23]. Another example occurs with alkali halide clusters; $(NaCl)_4$, $(NaCl)_5$, $(KCl)_4$ and $(KCl)_5$ all have locally stable planar rings and rectangles (or "ladders") as well as their global energy minima which are cuboidal. These clusters have temperature or energy ranges within which they can remain in cuboidal form for long periods and then pass to planar regions of their potential surfaces, where they pass rather easily between rings and rectangles for long periods before returning to the cuboidal regions [24, 25]. This is an example like the M_6 but more elaborate. Both these cases show a sort of softening without melting because there are several relatively low-lying minima above the lowest, solid-like minimum and these are connected to each other and to the well of the lowest minimum by saddles of lower energy (or free energy) than the saddles that connect them to the rest of the potential surface. We call this the graded-step case.

A second condition yielding no distinguishable solid-like or liquid-like forms in the transition range of temperature occurs if all the free energy barriers between catchment basins are low and wide enough that the clusters never spend long periods trapped in liquid-like regions of their potential surfaces. Possibly Ar_{17} belongs to this class, but it seems much less common than the graded-step case.

2.2.IMPLICATIONS

This brings us to the first of the implications for bulk phase transitions of the phase behavior of clusters. Because the free energy differences between *locally stable* solid and liquid clusters are relatively small at temperatures somewhat away from T_{eq}, the two forms may well coexist in measurable amounts at any pressure and temperature for which ΔF has those two minima as a function of γ. The two forms are as much like two chemical isomers, or, in terms used for the phase rule, like two components, as they are like phases. This finding is not really new [26, 27] only expressed in a somewhat new way and context which might make its content more intuitive than before.

There are some new insights, nonetheless. First is the question of what happens to the temperature difference $\Delta T_c \equiv T_m - T_f$. This was addressed [14, 15, 28] with a generalization of Stillinger and Weber's version of the defect model for melting [29]. The nature of the defects need not be specified; the only requirement for the approach is that the free energy of the liquid phase be expressible as a power series in the density or number of defects. The finding of this work is that if the defects either stabilize each other, i.e. attract each other, or if they lower the frequencies of the cluster's lattice vibrations, which is the most likely case, then ΔT_c remains finite as N grows arbitrarily large. If the defects do not interact with the lattice, raise the phonon frequencies or repel each other, then ΔT_c can go to zero as N grows very large. Because defects tend to lower phonon frequencies, we can expect that most materials whose clusters have nonzero ΔT_c 's will also have nonzero ΔT_c for bulk matter.

The next question, and the last before we go on to phase diagrams and surface melting, is that of observability of T_f, T_m or ΔT_c in bulk materials. In equilibrium bulk systems, these are simply unobservable. The discontinuities at T_f and T_m are far, far too tiny to be seen, and in a bulk system at equilibrium, ΔT_c is unobservable because the transition in D_{eq} is so sharp as to appear discontinuous. In fact the first order melting transition has this amusing irony about it: the transition at T_{eq} looks discontinuous but is really continuous for all finite systems, however large, because neither phase becomes truly unstable locally there. On the other hand, at T_f and T_m there are real discontinuities but they are far, far too tiny to be observed in a system that comes to equilibrium.

There is one way, in principle, to observe T_f, the sharp lower limit of local stability of the liquid form. A bulk liquid may be undercooled (or "supercooled") by keeping it in its local equilibrium state and preventing it from reaching global equilibrium. These results imply that there is a sharp lower limit to the temperatures to which any real liquid may be undercooled. The question arises as to whether fluctuations will hide the point of discontinuity [30]. However even if there are fluctuations, the state of local stability becomes the state with respect to which the fluctuations are defined. In any event, it becomes a challenging question now to decide whether a sharp temperature limit of undercooling can be found.

Undercooling has its counterpart in superheating. In cases in which a solid is not wet by its own melt, or in which the solid or solid cluster is trapped in a matrix so that it has no free surface, it may be possible to study superheating as well as undercooling[31]. However the question of possible superheating of other materials brings us to the topic of surface melting.

3. Surface Melting

3.1. CLUSTERS: SIMULATIONS AND ANALYTICS

The early findings of surface melting [4, 12] were not altogether surprising. We undertook to study surface melting of clusters in a systematic way in connection with a study of sintering, and in the process found some quite interesting results which seem to shed light on some of the puzzles of surface melting of bulk matter [32]. By keeping track of which atoms are in each shell, we were able in simulations [33, 34] to distinguish the behavior of the surface of a cluster from that of its interior. These simulations were done for Cu_{55} andfor a variety of argon clusters, magic and nonmagic. In particular, the mean square displacements of atoms in each shell proved an excellent index for the solid or liquid character of that shell. Some typical values are given in Table 1, for two clusters simulating Ar_{55} and Ar_{120}. Bulk argon has a self-diffusion coefficient of $1.84 \times 10^{-5} cm^2 s^{-1}$ at 85K, which would extrapolate to about 1.23 at 38K, the approximate (homogeneous) melting temperature for Ar_{147}.

Table 1. Sample values, shell by shell, of diffusion coefficients D, in units of 10^{-5} $cm^2 s^{-1}$; energies E are in eV; simulations were done at constant E and yielded the mean temperatures (K) as shown.

N	E	T	Shell	D
55	−0.328	29.6	1	0.001
			2	0.001
			3	0.001
	−0.321	30.3	1	0.000
			2	0.003
			3	0.035
			outermost 12 atoms	0.135
	−0.3175	30.7	1	0.001
			2	0.008
			3	0.083
			outermost 12 atoms	0.150
	−0.314	28.3	1	0.003
			2	0.063
			3	0.125
			average over all atoms	0.109
120	−0.351	31.1	1	0.005
			2	0.005
			3	0.010
			4	0.180
	−0.345	32.0	1	0.006
			2	0.008
			3	0.109
			4	0.297

Snapshots of clusters in energy ranges where their surfaces have liquid-like diffusion coefficients show amorphous-looking outer shells and regular, polyhedral inner shells. Other diagnostics, such as the radial distributions, also show the sharp inner peaks and wide outermost peak one expects for a cluster with a mobile, amorphous surface and a regular, solid-like core. However this turns out not to be an accurate representation of the behavior of a cluster with a "molten" surface. *Animations* revealed a kind of behavior that is verified by closer examination of the radial distributions, for example.

The animations show that an argon cluster with a melted surface has surface atoms most of which undergo large-amplitude, highly anharmonic*collective motions about equilibrium positions in the polyhedral geometry determined by the core*. While these vibrations go on, a few atoms are promoted out of the outermost shell, leaving corresponding vacancies. The promoted atoms, which we call "floaters," move around the surface rather freely, yet remain well bound to the cluster. The floaters contribute significantly to the diffusion in the outer layer and to the soft modes that characterize liquid-like behavior. They exchange with other atoms in the outer layer within about every 10,000 time steps of 10^{-14} s. In short, the "molten surface" of a cluster is indeed a mobile layer of material but not an amorphous swarm of atoms in a swollen shell. Rather, it is a loosely bound polyhedral shell with a few of its members popped out to form a sort of planetary atmosphere. Typically, about 2 atoms, sometimes 1 and sometimes 3, are floaters in an Ar_{147} cluster with a "molten" surface.

With this picture, we constructed an analytic model that describes the statistical thermodynamics of surface melting for homogeneous clusters [33]. In its most refined form, the partition function is grand-canonical, a sum over terms, each corresponding to a specific number of floaters. At the present stage, we have simplified the problem by making the partition function a function of the number of floaters i and evaluated the corresponding free energies, looking for value of i that yields the minimum free energy.

The vibrations of the core and outer shell are represented by Debye models. The cutoff frequency for the outer shell was taken to be 1/9 of that of the core, on the basis of the amplitudes of motion revealed in the simulations. The floaters are treated as particles free to move in a spherical shell outside the outer shell of the cluster. The energy of formation of a floater is the net loss of nearest-neighbor (attractive) interactions when the floater is taken out of the equilibrium geometry of the initial cluster and nested onto three surface atoms; for a close-packed system, this is a loss of 6 such interactions per floater. Two slightly adjustable free volumes enter as parameters: the free volume per atom in the outer shell of the cluster when the vacancies are formed and the free volume of a floater. The free volume of an atom bound in the outer shell prior to promotion of floaters also enters but, at least for magic-number clusters, this is almost not adjustable, being determined by the amplitude of vibration of the outer atoms in the closed-shell polyhedron. Higher-order terms include floater-floater and vacancy-vacancy interactions.

The procedure is to treat the number of floaters i as an order parameter and plot the cluster's free energy F(T,i) as a function of i, for a series of fixet temperatures. A more refined treatment would have F be a function of T and the most probable or mean number of floaters, given that the number of floaters changes a little with time. However in our animations, it seemed that the number of floaters does not vary much, and that the approximation of a fixed number of floaters is quite plausible.

The issues are a) whether the entropy associated with a few floaters can stabilize a cluster with mobile surface atoms, and b) how this stability depends of floaters. Homogeneous melting is described by a free energy with two minima and therefore two stable forms of cluster over a range of temperatures (and pressures as well [35]). The analytic model indicates that F(T,i) shows *only a single minimum as a function of i*. Below a temperature $T_s(N)$, that minimum occurs at i=0, implying that the surface of the cluster is solid. From T_s to the temperature at which the cluster can be a homogeneous liquid, i.e. to T_f, the minimum of F(T,i) increases to 1, then 2 and just about reaches 3 for Ar_{147}. In short, surface melting of clusters, from the combination of simulation and analytic theory, is quite different from homogeneous melting of clusters. Up to T_f, only one stable form of cluster is possible, the solid below T_s and the melted-surface cluster above.

3.2. IMPLICATIONS FOR BULK MATTER

If we extrapolate these results to large N, we infer several things about surface melting of bulk matter. We must assume that the same floater-vacancy mechanism is responsible for surface melting of macroscopic samples, of course. We infer first that surface melting is accompanied by a small but nonzero latent heat associated with the creation of the floaters and vacancies. The magnitude of this can be estimated from the fact that the Ar_{147} cluster, with 92 surface atoms, typically has only two floaters when its surface appears molten. Let us assume that 6 nearest-neighbor interactions or bonds are lost with the formation of each floater, so that about 1 bond is lost per 7 or 8 atoms in the surface. If each atom in the surface shares 9 bonds, so that there are about 4.5 bonds per atom holding the surface layer to the substrate, the latent heat of formation of the melted surface must be only about 1/30 of the mean binding energy per surface atom. It may be easy to overlook such a small latent heat unless one is expecting to find so small a quantity. Typical latent heats of fusion for nonmetallic bulk matter are of order 1/6 to 1/3 the binding energy of the solid, an order of magnitude larger than our estimate for the ratio of surface "latent heat of fusion" to surface binding energy.

Finally, the occurrence of a single minimum in F(T,i) has an important implication for bulk matter. It means that there is no metastable solid or liquid surface analogous to the metastable supercooled solid and undercooled liquid that can be made by suitable, careful preparation from an equilibrium sample by heating or cooling. It implies further that the surface of a solid, in a temperature range in which that surface is mobile, has a single surface-vapor tension, and that (unless the melt does not wet the underlying solid, of course) a solid undergoing a process such as sintering should be treated as having a uniform surface, not a surface with dry and wet areas.

These conclusions will illustrate the kind of insight into bulk phenomena that can be obtained through investigation of clusters.

4. Acknowledgments

This research was supported by the Office of Naval Research under Contract 88–00091.

References

1. D. J. McGinty, J. Chem. Phys. **58**, 4733 (1973).

2. W. Damgaard Kristensen, E. J. Jensen and R. M. J. Cotterill, J. Chem Phys. **60**, 4161 (1974).

3. R. M. Cotterill, W. Damgaard Kristensen, J. W. Martin, L. B. Peterson and E. J. Jensen, Comput. Phys. Comm. **5**, 28 (1973).

4. C. L. Briant and J. J. Burton, J. Chem. Phys. **63**, 2045 (1975).

5. R. D. Etters and J. B. Kaelberer, Phys. Rev. A. **11**, 1068 (1975).

6. R. D. Etters and J. B. Kaelberer, J. Chem. Phys. **66**, 5112 (1977).

7. J. B. Kaelberer and R. D. Etters, J. Chem. Phys. **66**, 3233 (1977).

8. R. S. Berry, T. L. Beck, H. L. Davis and J. Jellinek. "Solid-Liquid Phase Behavior in Microclusters." in *Evolution of Size Effects in Chemical Dynamics, Part 2*, I. Prigogine and S. A. Rice ed. (John Wiley and Sons, New York, 1988), p. 75

9. T. L. Beck, J. Jellinek and R. S. Berry, J. Chem. Phys. **87**, 545 (1987).

10. T.L. Beck and R. S. Berry, J. Chem. Phys. **88**, 3910 (1988).

11. H. L. Davis, D. J. Wales and R. S. Berry, J. Chem. Phys. **92**, 4473 (1990).

12. V.V. Nauchitel and A. J. Pertsin, Mol. Phys. **40**, 1341 (1980).

13. R. S. Berry, J. Chem. Soc. Faraday Trans. **86**, 2343 (1990).

14. R. S. Berry and D. J. Wales, Phys. Rev. Lett. **63**, 1156 (1989).

15. D. J. Wales and R. S. Berry, J. Chem. Phys. **92**, 4473 (1990).

16. J. Bösiger and S. Leutwyler, Phys. Rev. Lett. **59**, 1895 (1987).

17. J. Bösiger, R. Knochenmuss and S. Leutwyler, Phys. Rev. Lett. **62**, 3058 (1989).

18. U. Even, N. Ben-Horin and J. Jortner, Phys. Rev. Lett. **62**, 140 (1989).

19. U. Even, N. Ben-Horin and J. Jortner, Chem. Phys. Lett. **156**, 138 (1989).

20. L. E. Fried and S. Mukamel, Phys. Rev. Lett. **66**, 2340 (1991).

21. R. J. LeRoy, J. C. Shelley and D. Eichenauer. "Spectra, Structure and Dynamics of SF_6-$(Ar)_n$ Clusters." in *Large Finite Systems*, J. Jortner, A. Pullman and B. Pullman ed. (D. Reidel, Dordrecht, 1987), p. 165

22. P. A. Braier, R. S. Berry and D. J. Wales, J. Chem. Phys. **93**, 8745 (1990).

23. S. Sawada and S. Sugano, Z. Phys. D. **12**, 189 (1989).

24. J. Luo, U. Landman and J. Jortner. "Isomerization and Melting of Small Alkali-Halide Clusters." in *Physics and Chemistry of Small Clusters*, P. Jena, B. K. Rao and S. N. Khanna ed. (Plenum Press, New York, N. Y., 1987), p. 201

25. J. Rose and R. S. Berry, J. Chem. Phys. (in press) (1991).

26. T. L. Hill, *The Thermodynamics of Small Systems, Part 1*, (W. A. Benjamin, New York, 1963).

27. T. L. Hill, *The Thermodynamics of Small Systems, Part 2*, (W. A. Benjamin, New York, 1964).

28. D. J. Wales and R. S. Berry, J. Chem. Phys. **92**, 4283 (1990).

29.F. H. Stillinger and T. A. Weber, J. Chem. Phys. **81**, 5095 (1984).

30.R. K. Pathria, J. Math. Phys. **24**, 1927 (1983).

31.J. Bohr, Z. Physik D. **20**, 215 (1991).

32.K. Strandburg, Revs. Mod. Phys. **60**, 161 (1988).

33.H.-P. Cheng and R. S. Berry, (submitted) (1991).

34.H.-P. Cheng and R. S. Berry, in Proc. Symposium on Clusters and Cluster-Assembled Materials, Conference Location, 1991, R. Averbach, ed. (Materials Research Society, **206**), p. 241.

35.H.-P. Cheng, X. Li, R. L. Whetten and R. S. Berry, (submitted) (1991).

QUANTUM MOLECULAR DYNAMICS OF CLUSTERS

J. Bernholc, Jae-Yel Yi, Q.-M. Zhang, D. J. Sullivan, C. J. Brabec,
S. A. Kajihara, E. B. Anderson, and B. N. Davidson
Department of Physics, North Carolina State University
Raleigh, NC 27695-8202, USA

ABSTRACT. Recent quantum molecular dynamics studies of Al and carbon clusters are described. For Al, we focused on the 13- and 55-atom clusters, which can assume perfect icosahedral and cubic structures. However, the distortions from these ideal structures are substantial. For the 55-atom cluster, several inequivalent but nearly energetically degenerate structures are found, due to the short range of the screened interatomic interactions. For solid C_{60}, it is found that the soccerball structure is well-preserved in the solid. The intermolecular interactions are so weak that the individual C_{60} can rotate at relatively low temperatures. At high temperatures vibrations cause large distortions, but the cage structure is still preserved. The C_{60} isomer containing two pairs of adjacent five-fold rings has a binding energy only 1.6 eV smaller than that of perfect C_{60}, but the transformation between these two structures is hindered by a 5.5 eV barrier. It thus requires high temperatures and long annealing times. High temperatures are also needed for the transformation of the lowest energy C_{20} isomer, a dodecahedron, to a corannulene structure, which can be thought of as a fragment of C_{60}. The corannulene structure is a natural precursor for the formation of C_{60}. These results are consistent with the experimental findings that high temperatures are necessary for the formation of substantial quantities of C_{60}. A formulation and the first applications of a new, real space quantum molecular dynamics method, particularly suitable for cluster calculations, are also described.

1. Introduction and Summary

This paper describes quantum molecular dynamics studies of Al clusters and of the structure, dynamics, and formation of C_{60}. We also discuss briefly a development of a new, real-space method for quantum molecular dynamics. The Al clusters were chosen as a paradigm for studies of structural transformations in small and medium size metal clusters. The interest in C_{60} was of course stimulated by the recent discovery of an efficient method for its production [1].

For small metallic clusters, the splittings between the energy levels are still large and the symmetries of the occupied atomic orbitals allow only few well-defined structures. Such clusters can be considered molecular. As the size of the cluster increases, however, the number of possible structures increases dramatically. The work described below focuses on Al_{13} and Al_{55}, where both perfect cubic and icosahedral structures can exist. The results, which we regard as paradigmatic and thus applicable to other metal clusters, show clearly how the dominance of the surface and the lack of structural constraints lead to several nearly energetically degenerate structures for Al_{55}. Such structural flexibility can also lead to a distribution of cluster structures as well as floppiness at relatively low temperatures.

P. Jena et al. (eds.), Physics and Chemistry of Finite Systems: From Clusters to Crystals, Vol. I, 287–297.
© 1992 *Kluwer Academic Publishers.*

We have performed symmetry-unconstrained structure optimization and finite temperature quantum molecular dynamics simulations of solid C_{60}. It was found that the intermolecular interactions are weak and do not substantially affect the structure of the individual fullerenes. The T=0 °K structure consists of alternating single and double bonds, but this distinction becomes blurred at relatively low temperatures. The simulations show that the C_{60} molecules indeed rotate at relatively low temperatures, as indicated by the NMR data. At high temperatures the vibrational amplitudes are substantial but the cage structure is preserved, indicating excellent high temperature stability.

The perfect buckminsterfullerene is the lowest energy structure for a 60-atom carbon cluster, but a "defect" fullerene structure which contains two pairs of adjacent five-fold rings is only 1.6 eV higher in energy, according to our calculations. However, the activation energy for the transformation from the defect to the perfect structure is substantial (5.5 eV). This transformation thus requires high temperatures and long annealing times. The high barrier explains the results of various (usually constrained) simulations of the formation of C_{60}, which always lead to defective structures. The need for high temperatures is clearly demonstrated in the experiments, since such conditions lead to a much increased yield of C_{60}.

We have also examined the structural transformations of the C_{20} cluster since it is a likely precursor to the formation of C_{60}. The C_{20} is the smallest cluster which can attain a fullerene structure, consisting of twelve five-fold rings. However, our calculations find that although the fullerene structure has the lowest energy at 0 °K, entropy effects lead to an increased abundance of an open cap structure at high temperatures. The cap structure, which consists of a five-fold ring surrounded by six-fold rings, can be thought of as a fragment of C_{60} fullerene and is thus a perfect precursor for its formation. These findings are again consistent with the need for high temperatures in fullerene sources.

Although the quantum molecular dynamics methodology is quite general, its most efficient implementations rely on the use of a plane wave basis set. Periodic boundary conditions are necessarily imposed, and calculations involving first-row or transition metal atoms become quite expensive, sometimes prohibitively so. In order to reduce the computational effort for such clusters, we have developed a new real space method based on multigrid techniques which can be used with or without periodic boundary conditions. It can also handle first-row and transition metal atoms without a substantial increase in computational time. This method and the first tests of it are briefly described in Section 7.

2. Methodology

The quantum molecular dynamics (Car-Parrinello) solves the density functional equations in the local density approximation by introducing the Lagrangian [2]

$$L = \sum_i \frac{1}{2} \mu_i \int_\Omega d^3r \, |\dot{\psi}_i|^2 + \sum_I \frac{1}{2} M_I \dot{\mathbf{R}}_I^2 - E \left[\{\psi_i\}, \{\mathbf{R}_I\} \right] \tag{1}$$

where μ_i are fictitious masses associated with the dynamics of the electronic wavefunctions ψ_i, M_I and $\mathbf{R}_I$ are the masses and the positions of the atoms, and E is the total energy of the system in the local density approximation. After imposing the constraints of orthonormality of the wavefunctions, the Euler-Jacobi equations resulting from (1) become coupled equations of motion for electrons and ions. At zero temperature, the time evolution of the system reduces the time derivatives of the electronic wavefunctions to zero, and the equations of electronic wavefunctions become identical to the usual local density equations within a unitary transformation. However, one of the main advantages of the CP method is the explicit inclusion of time dependence, which opens new avenues for first-principles calculations, such finite temperature or trajectory simulations, or structure optimization based on simulated annealing.

The technical details of our QMD calculations are described in papers [3] and [4], for Al and C clusters, respectively. Specially constructed soft pseudopotentials were used in both cases, which allowed calculations with reduced plane wave cutoffs without sacrificing accuracy. The cutoffs were 4-6 Ry for Al clusters and 26-35 Ry for C clusters. Due to the use of plane waves, the clusters had to be placed in periodically repeated unit cells, but these cells were large enough to prevent significant cluster-cluster interactions, except for the solid C_{60} simulations, where the experimental fcc structure with the intercluster distance of 10.04 Å was used.

For the C_{20} calculations, the tight-binding method was used to estimate the entropies of its alternative structures [5]. The tight binding parameters were fitted to local density calculations [6].

3. Structural Distortions in Al Clusters

Our calculations focused on 13- and 55-atom clusters, for which two competing highly symmetric structures exist, *i.e.*, icosahedra and cuboctahedra. Previous calculations [7, 8 ,9], which considered only symmetry-preserving relaxations, found that the ideal icosahedron is lower in total energy than the ideal cuboctahedron for the 13-atom cluster, while the reverse is true for the 55-atom cluster. However, due to the presence of a large number of surface atoms, significant distortions from the ideal structures can be expected. We have carried out several constant thermodynamic speed annealing runs using both icosahedra and cuboctahedra as starting configurations. For the 13-atom cluster all annealing runs terminated with the same distorted icosahedral structure. The energy gain was only 0.02 eV per atom. For the 55-atom cluster three annealing runs were performed. In all cases the clusters were briefly heated to 2000 °K and then gradually cooled using the algorithm described above. One run, which started from the cuboctahedron, included 6 additional levels in addition to the occupied ones in order to allow for level crossing during annealing. The remaining runs included only the occupied levels and

Table I. Structural energies relative to the radially-relaxed Mackay icosahedra, ionization potentials (IP), and electron affinities (EA) for various alternative structures of Al_{13} and Al_{55}. For Al_{55} different simulated annealing runs led to somewhat different structures.

	E_{coh} (eV)	IP (eV)	EA (eV)
13 atoms			
ideal cub.	0.6	6.62	3.22
ideal ico.	0.0	7.15	3.81
annealed struct.	-0.3	6.56	3.04
experiment		6.42[a]	2.6[b], 2.86[c]
55 atoms			
ideal cub.	-2.2	5.50	3.29
ideal ico.	0.0	5.10	2.87
weak. ann. cub.	-5.3	5.45	3.23
weak. ann.	-6.5	5.35	3.19
fully ann. struct.	-6.0 – -6.6	5.30 - 5.38	3.14 - 3.26
experiment		4.95[d]	

(a) from [12]; *(b)* from [13]; *(c)* from [14]; *(d)* from [15].

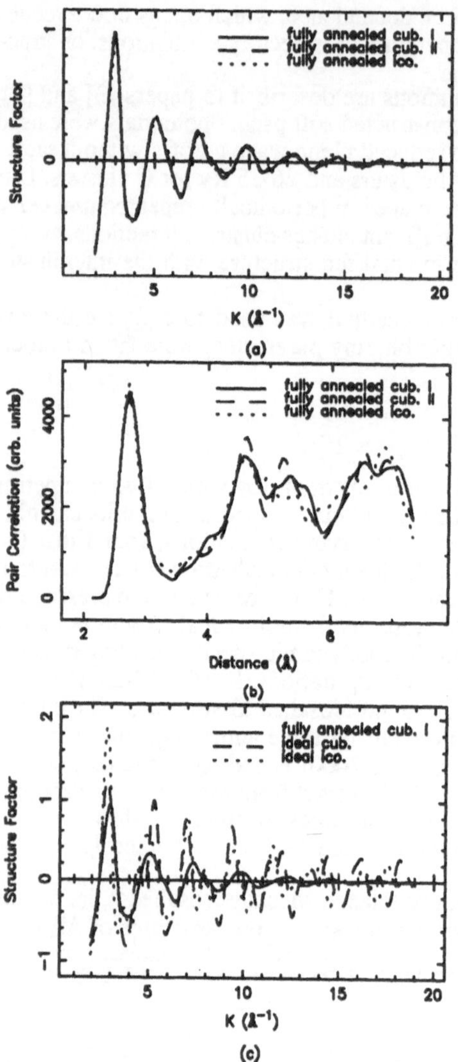

Fig. 1. Comparisons of (a) the structure factors and (b) the pair correlation functions for the fully annealed structures of Al$_{55}$. (c) Comparison of the structure factors of the radially relaxed icosahedron and cuboctahedron with the lowest energy fully annealed structure.

started from the cuboctahedron and the icosahedron, respectively. The annealing always led to severely distorted icosahedral structures with very similar structure factors but different pair correlation functions, see fig. 1a-b. The lowest energy structure is shown in fig. 2 and clearly is icosahedral in origin. The energy gain for Al$_{55}$ upon annealing, when compared to the lowest energy undistorted structure (the cuboctahedron) was about 0.1 eV/atom for the 55-atom cluster, in contrast to 0.02 eV/atom for Al$_{13}$. Since the level spacing near the Fermi level is smaller than 0.2 eV even for the annealed Al$_{55}$, this gain could not have originated from the Jahn-Teller effect and must be due to rebonding. The binding energy differences for the clusters described above are shown in Table I. Since the energetic differences between the three annealed structures are less than 0.01 eV per atom for Al$_{55}$, it is clear that this cluster can assume quite a few inequivalent structures with very similar energies. This should lead to floppiness and a low melting point. Indeed, it is well-known that metal clusters have melting points far below the bulk values [10].

In order to investigate the importance of annealing, we have also performed "weak annealing" computer experiments, in which the 55 atoms in a perfect cuboctahedral or icosahedral structure were given initial velocities corresponding to T = 500 °K in order to break possible symmetry barriers and were then dynamically relaxed. The resulting structures were still clearly cuboctahedral and icosahedral, respectively, and their structure factors were quite similar to those of the fully annealed clusters. However, the pair correlation and angular distribution functions differed substantially. Finally, in fig. 1c we compare structure factors of the ideal radially relaxed cuboctahedron and icosahedron with one of the fully annealed structures. It is clear that the three structures are *very* different. The pair correlation and angular distribution plots display very large differences as well.

The reason for the occurrence of several inequivalent but energetically almost degenerate structures becomes apparent when analyzing the angular distribution and pair correlation functions. They demonstrate that the local topology of each structure is almost the same, i.e. the nearest neighbor and the most probable angle peaks occur at exactly the same positions for the three annealed clusters, and there are strong similarities in the positions of second neighbor peaks. It follows that the energetics of Al55 is mainly determined by the first and possibly second nearest neighbors. The multitude of structures for the 55-atom cluster is possible because most of the atoms reside on the surface and are the therefore free from the constraints of bulk packing. Due to the size of this cluster, slight changes in the positions of distant neighbors can lead to completely inequivalent structures. However, a careful analysis of the annealed structures on a graphics workstation, capable of manipulating three-dimensional images shows that these structures cannot be simply described as consisting of the same solid core and differently arranged surfaces, since the positions of their inner atoms are substantially different.

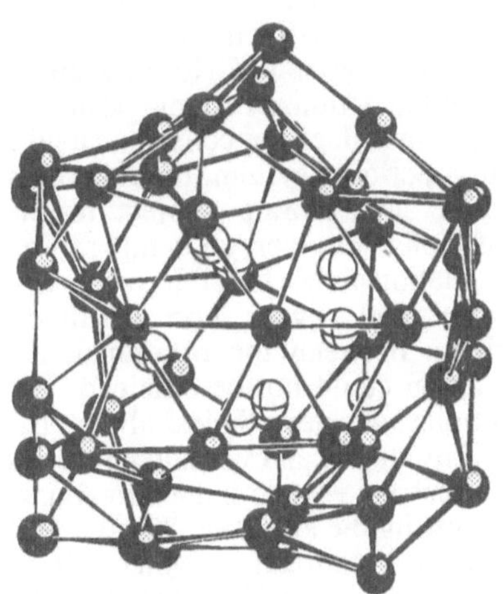

Fig. 2. The lowest energy annealed structure of Al55.

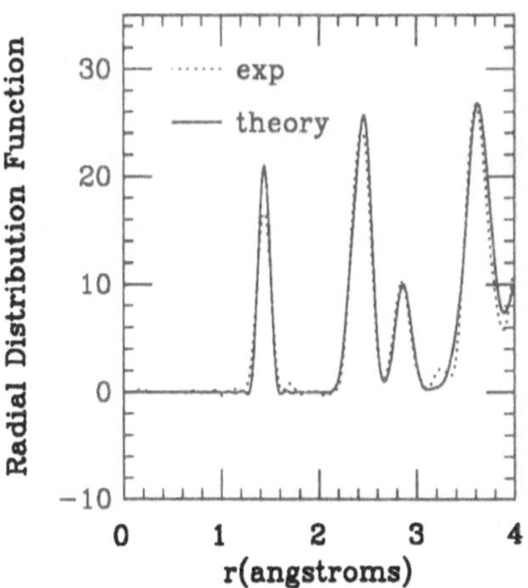

Fig. 3. Comparison of the experimental (solid line) and theoretical (dashed line) radial distribution functions for solid C60. The data are from Ref [20].

Cluster structures can also be inferred by comparisons between experimental and calculated Ionization Potentials (IPs) and Electron Affinities (EAs). We have calculated both IPs and EAs for isolated Al clusters in the structures described above, using a procedure described in Ref. [7]. The results are listed in Table I together with the experimental data. For Al13, the annealing brings the computed values much closer to the experimental ones. For Al55, the differences between the IPs and EAs computed for the various structures are too small to allow for the determination of the structure solely by comparison to the experimental data. Furthermore, since the various isomers are energetically almost degenerate, the experiments may measure an average over the possible structures. Indeed, for Co clusters experimental conditions lead to several magnetically resolved isomers in the N=55-66 range [11].

4. Structure and Dynamics of Solid C60

This section describes the results of symmetry-unconstrained structure optimization and finite temperature ab-initio molecular dynamics simulations of solid C60. The symmetry-unconstrained structure optimization led to an almost ideal fullerene with an average diameter of 7.1 Å. Similarly to what has been predicted for free C60 [16-19], two carbon-carbon bond distances were found, corresponding to nominally single and double carbon-carbon bonds. The average lengths in the optimized structure for the single and double bonds are 1.40 and 1.45 Å, respectively. As expected, these values straddle the nearest neighbor bond length in graphite which is 1.42 Å. In hexagonal rings the single and double bonds alternate, while only single bonds are present in pentagonal rings. The intermolecular interactions in the solid introduce

292

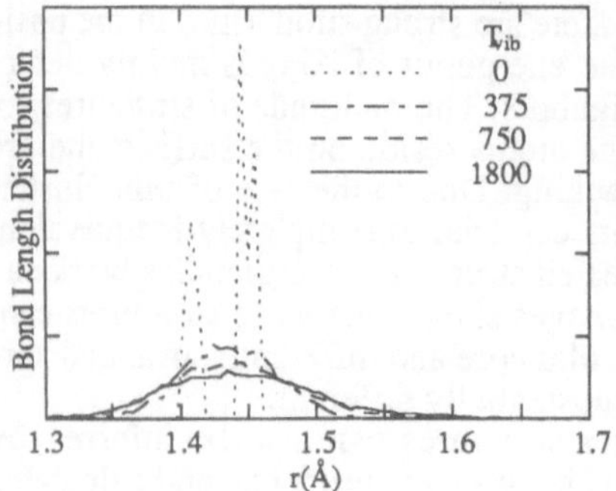

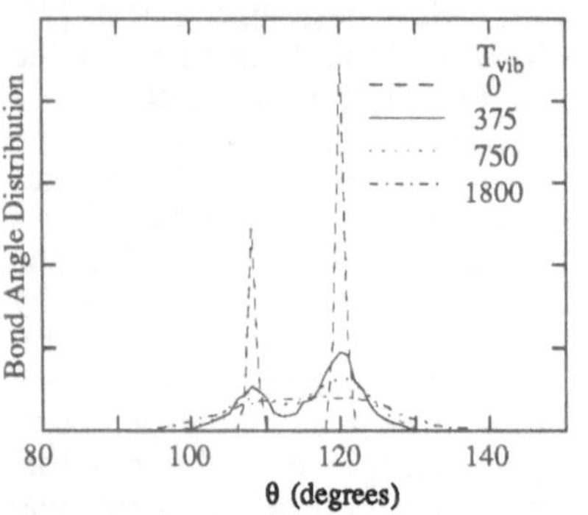

Fig. 4. Bond length and bond angle distribution in solid C_{60} at various temperatures. See text.

small variations in bond lengths, however, not exceeding 0.01 Å at T=0 °K. Similarly, deviations in the bond angles from the ideal pentagon and hexagon values do not exceed 1°. The computed radial distribution function agrees very well with the experimental distribution obtained by neutron diffraction [20], see fig. 3. Our results agree also very well with photoemission spectra and EXAFS data, see [4].

Constant temperature simulations were carried out for a number of representative temperatures. They utilized the quantum molecular dynamics version of Nosé dynamics and started from the optimized atomic positions. Since Nosé dynamics does not conserve angular momentum in a periodic system, the fullerenes can both rotate and vibrate subject to the constraint of constant average temperature. Since the simulation times possible on todays computers correspond to very short real times, the equilibriums between the rotational and vibrational motion could not be obtained. Our starting conditions have nevertheless always led to a mixture of rotational and vibrational modes and it was particularly easy to induce rotations. For example, heating the system to $T_{Nosé} = 300$ °K already led to rotational motion. This is consistent with NMR data, which show a single narrow line at this temperature [21, 22].

The bond length and bond angle distribution at various temperatures are shown in fig. 4. At T=0 °K one can clearly distinguish between single and double bonds. Since each carbon atom forms two single and one double bond, the single bond peak at 1.45 Å is twice as strong as the double bond peak at 1.41 Å (fig. 4a). However, at vibrational temperatures as low as 375 °K the vibrational amplitudes are sufficiently strong to wash out the small difference between the single and double bond lengths and only a single broad peak is present. At higher temperatures further broadening and bond softening is observed, but the fullerene structure remains intact. Similar trends are observed in the bond angle distribution (fig. 4b).

A more quantitative measure of the distortions from the ideal structure is provided by the ratio of the maximum distance to the corresponding minimum distance of any of the carbon atoms from the center of mass of the C_{60}. At T=0 °K this ratio is 1.01. The slight deviation from unity is caused by the weak crystal field. At T=375, 750, and 1800 °K, this ratio, averaged over the duration of the simulations, increases to 1.08, 1.09, and 1.22, respectively. However, the average diameter of the C_{60} remains almost constant, increasing only by 0.02 Å upon heating from T=0 to 1800 °K.

The vibrational modes of C_{60} have also been analyzed by QMD techniques, see [24, 25].

5. Annealing of C60 Isomers

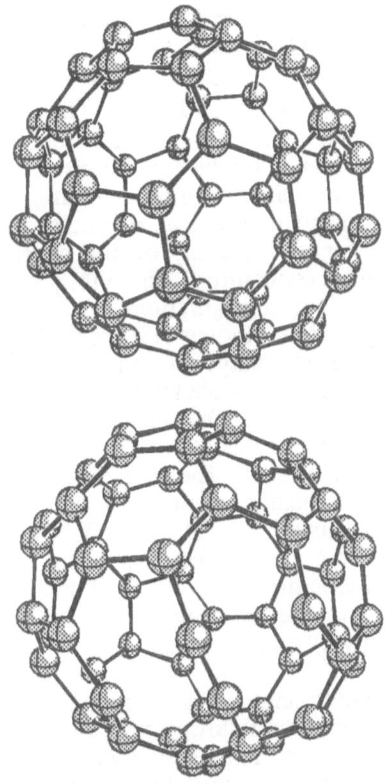

Fig. 5 (a) The structure of defect C_{60}. Note the adjacent five-fold rings. (b) The structure of the transition state. See text.

The isomer energetically closest to the C_{60} buckminsterfullerene contains two pairs of adjacent five-fold rings [23] (see fig. 5a). Our computed binding energy of the fully relaxed structure of this isomer is only 1.6 eV smaller than that of the buckminsterfullerene. We have thus proceeded to investigate the transformation between the two structures. The simplest path from the "defect" to the "perfect" C_{60} involves the rotation of two atoms around their bond center. This is the well-known concerted exchange path, which has an activation energy of 10.4 eV in graphite [26]. Starting from the concerted exchange initial geometry, we determined the relaxed saddle point configuration by a combination of an adiabatic trajectory simulation and a modified steepest descent procedure. The geometry of the saddle point is shown in fig 5b. Its energy is 5.5 eV above that of the defect C_{60}. Although it is substantially smaller than in graphite, it is still very large. Since the transformations between other C_{60} isomers are also likely to have large activation barriers, it follows that annealing of C_{60} isomers to a perfect buckminsterfullerene structure requires very high temperatures and long annealing times. This explains why all molecular dynamics simulations to date, both quantum and classical, resulted in the formation of highly defected C_{60}. The annealing times necessary to produce a perfect C_{60} are simply unattainable on today's computers.

The experimentally observed predominance of perfect C_{60}, on the other hand, is easily explained by heat of fusion arguments. For example, the reaction of C_{58} and C_2 to form a defect C_{60} releases 12 eV of energy. Provided that the cooling is not too fast, *i.e.*, the He pressure is moderate and/or a high temperature is maintained, the transformation to a perfect structure will occur. Annealing of the isomers is thus one of the reasons for the need to maintain high temperatures for a high-yield formation of C_{60}.

6. C20 Clusters as Precursors to the Formation of C60 Fullerenes

Since C_{60} is formed at high temperatures in graphite arc or laser vaporization sources, its formation process likely involves the fusion of smaller carbon clusters. C_{20} is the smallest carbon cluster which can have either a fullerene structure, consisting of twelve five-fold rings, or a much more open "cap" structure, consisting of a five-fold ring surrounded by five six-fold rings. The cap structure is curved, resembling a fragment of C_{60} (see fig. 6), and has been hypothesized to be the initial building block for the growth

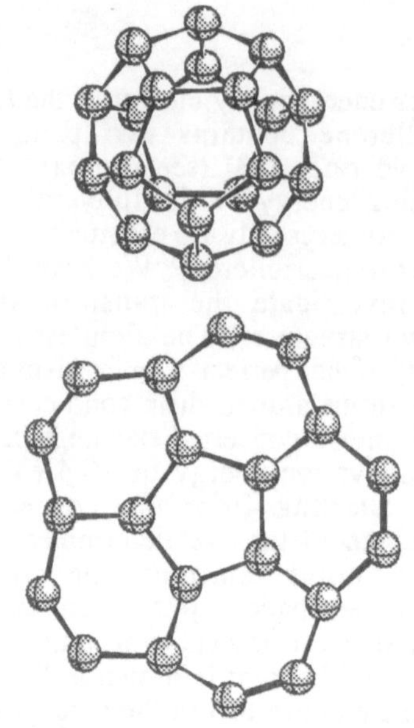

Fig. 6. The structures of (a) A C_{20} fullerene; and (b) A C_{20} pentagonal cap. See text.

of spiral carbon cluster cages. Other possible structures of C_{20} are linear chains and monocyclic rings, which are the preferred configurations for small carbon clusters. Table II gives our results for the relative binding energies calculated for the four relaxed configurations at T=0 °K. The cage structure is found to be the minimum energy configuration, followed by the cap, the ring, and the chain. However, since high temperatures are involved in the formation of C_{60}, one must also consider the effects of entropy on the relative stability of these structures. The vibrational entropies of the four configurations of C_{20} are obtained within a tight-binding (TB) formalism, which offers the advantage of requiring less computational time to calculate total energies and forces than the CP method. The energies obtained from the tight-binding calculations are compared to the CP results in Table II. The predictive power of the *ab initio* CP method is superior to the more economical parametrized TB method, which explains the minor discrepancies between the results. Both methods found the total binding energy of the structures (with respect to isolated atoms) to be larger than 100 eV, so the results differ only by a few percent.

Fig. 7 shows the free energy difference of the four structures as a function of temperature. At T=0 °K, the closed cage is the preferred geometry, followed by the cap, the ring and the chain. As the temperature is increased, however, the free energy of the cap structure approaches that of the cage. To a large degree, this is caused by the larger vibrational entropies of the more "floppy" structures. Thus the cap structure is more likely to form than the cage structure at higher temperatures. Since the cap arrangement of 20 atoms is contained in the C_{20} structure, this is a logical first step in the formation of C_{60}. The C_{20} cap could be the seed upon which the other fragments attach to form the complete closed C_{60} shell. Indeed, a recent "hot shrinking sphere" simulation of C_{60} formation showed that five-fold rings are the first to form out of the carbon vapor [27]. If C_{20} or a larger carbon cluster cap structure is a precursor to C_{60} formation, then ligands attached to the concave side of the cap might survive the formation of a larger fullerene. The introduction of a suitable ligand during the formation process might thus lead to trapping of atoms or small molecules inside the larger fullerene.

The present results have important implications on the models describing the formation of C_{60} and other cage structures. A previous study suggested that the growth of carbon clusters proceeds by the formation of graphitic sheets that minimize the number of dangling bonds, while maximizing the number of non-adjacent pentagonal rings [28]. The calculations presented here, however, show that

Table II. Relative cohesive energies of the four relaxed structures. The Car-Parrinello (CP) results are more accurate than the tight-binding (TB) results.

Structure	CP (eV)	TB (eV)
Cage	0.0	0.0
Cap	1.7	-3.3
Ring	5.4	1.6
Chain	10.0	0.6

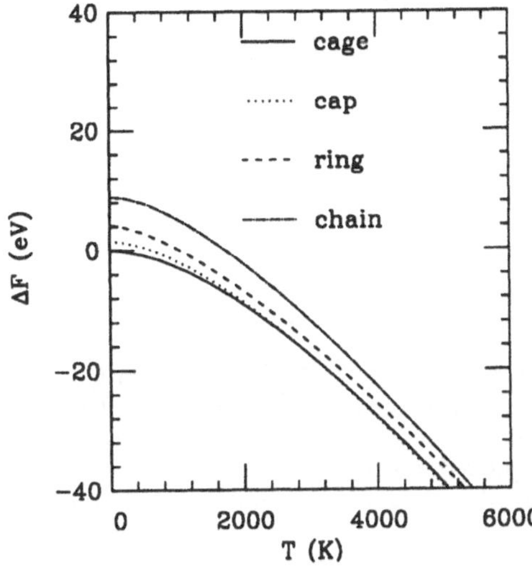

Fig. 7. Free energy difference ΔF with respect to the cage at T=0 °K as a function of temperature.

the structure of C_{20} with the lowest internal energy is a closed cage, even though it consists entirely of adjacent pentagonal rings. Therefore, if this cluster is able to assume its closed cage structure with minimum internal energy, the formation of C_{60} would be inhibited, since the closed cage structures should be less reactive than the open structures. Clearly, considerations based on the internal energy alone cannot explain the high 30-40% yield of C_{60} observed in the laboratory. As shown in Figure 7, entropic contributions favor the more open cap structure. Therefore, the fraction of carbon clusters in the cap structure should increase with temperature. This open structure is more prone to react and form larger structures than the cage structure. This result is also consistent with the high temperature growth conditions which are needed for a high-yield formation of C_{60}.

The above mechanism is applicable to other clusters of similar sizes. It is a reasonable conjecture that such clusters prefer closed cages at T=0 °K, but convert to open flakes at higher temperatures. For C_{60}, C_{70} and other "perfect" fullerenes, the energetic preference for the closed cage structure is so large that high temperatures alone are not enough to crack them open. C_{60} is the smallest of these structures, and is acting as a growth terminator. Assuming that the fullerenes form from smaller carbon clusters, C_{60} would form first and therefore in high yield, as observed experimentally.

Considering larger fullerenes, Smoluchowski's aggregation equations for clusters show that the peak in cluster distribution depends on the residence time [29]. Consequently, further increases in the yield of specific clusters can be achieved by optimizing the residence time and/or cluster density. This explains the ability of optimizing the experimental conditions for an increased yield of C_{60}, C_{70}, or C_{84}, although the yield of C_{60} is always substantial.

7. Real Space Quantum Molecular Dynamics

For a number of systems, the plane wave quantum molecular dynamics formalism suffers from several disadvantages. For clusters, the greatest is the need to maintain periodic boundary conditions, which leads to very large supercells in order to separate the periodically repeated clusters, thereby substantially increasing the computational cost. The plane wave basis is also very expensive to use for first row and transition metal atoms, since their hard pseudopotentials require very large plane wave cutoffs. Furthermore, the inclusion of even one "difficult" atom increases the number of plane waves *everywhere*. Although recent developments in optimized pseudopotentials decrease substantially the number of required plane waves, other limitations remain, and a calculation for a transition metal cluster would be very expensive.

We are developing a new real-space method for electronic structure calculations which does not use plane waves. Instead, real space grids are employed as a basis set, and all calculations occur in real space. Our method is based on multigrid techniques [30], which are utilized in iteratively solving both the Schrödinger and the Poisson equations. Briefly, in a multigrid or multilevel approach, a differential equation is solved on several grids at a time. The final solution is obtained on a sufficiently fine grid so that it represents well the pseudo-wavefunctions and the pseudopotentials in the problem. However, the convergence is accelerated by the use of several auxiliary grids, each with the spacing increased by a factor of two over the previous grid. Denoting the grid spacing by Δx, one can show that iterations on any given grid reduce quickly those Fourier components of the error with wavelengths $\sim \Delta x$ but are ineffective when dealing with substantially longer wavelengths. The role of the extra grids is thus to converge quickly and systematically *all* wavelength components of the error. One important advantage of the algorithm is that the grids need not be uniformly spaced so that the grids can be enhanced locally near, *e.g.*, transition metal atoms.

The multigrid algorithm is very stable and efficient, scaling as $O(N)$. It has long been used for solving Navier-Stokes and Poisson equations, but has not before been applied to the electronic structure problem. We have adapted it for the solution of local density equations and in particular developed procedures for handling the near-singularities which occur near the atomic origins, even when pseudopotentials are used. Substantial programming and some algorithm development efforts are still needed, but we already obtained test results for H, H_2, and Li. In Table III we compare the results for H_2 using a uniform and a non-uniform cubic grids, with zero boundary conditions applied at the faces of a cube. The cube side is 20 a.u. The substantial improvement upon the use of nonuniform grids is evident. Finally, note that the speed of these calculations is comparable to the Car-Parrinello method, both in terms of the number of iterations and CPU time.

Table III. Results of multigrid calculations for H_2 using uniform and non-uniform grids. See text.

H_2 Molecule	r_e (Bohr)	E_{bind} [eV]	ω (cm^{-1})	Eigenval. [eV]
Uniform Grid	1.35	4.95	7310	-9.30
Nonuniform Grid	1.41	5.06	4320	-9.64
Local orb. [31]	1.45	4.91	4210	-10.15
Experiment	1.40	4.75	4401	N/A

8. Acknowledgements

This work was supported by the Office of Naval Research, grant numbers N00014-88-K-0491 and N00014-91-J-1516. The supercomputer calculations were carried out at the North Carolina and Pittsburgh Supercomputing Centers.

References

1. W. Krätchmer, L. D. Lamb, K. Fostiropoulos, and D. R. Huffman, Nature **347**, 354 (1990).
2. R. Car and M. Parrinello, Phys. Rev. Lett. **55**, 2471 (1985).
3. J.-Y. Yi, D. J. Oh, and J. Bernholc, Phys. Rev. Lett. **67**, 1594 (1991).
4. Q.-M. Zhang, J.-Y Yi, and J. Bernholc, Phys. Rev. Lett. **66**, 2633 (1991).
5. C. J. Brabec, E. Anderson, B. N. Davidson, S. A. Kajihara, Q.-M. Zhang, J. Bernholc, and D. Tománek, to be published.
6. D. Tománek and M. Schlüter, C. Sun, N. Sharma, and L. Wang, Phys. Rev. B **39**, 5361 (1989).
7. J.-Y. Yi, D. J. Oh, J. Bernholc, and R. Car, Chem. Phys. Lett. **174**, 461 (1990).
8. H.-P. Cheng, R. S. Berry, and R. L. Whetten, Phys. Rev. B **43**, 10647 (1991).

9.   L. L. Boyer, M. R. Pederson, K. A. Jackson, and J. Q. Broughton, in *Clusters and Cluster-Assembled Materials*, edited by R. S. Averback, J. Bernholc, and D. L. Nelson, MRS Proc. **206**, 253 (1991).
10. Ph. Buffat and J.-P. Borel, Phys. Rev. A **13**, 2287 (1976).
11. J. P. Bucher, D. C. Douglass, and L. A. Bloomfield, Phys. Rev. Lett. **66**, 3052 (1991).
12. D. M. Cox, D. J. Trevor, R. L. Whetten, and A. Kaldor, J. Phys. Chem. **92**, 421 (1988).
13. G. Ganteför, M. Gausa, K. H. Meiwes-Broer, and H. O. Lutz, Z. Phys. D **9**, 253 (1988).
14. K. J. Taylor, C. L. Pettiette, M. J. Craycraft, O. Chesnovsky, and R. E. Smalley, Chem. Phys. Lett. **182**, 347 (1988).
15. K. E. Schriver, J. L. Persson, E. C. Honea, and R. L. Whetten, Phys. Rev. Lett. **64**, 2539 (1990).
16. D. S. Marynick and S. Estreicher, Chem. Phys. Lett. **132**, 383 (1986).
17. S. Satpathy, Chem. Phys. Lett. **130**, 545 (1986).
18. M. L. McKee and W. C. Herndon, J. Mol. Struct. (Theochem) **153**, 75 (1987).
19. B. I. Dunlap, Intern. J. Quant. Chem., Quant. Chem. Symp. **22**, 257 (1988).
20. F. Li, D. Ramage, J. S. Lannin and J. Conceicao, to be published.
21. C. S. Yannoni, R. D. Johnson, G. Meijer, D. S. Bethune, and J. R. Salem, J. Phys. Chem. **95**, 9 (1991).
22. R. Tycko, R. C. Haddon, G. Dabbagh, S. H. Glarum, D. C. Douglass, and A. M. Mujsce, J. Phys. Chem. **95**, 518 (1991).
23. A. J. Stone and D. J. Wales, Chem. Phys. Lett. **128**, 501 (1986).
24. G. B. Adams, J. B. Page, O. F. Sankey, K. Sinha, J. Menandez, and D. R. Huffman, Phys. Rev. B **44**, 4052 (1991)
25. B. P. Feuston, W. Andreoni, M. Parrinello, and E. Clementi, Phys. Rev. B **44**, 4056 (1991)
26. T. Kaxiras and K. C. Pandey, Phys. Rev. Lett. **61**, 2693 (1988).
27. C. Z. Wang, C. H. Xu, C. T. Chan, and K. M. Ho (submitted for publication).
28. R. E. Haufler, Y. Chai, L. P. F. Chibante, J. Conceicao, Changming Jin, Lai-Sheng Wang, Shigeo Maruyama and R. E. Smalley, in Ref. [9] p. 627 (1991).
29. J. Bernholc and J. C. Phillips, J. Chem. Phys. **85**, 3258 (1986).
30. A. Brandt, Math. Comp. **31**, 33 (1977); A. Brandt, S. McCormick, and J. Ruge, Siam J. Sci. Stat. Comp. **4**, 244 (1983).
31. I. Moullet and J. L. Martins, J. Chem. Phys. **92**, 527 (1990).

RANDOM CLUSTER MODEL FOR ICOSAHEDRAL PHASE AlMnSi

J. L. Robertson

Reactor Radiation Division
National Institute of Standards and Technology
Gaithersburg, Maryland 20899
U. S. A.

ABSTRACT.   Results are presented on a computer generated model of AlMnSi. Atomic clusters consisting of 12 Mn and 42 Al(Si) atoms are randomly packed to form a continuous random network while placing constraints on the allowed local cluster configurations. The topological characteristics of the model are explored in both $R^\parallel$ and $R^\perp$ space including an examination of the phason disorder. Preliminary X-ray and neutron diffraction patterns are calculated for comparison with experimental intensity data. The origin of the peak at $Q=1.62\text{Å}^{-1}$, associated with the prepeak found in "amorphous" AlMnSi, as well as the "diffuse" scattering seen experimentally under the groups of strong peaks, are revealed selectively in the calculated partial intensities for the Al-Al, Mn-Mn, and Mn-Al correlations.

There is substantial experimental evidence[1] that some icosahedral phase (i-phase) alloys consist for the most part of large icosahedrally symmetric atomic clusters. Indeed, such clusters are the primary structural unit of certain crystalline phases closely related to the i-phase[2] and are consistent with the three most widely discussed structural models, the three dimensional Penrose tiling (3DPT),[3] the random tiling model (RTM)[4] and the icosahedral glass model (IGM).[5,6] Ideal quasicrystals, e.g. the 3DPT, may be realized mathematically by various geometrical constructions,[1,7] but it is difficult to imagine how they can exist in nature. The difficulty is that when growing a quasicrystal solely by local attachment rules[8] there is no apparent physical mechanism by which the long range quasiperiodic order can be imposed. The RTM and the IGM offer alternative interpretations of the observed diffraction patterns without the strict requirement of perfect quasiperiodicity. In these two models entropy, rather than energy, is the dominant stabilizing factor[4,9] whereas the quasicrystal represents a ground state energy configuration in accordance with its matching rules.

The IGM provides a very natural growth scheme in which the atomic clusters that would normally pack together periodically to form a crystal instead form a more disordered array because of the large number of nearly equivalent cluster configurations.[6,9] This paper summerizes results on computer generated random cluster models (RCM)[10] indicating how the application of physically motivated local constraints can produce a model which agrees well with the observed diffraction pattern

299

P. Jena et al. (eds.), Physics and Chemistry of Finite Systems: From Clusters to Crystals, Vol. I, 299–308.
© *1992 Kluwer Academic Publishers.*

of simple i-phase alloys such as AlMnSi, AlTi or AlLiCu. A more detailed treatment of the RCM can be found in references 11 and 12.

Icosahedral phase AlMnSi (i-AlMnSi) has a closely related crystalline phase, α-AlMnSi,[2] which is essentially a bcc packing of 54-atom icosahedrally symmetric clusters.[13] Each cluster consists of two concentric shells of atoms; the inner shell is composed of 12 Al(Si) atoms on the vertices of an icosahedron. The second shell, also an icosahedron, has 12 Mn atoms on the vertices (directly above the Al atoms of the inner shell) and 30 Al(Si) atoms on the edges. There are additional Al(Si) atoms that fill the space between clusters and can be identified with a third shell of atoms,[14] but this shell is incomplete and each atom belongs to the third shell of more than one cluster. In the α-phase the clusters are connected face to face (along 3-fold axes) in the <111> directions of the cubic lattice. This type of connection ensures that the orientational order is preserved throughout the structure.

The RCM, in common with previous models[5,6,9] constructed by randomly packing icosahedra, retains the identity of the cluster linkages (face to face) thus preserving the orientation of the icosahedral symmetry axes throughout the structure; but unlike the α-phase, the faces connecting clusters are chosen at random, ignoring the bcc lattice. The simplest model is constructed by starting with a single "seed" icosahedron and adding clusters by randomly choosing a face on one of the existing icosahedra to which the next cluster is attached with the only constraint being that it does not interpenetrate any of its neighbors; if it does it is discarded and another face is chosen. This process is continued until it has "grown" to fill some predetermined volume. The basic procedure described above is identical to that first employed by Schectman and Blech[5], and later by Stephens and Goldman[6] in a more detailed study. Models produced in this way are in poor agreement, however, with the observed X-ray and neutron powder diffraction patterns (see Fig. 1). For convenience the center to center distance between face connected clusters is assigned a value of 1.0 which I define to be a model unit (mu). Thus in order to compare with experiment the scale must be multiplied by a factor of 10.984Å/mu as determined by fitting the calculated diffraction pattern to experimental i-AlMnSi diffraction data. A factor of 10.956 is calculated from the lattice parameter of the α-phase, but the clusters in the crystal are slightly elongated in the <111> directions of the bcc lattice.[13]

In order to improve the model, three adjustable parameters are introduced which are used to impose additional constraints on the near neighbor cluster configurations produced by the random packing algorithm. Specifically, the goal is to suppress the formation of tears[7,9] in the structure and increase the connectivity (average number of linkages per cluster) thus extending the long range translational order in the model. To reduce the tendency for isolated "branches" to grow outward from the seed, two parameters, ΔS and P, are introduced which are used to constrain the addition of new icosahedra to the surface of the model. Depending on the values of ΔS and P this constraint drastically alters the behavior of the model as it grows. The constraint is implemented as follows: the seed icosahedron is initially placed at the center of a small bounding sphere of radius S (~3mu) and icosahedra are added by the random packing algorithm until the probability of selecting an allowable site, i.e. a face for attaching the next icosahedron where it doesn't interpenetrate any existing clusters and is inside the bounding sphere, falls below some cutoff probability, P, which is computed continuously by monitoring the success rate of adding new clusters. Once

Table 1. Neighbor distances ≤2.0mu between clusters. The bcc neighbor distances and M3DPT vertex separations are indicated and the 1.229mu and 1.686mu distances are not allowed as discussed in the text.

distance in model units	(in Å)	neighbor type
1.000	(10.948)	M3DPT and bcc($\frac{1}{2}$, $\frac{1}{2}$, $\frac{1}{2}$)
1.155	(12.645)	M3DPT and bcc(1,0,0)
1.229	(13.455)	not allowed
1.528	(16.729)	M3DPT
1.633	(17.878)	M3DPT and bcc(1,1,0)
1.686	(18.458)	not allowed
1.777	(19.455)	M3DPT
1.868	(20.451)	M3DPT
1.915	(20.965)	bcc($\frac{1}{2}$, $\frac{1}{2}$, $\frac{3}{2}$)
2.000	(21.896)	bcc(1,1,1)

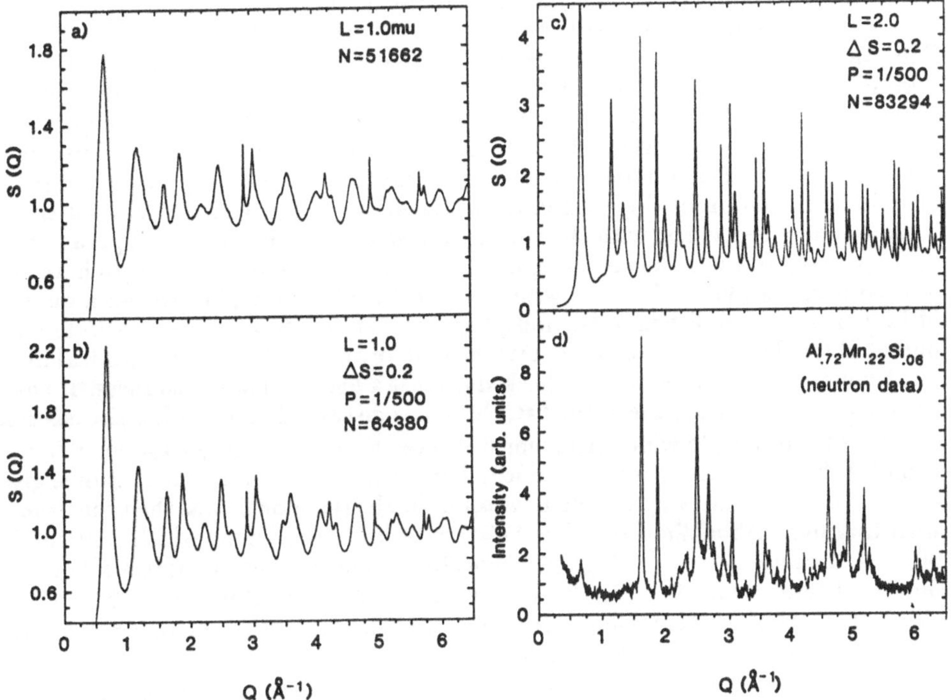

Figure 1. Powder diffraction patterns: (a) calculated S(Q) for a model constructed with only the constraint that clusters not interpenetrate; (b) calculated S(Q) for a model grown with the surface growth constraint ($\Delta S=0.2$mu and P=1/500) and the constraint that clusters not interpenetrate (L=1.0mu); (c) calculated S(Q) for a model with L=2.0mu, $\Delta S=0.2$mu and P=1/500; (d) neutron intensity data on i-AlMnSi.

the cutoff probability is reached, the radius of the bounding sphere, S, is increased by ΔS, thus presenting a "fresh" surface of available sites. Clusters are again added randomly until the probability of finding an allowable site once again falls below P; S is again increased by ΔS and the process continues until the model reaches the desired size, in the present case 50.0mu or ~550Å in diameter. After constructing a series of small trial models it was determined that the resulting structure has only a weak dependence on the value of the cutoff probability, P, in the range 1 in 200 to 1 in 2000; P = 1 in 500 was chosen for the present study.

The effect of varying ΔS was also explored. On decreasing ΔS, no significant improvement in the connectivity or the calculated diffraction pattern was found for ΔS greater than 2.0mu with the best results occurring when ΔS is less than 0.5mu. For ΔS less than 0.2mu, however, a faulted bcc or diamond cubic structure is likely to occur, presumably depending on the the configuration of the first few clusters; thus ΔS = 0.2mu was deemed the smallest "safe" value. It should be noted that this surface growth constraint also imposes some degree of non-locality on the growth algorithm in the sense that because of it, the model grows as a sphere which is a global rather than a local effect.

In order to produce a well-connected network of clusters and achieve a higher packing fraction, it is additionally necessary to discriminate against local cluster configurations which lead to the formation of large voids in the structure. It is also desirable that the path connecting nearby clusters consist of a small number of face-face linkages. Rather than discriminate against actual local cluster configurations, which would require a considerable amount of computer time for large models, neighbor distances are discriminated against, a somewhat less strict criterion. Here we introduce the notion of the local environment of a cluster, which is defined simply to be a sphere concentric with the cluster, whose size is controlled by a parameter L corresponding to the radius of the sphere. The idea is to allow a cluster to have only a discrete set neighbor distances within its local environment, i.e. within a distance of L. This is accomplished in the following way: after a face has been selected as a potential for the next cluster to be attached, all of the neighbor distances within the local environment of this site are checked against a list of allowed distances. If any of the distances do not appear in the list, the site is discarded and another site selected.

In order to decide which neighbor distances to allow, a list of all possible near neighbor distances for six face-linked icosahedra was computed and the result is given in Table 1 out to a distance of 2.0mu. A set of small trial models was then constructed with L=2.0mu, ΔS=0.2mu and P = 1/500 in order to determine which combination of these distances gives the highest connectivity and packing fraction while retaining icosahedral symmetry. By this process neighbors at 1.229mu and 1.686mu were eliminated. All those remaining correspond to either an α-phase distance or a vertex separation fond in the 3DPT or both.

The calculated S(Q) for the simple model, where the only constraint is that neighboring clusters not interpenetrate which corresponds to L=1.0mu and no surface growth constraint, is shown in Fig. 1a. Fig. 1c shows the calculated S(Q) for the best model, L=2.0, ΔS=0.2mu and P=1/500. By comparison of these S(Q)'s with the neutron powder intensity data[14] shown in Fig. 1d, it is clear that the pattern shown in Fig. 1c is in much better agreement with experiment; i.e. the peak positions and peak shapes are in good agreement whereas they are not for the simple models. (The model

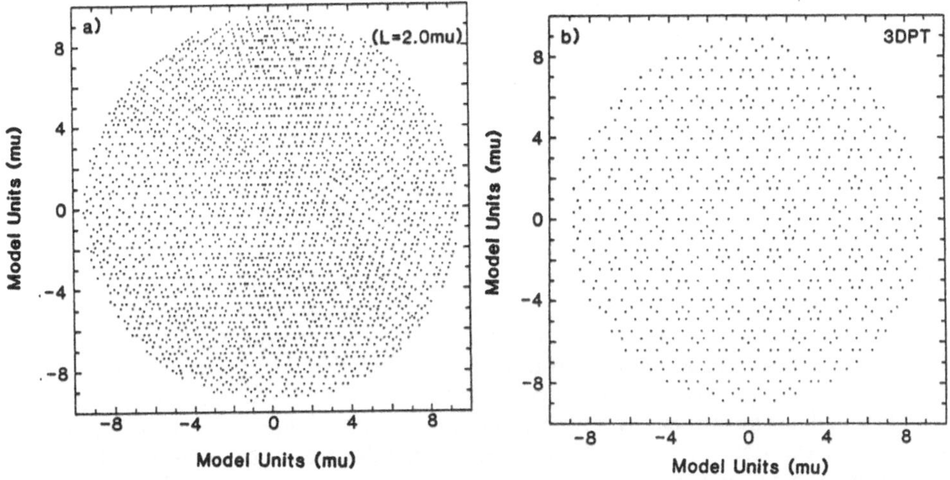

Figure 2. 20.0mu cross sections taken in R$^{\parallel}$ through (a) RCM (L=2.0mu, ΔS=0.2mu, P=1/500) and (b) the 3DPT. Each slice represents a cylinder with its axis parallel to an icosahedral basis vector and diameter equal to its length ($\sim$ 220Å), projected onto a plane perpendicular to its axis. The dots represent projected cluster center.

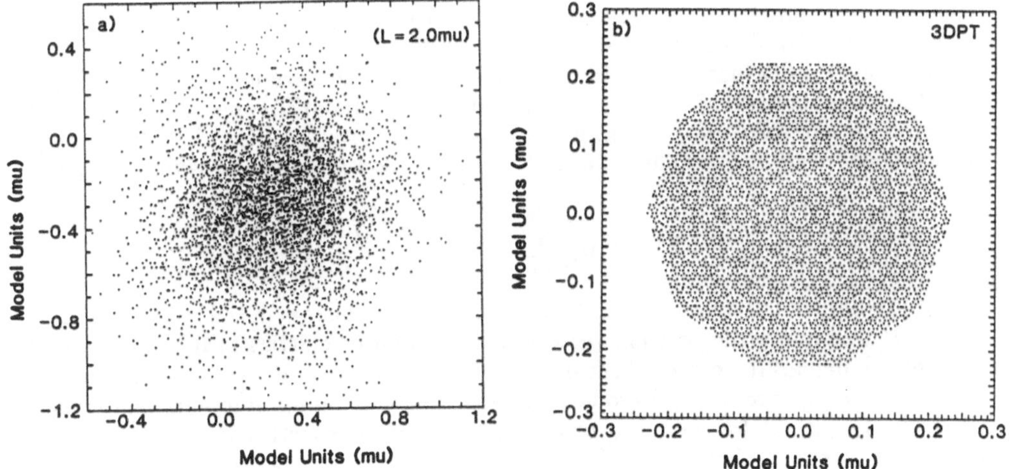

Figure 3. 0.04mu cross section of (a) RCM (L=2.0mu, ΔS=0.2mu, P=1/500) and (b) the 3DPT in R$^{\perp}$ taken perpendicular to an icosahedral basis vector (5-fold axis). Each dot represents a cluster center projected onto the plane of the slice. Note that the row structure, easily seen by viewing (b) at a glancing angle, is also visible in (a) but is much less distinct.

has not yet been decorated with atoms, each cluster is treated as a point scatterer in the calculation, so the peak intensities are not directly comparable.) Actually the the peaks in the calculated S(Q) of Fig. 1c are somewhat narrower than those in the experimental data in spite of the small size of the model ($\sim$550Åin diameter) indicating that the correlation length associated with the long range translational order is greater than in the actual i-AlMnSi alloy.

Further insight can be gained by examining directly a cross section of the grown model. While this does not yield any new quantitative results, it does provide us with a qualitative sense of the amount of disorder in the model. Fig. 2a shows a slice in $R^{\parallel}$ taken through the center of the best model (L=2.0mu) perpendicular to one of the icosahedral basis vectors. Each dot represents a cluster center projected onto the plane of the slice. Fig. 2b shows the same slice for the 3DPT. The slices are 20.0mu ($\sim$220.0Å) thick. This, the most ordered of the RCM's, is clearly far more disordered than the 3DPT. By viewing Fig. 2 at a glancing angle, rows of dots are easily identified that repeat every 72° in both structures, whence the 5-fold symmetry. The reason the density in Fig. 2a appears so much greater than that in Fig. 2b is because the row structure extends perpendicular to the plane of the slice as well and the disorder found in the model means that fewer clusters superpose.

Slices through the $R^{\perp}$ structure of model II (L=2.0mu) and the 3DPT are shown in Fig. 3 and are taken normal to an icosahedral basis vector (5-fold axis). In this case the thickness is 0.04mu. The density distribution in $R^{\perp}$ for the RCM (Fig. 3a) is well localized but is not centered (nor need it be) about the origin. The $R^{\perp}$ density distribution for the 3DPT (Fig. 3b), where the faceting of the acceptance domain can be seen, is clearly far more ordered than the RCM. Here the differences between the models appear much greater than is evident in $R^{\parallel}$. But, as before, a row structure similar to the 3DPT can be discerned in the RCM.

The phason fluctuations for models constructed with various values for the parameter L are shown in Fig. 4a vs. the model size, R. Here the same square-gradient form for the phason entropy used by Strandburg et al[4] is assumed. The long range translational order is then characterized by the phason fluctuations which are given by

$$F(R) = \langle \frac{1}{N_R}\sum_{i=1}^{N_R}|\vec{r}_i^{\perp} - \frac{1}{N_R}\sum_{j=1}^{N_R}\vec{r}_j^{\perp}|^2 \rangle,$$

where R is the model radius in $R^{\parallel}$, N_R is the number of clusters in the model of radius R, and $\vec{r}_i^{\perp}$ and $\vec{r}_j^{\perp}$ are the position vectors of the ith and j^{th} clusters in $R^{\perp}$. The second sum, over j, defines the center of the distribution of clusters in $R^{\perp}$ and $\langle\rangle$ represents an ensemble average over several models. As expected the phason fluctuations increase more slowly with R for models grown with the surface growth constraint, as L increases. This means that there is less phason strain, and thus a greater translational correlation range, as L increases. F(R) for the 3DPT is shown for reference. For R>15.0mu, F(R) is roughly linear, at least for the model sizes considered here. The slope, dF(R)/dR, is plotted in Fig. 4b as a function of L for R>16.0mu. The points could not be fit to any simple functional form but the slopes appear to approach a limiting value which is greater than zero. Thus, even for very large L, it is not likely that one can construct an equilibrium RCM free of phason strain.

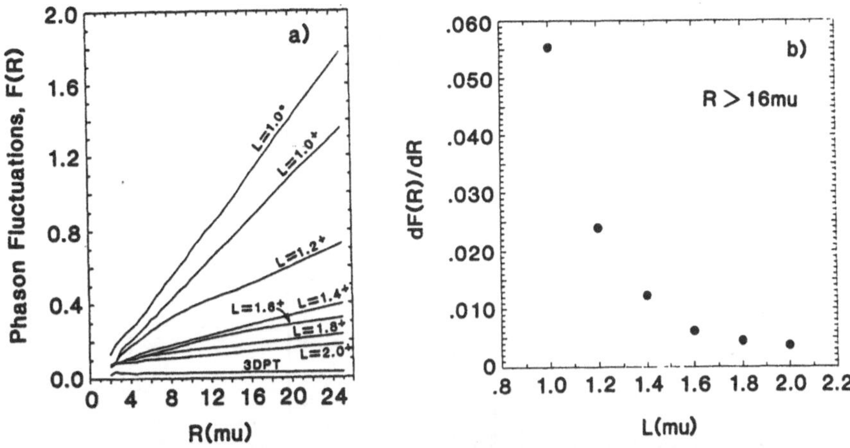

Figure 4. Phason fluctuations, F(R), as a function of model size, R. Note that the fluctuations decrease with increasing L (+). The F(r)'s for the simple model (*) and the 3DPT are shown for comparison. F(r) is approximately linear for R>15.0mu. The slope, dF(R)/dR, is shown in (b) where it appears that a limiting value (greater than zero) is approached for L>2.0mu. This implies that little improvement is expected for larger values of L and that a structure free of phason strain can not be achieved by simply increasing L.

In the case of AlMnSi the density and stoichiometry[1,15] of the i-phase are nearly the same as in the α-phase. Thus for the best model (L=2.0mu) where the density of clusters is $\sim$90% os that in the α-phase, the remaining $\sim$10% must be part of the "glue." In the α-phase there are 15 Al(Si) atoms (22%) per cluster that comprise the "glue." For the model the "glue" atoms will comprise 30% of the total and 5% of them must thereby be Mn; i.e. 10% of the Mn atoms must be part of the glue. The calculated diffraction intensities discussed here then represent only 70% of the total number of atoms because the glue atoms have not yet been included. In this sense the results presented here are only preliminary; none-the-less one can learn a great deal from this calculation.

In Fig. 5 is shown the partial neutron intensities, partial X-ray intensities, and the combined total neutron and X-ray intensities together with experimental neutron and X-ray data. Powder patterns have been calculated because we are interested in the total scattered intensity in order to include both structural disorder as well as the envelope of all the weak peaks (large $Q^{\perp}$). A small sphere ($\sim$300Å) was used taken from the larger model to bring the amount of computer time used on the decorated model to within a reasonable amount. The undecorated S(Q), Fig. 1c, has peak shapes and peak positions that are in good agreement with the neutron (Fig. 5e) and X-ray (Fig. 5j) intensity data; every peak in each experimental powder pattern has a corresponding peak in the undecorated S(Q). Note that there are corresponding peaks that are strong/weak in the X-ray data and weak/strong in the neutron data (for example, the peaks at $Q\approx2.89$Å^{-1}, 3.04Å^{-1}, 5.72Å^{-1} and 5.80Å^{-1}). These differences are due to the difference in scattering contrast between Mn and Al for neutrons ($b_{Mn} = -0.38 \times 10^{-12}$cm and $b_{Al} = 0.36 \times 10^{-12}$cm) and X-rays (12 electrons difference) which means that the neutron data emphasize the "difference lattice" while the X-ray pattern emphasizes the "average lattice".

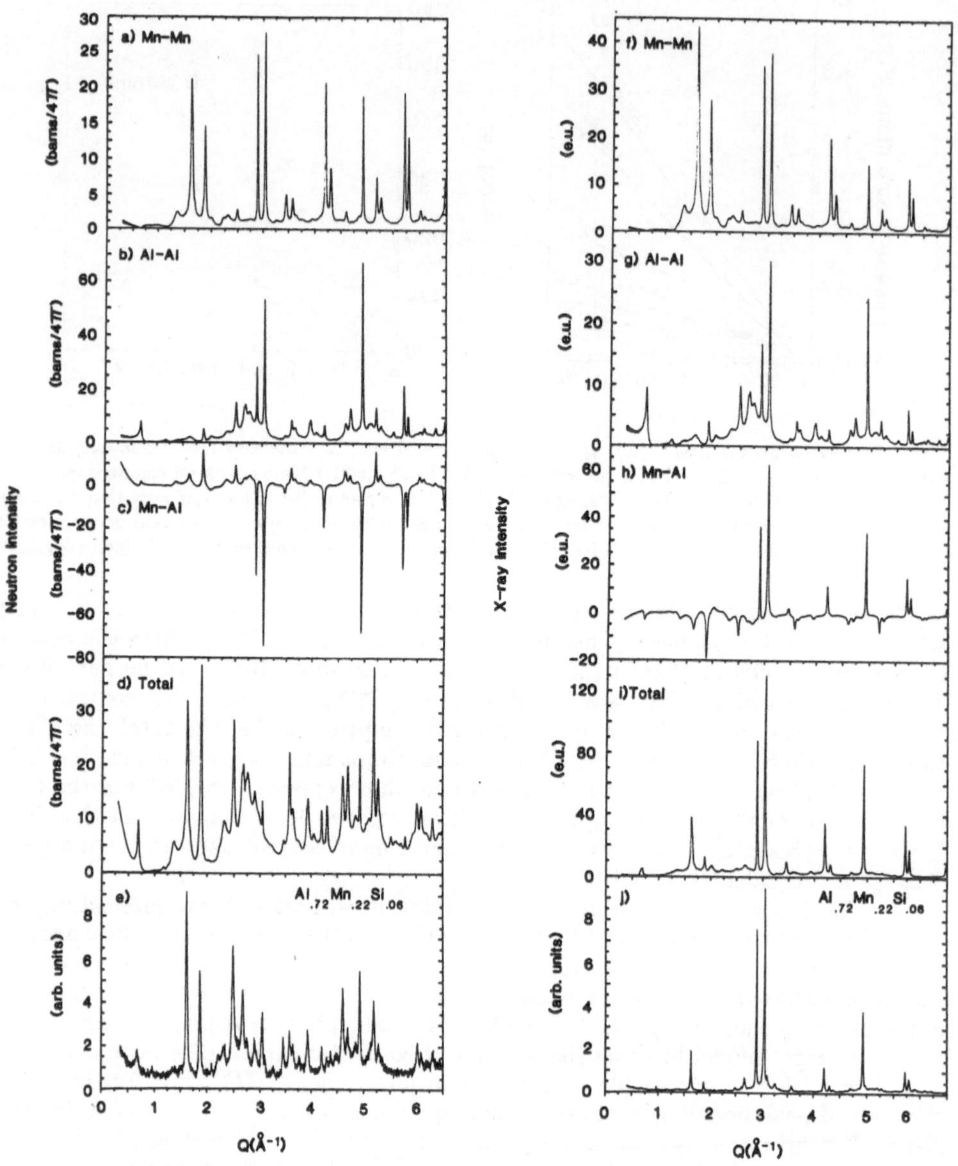

Figure 5. Partial neutron scattering intensity calculated for (a) Mn-Mn, (b) Al-Mn and (c) Mn-Al correlations; total neutron intensity (d) and experimental neutron intensity data (e) on i-AlMnSi; partial X-ray intensity for (f) Mn-Mn, (g) Al-Al and (h) Mn-Al correlations; total X-ray intensity (i) and experimental X-ray intensity data (j) on the same i-AlMnSi material as in (e).

Note also the "diffuse" intensity found under the groups of strong peaks near $Q\approx2.8\text{Å}^{-1}$ and 5.0Å^{-1} with weaker diffuse scattering at $Q\approx3.5\text{Å}^{-1}$ and 5.8Å^{-1} all of which seems to resemble closely the main features in the "amorphous" diffraction pattern.[16,17] Such diffuse scattering, which is common to all of the i-phase alloys studied thus far, including those thought to be free of phason strain (e.g. i-AlFeCu or i-AlRuCu), is clearly visible in the total intensity plots although it does not seem to be in the undecorated $S(Q)$. By examining the partial intensities we see that the diffuse scattering appears predominantly in the <u>Al-Al</u> correlations.

The partial intensities are quite instructive in determining which atomic correlations give rise to various features in the experimental diffraction patterns. For instance, consider the two peaks at $Q\approx5.72\text{Å}^{-1}$ and 5.80Å^{-1} which have relatively small values of $Q^{\perp}$ but are absent in the neutron data. The Mn-Mn and Al-Al partial intensities both have a positive contribution for these peaks while the Mn-Al partial intensity is negative and nearly cancels the other contributions. For X-rays all three partials give a positive contribution at these positions. Of even greater interest is the peak at $Q=1.62\text{Å}^{-1}$ which is almost entirely due to Mn-Mn correlations. In the "amorphous" diffraction pattern of AlMnSi of the same composition there is a prepeak at this position.[16] Such a prepeak is usually associated with correlations in the minority atomic species[18] and in this case has been identified with the Mn-Mn distance of $\sim4.6\text{Å}$.[16] The partial intensities shown here demonstrate that this remains the case in the quasicrystal which is of particular interest because the broadened quasicrystal pattern agrees so well with the "amorphous" pattern.[16]

We also note the "diffuse" intensity found under the groups of strong peaks near $Q\approx2.7\text{Å}^{-1}$ and 5.0Å^{-1} with weaker diffuse scattering at $Q\approx3.5\text{Å}^{-1}$ and 5.8Å^{-1} all of which seems to resemble the main features in the "amorphous" diffraction pattern.[16,17] Such diffuse scattering, which is common to the powder patterns of all of the i-phase alloys studied thus far, including those thought to be free of phason strain (e.g. i-AlFeCu or i-AlRuCu), is clearly visible in the total intensity plots although it does not seem to be in the undecorated $S(Q)$. By examining the partial intensities we see that the diffuse scattering appears predominantly in the Al-Al correlations. It is interesting that this feature is reproduced without either the "glue" atoms or any substitutional disorder within the atomic clusters. At least for i-AlMnSi, this must result from weak peaks with large $Q^{\perp}$ that are part of the background in the undecorated $S(Q)$[19] but are more heavily weighted in the Al-Al partial intensity. This has been confirmed by a calculation of the structure factor for a single 54-atom icosahedron in which strong broad peaks appear in the Al-Al partial at $\sim2.8\text{Å}^{-1}$ and $\sim5.0\text{Å}^{-1}$ which are not in the Mn-Al or Mn-Mn functions. At these $|Q^{\parallel}|$ positions, however, these strong Al-Al contributions appear about symmetry axes where the Bragg peaks due to the model are weak (i.e. large $|Q^{\perp}|$). To be specific, the prominent experimental peaks at $|Q^{\parallel}|$ $=2.89\text{Å}^{-1}$ and 3.04Å^{-1} are on the 5-fold and 2-fold axes, respectively, while the strong Al-Al contribution at $|Q^{\parallel}|\simeq3.0\text{Å}^{-1}$ is along the 3-fold axis of the 54-atom icosahedron.

The agreement with the calculated total intensities cannot be perfect, of course, because 30% of the atoms are missing. In addition, the X-ray patterns are in better agreement than are the neutron patterns because neutrons will be more sensitive to the decoration in the "glue" than X-rays. There are also other, more curious, differences such as the peak at $Q\sim3.45\text{Å}^{-1}$ which is much weaker in the calculated neutron pattern than in the data but is much stronger in the calculated X-ray pattern.

These partial intensities nonetheless compare well with the neutron study of contrast variation in Al-transition metal alloys by Janot and co-workers.[20]

An important theme in structural studies of i-phase alloys has been the 6D Patterson analysis[21] from which an average 6D lattice can be determined and the real 3D structure inferred. Criticism by its practitioners[22] of the sole use of this procedure has, however, recently appeared. The present results are thus offered as a preliminary attempt to provide an alternative in which the i-phase alloy may be viewed as a decorated RCM with quenched in phason strain.

The author wishes to thank S. C. Moss for his invaluable help and encouragement and B. Mozer for kindly supplying the neutron data. We also gratefully acknowledge the support of the Department of Energy, Division of Basic Energy Sciences on grant No. DE-FG05-87ER45325 and the use of the computer resources of the National Energy Research Supercomputer Center at the LLNL.

REFERENCES

1. M. A. Marcus, H. S. Chen, G. P. Espinosa and C.-L. Tsasi, Solid State Comm. **58**, 227 (1986); R. J. Schaefer and L. A. Bendersy, *Aperiodicity and Order 1: Introduction to Quasicrystals*, ed. M. V. Jarić, Academic Press Inc., San Diego (1988).
2. P. Guyot and M. Audier, Phil. Mag. **52**, L15 (1985); M. Audier and P. Guyot, Phil. Mag. B **53**, L43, (1986); M. Audier, P. Sainfort and B. Dubost, Phil. Mag. B **54**, 105 (1986); C. L. Henley and V. Elser, Phil. Mag. **53**, L59 (1986).
3. D. Levine and P. J. Steinhardt, Phys. Rev. Lett. **53**, 2477 (1984); D. Levine and P. J. Steinhardt, Phys. Rev. B **34**, 596 (1986); J. E. S. Socolar and P. J. Steinhardt, Phys. Rev. B **34**, 617 (1986); V. Elser, Phys. Rev. B **32**, 4892 (1985).
4. K. J. Strandburg, L.-H. Tang and M. V. Jarić, Phys. Rev. Lett. **63**, 314 (1989).
5. D. Shechtman and I. A. Blech, Metallurgical Transactions **16A**, 1005 (1985).
6. P. W. Stephens, A. I. Goldman, Phys. Rev. Lett. **56**, 2331 (1986); P. W. Stephens, *Aperiodicity and Order 3: Extended Icosahedral Structures*, ed. M. V. Jarić and D. Gratias, Academic Press Inc., San Diego (1989); A. I. Goldman and P. W. Stephens, Phys. Rev. B **32**, 2826 (1988).
7. V. Elser, Acta Cryst. /bf42A, 36 (1986); A. Katz and M. Duneau, J. Phys. (Paris) **47**, 181 (1986); N. G. de Bruijn, Ned. Akad. Weten. Proc. Ser. A **43**, 39 and 53 (1981).
8. G. Y. Onada, P. J. Steinhardt, D. P. DiVincenzo and J. E. S. Socolar, Phys. Rev. Lett. /bf60, 2653 (1988); J. E. S. Socolar, Comm. in Math. Phys. **129**, 599 (1990).
9. V. Elser, *Aperiodicity and Order 3: Extended icosahedral structures*, ed. M. V. Jarić and D. Gratias, Academic Press Inc., San Diego (1989).
10. The name RCM has been adopted for this model although it is a natural extension of the IGM. This is to distinguish it from earlier work and, more importantly, because the use of the term "glass" does not seem appropriate to describe structures with long range translational and orientational order.
11. J. L. Robertson and S. C. Moss, Zeit. für Phys. B – Condensed Matter **83**, 391 (1991).
12. J. L. Robertson and S. C. Moss, Phys. Rev. Lett. **66**, 3 (1991).
13. M. Cooper and K. Robinson, Acta Cryst. **20**, 614 (1966); C. Romming and J. E. Tibbals (preprint).
14. Neutron data supplied by B. Mozer. The sample was prepared by J. Bigot and coworkers at CECM, Vitry-sur-Seine, France.
15. P. Guyot, M. Audier and R. Lequette, J. Phys. (Paris), colloque C3, supp. n°7 **47**, C3-389 (1986); Chr. Janot, J. Pannetier, J. M. Dubois, J. P. Houin and P. Weinland, Phil. Mag. B **58**, 59 (1988).
16. J. L. Robertson, S. C. Moss and K. G. Kreider, Phys. Rev. Lett. **60**, 2062 (1988).
17. J. M. Dubois and K. Dehghan, J. Non-Cryst. Solids **93**, 179 (1987).
18. S. C. Moss and D. L. Price, *Physics of Disordered Materials*, ed. D. Alder, H. Fritzsche and S. R. Ovshinsky, Plenum Publ., New York (1987).
19. Y. Ma, E. A. Stern, X.-O. Li and Chr. Janot, Phys. Rev. B **40**, 8053 (1989); X.-O. Li, E. A. Stern and Y. Ma, Phys. Rev. B (in press).
20. Chr. Janot, J. Pannetier, M. de Boissieu and J.-M. Dubois, Europhys. Lett. **3**, 995 (1987).
21. J. W. Cahn, D. Gratias and B. Mozer, J. de Phys. (Paris) **49**, 1225 (1988); S.-Y. Qiu and M. V. Jarić, *Quasicrystals*, ed. M. V. Jarić and S. Lundqvist, World Scientific, Singapore (1990).
22. M. de Boissieu, Chr. Janot and J.-M. Dubois, J. Phys. Cond. Matter.

SIZE-DEPENDENT THERMODYNAMIC AND ELECTRONIC PROPERTIES OF INDIVIDUAL NANOMETER-SIZE SUPPORTED GOLD CLUSTERS

M.E. LIN,* A. RAMACHANDRA,** R.P. ANDRES** and
R. REIFENBERGER*
Purdue University
* Department of Physics
** School of Chemical Engineering
W. Lafayette, IN 47907
U.S.A.

ABSTRACT. Field emission techniques devised to measure the melting temperature and electronic structure of individual, nanometer-size clusters supported on electrically conducting substrates are reviewed. Data on the size-dependent reduction in melting temperature of Au clusters are compared to existing thermodynamic descriptions and molecular dynamic calculations. Data on the electronic structure of an individual 1nm Au cluster are compared to the predictions of simple electron shell models for cluster electronic states.

1. Introduction

A recent focus in cluster research is the novel structural, electronic, and optical properties expected for cluster-assembled, nanophase materials.[1-3] To date most information on the properties of small clusters has been obtained from cluster-beam experiments in which isolated clusters are studied while in flight. This research has established that in the nanometer size range, unsupported or free-space clusters exhibit unique, size-dependent, structural and electronic features different from bulk samples of the same atoms, but cannot answer important questions regarding their properties for cluster-assembled materials. An important research approach that bridges the gap between studies of free clusters and cluster-assembled materials is the study of an individual cluster supported on a solid surface. The information obtained from such single-cluster studies is of fundamental importance to an understanding of the interaction between clusters and substrates and for determining whether the unique properties of free clusters survive when they are deposited on a solid surface. By performing experiments on a single supported cluster, uncertainties associated with

P. Jena et al. (eds.), *Physics and Chemistry of Finite Systems: From Clusters to Crystals, Vol. I,* 309–322.

the size distribution of clusters, that plague experiments requiring an ensemble of supported clusters, can be circumvented.

The following paper reviews experimental techniques recently developed for studying individual supported clusters and summarizes results that have been obtained in our laboratory using these techniques.[4,5] Controlled-size clusters grown in a multiple expansion cluster source (MECS),[6] and subsequently deposited on clean substrates, have been used to study melting temperature as a function of cluster size,[7-9] the shape of supported Au clusters on flat substrates,[10,11] and the size-dependent electronic structure of supported Au clusters.[12]

2. Field Emission Techniques

The field emission process is a well known phenomenon that relies on the point-projection imaging of electrons emitted from a small apex of an electrically conducting tip subjected to a high electric field. The field emission process has been well studied from a theoretical context and has been used to provide a wide variety of information about the work function of metal surfaces,[13] surface diffusion of adsorbates on metallic surfaces,[14,15] the electronic band structure of metal surfaces near the Fermi energy,[16] and the interaction of light with sub-micron metallic surfaces.[17,18]

Field emission experiments are versatile and relatively simple to conduct. The common ingredients in any field emission experiment include a sharp metallic tip positioned in front of a light emitting screen located inside an ultra-high vacuum (UHV) chamber. The tip can be constructed from a large variety of high melting-point elements, with tungsten being the material of choice for most studies. The tips are usually electro-chemically etched from a wire in much the same way that scanning tunneling microscope tips are often prepared.[19] This etching produces a rough, irregular tip with an end form of radius $\sim 10\text{-}50\text{nm}$. An important modification occurs to the tip morphology when it is mounted in a UHV chamber and heated repeatedly to high temperatures to produce an atomically clean surface. The heating procedure produces a well-annealed end form of radius ~ 100 nm that is single crystal faceted along crystallographic directions with low surface free energy. Thus, the tip that supports the clusters in our studies is a well annealed single crystal whose orientation can be readily discerned from the symmetry of the resulting field emission pattern.

The key that makes field emission techniques useful to study supported clusters is the small radius of curvature of the supported cluster when compared to the underlying field emission tip. This simple geometric consideration results in an electric field enhancement around the cluster with respect to the substrate and in turn results in electron emission from the cluster alone.

To produce a measurable field emission current usually requires a few thousand volts potential difference between the tip and a nearby phosphor-coated screen (~ 10

cm diameter) located in front of the field emission tip. The image of the field emitted electrons can then be viewed directly from the screen. Often, it is convenient to store the field emission images on a video recorder for further analysis. More quantitative measurements can be made by measuring the current as a function of the applied voltage or by energy resolving the emitted electrons with a suitable field emission energy analyzer.

The field emission process has been well studied and is concerned with the calculation of j(E), the number of electrons of mass, m, emitted per unit area per unit time with energy between E and $E + dE$, due to an electric field F applied in the z-direction.[16] This calculation requires an integration over all possible energies E and wavevectors $\vec{k} = (k_z, k_{\parallel})$

$$j(E)dE = \int_E^{E+dE} N(\vec{k})|T(k_z, E)|^2 \frac{d^3k}{(2\pi)^3} \qquad (1)$$

where, $N(\vec{k})$ is the number of electrons with wavevector $\vec{k}$ incident on the barrier per unit time per unit area and $|T(k_z, E)|^2$ is the transmission probability through the surface potential barrier. The metal-vacuum interface is often described in 1-dimension by an image rounded barrier given by

$$\begin{aligned} V(z) &= E_F + \phi - \frac{e^2}{4z} - eFz \qquad && \text{for } z > z_c \\ &= 0 && \text{for } z < z_c \end{aligned} \qquad (2)$$

where z_c is determined by $V(z_c) = 0$, E_F is the Fermi-energy of the metal and ϕ is the metal work function (see Figure 1).

Using the WKB approximation to estimate $T(k_z, E)$ and a free electron approximation to evaluate $N(\vec{k})$, the distribution of emitted electrons is:

$$j_o(E) = \frac{m d_E f(E)}{2\pi^2 \hbar^3} e^{-\frac{4}{3}\left(\frac{2m}{\hbar^2}\right)^{\frac{1}{2}} \frac{(\phi+E_F-E)^{\frac{3}{2}}}{eF} v(y)} \qquad (3)$$

where

$$\frac{1}{d_E} = 2\left(\frac{2m}{\hbar^2}\right)^{\frac{1}{2}} \frac{(\phi + E_F - E)^{\frac{1}{2}}}{eF} t(y) \qquad (4)$$

$v(y)$, $t(y)$ are slowly varying, tabulated[20] functions of $y = (e^3 F)^{\frac{1}{2}}/(\phi)$ that account for the static image-rounded term in the potential given in Eq. 2, and f(E) is the Fermi-Dirac function describing the distribution of filled states within the metal.

The shape of $j_o(E)$ is roughly governed by the temperature of the emitter for $E > E_F$ and the rapid decrease in the transmission probability through the metal-vacuum interface for $E < E_F$. The exponentially rapid decrease in the transmission probability with energy below E_F strongly favors electron emission from filled states nearest the Fermi energy. An experimental measurement of this energy distribution

Metal-Vacuum Potential Barrier

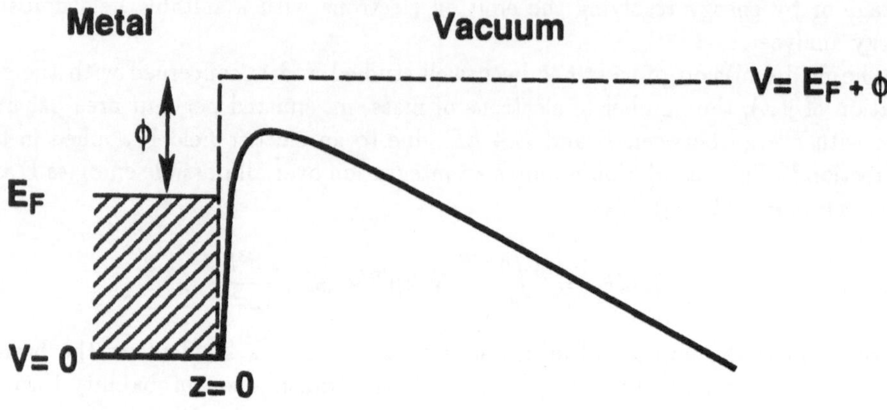

Figure 1: A simple model for the surface potential barrier of a metal-vacuum interface described by Eq. 2. The barrier of finite width is a result of an applied electric field. Electrons tunneling from the metal through the barrier into the vacuum produce a field emission energy distribution given by Eq. 3. When a nanometer-size metallic cluster is deposited on the metal substrate, electrons tunneling from the quantized cluster states can be measured.

involves further convolution of Eq. 3 with a symmetric gaussian window function that accounts for the finite energy resolution of the energy analyzer. It is an experimental fact that Eq. 3 well describes the overall shape of the distribution of electrons emitted from field emission tips fabricated from a variety of refractory metals.[16]

Eq. 3 can be further integrated to give the total current

$$
\begin{aligned}
J(F) &= e \int_{-\infty}^{\infty} j_o(E)dE \\
&= A'F^2 e^{-(B'\phi^{\frac{3}{2}}/F)}
\end{aligned}
\tag{5}
$$

where A' and B' are known constants.[21] This result, known as the Fowler-Nordheim (F-N) equation, predicts a plot of $\ell n(J(F)/F^2)$ $vs.$ $1/F$ will give a straight line with a slope related to the work function ϕ of the emitting surface. Use of this equation has provided much of the systematic information now available on the work function of the various faces of pure metals.

These basic ideas of field emission can be used to learn more about the shape and electronic structure of a supported cluster. Information about the shape of the supported cluster comes through a geometrical factor that relates the applied voltage to the electric field F that appears in the equations discussed above. Since the electron emission current depends exponentially on the field strength, and since the field strength is determined by the shape (i.e. radius of curvature) of the supported cluster, small changes in shape produce large changes in the tunneling current. Information about the electronic structure of a supported cluster is contained in $N(\vec{k})$. When considering electron emission from a cluster, the free-electron density $N(\vec{k})$ in Eq. 1 must be replaced with the quantized density of states for the cluster.

3. Cluster synthesis and sample preparation

A description of the design and operation of the multiple expansion cluster source (MECS) used in this study can be found in the literature.[6,22,4] The clusters are grown as aerosol particles in a room temperature inert gas and introduced into a vacuum chamber (10^{-6} torr) to form a cluster beam. In each of the experiments described below, cluster samples captured on thin amorphous carbon substrates were analyzed by TEM to determine their distribution in size. Their size distribution studied in this way confirm the ability of the MECS to produce metal clusters having an approximately spherical shape, a controlled mean size, and a narrow size distribution.[6,22]

To perform the field emission studies discussed below, a clean tungsten tip is briefly exposed to the cluster beam until a single cluster is observed by its field emission signature. The tip is then transferred in a vacuum transfer cell at pressures of $\sim 5 \times 10^{-8}$ torr to a UHV field emission apparatus. Experiments are performed in

314

the field emission chamber at a pressure of $\sim 2 \times 10^{-10}$ torr.

4. Size-Dependent Melting of Au Clusters

In order to investigate the melting transition in nanometer-size clusters, we have developed a technique that uses the field emission current from an individual cluster to measure the melting temperature of that cluster. This approach when coupled with the ability of the MECS to grow clusters of controlled size enables the study of a wide range of cluster sizes ($\sim 1.0 - 20$ nm in diameter). Moreover, the sensitivity of the technique appears to be unaffected by cluster size. The details of these measurements have been reported previously[8,9] and we will not repeat them here. Rather, we will discuss the experimental results for Au clusters within the context of various theoretical treatments of size-dependent melting.

The classical thermodynamic treatment of melting holds that the melting transition occurs when the chemical potentials of the liquid and solid phase are equal.[23,24] In this context, the melting depression for a small cluster results because the chemical potential for a cluster is modified (compared to the bulk value) by a differential pressure drop across the surface. For a spherical cluster (solid or liquid) of radius r, and having a surface tension γ the pressure difference is given by[25]

$$P - P_{ext} = 2\gamma/r. \tag{6}$$

A thermodynamic treatment of the melting temperature of small clusters based on this premise was first proposed by Pawlow[26] and has been more recently reevaluated by Hanszen.[27] This classic analysis of the problem predicts that a reduced plot of $T_m(r)/T_o$ (where T_o is the melting temperature of the bulk solid) vs. the inverse of the cluster radius, (1/r), will yield a straight line. If the depression in melting temperature is large (as is the case when r is small), then the change in bulk thermodynamic variables with temperature must be taken into account and the straight line predicted by Pawlow becomes slightly curved.[28] The predictions of this corrected thermodynamic model, are compared to the experimental data for Au clusters in Figure 2. A reasonable fit is obtained provided the cluster radius is greater than $\sim$ 2 nm.[9] In fitting the model to the experimental results, γ_s is the only free parameter that is varied. The two curves in Figure 2 demonstrate the sensitivity of the cluster melting temperature on this parameter.

Other thermodynamic theories for cluster melting exist. For example, Couchman and Jesser[29] have proposed a dynamic model that treats surface atoms differently than interior atoms. They assume cluster melting occurs via transformation of a surface layer of atoms from solid to liquid (i.e. surface melting) followed by spontaneous growth of this liquid region to consume the entire cluster. The Couchman and Jesser model predicts the same qualitative behavior as the Pawlow model, a monotonic decrease in melting temperature with decreasing cluster size.

The experimental data obtained by Buffat and Borel,[28] who studied cluster melting by observing the disappearance of electron diffraction rings from ensembles of Au clusters (grown by depositing Au atoms on carbon substrates,), are plotted as the shaded area in Figure 2. Although these data are of questionable accuracy as the cluster radius becomes smaller than ~ 2.5 nm, they seem to indicate a leveling of $T_m(r)$ as r becomes small, similar to that measured by field emission. The interesting problem is to theoretically predict this leveling of $T_m(r)$.

Several authors have studied cluster melting by means of molecular dynamics (MD) simulation. Most of these studies have been for atoms that interact *via* a Lennard-Jones 12-6 potential. Using best fit Lennard-Jones parameters for Au, the absolute predictions of these calculations for $T_m(r)$ are poor. However, if the $T_m(r)$ obtained in the MD simulation is normalized by a T_o also obtained by means of MD simulation, the comparison between theory and experiment for $r \leq 1$ nm is surprisingly good. An example of this agreement is the theoretical calculations of Honeycutt and Anderson[30] which are plotted on Figure 2.

MD calculations by Ercolessi et al.[31] and by Jellinek and Garzón[32] employ empirical potential functions designed to better model the bulk properties of gold. The predictions of these two studies are also plotted on Figure 2. Unfortunately, the calculations of Ercolessi et al. do not extend to clusters with a radius smaller than $\sim$ 1 nm, while the calculations of Garzón and Jellinek were performed for 12-14 atom clusters only. Furthermore, it is not clear that the empirical potentials used by these authors have validity outside of the respective ranges of their calculations.

The experimental data presented in Figure 2 represent a severe test for any empirical potential function of Au. One caution with regard to direct comparison between these experimental data and MD calculations must be noted. The Au clusters studied by field emission were supported on a W substrate while the MD calculations are typically for free clusters. As cluster size decrease, the effects of cluster-substrate interaction on the melting transition can be significant. Garzón et al.[33] have simulated the melting of argon clusters on fixed solid lattices and find that T_m depends o· the strength of the cluster-substrate interaction. Further experiments (with differeı metals and with a variety of substrate materials) and theoretical calculations, (ı the size range below $r \simeq 2$ nm and for supported as well as free clusters) are clearly needed.

5. Field Emission Energy Distributions from Au Clusters

It is useful to consider simple models for the quantized electronic states of a cluster to learn more about the feasibility of detecting these states using field emission. Since, the electronic states of free space (*i.e.*, unsupported) clusters can be described using simple models, systematic case studies illustrating important trends can be made. In what follows, we calculate the expected energy distribution emitted from a cluster

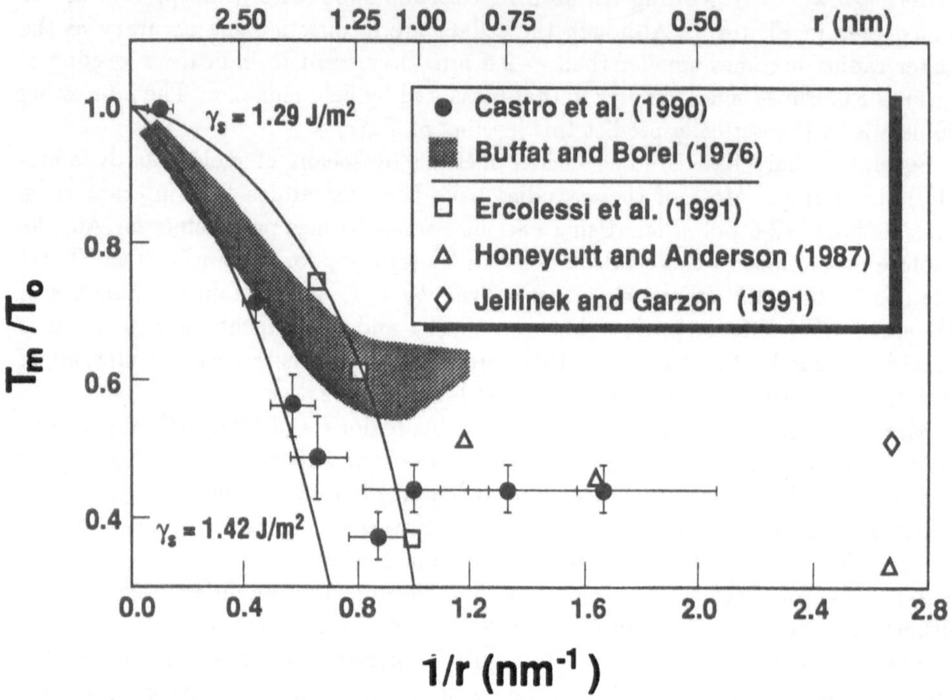

Figure 2: A compilation of the size-dependent melting temperature of supported Au metal clusters. The cluster melting temperature (T_m) is normalized to the bulk melting temperature (T_o) and plotted as a function of $1/r$ where r is the cluster radius in nm. The solid dots are data obtained from the field emission technique developed by Castro et al. (Ref. 7). The data from Buffet and Borel (Ref. 26) encompass the shaded region. The solid lines are calculated from the modified thermodynamic model for two different values of γ_s (ref. 7). The results of molecular dynamics calculations by Ercolessi et al.(Ref. 29), Honeycutt and Andreson (Ref. 28), and Jellnick and Garzón (Ref. 30) are also shown.

on the basis of these idealized models and compare this theoretical prediction to a measured energy distribution from a 1 nm diameter Au cluster. The calculations serve as a useful guide for determining the size of clusters which might show structure in field emission experiments due to their discrete energy states, and the comparison between theory and experiment serves as an indication of how field emission from a single supported cluster can yield valuable information about a cluster's quantized electronic states.

The electronic structure reflected in the mass spectra of alkali clusters[34,35] indicate the suitability of a shell model for describing cluster states. It is customary to discuss this shell structure using either the infinite square well potential or the infinite harmonic potential well model. In the infinite square well model, the electrons are confined by a potential V(r) given by

$$V(r) = \begin{cases} 0 & \text{if } r < a \\ \infty & \text{if } r > a, \end{cases} \tag{7}$$

where a is the radius of the cluster. The radial Schrödinger equation for $r < a$ gives the well known solution

$$R_l(r) = A_l j_l\left(\frac{\sqrt{2mE}}{\hbar}r\right), \tag{8}$$

where j_l is the spherical Bessel function. Since this is an infinite well, the eigenfunctions must vanish for $r > a$. Thus, Eq. 8 must equal zero at r = a giving the energies of the n'th shell

$$E_{nl} = \frac{\hbar^2 Z_{nl}^2}{2ma^2}, \tag{9}$$

where Z_{nl} are the zeros of the spherical Bessel function.

In the infinite harmonic well model, the electrons are confined by a potential V(r) specified by

$$V(r) = \frac{1}{2}m\omega^2 r^2, \tag{10}$$

where m is the mass of an electron. The parameter ω is determined for a given cluster by setting the width of the potential well at the vacuum level equal to the diameter of the cluster. A solution of Schrödinger's equation for spherical harmonic oscillator produces complicated eigenfunctions but the eigenvalues are well-known and given by

$$E_{nl} = (2n + l + \frac{3}{2})\hbar\omega, \tag{11}$$

where n,l = 0,1,2,$\cdots$. In general, if $N = 2n + l$, then for N even, there are $\frac{1}{2}N + 1$ degenerate solutions while, if N is odd, $\frac{1}{2}(N + 1)$ degeneracies occur.

A realistic potential well for a cluster most likely is an interpolation between the previous examples[36] and the actual energy levels should lie somewhere between the corresponding levels derived from these simple models. Further elaboration can remove the degeneracies inherent in Eq. 11. For instance, if we introduce a spin-orbit coupling term $\alpha \hat{l}\cdot\hat{s}$, where α is the spin-orbit splitting parameter, we can remove degeneracies. At the present, little is known about realistic values of α to warrant further calculations. Yet another modification is the deviation of the cluster from spherical symmetry. An ellipsoidal shell model was developed for nuclei by Nilsson.[37] A similar idea can be applied to metal clusters. However, these elaborations will not be applied here.

Based on the electron energy distribution predicted using these two models (Eqs. 9 and 11), we can ascertain the extent that observable structure in the field emission energy distributions can be measured. We find that resolvable splittings should be observed for Au clusters smaller than ~ 3 nm. Clusters with diameters smaller than this value should exhibit energy distributions with experimentally resolvable splittings and these distributions will have a shape that deviates significantly from the standard results of field emission theory (see Eq. 3).

An example is shown in Figure 3 where the electron energy distributions predicted for a 1nm Au cluster are compared to experiment. In performing these calculations, the transmission probability through the interface potential barrier was calculated using the Lambin-Vigneron algorithm[38,39] and combined with the discrete density of states predicted by the models discussed above to yield the field emission energy distributions. These results are compared to the field emission energy distribution obtained from a nominal 1 nm diameter Au cluster.[12] The calculated distributions are all convoluted with a gaussian function with a full-width at half-maximum of 70 meV to take into account the finite energy resolution of our electron energy analyzer.[40]

These calculations show that for clusters this small, there is a significant deviation from bulk behavior. The calculations based on the two separate shell models for the clusters roughly predict the same structure since the calculated spectrum is strongly weighted by the highest filled states. For clusters in this size range, measurements of the electron energy distribution extending to lower energies are required to distinguish between the two models. At the present time, this structure is not accessible experimentally.

The curve showing results obtained from experiment reveals a number of important differences from the model calculations. First, the entire distribution is shifted below the Fermi energy of the substrate, indicating that the electronic levels of the cluster need not be pinned to the Fermi energy of the substrate. Second, the significant structure in the energy distribution is observed to shift to lower energies with applied electric field, indicating that a more complete theory involving field penetration into the cluster is required. These observations require modifications to the simple models discussed above which are discussed elsewhere.[12,41]

The most important result indicated by Figure 3 is that the electronic structure

Energy Distributions for 1 nm Au Cluster

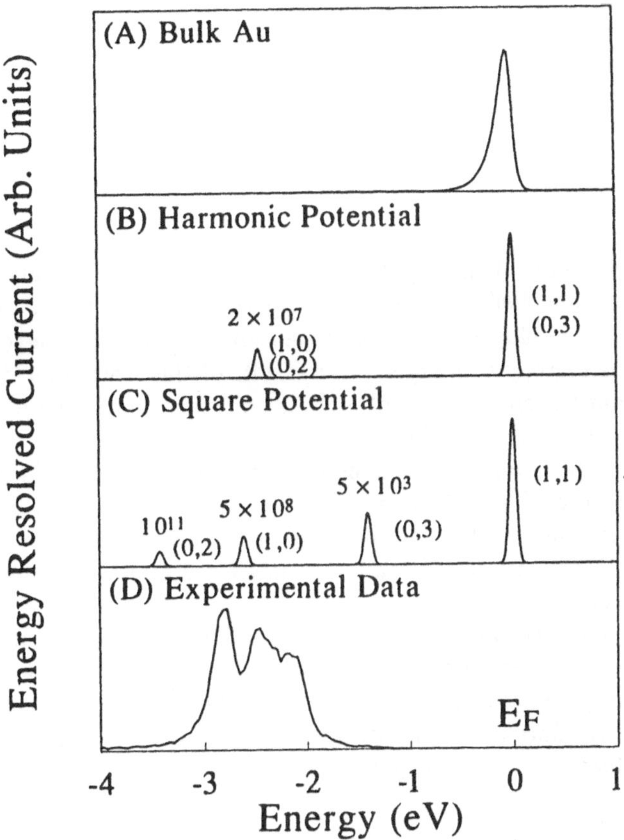

Figure 3: The energy distributions for a 1 nm diameter Au cluster containing ~ 40 atoms. The three calculated curves represent the predictions based on treating the cluster atoms as (a) bulk part of a metal (Eq. 3) with work function $\phi = 5.4$ eV and an applied electric field of 3×10^9 V/m, (b) a separate species described by the infinite harmonic model, and (c) a separate species described by the infinite square potential discussed above. The three model calculations are all plotted with their highest filled states set equal to E_F. Before plotting, the features at low energy were magnified by the multiplicative factors shown. The (n,l) values for each feature are also indicated. In (d), an experimental data curve obtained from a nominal 1 nm Au cluster supported on a W substrate is also shown. The zero of energy for this curve is the Fermi energy of the substrate.

of a supported nanometer-size cluster can now be probed. Studies performed as a function of cluster size can be undertaken on a wide variety of different substrates and for different regimes of coupling between the substrate and cluster. We believe that a systematic exploitation of this technique will provide an important set of data regarding the factors that influence the quantized electronic structure of small supported clusters.

6. Summary

We have reviewed some of the results that have been recently obtained on *individual* metal clusters of controlled size that were supported on a substrate. The experiments summarized above employed a multiple expansion cluster source in conjunction with field emission microscopy techniques. The results of this work indicate that detailed information about the size-dependent melting and electronic structure of nanometer-size supported clusters can be obtained.

The utility of this approach was illustrated by two examples. The size-dependent melting temperature of Au clusters has been measured and was discussed within the context of existing models for the melting transition. The energy distribution emitted from a 1 nm Au cluster has been measured and was compared to simple model predictions.

In conclusion, based on the results already obtained, it now seems feasible to consider a sequence of experiments in which free-space clusters of known size and composition are systematically prepared and studied on a wide variety of different substrates to learn more about the structural, thermodynamic, and electronic properties of individual, supported clusters.

Acknowledgements

This work was partially supported by the U.S. Department of Energy under grant DE-FG02-88ER45162. M.E. Lin thanks the David Ross Foundation of Purdue University for financial support during the course of this study. A. Ramachandra thanks the Dow Chemical Corporation and Arco for financial support during the course of this study.

References

[1] R.P. Andres, R.S. Averback, W.L. Brown, L.E. Brus, W.A. Goddard, A. Kaldor, S.G. Louie, M. Moscovits, P.S. Peercy, S.J. Riley, R.W. Siegel, F. Spaepen, and Y. Wang , J. Mater. Res. **4**, 704 (1989).

[2] R.W. Siegel, MRS Bulletin **XV; No. 10**, 60 (1990).

[3] R.S. Averback, J. Bernholc, and D.L. Nelson, editors, *Clusters and Cluster-Assembled Materials*, Materials Research Society; Vol. 206, Pittsburgh, PA, 1991.

[4] T. Castro, Y.Z. Li, R. Reifenberger, E. Choi, S.B. Park, and R.P. Andres, J. Vac. Sci. Technol. **A7**, 2845 (1989).

[5] T. Castro, E. Choi, Y.Z. Li, R.P. Andres, and R. Reifenberger, Proc. Mat. Res. Soc. **206**, 159 (1991).

[6] S.B. Park, PhD thesis, 1988, Purdue University (unpublished).

[7] T. Castro, PhD thesis, 1989, Purdue University (unpublished).

[8] T. Castro, R. Reifenberger, E. Choi, and R.P. Andres, Surf. Sci. **234**, 43 (1990).

[9] T. Castro, R. Reifenberger, E. Choi, and R.P. Andres, Phys. Rev. B **42**, 8548 (1990).

[10] Y.Z. Li, PhD thesis, 1989, Purdue University (unpublished).

[11] Y.Z. Li, R. Reifenberger, E. Choi, and R.P. Andres, Surf. Sci. **250**, 1 (1991).

[12] M.E. Lin, R.P. Andres, and R. Reifenberger, Phys. Rev. Lett. **67**, 477 (1991).

[13] J. Hölzl and F.K. Schulte, Work function of metals, in *Springer Tracts in Modern Physics, Vol 85*, edited by G. Hohler, Springer Verlag, 1979.

[14] R. DiFoggio and R. Gomer, Phys. Rev. B **25**, 3490 (1982).

[15] C. Dharmadhikari and R. Gomer, Surf. Sci. **143**, 223 (1984).

[16] J.W. Gadzuk and E.W. Plummer, Rev. Mod. Phys. **43**, 487 (1973).

[17] D. Venus and M.J.G. Lee, Surf. Sci. **125**, 452 (1983).

[18] Y. Gao and R. Reifenberger, Phys. Rev. **B35**, 8301 (1987).

[19] A.J. Melmed, J. Vac. Sci. Technol. **B9**, 601 (1991).

[20] A. Modinos, *Field, Thermionic and Secondary Electron Emission Spectroscopy*, Plenum Press, New York, 1984.

[21] R.H. Fowler and L.W. Nordheim, Proc. Roy. Soc. Lond. **A119**, 173 (1928).

[22] E. Choi and R.P Andres, in *Physics and Chemistry of Small Clusters*, edited by P. Jena, B.K. Rao, and S.N. Khanna, page 61, Plenum Press, New York, 1987.

[23] Charles Kittel and Herbert Kroemer, *Thermal Physics*, W.H. Freeman and Company, 1980.

[24] F. Reif, *Fundamentals of Statistical and Thermal Physics*, McGraw-Hill, 1965.

[25] I. Prigogine R. Defay and A. Bellemans, *Surface Tension and Adsorption*, John Wiley and Sons Inc., 1966.

[26] P. Pawlow, Z. Phys. Chem. **65**, 1 (1909).

[27] K. J. Hanszen, Z. Phys. **157**, 523 (1960).

[28] Ph Buffat and J.P. Borel, Phys. Rev. **A13**, 2287 (1976).

[29] P.R. Couchman and W.A. Jesser, Nature **269**, 481 (1977).

[30] J.D. Honeycutt and H.C. Anderson, J. Phys. Chem. **91**, 4950 (1987).

[31] F. Ercolessi, W. Andreoni, and E. Tosatti, Phys. Rev. Lett. **66**, 911 (1991).

[32] J. Jellinek and I.L. Garzón, Z. Phys. D **20**, 235 (1991).

[33] I.L Garzon et al., in *Physics and Chemistry of Small Clusters*, edited by P. Jena, B.K. Rao, and S.N. Khanna, page 193, Plenum Press, New York, 1987.

[34] W.D. Knight, K. Clemenger, W.A. de Heer, W.A. Saunders, M.Y. Chou, and M.L. Cohen, Phys. Rev. Lett. **52**, 2141 (1984).

[35] W.D. Knight, W.A. de Heer, K. Clemenger, and W.A. Saunders, Solid State Commun. **53**, 445 (1985).

[36] M. G. Mayer and J. H. D. Jensen, *Elementary Theory of Nuclear Shell Structure*, John Wiley and Sons Inc., 1955.

[37] S.F. Nilsson, Mat.-Fys. Medd. Dan. Vidensk. Selsk. **29**, 1 (1955).

[38] Ph Lambin and J P Vigneron, J. Phys. A: Math. Gen. **14**, 1815 (1981).

[39] J P Vigneron and Ph Lambin, J. Phys. A: Math. Gen. **13**, 1135 (1980).

[40] Y. Gao, R. Reifenberger, and R.M. Kramer, J. Phys. E: Sci. Instrum. **18**, 381 (1985).

[41] M.E. Lin, R.P. Andres, and R. Reifenberger, (to be published) (1991).

INELASTIC NEUTRON SCATTERING FROM SINGLE GRAINS OF ICOSAHEDRAL AND CUBIC Al-Li-Cu

A. I. Goldman
Ames Laboratory, U.S.D.O.E.
and Department of Physics and Astronomy
Iowa State University
Ames, IA 50011

ABSTRACT. The results of inelastic neutron scattering measurements on single grains of icosahedral Al-Li-Cu and the related cubic R-phase are described. Measurements along the high symmetry two-, three- and five-fold axes of the icosahedral phase clearly show the existence of well-defined propagating modes, with isotropic dispersion, near strong Bragg peaks. In several instances "quasi"-Brillouin zones, with stationary points at the "special" high-symmetry points of a simple icosahedral lattice, can be identified.

1. Introduction

Over the past six years there have been extensive investigations into the structure of quasiperiodic alloys[1]. While studies of the dynamical properties have lagged, there has been a recent surge of interest in both theoretical and experimental determinations of the electronic and vibrational spectra of these novel materials. The small grain size of most icosahedral phases has limited neutron scattering studies to measurements of the vibrational density of states (VDOS) for powder samples. Inelastic neutron scattering measurements of phonons in icosahedral Al-Li-Cu[2,3] and Al-Cu-Fe[4] are now possible because of the availability of relatively large single grains of these stable phases.

The details of elementary excitations in quasicrystals are expected to differ from those found for periodic crystals. First, the absence of elastic anisotropy is expected for icosahedral alloys, since the symmetry of the icosahedral point group $(m\bar{3}5)$ is higher than cubic. Invariance of the cubic elastic constants under rotations by $\frac{2\pi}{5}$ about a five-fold axis leads to the isotropic relation, $2c_{44} = c_{11} - c_{12}$.[5] Secondly, differences arise from the fundamental incommensurability of quasiperiodic structures. One consequence of this is that the concept of Brillouin zones in the reciprocal lattice is ill-defined. Reciprocal space is, in principle, densely filled with Bragg peaks although most of these peaks have negligible intensity. Incommensurability also leads to the existence of new dynamical modes[6,7] (phason modes) which, for most icosahedral alloys, are believed to be pinned at room temperature.[8]

P. Jena et al. (eds.), Physics and Chemistry of Finite Systems: From Clusters to Crystals, Vol. I, 323–332.
© 1992 *Kluwer Academic Publishers.*

Theoretical calculations of the dynamical reponse of these sysems is complicated by the absence of periodicity and the lack of specific knowledge concerning the atomic structure, in three-dimensions, of real icosahedral alloys. Several groups have, however, attempted calculations of the full wavevector- and frequency-dependent response function $S(\vec{Q}, \omega)$ for one-, two- and three-dimensional quasilattices. Although most of this work describes calculations of spin dynamics (magnons) on a quasilattice, the results carry over directly to vibrational excitations (phonons) by the transformation $\omega \to \omega^2$. These studies have resulted in several specific predictions for the behavior of phonon modes in quasicrystals that may be tested by inelastic neutron scattering measurements.

Patel and Sherrington calculated $S(\vec{Q}, \omega)$ for a system of ferromagnetically coupled spins on 4181 sites of a two-dimensional Penrose tiling.[9] They found well-defined propagating spin-waves with isotropic dispersion close to the zone centers (Bragg peaks) of the quasilattice. The behavior of $S(\vec{Q}, \omega)$ beyond the long-wavelength $(\vec{q} \to 0)$ limit has been studied by several groups. For the one dimensional Fibonacci sequence, Ashraff and Stinchcombe derived an analytical expression for $S(\vec{Q}, \omega)$ in the equal coupling limit.[10,11] In this case, the dispersion curves consist of main branches accompanied by many satellite branches of much lower intensity. The strongest branch is associated with the strongest zone center. The satellite branches originate from other nearby zone centers of weaker intensity. In principle, there are an infinite number of satellite dispersion curves, since the reciprocal space is densely filled with zone centers, although most of these carry negligible spectral weight. In the equal coupling limit these branches cross without opening gaps. However, once the coupling between sites separated by different lengths are assigned different strengths, gaps open at the crossing points. The resulting picture for the magnon (or equivalently phonon) dispersion curves consists of propagating modes at small wavevectors separated by a set of gaps from stripes of dispersionless modes at higher frequency. Similar features are observed in the phonon $S(\vec{Q}, \omega)$ for a one-dimensional Fibonacci sequence obtained by the spectral moments method by Benoit, Poussigue and Azougarh[12]. In addition, they studied the Fourier transform of the time dependent displacement-displacement correlation function and found that the higher frequency modes are significantly more localized than the acoustic modes.

All of the calculations described above ignore the atomic decoration of the quasilattice. Therefore, while qualitative trends in the calculated $S(\vec{Q}, \omega)$ may be compared with the results of neutron scattering experiments, quantitative comparisons can not be made. One approach to the determination of $S(\vec{Q}, \omega)$ for real quasicrystals is suggested by the existence of so-called crystalline approximants to the quasicrystalline alloys. These are large unit cell periodic structures with internal icosahedral symmetry. For instance, the cubic R-phase of Al-Li-Cu may be characterized as a body-centered cubic packing of icosahedral clusters of atoms[13−15]. The icosahedral phase of Al-Li-Cu may be viewed as an aperiodic[16,17] or disordered[18] packing of these same clusters. Knowledge of $S(\vec{Q},\omega)$ for the R-phase then, can provide a good starting point for realistic calculations of the response function for the icosahedral alloy[19,20].

2. Experimental details

Inelastic neutron scattering measurements were made on a 0.3 gram single grain of $Al_{60.3}Li_{29.2}Cu_{10.5}$ oriented so that a two-fold plane, shown schematically in Figure 1, was coincident with the scattering plane. This geometry facilitates measurements along the high-symmetry two-, three-, and five-fold axes. Inelastic scattering measurements were also made on a 1.3 gram grain of the cubic R-$Al_{56}Li_{32}Cu_{12}$. The R-phase sample was oriented so that the analogous two-fold plane was coincident with the scattering plane (also shown in Figure 1). The areas of the circles in these plots are proportional to the measured or calculated[21], intensities of the respective elastic scattering peaks. In the schematic of the two-fold plane for the cubic R-phase, arrows denoting the three- and five-fold axis of the icosahedral phase have been included to emphasize the close similarity of the two diffraction patterns. Although the pattern of diffraction spots for the R-phase is periodic, the intensity modulations of the pattern bear a strong resemblence to those of the icosahedral phase. This results from the fact that the dominant scattering clusters in the cubic unit cell of the R-phase have icosahedral symmetry.

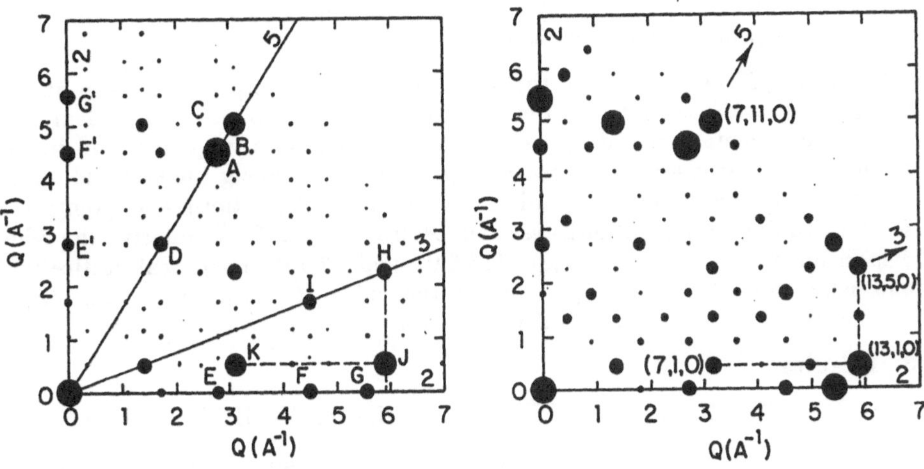

Figure 1.The two-fold planes for the icosahedral (left) and R-phase (right) of Al-Li-Cu investigated in these measurements. The dashed lines denote the region of inelastic scattering measurements for the two-fold longitudinal and transverse modes as described in the text. The arrows in the right panel denote the 3- and 5-fold axes of the icosahedral phase for comparison.

The phonon wavevector corresponding to inelastic measurements is defined by the relation $\vec{q} = \vec{Q} - \vec{G}$, where $\vec{Q}$ is the momentum transfer, $\vec{k}_F - \vec{k}_I$, and $\vec{G}$ is the reciprocal lattice vector of the zone center (Bragg peak) chosen as the origin for the measurement. Since the reciprocal space of the icosahedral phase is aperiodic, all wavevectors are labelled in units of $\overset{\circ}{A}{}^{-1}$, rather than reciprocal lattice units as is typically done for periodic crystals. For reference, the Miller indices for a few of the diffraction spots of the R-phase are given in Figure 1. Some icosahedral phase peaks are labelled by letters A-K.

Most of the inelastic neutron scattering measurements described here were made on the IN8 triple axis spectrometer at the ILL using a vertically bent copper (111) monochromator and a vertically bent pyrolytic graphite (PG) (002) analyzer. Constant-q scans were performed at a fixed final neutron energy of 44.8 meV, and collimations of 50'-40'-40'-40', numbers which refer to the horizontal collimations of the beam between the reactor and monochromator, monochromator and sample, sample and analyzer, and analyzer and detector respectively. This configuration was chosen so that the (E-q) slope of the instrumental resolution function was as close as possible to the slope of the transverse acoustic phonon branch to maximize the effect of focusing. Measurements of the low energy acoustic modes, at small $\vec{q}$, were made at IN8 using a fixed final energy of 14.7 meV, the same collimation, and a pyrolytic graphite filter to eliminate higher harmonic contamination of the scattered beam. In addition to the measurements made at the ILL, data were taken on the thermal neutron 2T and cold neutron 4F2 triple axis spectrometers of the Laboratoire Léon Brillouin at Saclay, France.

Constant-$\vec{q}$ energy scans of the icosahedral phase sample were taken along the two-, three- and five-fold directions. Frequently, excitations were measured from several zone centers in order to check for consistency in frequency and intensity. A more complete map of the low-lying phonon modes along the two-fold direction was taken between points J and H (transverse polarization) and points J and K (longitudinal polarization), as well as from point G. Similarly, the phonon dispersion between the (13,1,0), (13,5,0) and (7,1,0) reciprocal lattice points and near the (12,0,0) Bragg point in the cubic R-phase were determined for comparison with the icosahedral phase data.

3. Results

3.1 LOW-$\vec{q}$ PHONON DISPERSION FOR THE ICOSAHEDRAL ALLOY

The data in Figure 2 represent the peak centroids of phonon groups taken along the high symmetry two-, three- and five-fold axes of the icosahedral phase sample. Within experimental error, the dispersion relations for both longitudinal and transverse modes out to phonon wavevectors of $0.2\text{-}0.3\overset{\circ}{A}{}^{-1}$ are independent of direction. Furthermore, measurements of the transverse acoustic phonons along the pair of two-fold directions, separated by 90° in the two-fold plane, probe modes of different polarizations and were found to be indistinguishable.

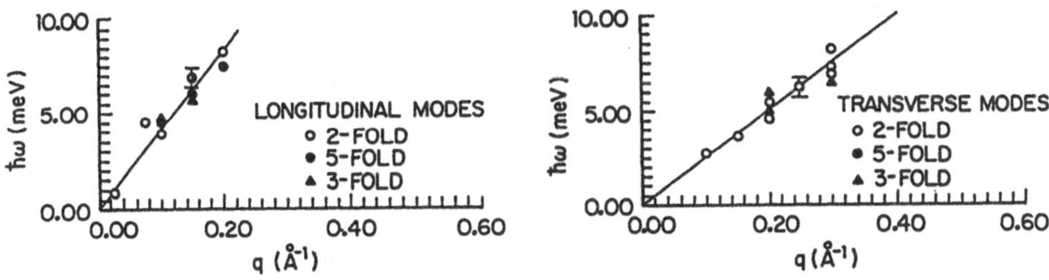

Figure 2.Low energy acoustic modes along the two-, three- and five-fold axes for icosahedral Al-Li-Cu. The solid lines describe the extrapolated initial slopes of the dispersion curves using sound velocities from ultrasonic measurements as described in the text.

In the long wavelength limit ($|\vec{q}| \to 0$) the slopes of the dispersion curves may be related to sound velocities determined by ultrasonic measurements using:

$$\omega = \frac{E}{\hbar} = v_j |\vec{q}|,$$

where v_j is the appropriate longitudinal or transverse sound velocity. Ultrasonic measurements by Reynolds et al. on single grains of Al-Li-Cu found that $v_L = 6.4 \pm 0.1 X 10^5 \mathrm{cm/s}$ and $v_T = 3.8 \pm 0.1 X 10^5 \mathrm{cm/s}$ along both the two-fold and five-fold axes.[22] The solid lines in Figure 2 represent the extrapolated linear dispersion calculated from the sound velocities determined by Reynolds et al. There is clearly excellent agreement between the initial slope of the dispersion curve extrapolated from the ultrasonic data and the neutron scattering data.

Unfortunately, we were not able to complete a full set of similar measurements, at small $\vec{q}$, on the R-phase sample. In previous work[23] on the IN14 triple axis spectrometer, however, it was found that even for the cubic crystal no significant anisotropy could be detected. The measured transverse sound velocity for the cubic phase is $3.8 X 10^5$ cm/s; the same as for the icosahedral phase.

3.2 DISPERSION RELATIONS FOR THE ICOSAHEDRAL AND CUBIC PHASES

In order to compare the dispersion relations for the icosahedral and R-phase samples, scans of transverse and longitudinal modes along the two-fold axis were taken along the directions indicated by dashed lines in Figure 1. These particular regions were selected because they cover approximately the same range of reciprocal space for both samples and the sequence of elastic scattering peaks, both in position and intensity, are quite similar.

328

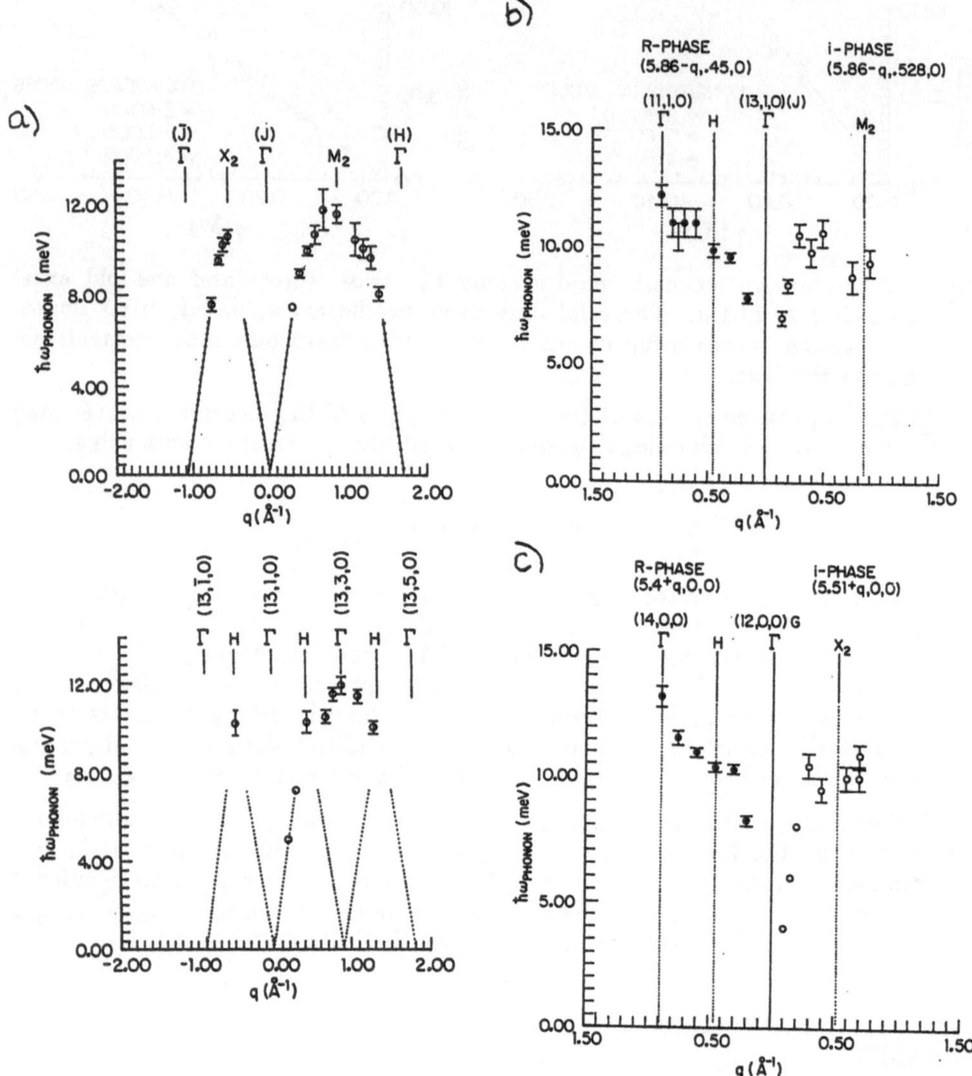

Figure 3.a) Transverse phonon modes along the two-fold directions for icosahedral
(top panel) and the cubic R-phase (bottom panel) of Al-Li-Cu. The Γ points
denote the zone center positions, while other special points are described in
the text. The dotted lines indicate the low energy acoustic mode dispersion.
Zone centers are labelled by their miller indices, or letters; b) the longitudinal
phonon dispersion along a two-fold direction for the icosahedral phase measured
from point J (right side) and R-phase, measured from the (13,1,0) (left side); c)
The longitudinal phonon dispersion along a two-fold direction for the icosahedral
phase measured from point G (right side) and R-phase, measured from the
(12,0,0) (left side).

Figure 3 shows the transverse phonon dispersion along the two-fold direction for the icosahedral and R-phase samples as well as the longitudinal phonon dispersion along the two-fold axes for both samples. Symmetry points for the cubic R-phase are labelled using the conventional group theoretic notation, along with the identification of the Bragg point corresponding to the zone center. For the icosahedral phase sample, the behavior of the transverse phonon dispersion along the two-fold axis, as well as previously reported results for longitudinal modes along the five-fold axis,[2] suggest that one can define quasi-Brillouin zones in the reciprocal space of the icosahedral phase between pairs of strong Γ points as predicted by several theoretical claculations of $S(\vec{Q},\omega)$. Although, in principle, the concept of Brillouin zones is ill-defined for quasiperiodic structures, one may identify "quasi"-zones between the most intense zone-centers.

The positions denoted as Γ, M_2 and X_2 in Figure 3 for the icosahedral phase, represent the projection of the special points of a six-dimensional simple hypercubic lattice down onto the two-fold plane of the icosahedral phase as described by Niizeki and Akamatsu.[24,25] Briefly, these special points describe positions of high symmetry in the six-dimensional lattice of the form $\frac{1}{2}(n_1,n_2,n_3,n_4,n_5,n_6)$ where n_1 through n_6 are integers. For example, the Γ label denotes points where all n_i are even (zone centers), while the X_2 points are of the form $\frac{1}{2}(1,1,0,0,0,0)$. The physical significance of these special points is analogous to that for the corresponding special points in the reciprocal lattice of periodic crystals. They describe critical, or stationary, points of the dispersion relation for electronic or vibrational states where one would expect maxima in the dispersion curve, as are found in our measurements for icosahedral Al-Li-Cu.

Generally speaking, the longitudinal and transverse phonon dispersion curves for the icosahedral phase and cubic R-phase are quite similar. Indeed, as noted before, the long wavelength acoustic dispersion along the two-fold directions are essentially indistinguishable. This is not unexpected since at the long-wavelength (small $\vec{q}$) continuum limit, excitations are relatively insensitive to the details of the atomic structure of materials. At higher energy, at least for the transverse modes, there are also similarities in the form of the phonon dispersion for the icosahedral and R-phase samples. This result is encouraging since it validates one approach to the determination of the inelastic structure factor for real quasicrysatlline sytems; the extension of theoretical calculations of $S(\vec{Q},\omega)$ for the crystalline approximants.

On the other hand, differences in the phonon dispersion between the two alloys are found at higher energy for phonon wavevectors larger than about 0.4 to $0.5 \mathring{A}^{-1}$. The longitudinal modes of the R-phase exhibit low-lying dispersive optical branches in the region $\Gamma \rightarrow H \rightarrow \Gamma$. In contrast, for the icosahedral phase sample, the higher energy longitudinal phonon excitations along the two-fold axes appear as dispersionless branches which are quite similar to the localized excitations at higher energy found in the one- and two-dimensional calculations of $S(\vec{Q},\omega)$ for Fibonacci lattices. In fact, there is an interesting connection between the two-fold direction of the icosahedral phase and the one-dimensional Fibonacci sequence. The two-fold planes of the icosahedral phase are dense-packed layers and, as high resolution electron microscopy images have shown, the atomic configurations along the two-fold axis closely resemble a Fibonacci sequence of "atomic planes." This has also been proposed as an alternative way of viewing the quasicrystalline structure of Al-Li-Cu[21]. In terms of diffraction properties, it is interesting that the sequence of diffraction peaks along a two-fold axis of the icosahedral phase is identical to the diffraction pattern from a one-dimensional Fibonacci sequence. Indeed, the close similarity between the diffraction patterns itself motivates a comparison of the longitudinal modes probed along a two-fold direction with dynamical calculations for one-dimensional Fibonacci sequence.

4. Discussion

While phonon DOS measurements of icosahedral Al-Li-Cu by Suck et al.[26] found maxima in the inelastic spectrum at 13, 16 and 21 meV, our own measurements on the single grain of icosahedral Al-Li-Cu found no distinct excitations at these energies. Interestingly, the flat 10 meV mode found in our measurements does not appear as a distinct peak in the phonon DOS measurements. This means that the spectral weight carried by this mode must be smaller than that carried by other, perhaps more localized, modes at higher energy. Excitations which are spatially localized give rise to dispersionless scattering which is distributed over a correspondingly large region in reciprocal space. The density of states measurement is more sensitive to the presence of localized modes since it makes use of powder samples, thereby averaging over all directions and phonon wavevectors, for some particular energy range. Other recent measurements of inelastic excitations in quasicrystals have had little success in finding the higher energy optic-like modes in these systems. In icosahedral Al-Cu-Fe, for instance, excitations above 10 meV could not be observed.[4] Similarly, atomic beam scattering experiments on the decagonal phase of Al-Co-Ni found that surface phonon peaks broadened and diminished at higher energy (on the order of 10 meV).[27] All of the evidence cited above appears to point to an enhanced degree of localization for the higher energy modes in quasiperiodic systems.

Independent of any specific interpretation of behavior of the phonons in the icosahedral phase at higher energy and wavevector, we point out that it is precisely in this region that we should expect to see differences between the icosahedral and R-phase samples. First consider the measurements made on the R-phase sample. Inelastic neutron scattering experiments probe dynamical correlations between atoms over distances which vary inversely with $|\vec{q}|$. Very close to the zone center dynamical effects over very large length scales compared to the size of the unit cell are probed. As $\vec{q}$ increases, dynamical correlations over smaller distances become more important until, at the zone boundary, we essentially probe the correlated motions between atoms on neighboring icosahedral clusters of atoms in the cubic unit cell of the R-phase. As mentioned in the Introduction, some models for the icosahedral phase of Al-Li-Cu view the structure as an aperiodic or disordered packing of these same icosahedral clusters of atoms. Therefore, differences in the behavior of excitations at phonon wavevectors ($\approx 0.4 - 0.5 \text{\AA}$) approaching intercluster separations ($\approx 12 \text{\AA}$)may be attributed to differences in the details of how these clusters are packed together.

The work described here was carried out in collaboration with C. Stassis, R. Bellissent, H. Moudden, M. deBoissieu, C. Janot and R. Currat. I also wish to thank C. Benoit, B.N. Harmon, K.M. Ho and M. Widom for their assistance and useful discussions; F.W. Gayle for the preparation of the icosahedral phase grain and B. Dubost for preparation of the R-Al-Li-Cu phase. Ames Laboratory is operated for the U.S. Department of Energy by Iowa State University under Contract No. W-7405-ENG.

References

1. For a recent review, see A.I. Goldman and M. Widom, Ann. Rev. Phys. Chem **42**, 685 (1991).

2. A.I. Goldman, C. Stassis, R. Bellissent, H. Moudden, N. Pyka and F.W. Gayle, Phys. Rev. **B43**, 8763 (1991).

3. A.I. Goldman, C. Stassis, M. de Boissieu, R. Currat, C. Janot, R. Bellissent, H. Moudden and F.W. Gayle, "Phonons in Icosahedral and Cubic Al-Li-Cu," (submitted to Phys. Rev. B, August 1991).

4. M. Quillichini, G. Heger, B. Hennion, S. Lefebvre and A. Quivy, J. Phys. France **51**, 1785 (1990).

5. L.C. Chen, S. Ebalard, L.M. Goldman, W. Ohashi, B. Park and F. Spaepen, J. Appl. Phys. **60**, 2638 (1986).

6. P. Bak, Phys. Rev. Lett. **54**, 1517 (1985); Phys. Rev. **B32**, 5764 (1985).

7. D. Levine, T.C. Lubensky, S. Ostlund, S. Ramaswamy, P.J. Steinhardt and J. Toner, Phys. Rev. Lett. **54**, 1520 (1985).

8. For example, see T.C. Lubensky in, *Introduction to Quasicrystals*, M.V. Jaric, ed., Academic Press, San Diego, 1988, pp. 199-280.

9. H. Patel and D. Sherrington, Phys. Rev. **B40**, 11185 (1989).

10. J.A Ashraff and R.B. Stinchcombe, Phys. Rev. **B39**, 2670 (1989).

11. J.A. Ashraff, J-M. Luck and R.B. Stinchcombe, Phys. Rev. **B41**, 4314 (1990).

12. C. Benoit, G. Poussigue and A. Azougarh, J. Phys. Cond. Matter. **2**, 2519 (1990); J. Phys.: Condens. Matter **1**,335 (1989).

13. E.E. Cherkashin, P.I. Kripyakevich and G.I. Oleksiv, Sov. Phys. Crystallogr. **8**, 681 (1964).

14. G. Bergman, J.T. Waugh and L. Pauling, Acta Crystallogr. **10**, 154 (1957).

15. C.A. Guryan, P.W. Stephens, A.I. Goldman and F.W. Gayle, Phys. Rev. B37, 8495 (1988).

16. V. Elser and C.L. Henley, Phys. Rev. Lett. 55, 2883 (1985).

17. M. Audier and P. Guyot, Phil. Mag. B53, L43 (1986).

18. P.W. Stephens and A.I. Goldman, Phys. Rev. Lett. 56, 1168 (1986).

19. J. Los and T. Janssen, J. Phys.:Condens. Matter 2, 9553 (1990).

20. T. Janssen, preprint (1991).

21. M. de Boissieu, C. Janot, J.M. Dubois, M. Audier and B. Dubost, J. Phys.: Condens. Matter 3, 1 (1991).

22. G.A.M. Reynolds, B. Golding, A.R. Kortan and J.M. Parsey Jr., Phys. Rev. B 41, 194 (1990).

23. M. deBoissieu, R. Currat and C. Janot, ILL Report 4.01.387, 1990.

24. K. Niizeki and T. Akamatsu, J. Phys.: Condens. Matter 2, 2759 (1990).

25. K. Niizeki, J. Phys. A: Math. Gen. 22, 4295 (1989).

26. J.-B. Suck, C. Janot, M. De Boissieu and B. Dubost, in *Phonons '89*, S. Hunklinger, W. Ludwig and G. Weiss, eds., (World Scientific, Singapore, 1990), pp. 576-578.

27. B.R. Doak, private communication.

STRUCTURAL AND ELECTRONIC PROPERTIES OF PURE AND DOPED FULLERENES

Wanda ANDREONI, François GYGI and Michele PARRINELLO
IBM Research Division
Zurich Research Laboratory
8803 Rüschlikon
Switzerland

ABSTRACT. We present the results of ab-initio molecular dynamics studies of pure and doped fullerenes. In particular, we have determined the electronic and ionic properties of pure C_{60} and C_{70}. For C_{60}, we have also studied finite temperature and dynamical properties, which are found to be in excellent agreement with experimental results. In view of the recent synthesis of doped C_{60} cages, we have also investigated the effect of chemical doping. The systems investigated are $C_{59}B$, $C_{59}N$, $K@(C_{59}B)$ and $C_{58}BN$.

1. Introduction

An enormous amount of experimental and theoretical work has recently been done on fullerenes, both in the molecular and in the solid phases [1, 2]. Pure C_{60} clusters and crystals have been widely investigated experimentally, so that by now the molecular structure [3] as well as several electronic properties (see e.g. [4, 5, 6, 7]) and vibrational frequencies [8] have been measured. The study of C_{70} is less advanced, due to the fact that pure C_{70} is much harder to obtain. In particular, the molecular structure, although proven by nuclear magnetic resonance (NMR) experiments to be compatible with a D_{5h} symmetry [9, 10], is not known experimentally. More recently, a new class of fullerenes has started to be synthesized in the molecular phase by Smalley's group, which are obtained by doping C_{60} as well as higher fullerenes [11]. Reactivity and photolysis experiments are used as a tool to identify whether the impurity is internal (endodoping) to the cage or not (exodoping). The result is that new fullerene cages are formed by either replacing carbon with boron ($C_{59}B$, $C_{58}B_2$) [12] or by inserting in the hollow cages atoms such as K and La [13].

In this paper, we present some very recent results on C_{60}, C_{70} as well as some doped C_{60} clusters, obtained by using the Car-Parrinello (CP) method [14]. The work on C_{60}, which has already appeared in Ref. [15], can be considered as a test of the accuracy of the whole computational scheme in predicting structural parameters, ground state electronic properties and vibrational frequencies. These results will be reviewed in Sect. 2. Sect. 3 contains the extension of these calculations to C_{70}, for which we provide a structure optimized in the local density approximation (LDA) of the density functional theory and discuss the one-electron states and energy spectrum. In Sect. 4 we shall discuss our predictions for the equilibrium structure of various doped C_{60} cages, namely $C_{59}B$, $C_{59}N$, $K@(C_{59}B)$ [16] and $C_{58}BN$. We shall see that in all cases doping induces significant but spatially localized deformations of the cage.

P. Jena et al. (eds.), Physics and Chemistry of Finite Systems: From Clusters to Crystals, Vol. I, 333–338.
© *1992 Kluwer Academic Publishers. Printed in the Netherlands.*

In the first three systems such deformations are accompanied by the formation of localized impurity states, which lie in the middle of the energy gap of the undoped cluster, in analogy with the case of deep impurities in semiconductors.

2. Method and Application to C_{60}

An extensive description of the Car-Parrinello method [14] is contained in Refs. [17] and [18]. Unique to this method is the possibility of treating simultaneously electronic and ionic variables, thus allowing efficient procedures for the optimization of geometrical structures, and to study finite temperature properties and dynamics of complex systems "ab-initio". So far, in the application to fullerenes, we have used the CP method mainly to determine the equilibrium structure with a combined ionic and electronic steepest-descent procedure. In the case of C_{60} we have also studied average properties and dynamics at finite temperature.

In all these calculations we have used a plane wave basis set for the expansion of the one-electron orbitals, with a cutoff of 35 Ry, a repeated cell scheme with a fcc unit cell with a lattice constant of 17.46 Å, and a fully nonlocal pseudopotential [19] to represent the core-valence interaction.

The calculations on C_{60} [15] can be summarized as follows. Starting from a buckminsterfullerene structure with a unique bond length, and relaxing simultaneously ionic and electronic variables, we have obtained the desired structure with two bond lengths corresponding to the frontier between two hexagons ("double bond") and to the frontier between two pentagons ("single bond"). The calculated values of 1.39 Å and 1.45 Å respectively turned out to be in very good agreement with the data obtained from NMR experiments [3]: 1.40 $\pm$0.015 Å and 1.45 $\pm$ 0.015 Å, and also with other LDA results [20, 21]. For comparison, the Hartree-Fock (HF) values are 1.365 Å and 1.453 Å [22, 23] and the MNDO values are 1.400 Å and 1.474 Å [24]. Subsequent simulations at ~450 K revealed very slight changes in the average structural parameters and provided a distribution of the interatomic distances closely resembling the radial distribution function derived from neutron scattering experiments [25]. In particular, in spite of the existence of two different bond lengths, at ~450 K thermal broadening already masks all the splittings derived from it. Comparison of the time averaged density of one-electron Kohn-Sham states (DOS) with the photoelectron spectrum [4] revealed a globally good agreement in the position of the peaks and peak widths. This was particularly true for the higher ionization transitions, for which the difference between the observed spectrum and the calculated DOS is expected to be smaller [26]. However, the discrepancies we observe could be attributed in part to the LDA representation of the DOS, which typically leads to a narrowing of bandwidths and energy gaps. Indeed, other LDA calculations [21] including the effect of the transition matrix elements, do not improve the agreement significantly. We note that in this case the LDA energy gap turns out to be 1.7 eV. This value must be compared to the HOMO-LUMO gap of 1.9 eV obtained from photoelectron spectroscopy of the anion [27] and to the band gap of the solid of 1.7 eV [7b]. From the atomic trajectories we have also calculated the vibrational modes and in particular we have extracted the frequencies of the optically active ones, which could be compared with infrared and Raman spectroscopy. All these frequencies were reproduced within a few percent. Such a discrepancy, which amounts to few cm^{-1} for the lowest modes but reaches 90 cm^{-1} for the highest frequencies, must be primarily associated to the inaccuracy of LDA. It must be pointed out, however, that semi-empirical methods do not reproduce the vibrational spectrum with higher accuracy.

3. C_{70}

C_{70} is believed to assume a so-called rugby-ball structure with a D_{5h} symmetry, which is consistent with NMR experiments [9, 10] showing the presence of five non-equivalent sites. Structural parameters have not been measured yet. Our optimized structure has values which are intermediate between the HF ones [23, 29] and the ones obtained with MNDO [23], in full analogy with the results for C_{60}. The shortest bond length (1.386 A) is only very slightly reduced with respect to C_{60} and the largest bond length (1.467 A) corresponds to the smallest distance in the additional ring of 10 atoms at the equator of the structure. This 0.08 A bond length separation is comparable to the HF values of 0.11 A to 0.12 A [23, 29] and to the MNDO value of 0.095 A [23]. The LDA electronic spectrum differs significantly from that obtained in the HF calculations in the spacing of the energy levels. However, the symmetry of the states near the gap is the same: the HOMO is a doublet of e''_1 symmetry, close to a singly degenerate level of a''_2 symmetry, while the LUMO is a doublet of e''_1 followed by a singly degenerate level of a''_1 symmetry. In Fig. 1 we compare the DOS we have calculated for C_{70} with that of C_{60}, both broadened by Lorentzians of width 0.2 eV. For C_{60} such a broadening roughly corresponds to the one resulting from our finite temperature simulation [15]. We notice a more structured DOS in C_{70}, which is consistent with the lowering of symmetry with respect to C_{60}. Comparison of the DOS with the photoemission spectrum measured on the solid [30] reveals the same type of agreement than in the case of C_{60}. The photoelectron spectrum calculated in Ref. [21] also shows the same type of qualitative agreement, in spite of the significant deviation of the structural parameters from our optimized values. The same comment can be applied to the recent LDA DOS obtained in Ref. [28] for another different geometrical model. Such comparisons could be naively interpreted a sign of the weak dependence of the electronic spectrum from the details of the structure (in particular the bond distances in the additional ring) or of the relevance of energy band dispersion. Indeed, owing to the low resolution of the photoemission spectrum, no conclusion can yet be drawn. The LDA HOMO-LUMO gap (1.75 eV) turns out to be very close to that of C_{60}. A complete report on these calculations will be given elsewhere [31].

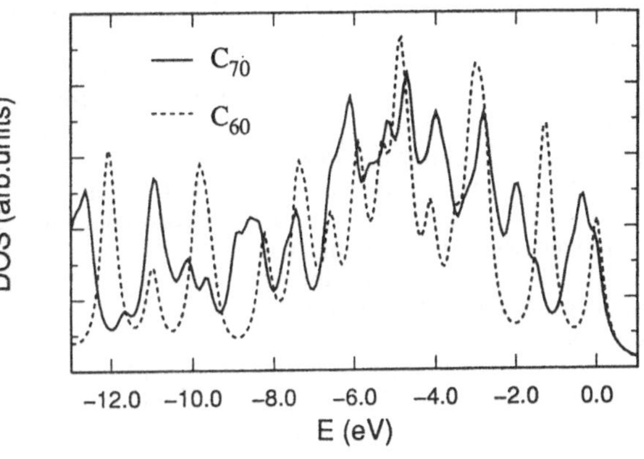

Figure 1. Density of states calculated in C_{60} and C_{70} (see text).

4. Doped C_{60}

Contrary to the most conventional approaches the CP method does not necessarily rely on preassigned symmetries. This makes it very useful to engineer without prejudices new materials on the computer. In the search for new fullerenes, the B- and

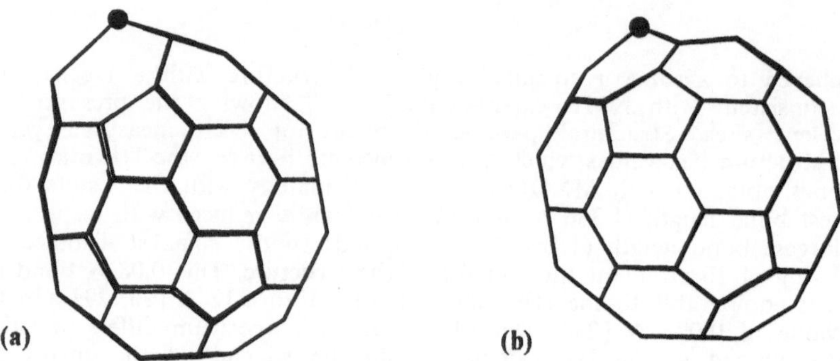

Figure 2. Structural deformations induced by doping (see text): (a) $C_{59}B$, (b) $C_{59}N$.

N-doped C_{60} are the most natural candidates. $C_{59}B$ has indeed been recently synthesized in beam, and the boron is believed to enter substitutionally in the carbon network [12]. Starting from the C_{60} atomic positions, and relaxing them, we have found [32] that in both $C_{59}B$ and $C_{59}N$ the optimum arrangements involve significant distortions of the network, mainly localized around the impurity atom. This is illustrated in Fig. 2 where the distorsions are amplified by a factor of 5. These are more important in the case of $C_{59}B$ (a) than in $C_{59}N$ (b), as can be expected on the basis of the larger core size mismatch between the impurity and the host, and also owing to the nature of the electron states mostly involved in the doping: the HOMO and the LUMO of C_{60} which are respectively bonding and antibonding on the double bonds. Especially interesting is the appearance of new features in the electron energy spectrum, i.e. impurity levels in the middle of the HOMO-LUMO gap of the undoped cluster, in analogy with the case of deep impurities in semiconductors. This is illustrated in Fig. 3. In both cases the doped fullerene is a radical, with one singly occupied state which can act either as an acceptor or as a donor in the case of B- and N-doping respectively. Indeed, $C_{59}B$ has been found to act as a Lewis acid in reactions with ammonia [12].

Starting from $C_{59}B$ and adding a K atom or alternatively replacing one more C atom with N, new fullerenes can be formed which are isoelectronic to C_{60}. We have optimized the structure of two of them: one with K inside the cage $K@(C_{59}B)$, following the suggestions from experiments, and one (still hypothetical) with BN replacing a CC double bond. Structural deformations are again significant but rather localized and can be easily deduced from those of the singly doped systems. Very interesting is the energy spectrum near the gap, which is shown in Fig. 3. The spectrum of $K@(C_{59}B)$ is very similar to $C_{59}B$, in spite of the fact that

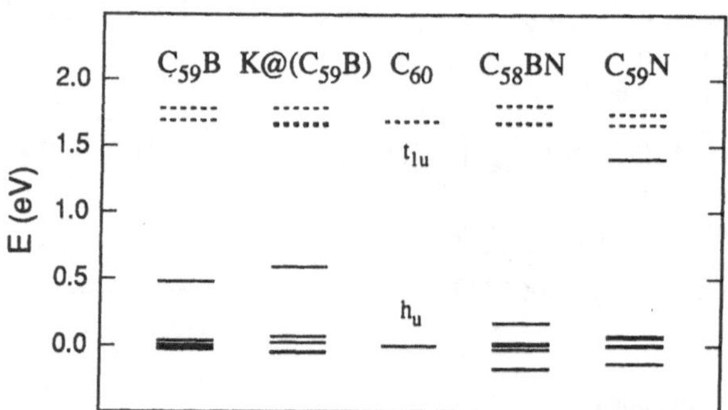

Figure 3. Energy level diagram of doped and undoped C_{60} (see text). Solid line = occupied states; dashed line = empty states.

the impurity state is now doubly occupied. Complete charge tranfer is observed from the K atom, which is found to reside only 0.1 Å away from the geometric center of the cage. The impurity state has a localized character similar to that of $C_{59}B$. As expected, the spectrum of $C_{58}BN$ more closely resembles that of the undoped system, with no specific impurity level in the gap, but only a global modification of the HOMO due to the difference in electronegativity of the two atoms replacing one of the 30 double bonds and consequent charge redistribution. This in particular causes a reduction in the energy gap of only 0.2 eV.

For all these systems we have also calculated other measurable quantities such as the dipole moments which range from 0.9D in $C_{58}BN$ to 3.5D in $K@(C_{59}B)$. In all cases we have observed a decrease of about 90% with respect to the pure ionic dipole moment. A complete discussion of these recent calculations will be reported elsewhere [33]. These compounds offer large freedom to change C_{60} properties like the HOMO-LUMO gap and also new pathways to doping. Of special relevance are the B-doped bucky-balls since they can be expected to lead to hole doped conductors.

References

1. R.E. Smalley, The Sciences, vol. 31, pp 22-30 (1991).
2. F. Diederich and R. Whetten, Ang. Chem. Int. 30, 678 (1991).
3. C. S. Yannoni, P. P. Bernier, D. S. Bethune, G. Meijer and J.R. Salem, J. Am. Chem. Soc. , 113:3190, 1991.
4. D.L. Lichtenberger et al. Chem. Phys. Lett. , 176:203, 1991.
5. J.P. Hare, H.W. Kroto and R. Taylor, Chem. Phys. Lett. 177, 394 (1991).
6. J.H. Weaver et al., Phys. Rev. Lett. 66, 1741 (1991).
7. (a) R.E. Haufler et al., Chem. Phys. Lett. 179, 449 (1991);
   (b) S. Krummacher, S. Cramm, K. Szot, W. Eberland and W. Krätzmer, Europhys. Lett. (in press) and references therein.
8. D.S. Bethune et al., Chem. Phys. Lett. 179, 181 (1991).
9. A.K. Abdul-Sada R. Taylor, J.P. Hare and H.W. Kroto. J. Chem. Soc. Chem. Commun. 1423, 1990.
10. R.D. Johnson, G. Meijer, J.R. Salem and D.S. Bethune, J. Am. Chem. Soc. 113, 3619, 1991.
11. R.E. Smalley, ACS Symposium Series, "Large Carbon Clusters", ed. G. Hammond and V. Kuck (in press).
12. Guo Ting, Jin Changming, and Smalley R.E. J. Phys. Chem. , 95:4948, 1991.
13. Yan Chai et al., J. Phys. Chem. 1991 (in press).
14. Car R. and Parrinello M. Phys. Rev. Lett. , 55:2471, 1985.
15. B.P. Feuston, W Andreoni, M. Parrinello and E. Clementi Phys. Rev. B44, 4506 (1991).
16. We use the recently introduced notation @ to indicate the inclusion of the dopant in the cage.
17. R. Car and M. Parrinello, in: Simple Molecular Systems at Very High Density, Vol. 186, Eds. A. Polian, P. Loubeyre, and N. Bocarra (Plenum, New York, 1989), p.455
18. G. Galli and M. Parrinello. Proc. Nato ASI "Computer Simulations in Material Science", Aussois, Fance 1991, Ed. M. Meyer and V. Pontikis, Kluwer Acad. Publisher
19. W. Andreoni, D. Scharf and P. Giannozzi Chem. Phys. Lett. 173:449, 1990.
20. Jae-Yel Yi Q. Zhang and J. Bernholc. Phys. Rev. Lett. , 66:2633, 1991.
21. J.W. Mintmire et al. Phys. Rev. , B43:14281, 1991.
22. G.E. Scuseria. Chem. Phys. Lett. , 176:423, 1991.
23. K. Raghavachari and C.M. Rohlfing, J. Phys. Chem. 95, 5768 (1991)

24. M.D. Newton and R.E. Stanton, J. Am. Chem. Soc. 108:2469, 1986.
25. J.S. Lannin, Fang Li, D. Ramage and J. Conceicao. submitted for publication.
26. see e.g. R. Clauberg, F.H. Frank, J.M. Nicholls and B. Reihl, Surf. Sci. 189/190, 44 (1987).
27. S.H. Yang, C.L. Pettiette, J. Conceicao, I. Cheshnovsky and R.E. Smalley, Chem. Phys. Lett. 176, 203 (1991).
28. S. Saito and A. Oshiyama, preprint.
29. G.E. Scuseria. Chem. Phys. Lett. 180:451, 1991.
30. M.B. Jost et al. Chem. Phys. Lett. , 1991. in press.
31. W Andreoni, F. Gygi, and M. Parrinello. (to be published)
32. W Andreoni, F. Gygi, and M. Parrinello. submitted for publication.
33. W Andreoni, F. Gygi, and M. Parrinello. (to be published)

VIBRATIONAL AND ELECTRONIC CONTRIBUTIONS TO THE SPECIFIC HEAT OF $Pd_{561}Phen_{36}O_{200}$.

J. Baak, H.B. Brom, L.J. de Jongh,
Kamerlingh Onnes Laboratory, Leiden University,
P.O. Box 9506, 2300 RA Leiden, the Netherlands.

G. Schmid,
Institut für Anorganische Chemie, Universität/GH Essen,
D-4300 Essen, Germany.

ABSTRACT. Specific heat measurements have been performed on the metal cluster compound $Pd_{561}Phen_{36}O_{200}$, which consists of Pd_{561} clusters surrounded by ligand shells. Above 1 K the specific heat can be successfully described by a contribution due to vibrations of the cluster as a whole, modeled via the Debye scheme, and a contribution due to vibrations within the cluster. The intra-cluster vibration frequencies are calculated in an elastic continuum approach. By solving the Navier equation, using stress free boundary conditions, eigenvalue equations for the spheroidal and toroidal oscillations are obtained. The results compare well with those of molecular dynamics calculations and experiment. The most salient feature of the specific heat below 1 K is a linear contribution $C = \gamma T$, which is interpreted as an electronic contribution. The observed value of γ is about one third of the value of bulk Palladium.

1 Introduction.

The study of metal cluster compounds has been an important field of interest in our group in recent years. These compounds consist of metal clusters of well defined size and structure (this in contrast to for example colloids), "chemically stabilized" by ligand shells. The ligands will not contribute to the specific heat at temperatures low compared to the vibrational frequencies involved (i.e. below 50 K [1]), making these compounds excellent model systems to study quantum size effects by specific heat experiments.

Smit et al. [2] were able to describe both Mössbauer and specific heat experiments on the $Au_{55}(PPh_3)_{12}Cl_6$ compound, containing Au_{55} clusters, by the sum of vibrations of the cluster as a whole (inter- cluster vibrations) and vibrations of the atoms within the cluster (intra- cluster vibrations). The inter-cluster vibrations were modeled by the Debye scheme, whereas the intra-cluster vibration frequencies were calculated by a brute force solution of the equations of motion. These so-called molecular dynamics (MD) calculations require the diagonalisation of a 3Nx3N matrix, N being the number of atoms in the cluster. Therefore an alternative approach may be profitable for the larger clusters. In an attempt to explain the experimental specific heat data on small Pb and In particles [3], Nishiguchi et al. [4] used an elastic continuum approximation (ECA) to calculate the vibrational spectrum. By

P. Jena et al. (eds.), Physics and Chemistry of Finite Systems: From Clusters to Crystals, Vol. I, 339–344.
© 1992 *Kluwer Academic Publishers.*

exactly solving the Navier equation with stress free boundary conditions, the intra-cluster vibrational frequencies were obtained. This method is expected to break down for smaller particles, since these can not be considered to be an elastic continuum.

Also the electronic spectrum for small metal particles is expected to be dominated by quantum size effects. In a simple model, such as the free electron gas in a box, the splittings of the energy levels can no longer be neglected, e.g. for a 600 electron system the splittings around the Fermi energy are of the order of 100 K. This implies that no linear contribution to the specific heat is expected at low temperatures. Indeed in specific heat measurements on $Au_{55}(PPh_3)_{12}Cl_6$ such a linear contribution was not seen by Smit et al. [2] down to 0.5 K, nor by Goll et al.[5] down to 60 mK.

In this paper we present specific heat measurements on the $Pd_{561}Phen_{36}O_{200}$ compound, synthesized by G. Schmid and coworkers [6]. The cuboctahedral Pd_{561} clusters consist of five closed shells of Pd atoms. The data are analyzed by using the MD and the ECA method. We show that, although the excitation spectra of both approaches are different, the calculated specific heat is similar and compares well with experiment.

Most important experimental result is the observation of a linear contribution to the specific heat at the lowest temperatures. As will be argued, the magnitude can well be understood by assuming an inner core fraction of the cluster atoms to be metallic. As far as we know this is the first experimental observation of such a linear term in a metal cluster molecule.

The specific heat measurements on the $Pd_{561}Phen_{36}O_{200}$ compound were performed in a dilution refrigerator apparatus, using a thermal relaxation method, on two different batches of approximately 35 mg each. The size of the Pd clusters was checked using high resolution electron microscopy. The (powder) samples were "glued" to a sapphire plate using approximately 7 mg of Apiezon N grease. The sapphire plate was connected to a heat sink via a small gold wire. A NiCr heater of 1257 Ω was sputtered on this sapphire plate. As a thermometer a 4.5 kΩ Philips RuO_2 resistor was used. The measured heat capacity was always corrected for the contributions of empty apparatus and the Apiezon N grease. Because of the high internal relaxation times and the large contribution of the empty apparatus we were not able to perform measurements at temperatures lower than 0.2 K.

2 Calculations of the phonon and electronic specfic heat.

The molecular dynamics approach will be discussed only very briefly. For an extensive discussion we refer to [2]. The Newtonian equations of motion for the cluster atoms are given by:

$$m \frac{d^2 Q_i}{dt^2} + \sum_{j=1}^{3N} F_{ij} Q_j = 0 \tag{2.1}$$

where m is the atomic mass and Q_i are cartesian displacements coordinates and F_{ij} the force constant between atom i and atom j. Only nearest neighbour interactions are taken into account and the force constant between them is taken as an adjustable fitting parameter. By introducing the trial solution:

$$Q_i = A_i \, cos(\omega t + \beta) \tag{2.2}$$

β being an unimportant phase factor, we arrive at the eigenvalue problem:

$$\sum_{j=1}^{3N} F_{ij} A_j = m\omega^2 A_i \tag{2.3}$$

The normal mode frequencies are obtained by numerical diagonalisation of the 3N × 3N matrix F_{ij}. Six of the 3N eigenvalues are zero, corresponding to the 3 translational and 3 rotational degrees of freedom of the cluster as a whole.

Turning now to the elastic continuum approximation, we follow Nishiguchi and Sakuma [4] to obtain the normal mode frequencies by solving the Navier equation for an isotropic and homogeneous sphere, using the spherical coordinates r, θ and ϕ:

$$\rho \, \ddot{\vec{Q}} = (\lambda + 2\mu) \, grad \, div \, \vec{Q} - \mu \, curl \, curl \, \vec{Q} \tag{2.4}$$

where $\vec{Q}$ is the displacement vector with components Q_r, Q_θ, Q_ϕ; ρ is the mass density and λ and μ are Lamé constants. An exact solution can be found by introducing one scalar and two vector potentials [4]. Due to the stress free boundary conditions only a discrete set of angular frequencies ω is found. For a detailed calculation we refer to [4]. The toroidal oscillation frequencies are determined by the solutions of the following equation:

$$j_{l+1}(\eta) - \frac{l-1}{\eta} j_l(\eta) = 0 \qquad (l \geq 1) \tag{2.5}$$

j_l being the spherical Bessel function of order l, $\eta = \omega R / v_t$, R the cluster radius and $v_t = \sqrt{\mu/\rho}$ the transverse sound velocity. Since there are two toroidal modes and the solutions are independent of the quantum number m they all have a 2(2l+1)-fold degeneracy. The spherical oscillation frequencies are determined by the solutions of the following equation:

$$2 \frac{v_t}{v_l} \eta \frac{j_{l+1}(\frac{v_t}{v_l}\eta)}{j_l(\frac{v_t}{v_l}\eta)} (\eta^2 + (l-1)(l+2)(\eta \frac{j_{l+1}(\eta)}{j_l(\eta)} - (l+1)))$$

$$-\frac{1}{2} \eta^4 + (l-1)(2l+1) \eta^2 + \eta \frac{j_{l+1}(\eta)}{j_l(\eta)} (\eta^2 - 2l(l-1)(l+2)) = 0 \tag{2.6}$$

$$(l \geq 0)$$

$v_l = \sqrt{\lambda + 2\mu/\rho}$ being the longitudinal sound velocity. Because the solutions are independent of the quantum number m they all have a (2l+1)-fold degeneracy. During the calculation the sound velocities were taken to be adjustable parameters, but the ratio of v_t and v_l was fixed at the bulk value, which we calculated to be 0.42.

Since there are infinitely many l-values, each giving infinitely many ω's, a cut-off frequency has to be introduced. Usually the Debye cut- off method is chosen, which means that only the 3N-6 lowest frequencies are used. A different cut-off method was introduced by Tamura et al. [7]. By arguing that the wavelength should be larger than two times the

342

interatomic distance, they introduce an upper limit for l. After setting this upper limit the lowest frequencies are selected. This cut-off method was used for our calculations.

Once the vibrational frequencies are known, the specific heat contributions of the (Einstein) oscillators can be obtained from:

$$\frac{C}{R} = \frac{(\frac{\hbar\omega}{k_B T})^2 \, e^{\frac{\hbar\omega}{k_B T}}}{(e^{\frac{\hbar\omega}{k_B T}} - 1)^2} \tag{2.7}$$

To calculate the total intra-cluster specific heat the contributions of all frequencies have to be summed.

The inter-cluster specific heat is calculated using the well known Debye formula:

$$\frac{C}{R} = 9 \left(\frac{T}{\theta_D}\right)^3 \int_0^{\theta_D/T} \frac{x^4 e^x}{(e^x - 1)^2} \, dx \tag{2.8}$$

where the parameters in the Debye temperature θ_D (the mass and springconstant) are to be related to the cluster as a whole.

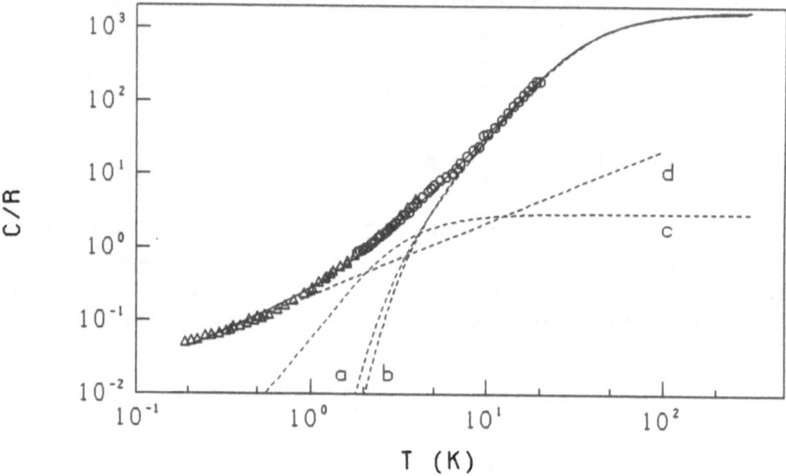

Fig. 1. Experimental specific heat data on the $Pd_{561}Phen_{36}O_{200}$ compound. The circles represent the data on batch 1, the triangles the data on batch 2. For an explanation of the solid and dashed lines see text.

Lastly, a possible electronic contribution to the specific heat should be considered, which for a degenerate electron gas would be a linear term $C_{el} \propto \gamma T$. The coefficient γ is directly related to the density of states at the Fermi level and is large ($\gamma = 9.42$ mJ mol^{-1} K^{-2}) for bulk Pd. This linear relation will break down when the energy splittings around E_F become larger than $k_B T$.

3 Results and discussion.

Figure 1 shows the experimental specific heat results on a double logarithmic scale. The dashed lines a and b represent the contributions to the specific heat due to intra-cluster vibrations, as calculated by the MD- and ECA methods respectively. Dashed line c represents the contribution of the inter-cluster vibrations. The specific heat above 1 K is well described by the sum of inter- and intra-cluster vibrations (resp. c and a,b in Fig.1). The fitted inter- cluster Debye temperature was 16 K. The intra-cluster vibrational frequencies were calculated using both the MD and ECA method. The calculated densities of vibrational frequencies are shown in fig. 2. Although the densities differ in details, the results of both methods for the specific heat are in good agreement. The adjustable parameters in the calculations, i.e. the transverse sound velocity for the ECA method and the force constant for the MD method, were taken to be 1.63 km/s and 19 N/m respectively. These values are below the values of bulk palladium (1.99 km/s and 34 N/m resp.). This "phonon softening" was also observed for the specific heat measurements on small Pb and In particles mentioned above.

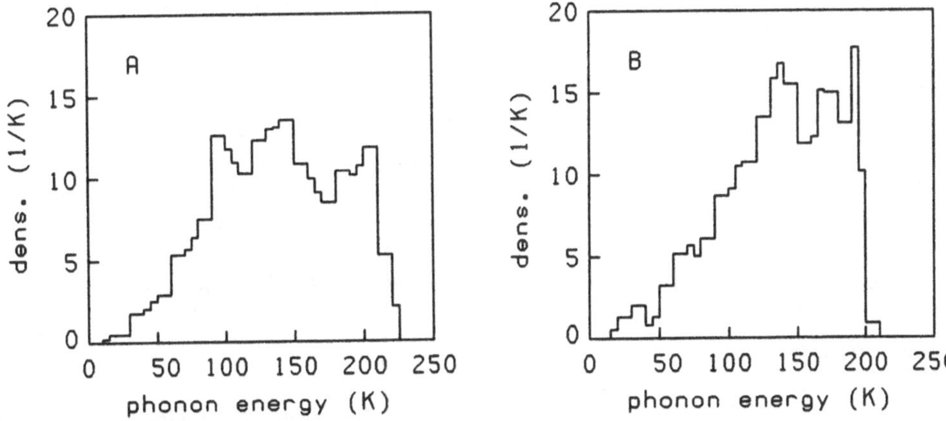

Fig. 2. Density of intra-cluster vibrational frequencies as calculated via the MD (fig. 2a) and the ECA (fig. 2b) methods.

Maybe the most important result is the observation that these two contributions are insufficient to describe the specific heat at the lowest temperatures. In order to fit the experimental data over the whole temperature range, a linear term ($C/R = 0.22$ T) was added to the calculated specific heat (dashed line d). The solid line in Fig. 1 represents then the sum of the contributions of intra- and inter-cluster vibrations and the linear contribution. The coefficient of the linear term amounts to approximately one third of the value of bulk Palladium ($C/R = (561 \cdot 9.4 \; 10^{-3}/R)$ T $= 0.64$ T). Such a reduction of the bulk value has to be expected. Only the inner-shells (309 atoms) might show metallic behaviour, the outer (surface) atoms being chemically bound by ligand molecules. Furthermore, the density of states near the Fermi energy of a metal particle as small as a few hundred atoms will always be less than the bulk value. Moreover the γ-value for bulk Pd is known to be enhanced, and it is not sure if this enhancement will remain unchanged for very small particles. These results compare well with data on the magnetic susceptibility of Pd_{561} [8], which show a (Stoner-enhanced) Pauli paramagnetic contribution corresponding to a similar fraction of the bulk value. Recently D. v.d. Putten et al., based on ^{195}Pt-Knight shift and relaxation data in the $Pt_{309}Phen_{36}O_{30}$ compound, also concluded that about half of the core atoms has a metallic character[9]. Specific heat measurements on this compound are presently underway.

It is of interest to note that at the lowest temperatures reached in the experiment (0.2 K), the thermal energy is already much lower than the estimated energy level splittings. For a large metal cluster the spacing would correspond roughly to $2E_F/N$, if degeneracy is discarded. Here E_F is the highest occupied orbital and N the number of (s and d) valence electrons in the cluster. Taking the Fermi energy to be a few eV and N = 561 x 10 = 5610, one has a level spacing of about 2 meV, or $\pm$ 20 K. This implies that level broadening of this order should be present to explain our experimental results. Several mechanisms for broadening can be found in the literature. Here we wish to point out that a weak inter-cluster electron transfer integral of 1 meV would be sufficient to produce a quasi-continuous spectrum. Another factor of importance is the high symmetry of the fcc-packing of the core, which will modify the equal level spacing, assumed above.

Acknowledgements.

This work is part of the research program of the Leiden Materials Science Centre (Werkgroep Fundamenteel Materialen Onderzoek). We would like to thank J.C.J.M. de Rooy and D.A. v.d. Straat for help with the measurements, D. Reefman for useful discussion and advice about numerical difficulties and M.W. Dirken for help with the MD calculations. In the first stage of this project we have largely profited from the contributions of P.R. Nugteren.

References

[1] N.N. Greenwood and E.J.F. Ross, "Index of Vibrational Spectra of Organic and Inorganic Metallic Compounds", Butterworth 1975, London.

[2] H.H.A. Smit, P.R. Nugteren, R.C. Thiel, L.J. de Jongh, Physica B 153 (1988) 33.

[3] V. Novotny, P.P.M. Meincke, Phys. Rev. B 8 (1973) 4186.

[4] N. Nishiguchi, T. Sakuma, Solid State Commun. 38 (1981) 1073.

[5] G. Goll, H. v. Löhneysen, U. Kreibig, G. Schmid, Z. Phys. D 20 (1991) 329.

[6] G. Schmid, Polyhedron 7 (1988) 2331.

[7] A. Tamura, K. Higeta, T. Ichinokawa, J. Phys. C 15 (1982) 4975.

[8] J.M. van Ruitenbeek, M.J.G.M. Jurgens, G. Schmid, D.A. van Leeuwen, H.W. Zandbergen and L.J. de Jongh, Z. Phys. D 20 (1991) 267.

[9] D. van der Putten, H.B. Brom, L.J. de Jongh, G. Schmid, this conference.

Vibrational States of Al Adsorbed on Ge Clusters

W.S. Bacsa and J.S. Lannin
Department of Physics,
Penn State University,
University Park, PA, 16802
U.S.A.

Interference enhanced Raman spectra of Al adsorbed on Ge clusters reveal an Al induced vibrational band. The resulting effective force constant implies a significantly weaker interaction than that of the similar reduced mass Si-Ge system. Consistent XPS core line shifts of Al and Ge (0.6eV) suggest the formation of a weak covalent bond between the metal adsorbate and semiconductor clusters.

1. Introduction

Chemisorption effects of H and O atoms on the dynamics of Ge clusters and ultrathin films have been recently observed by interference enhanced, in situ Raman spectroscopy [1]. It has been found that energy shifts of the optic phonon band of Ge, which are dependant on the cluster size, are reduced when H or O are chemisorbed. This has been attributed to the removal of dangling bonds on the surface of the semiconductor cluster upon adsorption of H and O. We report here on how the vibrational properties of Ge clusters are modified by adsorbtion of Al metal.

Metal-semiconductor interfaces have been studied for the past several decades because of their important applications and also because they are of fundamental interest. However, little is known about the interaction of metals and semiconductors right at the interface or in the monolayer range. The Al-Ge system is a prototype of such a metal - covalent semiconductor interface [2]. The effect of Al adsorption on the dynamics of Ge clusters is investigated by interference enhanced Raman spectroscopy which allows the observation of vibrational properties of semiconductors in the monolayer range. Simultaneously, photoemission spectroscopy (XPS) is used to study the atomic core lines and their chemical shifts with Al adsorption. The chemical shifts reflect the changes in the atomic charge distribution and therefore the atomic bonding between Al and Ge.

2. Experimental

Al of high purity (99.9999%) was evaporated from a BN crucible, heated by a tungsten filament. The evaporation rate was set to 2A/min and was calibrated with a quartz crystal thickness monitor. The pressure gauge, situated 15cm from the evaporating beam, indicated a pressure of

P. Jena et al. (eds.), Physics and Chemistry of Finite Systems: From Clusters to Crystals, Vol. I, 345–349.

2-4x10⁻⁸torr during deposition. It was found that lower evaporation rates (1A/min) give rise to a distinct XPS oxide signal despite a low background pressure (2-5x10⁻¹⁰torr).

Amorphous (a-) Ge of approximately 6A thickness was sputtered at room temperature with a DC magnetron onto a sputtered, disordered (d-) C substrate. This thickness forms a dense array of clusters of approximately 20A diameter [3]. For the Raman measurements, the d-C layer was the top layer of a three layer structure (d-C/SiOx/Al) needed to obtain interference enhancement. The enhancement due to the use of the trilayer is estimated to be one order of magnitute. Al and Ge were deposited in the preparation chamber of the UHV system. The d-C substrate was sputtered in the load lock of the same UHV system and cleaned by an ion source for 1-2 minutes.

The UHV system was such that it was possible to obtain XPS and Raman measurements in the same analysis chamber. Unpolarized Raman spectra were taken at 295K and at 514.5nm excitation with a Spex Triplemate (resolution: 6cm⁻¹) equipped with an ITT Mepsicron multichannel detector. For the XPS measurements, the Al K-α line of 1.1eV resolution was used. The accuracy of the HA100 electron analyser of Microscience Inc. was approximately 0.05eV. The large cross section of oxygen compared to that of Al made the measurements very sensitive to oxygen contamination. All the spectra have been smoothed over 25cm⁻¹ or 0.6eV for the Raman and XPS data, respectively.

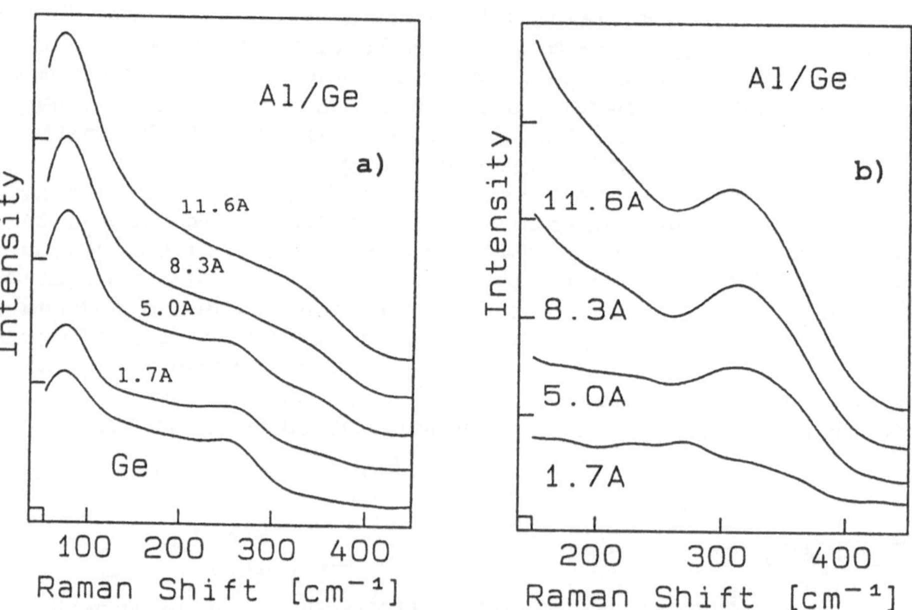

Figure 1. Interference enhanced Raman spectra of Ge clusters followed by Al depositions: 1.a) after subtraction of d-C background spectrum. 1.b) after subtraction of the spectrum due to the Ge clusters.

3. Results and Discussion

Figure 1.a) shows Raman spectra of 6A Ge deposited on d-C followed by Al evaporations. The d-C background from the substrate varies only slowly in this energy interval and has been subtracted. Apart from the phonon spectrum of 6A Ge in the 50-300 cm^{-1} range [4], a new phonon band appears at about 325 cm^{-1}. The Al induced band becomes stronger in the 2-8A range and merges with the Ge spectrum at higher Al thicknesses.

Figure 1.b) shows the same spectra after subtracting the pure Ge spectrum. The effect of the Al adsorption is more apparent. Due to the overlap of the Al induced band with the Ge spectrum the Al induced band shifts by approximately $10cm^{-1}$ to lower energies after subtraction of the Ge spectrum. As more Al is deposited, a small shift of $6cm^{-1}$ is observed, which reduces to $4cm^{-1}$ if the non-constant background is taken into account. This shift to lower energies is possibly due to reduced interaction of Ge with Al at higher Al coverages. However, it is noted that the shift is small compared to the broad Al induced band (halfwidth $70\text{-}90cm^{-1}$) and therefore, it is not a major effect. In Fig. 1.b) it is seen that the Ge spectrum with lowest Al coverage (1.7A) cannot be completely subtracted. This indicates that the Ge spectrum changes significantly in this thickness range [5]. At higher Al coverages (5-12A) the subtracted spectra do not show features characteristic of the Ge spectrum.

In order to test whether the Al induced band is due to an Al-Ge interaction, analogous measurements were carried out without depositing Ge. After evaporating Al on a d-C substrate, the Raman spectra showed no Al induced phonon band, while the elastic scattering at lower energy increased as more and more Al was deposited. The different behavior of Al on d-C compared to Al on Ge clusters indicates that that the Al induced band is due to interaction of Al with Ge. It is assumed here that possible variations of the Al morphology do not contribute to changes in the Raman spectra. When the Al covered d-C was exposed to 1atm oxygen for several minutes, the elastic scattering disappeared. In the case of Al adsorbed Ge clusters exposed to oxygen, the elastic background similarly disappeared; the Al induced phonon band persisted but tended to be weaker in intensity. The steep increase at lower energies in Figs. 1.a) and 1.b) are therefore attributed to appreciable elastic scattering.

Si has atomic mass similar to the atomic mass of Al. It is therefore interesting to compare the energy of the Si-Ge optical phonon band in crystalline and amorphous Si_xGe_{1-x} alloys ($407cm^{-1}$) [6-8] with the Al induced vibrational band in Fig.1 ($325cm^{-1}$). The substantially smaller energy of the Al induced band indicates that the Al is less strongly bonded to Ge, with an effective force constant approximately 40% smaller than Si is bound to Ge in Si_xGe_{1-x} alloys.

Figure 2 displays the line shapes and energy positions of the XPS core lines of Ge and Al. As Al is deposited on the Ge clusters, the core line 3d signal of Ge shifts to lower binding energies up to 0.6 eV from its bulk value. At the same time the Al 2p core line shifts also to lower binding energies by 0.5 eV towards its bulk value. Since the Ge core line appears first at its bulk value and the Al core line reaches the bulk value after the last deposition, the two core line shifts are opposite to each other. This is consistent with the formation of an Al-Ge bond. No chemical shift has been found in the the core line of the substrate (C 1s) as Al was deposited.

The Ge 3d core line starts to shift at low Al coverages. However, the shift ceases only after a rather large Al coverage (8A). The Al 2p core line does not shift sizably at lower coverages (1-4A) but finally shifts to its bulk value at about the same coverage (8A). Keeping in mind that the Ge forms an array of clusters on the d-C substrate, the surface area of the Ge clusters is clearly larger than the substrate surface. The surface area is, for example, twice as large if it is assumed that the clusters form uniform half spheres. In this case the Ge clusters would be covered by approximately 4A Al (3 monolayers) in order to shift the core lines by 0.5-0.6eV. Although the detailed morphology of the Ge clusters is not known, this estimate may explain the large Al depositions needed to shift the Ge and Al core lines. When the cluster size was reduced by sputtering half the amount of Ge (3A), it was observed, that the Al and Ge core lines shifted by 0.5-0.6eV at a smaller (6A) Al coverage.

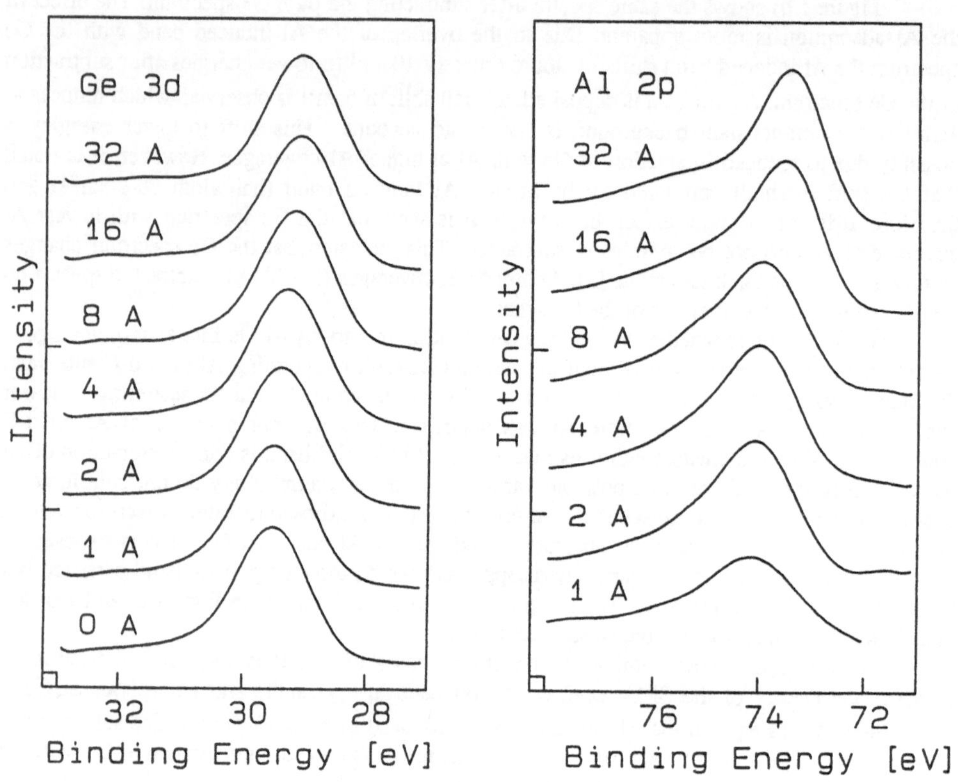

Figure 2 .   XPS spectra of Ge 3d and Al 2p core lines after Al depositon.

The widths and asymmetry of the Ge 3d core line signals are not changed with Al adsorption. However the Al 2p core line decreases in width and becomes more symmetric with increasing Al coverage. Since weak oxide core line signals were observed after Al was deposited, it is expected that the Al 2p core line shows a 2eV shifted component due to Al_2O_3 formation. It was found that the contribution of the oxide is larger in the case of lower growth rates. At a growth rate of 1A/min the component due to oxide formation dominated the core line signal at lowest coverage (1A). In order to reduce the influence of oxide formation on the Al 2p signal we have chosen a higher growth rate (2A/s). We therefore attribute the small changes of the line shape of the Al 2p core line signal in Fig. 2 as due to small contributions of Al_2O_3.

4. Conclusion

The appearance of a new vibrational band upon Al adsorption on Ge clusters has been observed with interference enhanced Raman scattering. This represents the first observation of metal-semiconductor vibration on a cluster surface. A comparison with the Si-Ge system shows that the effective force constant is 40% weaker in the case of Al adsorbed Ge clusters. It is found that oxidation of the evaporated Al reduces elastic scattering at low energies of the Raman spectra, but does not remove the Al induced phonon band. XPS photoelectron spectroscopy reveal consistent chemical shifts of the Ge 3d and Al 2p core lines which clearly indicate the formation of a bond between Ge and Al. The chemical shift of 0.5-0.6eV measured after evaporation of 8A Al on 6A sputtered Ge clusters is small. This indicates that the covalent interaction of the Ge cluster surface atoms and Al is weak and is consistent with the observed energy of the Al induced vibrational band.

5. Acknowledgement

The help obtained from S. Chase and R. Yu in performing the XPS measurements and M. Mitch with the UHV system is gratefully acknowledged. This work was supported by the National Science Foundation - Grant DMR-8922305-1.

References

1.   J. Fortner and J.S. Lannin (1991) 'Chemisorption effects on the dynamics of Ge clusters and ultrathin films', Surface Science, Volume 254, 251-60.
2.   A. Zangwill (1988) Physics at Surfaces, Cambridge University Press, p. 315.
3.   B.Y. Yang, T. Gu, R.Q. Yu, J.S. Lannin and R.W. Collins (1990) 'In situ ellipsometry of Ge clusters and ultrathin films', MRS Symposium Proceedings Volume 206, p. 417.
4.   J. Fortner, R.Q. Yu and J.S. Lannin (1990) 'Near-surface Raman scattering in germanium clusters and ultrathin amorphous films', Physical Review B 42, 7610-7613.
5.   W. Bacsa and J.S. Lannin, to be published.
6.   W.J. Brya (1973) 'Raman Scattering in Si-Ge alloys', Solid State Communication Volume 12, 947.
7.   J.S. Lannin (1977) ' Vibrational and Raman scattering properties of crystalline Si-Ge alloys', Physical Review B 16, 1510.
8.   J.S. Lannin (1974) 'Raman scattering in amorphous Ge-Si alloys', Proceedings of the 5th International Conference of Amorphous and Liquid Semiconductors, ed J. Stuke and W. Brenig, Taylor and Francis, p.1245.

FROM CLUSTERS TO LIQUIDS: DIFFUSION, STOKES-EINSTEIN BEHAVIOR, AND SOLVATION IN MIXED MOLECULE-RARE GAS CLUSTERS

THOMAS L. BECK, J. R. WALKER, AND T. L. MARCHIORO II
Department of Chemistry
University of Cincinnati
Cincinnati, OH 45221

ABSTRACT. Liquid-like behavior in small atomic and molecular clusters is analyzed from the standpoint of molecular hydrodynamic theory. A method for the computation of friction constants in clusters is presented which utilizes the Generalized Langevin Equation-Memory Function approach to liquid dynamics. The friction constant is determined by integrating the time correlation function for forces on a heavy mass Brownian particle inserted into the cluster. Preliminary molecular dynamics studies give relatively good agreement between the approximate friction constants obtained by this theory and exact results obtained from the Stokes-Einstein formula, for clusters larger than roughly $N = 30$ atoms. Molecular dynamics studies are also presented of structure, dynamics, and solvation in small benzene-argon complexes, systems of recent experimental interest.

1. Introduction

In the last five years, experimenters have begun to examine the possibility of observing melting and diffusion in small clusters[1]. One approach has been to study the spectra (IR or electronic) of a chromophore molecule imbedded in or placed on a small rare gas cluster. The experiments have given evidence for diffusive behavior, wetting-nonwetting transitions, surface-bulk effects, structural types for small clusters, and temperature effects indicative of melting. We do not review the extensive and growing literature concerning these experiments here, but simply raise some of the interesting theoretical questions motivated by the experiments.

First, the placement of a spectroscopically active probe molecule into the cluster, depending on the structure of the molecule and the strength of its interactions with the rare gas atoms, can severely perturb the low temperature geometries of the cluster. This behavior has been clearly elucidated in a comprehensive study of structural patterns in SF_6-argon complexes[2]. Second, different molecules may prefer to reside at the cluster surface as opposed to becoming fully imbedded or solvated; in addition, solvation may depend strongly on cluster size, temperature, and/or cluster phase[3]. Third, the observation of solvation and diffusion in small clusters presents basic theoretical challenges due to finite system effects. For example, to what extent can we speak of a cluster of $N = 10 - 50$ atoms as a liquid-like droplet which obeys macroscopically derived equations? Are diffusion constants even well-defined, and if so does the Stokes-Einstein equation apply in

351

P. Jena et al. (eds.), Physics and Chemistry of Finite Systems: From Clusters to Crystals, Vol. I, 351–356.
© 1992 *Kluwer Academic Publishers.*

the case of small systems? What is the temperature range and what are the time scales for stability of small liquid droplets?

In the present note, we present preliminary results addressing two areas discussed above: the statistical mechanics of diffusion in small clusters and molecular dynamics of benzene in argon complexes. We first present the theoretical methodology employed in this work. Then we present results pertaining to the applicability of molecular hydrodynamic theory in analyzing cluster diffusion. Finally, we discuss molecular dynamics studies of benzene-argon complexes where we have analyzed in detail cluster structural and dynamical transitions for several cluster sizes.

2. Theoretical Methods

The studies discussed below utilize molecular dynamics simulation. Details of the MD methodology are presented elsewhere[4]. In short, we prepare equilibrated clusters at a range of energies or temperatures and solve the classical equations of motion numerically via the Verlet algorithm. Simple Lennard-Jones potentials are assumed for argon-argon interactions, with parameters taken from previous work. The time step is taken as 10^{-14} sec. Recent extensive MD simulations have shown that there may be relatively sharp onset of transport in small clusters (especially for "magic number" sizes); however, diffusion constants are only defined over time scales appropriate for migration across the extent of the cluster[4]. Nevertheless, these time scales are on the order of 40 collision times for a cluster as small as Ar_7.

When the mass of a solute in a liquid approaches that of the solvent, as is the case for argon liquid or benzene in argon, the friction constant is not local in time, as assumed in the standard Langevin equation. An equation of motion for the velocity autocorrelation function or VAF ($C(t)$) of atoms in the liquid can be derived based on the Generalized Langevin Equation:

$$\frac{dC(t)}{dt} = -\int_0^t dt'\, M(t-t')\, C(t') \ . \tag{1}$$

$M(t-t')$ is termed the memory function. The friction constant is then given exactly by:

$$\zeta = m \int_0^\infty M(t)\, dt \ . \tag{2}$$

Hoheisel and coworkers[5] have found that a good approximation to the memory function for liquid argon is obtained by making one of the atoms in the liquid on the order of 50 - 100 times the mass of argon and computing the time correlation function of total forces on that particle (FAF). The *plateau* value of the integral (the infinite time integral goes to zero since the particle momenta are bounded) then gives the friction constant:

$$\zeta = \frac{1}{3kT} \int_0^{t_2} \langle\, F(t)\, F(0)\, \rangle\, dt \ . \tag{3}$$

Vogelsang and Hoheisel[5] found that this approximation gives very good agreement with exact values of the friction constant for liquid argon calculated via the Stokes-Einstein relation. Our work is aimed at carrying out a similar analysis for clusters to determine if and when such a hydrodynamic description might be applicable.

MD simulations of a molecule like benzene are somewhat problematic due to the planar structure. However, two approaches for carrying out MD calculations on such systems have been developed: quaternion dynamics and the generalized SHAKE algorithm[6]. We have employed the second approach in our work.

The benzene molecule is treated as a rigid object interacting with the argons via 12 Lennard-Jones interactions (one for each C and H of benzene). The site-site approach has been taken in studies of other rare gas-aromatic molecule systems[7]. Careful MD simulations were performed for a variety of initial configurations. We note that we are solving the classical equations of motion given our model for benzene; hence, we obtain essentially exact dynamical information within the confines of the model. Benzene is free to move, and in fact does move a great deal at the higher energies.

Two other recent studies have examined benzene-argon complexes: Adams and Stratt[8] carried out Monte Carlo simulations of the system and used a local normal mode approximation to probe dynamical transitions, while Fried and Mukamel[9] have employed quaternion methods for the dynamics and have also developed a semiclassical theory for the electronic lineshape of the benzene molecule.

3. Results

We have carried out a series of molecular dynamics studies on various cluster sizes ($N = 7$-55), and for masses of the Brownian solute particle ranging from equal mass to masses 100 times that of argon. In this section, we discuss initial results for the Ar_{33} cluster. The cluster temperature is 37 K, a value at which the cluster is liquid-like but at which evaporation events occur infrequently on the 10 ns time scale.

Integrals of the time correlation function of forces on the heavy mass particle are presented in Fig. 1. Three integrals are shown; curve b) represents the average integral computed from a long time trajectory (21 ns). While the curve exhibits no clear, flat plateau, it rises sharply, displays a relatively flat region, and then begins its decay towards zero. If the maximum value of the integral is taken (and multiplied by appropriate normalization constants) a value of $\zeta = 5.5 \times 10^{-13}$ kg / sec is obtained for the friction constant. If the exact diffusion constant is computed from molecular dynamics and the friction constant calculated via the Stokes-Einstein formula, a value of $\zeta = 9.3 \times 10^{-13}$ kg / sec is obtained. Hence, the memory function approach gives a value roughly 40% too low for the averaged friction constant. The quality of agreement is not as good as that obtained for liquid argon. However, we consider the agreement surprisingly good for a system of only 33 atoms.

The lack of a clear plateau value for the integral may result from several causes: 1) The system may be too small. 2) Structural ordering in the finite system may alter the nature of the force correlations. 3) There may exist surface-bulk distinctions in the cluster which differ markedly from liquid-argon (in the 33 atom cluster, on average 20 atoms are on the surface). To investigate these issues, we separated the trajectory into two bins based on the location of the heavy mass particle. This was done by monitoring the average number of

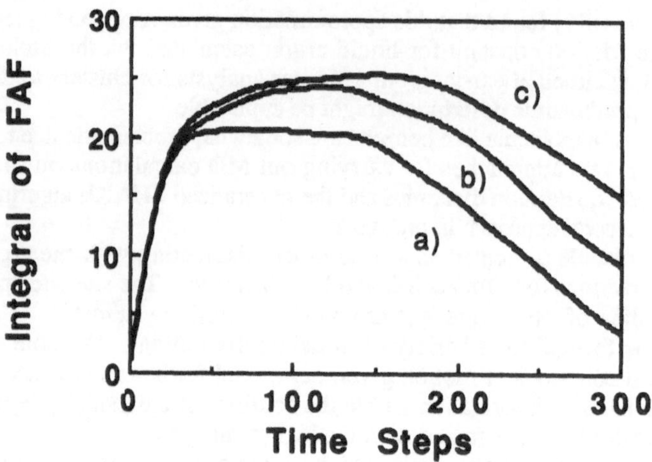

Figure 1. Time integral of the normalized force-force correlation function. a) Inner region b) average over full trajectory c) surface region.

nearest neighbors during a 300 time step interval. The integrals of the FAF are presented as curves a) (bulk) and c) (surface) of Figure 1. Clearly, the integral obtains a plateau value when the Brownian particle is located within the cluster, and the non-plateau behavior results from surface effects. The friction constants obtained are 8.2×10^{-13} kg / sec and 4.4×10^{-13} kg/sec for the bulk and surface cases, respectively. Whether these results are reflective of behavior expected of larger systems will be the subject of future study.

We now give preliminary results of our MD simulations of benzene-argon complexes. A range of cluster sizes has been studied ($N = 1 - 25$ argon atoms). Initial structures of the small complexes were prepared by addition of atoms to one or both sides of the benzene molecule. The system was then heated until large mobility was observed and cooled again to lower temperatures. Periodic steepest descent quenches were carried out along the trajectories as a "pattern recognition tool" to examine the structural bonding motifs in the clusters. Recent two photon ionization and UV absorption experiments suggest that experimental determination of low temperature structures is now possible[10,11]. The larger clusters ($N = 16$ and 25) were prepared by addition of the benzene molecule to a liquid-like cluster, followed by heating and cooling cycles. Initial states were prepared both with the benzene inside and on the surface.

We present first several low energy structures which dominate the structural patterns for small clusters (Figure 2). The lowest energy isomer for $N = 1$ places the argon on the center of benzene. The binding energy is 520 K in our model. The lowest energy structure for $N = 2$ is for the symmetric isomer--one argon on each side of benzene. The other isomer has two argons on one side of the benzene, one at the center and one displaced. The binding energies are 1048 K and 968 K, respectively. The $N = 3$ and $N = 4$ lowest energy isomers are both one-sided structures, and both depart from a "building up" pattern where the initial atom is placed at the center. These structures correspond to an equilateral triangle and a diamond-shaped cluster, respectively. However, near-lying isomers are easily accessible and argon mobility is observed at relatively low temperatures (10 K). Starting with N = 5, the most stable structures are comprised of incompletely filled "shells" of the N = 7 hexagonal structure, with one atom at the center of benzene. Intermediate sized clusters (such as N = 16) display a preponderance of "clamp" structures, with the benzene partially imbedded in the cluster, and argon atoms at the side of benzene. We have found an interesting structure for N = 25, where the argons provide a low energy "closed shell" around the benzene. Interestingly, while these structures are energetically favored, entropic effects can drive the system to other substantially less stable configurations at finite temperatures.

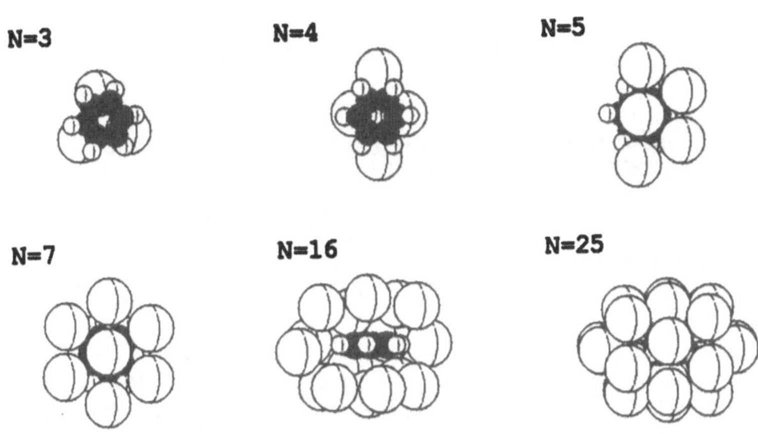

Figure 2. Structures of benzene-argon complexes determined by steepest-descent quenching along MD trajectories.

In addition to the structural information, various finite temperature quantities have been computed at a range of temperatures: bond length fluctuations (argon-argon and argon-benzene), power spectra of the VAF, mean square displacements, and principal moments of inertia. The instantaneous principal moments of inertia give a clear indication of the structural type for small N. As an example, a trajectory was run for N = 3 at an energy where a side flip occurred. The transition induced a substantial change in the three principal moments, implying different rotational band structures. For larger clusters (N = 5 - 9), various transitions were observed at 30 K, including side flips, atomic rearrangements in the plane above the benzene, and transitions where one atom hops from motion in the plane above benzene to a second argon layer.

4. Summary

Results have been presented concerning diffusion and solvation in clusters. The memory function approach implies that small clusters exhibit behavior remarkably like much larger systems in terms of a microscopic definition of the friction constant. Plateau behavior in the time integral of the force autocorrelation function is indicative of bulk hydrodynamic behavior; apparently this may occur in clusters when the Brownian particle is surrounded by one layer of "solvent" atoms. Also, the MD studies discussed here of solvation in benzene-argon clusters reveal a rich array of structural patterns and dynamical transitions. Extensions of this work will be presented in forthcoming papers.

5.   Acknowledgments

We acknowledge support from the donors of the Petroleum Research Fund, administered by the ACS, the Department of Chemistry of the University of Cincinnati, and a grant of computing time at the Ohio Supercomputer Center.

6.   References

[1]See, for example, X. J. Gu, D. J. Levandier, B. Zhang, G. Scoles, and D. Zhuang, J. Chem. Phys. **93**, 4898 (1990); S. Leutwyler and J. Bosiger, Chem. Rev. **90**, 489 (1990); M. Hahn and R. L. Whetten, Phys. Rev. Lett. **61**, 1190 (1988).
[2]D. J. Chartrand, J. C. Shelley, and R. J. LeRoy, J. Phys. Chem. (in press).
[3]L. Perera and F. G. Amar, J. Chem. Phys. **93**, 4884 (1990).
[4]T. L. Beck and T. L. Marchioro II, J. Chem. Phys. **93**, 1347 (1990); Phys. Rev. A **42**, 5019 (1990).
[5]C. Hoheisel, Comp. Phys. Repts. **12**, 29 (1990), and references therein.
[6]G. Ciccotti and J. P. Ryckaert, Comp. Phys. Rpts. **4**, 345 (1986).
[7]J. Bosiger, R. Bombach, and S. Leutwyler, J. Chem. Phys. **94**, 5098 (1991).
[8]J. E. Adams and R. M. Stratt, J. Chem. Phys. **93**, 1358 (1990).
[9]L. E. Fried and S. Mukamel, J. Chem. Phys. (in press).
[10]M. Schmidt, M. Mons, and J. L. LeCalve, Chem. Phys. Lett. **177**, 371 (1991).
[11]T. Weber, E. Riedle, and H. J. Neusser, Z. Phys. D **20**, 43 (1991).

PROTON SPIN RELAXATION AND THERMAL HISTORY EFFECTS IN ORGANIC MOLECULAR SOLIDS

PETER BECKMANN
Department of Physics
Bryn Mawr College
Bryn Mawr
Pennsylvania 19010, USA.

ABSTRACT. We present a preliminary report on a study of
the effect of thermal history on the solid state proton
spin relaxation (SSPSR) rate in organic molecular solids.
In the present case, the SSPSR technique uses the reorientation of methyl groups as a dynamic probe of the local
environment. We can learn about the effect of the local
electrostatic anisotropies as the state of the solid
changes with time and temperature. The solid used in this
study is 1,3,5-tri-t-butylbenzene (TTB). It is a highly
symmetric molecule and it serves as a model system for a
Van der Waals solid. The SSPSR rates are extremely sensitive to local and long-range structure (via the distribution of local structure) and the goal is to correlate the
observed SSPSR rate with the structure at the nearest-
neighbor-molecule level, the several-molecule level and the
macroscopic state of the solid. We are able to prepare TTB
in a reproducible manner in a variety of as yet unknown
states, presumably ranging from various crystalline states
to various amorphous states, by carefully controlling the
thermal history.

1. Introduction

We present an initial study of the use of the solid state
proton spin relaxation (SSPSR) technique to investigate the
states of organic solids. The technique exploits the dependence of the reorientational properties of intramolecular rotors like methyl (CH_3) groups on the local electrostatic environment. The observed relaxation rates are very
sensitive to the local structure of the solid. One can
also learn about distributions of single molecule properties throughout the macroscopic sample.

P. Jena et al. (eds.), Physics and Chemistry of Finite Systems: From Clusters to Crystals, Vol. I, 357–362.
© 1992 *Kluwer Academic Publishers.*

2. The Molecule

The molecule is 1,3,5-tri-*t*-butylbenzene (TTB). It has
three *t*-butyl groups, each of which has three methyl groups
as shown in references [1] and [2]. It is a highly symmetric organic molecule and it serves as a model system.
Because of the symmetry, the three *t*-butyl groups are
dynamically equivalent in the isolated molecule and the
three methyl groups in each *t*-butyl group are also dynamically equivalent. One can think of the molecule as a
system of nine rotors which is sensing the local electrostatic environment.

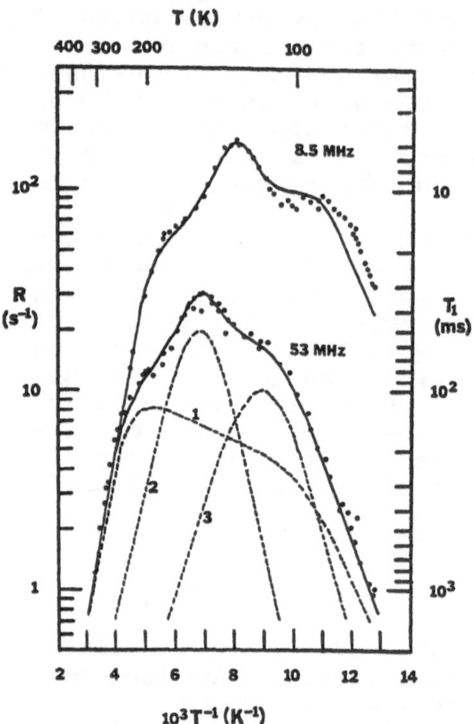

Figure 1. Proton spin zeeman relaxation rate R versus
temperature T at larmor frequencies of 8.5 and 53 MHz. The
fits are discussed in reference [1]. The decomposition of
the 53 MHz fit shows the contributions to R from the three
sets of rotors whose activation energies are shown in
figure 2. Number 1 corresponds to the broad 40% background
and numbers 2 and 3 correspond to the 40% and 20% spikes
respectively.

3. The Structure of the Solid

The perfect crystal structure is a characteristic inter-
locking herring-bone structure [3] which leads to two *t*-
butyl groups in each molecule seeing the same environment
and the third seeing a different environment. As suggested
below, there seems to be a variety of non-crystalline
phases, or, perhaps the SSPSR experiments are just *very*
sensitive to very subtle differences in structure which
result from different thermal histories. TTB melts at 343
K and has been appropriately purified.

4. The SSPSR Technique

The nuclear (zeeman) magnetization associated with the many
protons is perturbed and its recovery to equilibrium is
monitored [4]. The spin-spin dipolar interaction is
modulated by the rotation of the *t*-butyl groups and their
constituent methyl groups. The resulting time dependence
of the local dipolar field results in stimulated emission
and the spin system relaxes. (There is no spontaneous
emission at radio frequencies.) The rate of recovery is a
measure of a fourier component of the local fields at the
larmor frequency and, via appropriate models [4], this
information is translated into dynamical information about
the rotors.

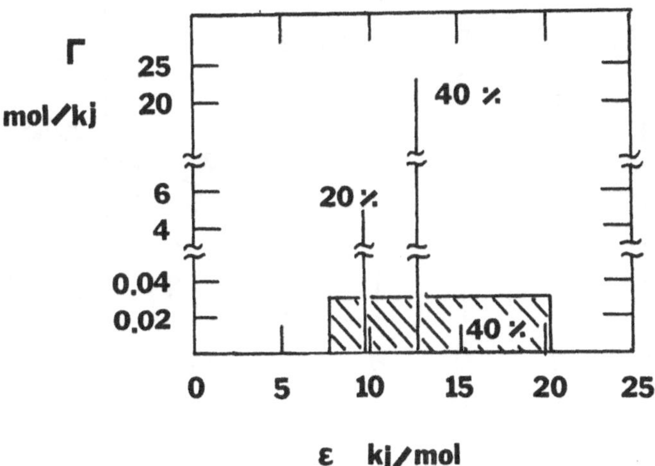

Figure 2. The distribution $\Gamma(\varepsilon)$ of *t*-butyl and methyl
reorientation activation energies ε corresponding to the
fit of the temperature dependence of the relaxation rate
shown in figure 1.

5. Previous experiments with TTB

The temperature T and larmor frequency f dependence of the relaxation rate R was investigated previously [1]. Figure 1 is taken directly from that work. In this study [1], the sample was always prepared with the same thermal history. The distribution of activation energies (electrostatic potentials) for methyl and *t*-butyl reorientation is shown in figure 2. The two-to-one ratio in the number of rotors found in the x-ray study [3] is seen. However, there is also a broad "background" of activation energies for 40% of the molecules. There is a hint of this in the x-ray study.

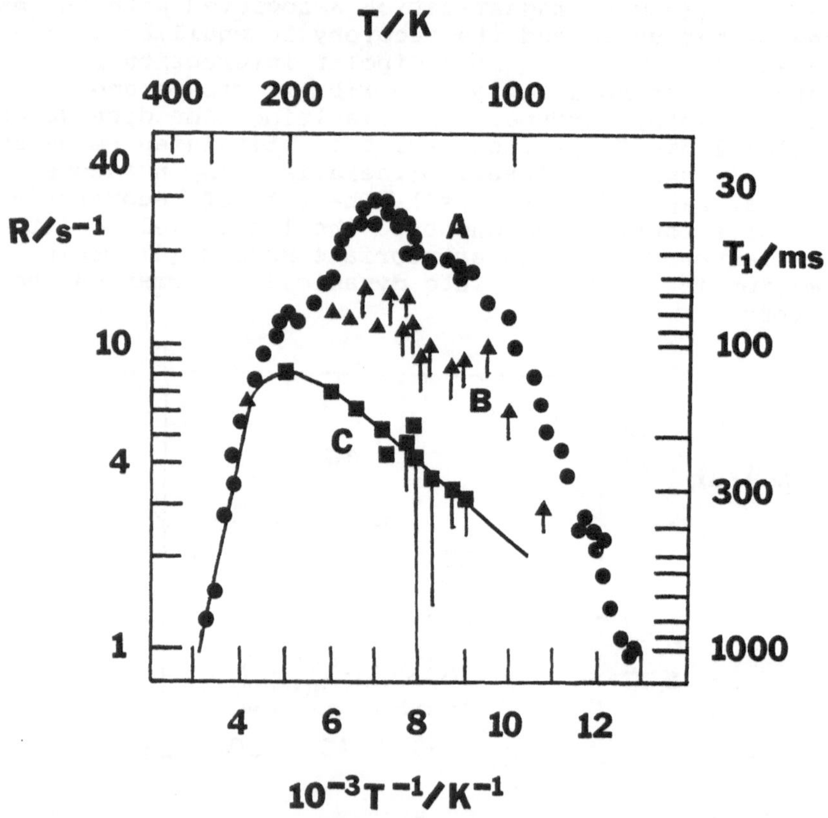

Figure 3. Proton spin zeeman relaxation rate R versus temperature T at a larmor frequency of 53 MHz. The states A, B and C result from different thermal preparations. The fit to state C is discussed in the text.

6. New Experiments with TTB

The f = 53 MHz data in figure 1 is reproduced in figure 3
and labelled state A. By performing different thermal
preparations of the solid, the T dependences of R labelled
B and C in figure 3 can be obtained. One can suggest with
some confidence that with appropriate preparation, any
"state" between A and C (and more?) can be obtained and
these letters are only used for convenient administrative
purposes.

7. Activation Energies

The states corresponding to lower R values in figure 3 are
characterized by larger distributions of activation energies. Figure 4 shows the distribution for state C. (See
references [5] and [6] for the fitting procedure.) The
cut-off energy of 14.2 kJ/mole in figure 4 (ε/E = 1) can be
compared with the energies in figure 2. The 40% spike in
figure 2 is at 12.7 kJ/mole.

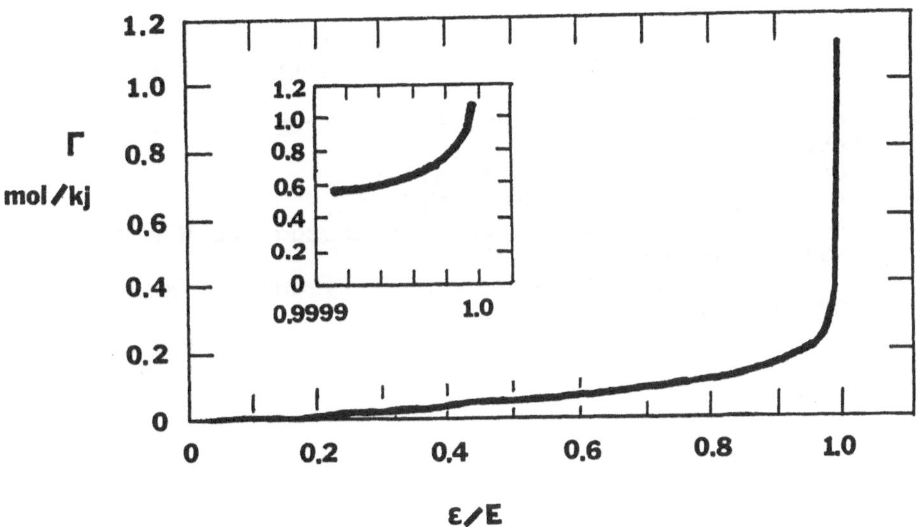

Figure 4. The distribution $\Gamma(\varepsilon)$ of t-butyl and methyl
reorientation activation energies ε corresponding to the
fit of the temperature dependence of the relaxation rate
shown in figure 3. The cutoff energy is E = 14.2 kJ/mole.

8. How Many Rotors are Involved?

Preliminary fits to the relaxation data for state C in
figure 3 suggest that only between 50% and 70% of the *t*-
butyl groups are reorienting on the nuclear magnetic resonance time scale. It is very difficult to understand how
these Van der Waals solids can allow the molecules to be so
close together that the activation energy is out the range
of the SSPSR technique. There are no molecular groups in
TTB that could "bond" into polymer-like systems.

9. Future Work

We need to carefully characterize the various states of TTB
by performing temperature-dependent x-ray studies in the
range from about 77 to 250 K. In addition, we must control
the thermometry more carefully in the SSPSR experiments.
Finally, more effort is needed in the development of the
models that link the observed relaxation rate R to the
properties that characterize the intramolecular reorientation process [2,5].

References

[1]   A M Albano, P A Beckmann, M E Carrington, E E Fisch,
      F A Fusco, A E O'Neill and M E Scott 1984 Phys Rev
      B, 2334.
[2]   P A Beckmann, R M Hathorn, F B Mallory 1990 Molec
      Phys 69, 411.
[3]   T Sakai 1978 Acta Cryst B 34, 3649.
[4]   P A Beckmann 1988 Phys Rep 171, 85.
[5]   P A Beckmann, L Happersett, A V Herzog and W M Tong
      1991 J Chem Phys 95, 828.
[6]   K G Conn, P A Beckmann, C W Mallory and F B Mallory
      1987 J Chem Phys 87, 20.

A MOLECULAR DYNAMICS STUDY OF THE SOLID PHASE TRANSITION IN SULPHUR HEXAFLUORIDE CLUSTERS

F. M. BENIERE, B. ROUSSEAU, A. H. FUCHS
Laboratoire de Chimie-Physique des Matériaux Amorphes,
URA 1104, Bât.490, Université de Paris-Sud, 91405 Orsay, France
M.-F. DE FERAUDY and G. TORCHET
Laboratoire de Physique des Solides, URA 2, Bât.510, Université de
Paris-Sud, 91405 Orsay, France

ABSTRACT. The structural phase transition occuring in SF_6 clusters is simulated in a Molecular Dynamics calculation performed on 229- and 137-molecule models. The variation of the potential energy obtained by first cooling and then warming the models shows that the transition is spread over roughly 20 K and that the transformation is reversible. The temperature of the end of the transition increases from 95 K to 105 K when going from 137 to 229 molecules. These results are consistent with the previous calculations on a 512-molecule model.

1. Introduction

Sulphur hexafluoride, SF_6, clusters of several hundreds of molecules, produced in a free jet expansion of a Ne-SF_6 mixture and observed by electron diffraction, experience a structural phase transition between a low temperature monoclinic phase and a body centred cubic structure [1]. These two phases are also known to be the stable ones in the solid polycrystalline material [2]. This phase transition has been recently reproduced in a Molecular Dynamics (MD) simulation performed on a 512-molecule model [3]. Starting from the low temperature structure, the transition is characterized by a range of temperature during which the cluster adopts progressively the high temperature bcc structure through molecular reorientations experienced first by surface and then by core molecules [4]. Intermediate experimental diffraction functions were in very good agreement with the calculated patterns, and have thus been attributed to the evolution of single clusters rather than to the coexistence of clusters with different structures as had been suggested previously [5].

In order to study systematically the reversibility of the transition and the effect of the cluster size, MD simulations have been performed by cooling and heating 229- and 137-molecule models in the range 50 to 120 K.

P. Jena et al. (eds.), Physics and Chemistry of Finite Systems: From Clusters to Crystals, Vol. I, 363–367.
© 1992 *Kluwer Academic Publishers.*

2. Computational Method and MD Model

As in the previous MD simulations on SF_6 clusters [3, 4, 6,7] a rigid molecule is assumed and a single Lennard-Jones F-F atom-atom potential is used to compute the intermolecular interactions. The model potential function is taken from an optimized fit between earlier MD simulations[8] and neutron diffraction measurements, and has been proved very efficient in modelling SF_6 behaviour, both in the free cluster and periodic boundary conditions [3, 4, 6, 7, 9]. A cut-off distance of 13 Å was used which corresponds approximately to four intermolecular distances. Two clusters were made by taking a plastic phase configuration and carving out spherical clusters containing 229 and 137 molecules, respectively. Both clusters have their outer shells completed. No boundary conditions were used. The position and quaternion coordinates are time-stepped using the Beeman algorithm which is accurate to the same level of approximation as the more commonly used Verlet algorithm.

Clusters were cooled down from 120 to 50 K by steps of 5 K and then warmed up again in the same way. A velocity rescaling is applied in order to change the mean temperature of the clusters. Except during the initial stabilization stage (where a very small time step is used), we have used here a time step of 5×10^{-3} ps. The equilibration stage lasts between ~100 to 300 ps, as indicated by the potential energy becoming stable, after which the simulations are being performed at constant energy for roughly 100 ps. The results shown and discussed below come out of the constant energy simulations.

The calculations have been performed on a massively parallel transputer-based computing surface. The details of the technique used for these simulations are given in reference [7]. A ring array of 6 transputers was used for most of the calculations presented here. The rate of calculated interatomic interactions for such a configuration is roughly equal to 14,000 per second, the cycle time being 2.3 and 1.7 s for the 229- and 137-molecule cluster, respectively.

3. Results and Discussion

3.1. TEMPERATURE SPREAD OF THE TRANSITION

We have sketched in figure 1 the evolution of the mean potential energy as a function of temperature for the 229-molecule cluster model. Structural information, as deduced from the calculated diffraction patterns, is represented in the following way: open symbols refer to the monoclinic structure, full symbols to a cubic structure and half-full symbols to the intermediate structures. The circles correspond to the cluster configurations obtained on cooling the initial bcc model and the squares to those obtained on reheating the cluster model.

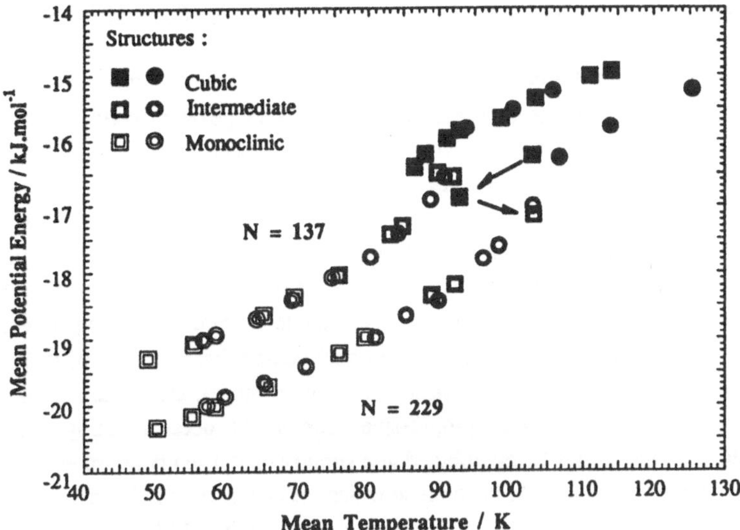

Figure 1. Mean potential energy as a fonction of mean temperature for the 137- and 229-molecule clusters, obtained on cooling(squares) or heating (circles) the models.

The initially cubic cluster can be cooled down to ~92 K without observing noticeable changes in the diffraction pattern. At this stage a spontaneous transformation occurs which leads to an intermediate pattern similar to those obtained in our previous work on the 512-molecule model. This irreversible transformation can be interpreted as due to a spontaneous nucleation of the monoclinic structure in the core of the cluster. However, the cluster model is not fully monoclinic at this stage. A further lowering of the temperature leads to a progressive transformation of the patterns from an intermediate to a purely monoclinic structure. On reheating the cluster, the same equilibrium thermodynamic path is followed and no "superheating" of the monoclinic structure is observed at low temperature, presumably because the transformation initiates at the surface and thus needs no real "seed". So, apart from the slight retardation observed on cooling the cubic cluster model, the phase transition, which is spread over a range of ~20 K, seems to be reversible. A very similar diagram is obtained in the case of the 137-molecule cluster model (figure 1), except that the phase transition takes place between ~75 and 95 K (instead of ~85 to 105 K).

3.2. CLUSTER SIZE EFFECT

It has been found earlier that the phase transition takes place between 90 K and 120 K for a 512-molecule model [4]. The present work gives then some evidence for a lowering of both the temperature limits ("beginning" and "end" of the transition) and the temperature spread of the transition, when decreasing the size of the cluster model.

3.3. STRUCTURAL CHANGE IN CALCULATED DIFFRACTION FUNCTIONS

Diffraction functions have been calculated [4] at every temperature. In order to account correctly for experimental patterns it has been necessary to make an average over 6 configurations for T < 90 K and 15 configurations for T > 90 K. The calculation has been made separately for the whole cluster and for the core defined as the set of molecules having at least their first coordination shell filled up. The ratio of two line heights, namely the (310) and (431) lines, measured on the diffraction patterns has been proved to be a good index of the progressive structural change, especially in the intermediate range [4, 10]. This ratio is shown in figure 2 for the 229-molecule cluster and the 89-molecule core. In the whole cluster, the lower limit of the temperature transition-range is roughly equal to 85 K but cannot be precisely determined because of the very smooth increase of the ratio of line-heights when heating the cluster model. This results from the fact that the monoclinic-to-cubic transition is a progressive disordering process initiating at the surface of the cluster, the core molecules being affected at a slightly higher temperature ~93 K. The upper limit of the transition corresponds to the point at which the cluster is entirely cubic and this point is much easier to detect. As shown in figure 2, the upper limit for the 229-molecule model is equal to 105 K, the uncertainty on the absolute temperature being +/- 2 K. This value is identical to the value provided by the potential energy curve in figure 1. The same computation made on the 137-molecule model showed the same qualitative behaviour of the line heights ratio, with an upper limit of the transition range of 95 K.

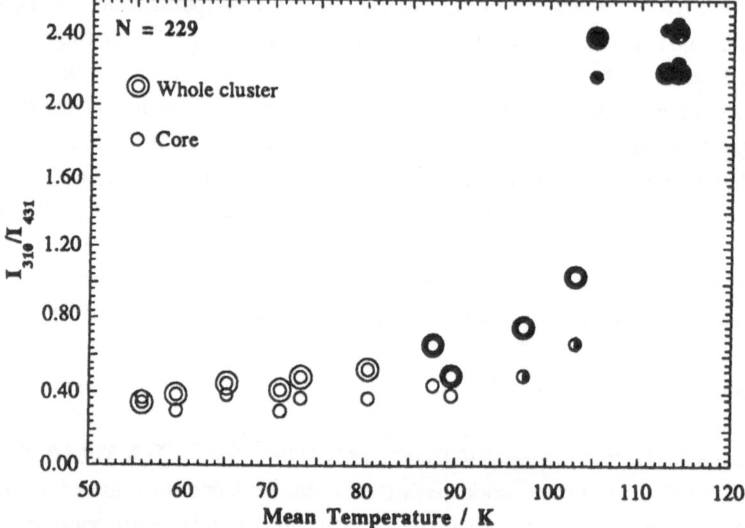

Figure 2. Ratio of two line heights measured on the calculated diffraction functions of the 229-molecule model as a fonction of mean temperature. Model structures are reproduced as in Fig. 1.

4. Conclusion

The study of two clusters containing 137 and 229 molecules has shown that the progressive transition between the monoclinic and the cubic phases is reversible. The spread of the transition is equal to roughly 20 K, which is significantly lower than the 30 K value found earlier for the 512-molecule cluster. It should be stressed however that only the upper limit of the temperature range of the transition is clearly detectable. Our findings that this upper limit increases with an increasing cluster size (95 K for 137 and 105 K for 229 molecules) is consistent with our previous results on a 512-molecule model [4] in which we have found an upper limit of 120 K. This is in agreement with the recent experiments suggesting that small cubic clusters are produced at lower temperatures than larger ones [10].

5. References

[1]. Farges, J., de Feraudy, M.-F., Raoult, B. and Torchet, G. (1986) 15th Rarefied Gas Dynamics Conference, Grado, Italy .

[2]. Dove, M. T. , Powell, B. M., Pawley, G. S. and Bartell, L. S. (1988) 'Monoclinic phase of SF_6 and the orientational ordering transition', Mol. Phys., 65, 353-358.

[3]. Fuchs, A. H. and Pawley, G. S. (1988) 'Molecular dynamics simulation of the plastic to triclinic phase transition in clusters of SF_6', J. Phys. France, 49, 41-51.

[4]. Torchet, G., de Feraudy, M.-F., Raoult, B., Farges, J., Fuchs, A. H. and Pawley, G. S. (1990) 'Cluster model for the monoclinic to cubic transition in SF_6 clusters', J. Chem. Phys., 92, 6768-6774.

[5]. Bartell, L. S., Valente, E. J. andCaillat, J. (1987) 'Electron diffraction studies of supersonic jets. 8. Nucleation of various phases of SF_6, SeF_6 and TeF_6', J. Phys. Chem., 91, 2498-2503.

[6]. Boyer, L. L. and Pawley, G. S. (1988) 'Molecular dynamics of clusters of particles interacting with pairwise forces using a massively parallel computer', J. Comp. Phys., 78(2), 405-423.

[7]. Rousseau, B., Boutin, A., Craven, C. J. and Fuchs, A. H., submitted to Mol. Phys.

[8]. Dove, M. T. and Pawley, G. S. (1983) 'A molecular dynamics simulation study of the plastic crystalline phase of sulphur hexafluoride', J. Phys. C, 16, 5969-5983, and *ibid* (1984) 'A molecular dynamics simulation study of the orientationally disordered phase of sulphur hexafluoride', J. Phys. C, 17, 6581-6599.

[9]. Boutin, A., Rousseau, B. and Fuchs, A. H., in preparation.

[10]. Torchet, G. (1991) 'Size effects in the solid phase transition of SF_6 clusters', Z. Phys. D, 20, 251-253.

OBSERVATION OF ELECTRONIC SHELLS IN LARGE LITHIUM CLUSTERS

C.BRECHIGNAC, Ph.CAHUZAC, M.de FRUTOS, J.Ph.ROUX and
K.BOWEN*
Laboratoire Aimé Cotton
CNRS II bât. 505
91405 Orsay cedex, France

ABSTRACT. Shell and supershell structures have been observed for lithium clusters
produced in a gas-aggregation type source, up to the size range $n \sim 2000$. Noteworthy
similarities are found as compared with observations performed on Na_n clusters.
Thermal effects are briefly investigated.

1. INTRODUCTION.

Recent experimental efforts to extend the study of clusters to large sizes have lead to the
observation of electronic shells in large Na-atom clusters [1-3]. Sequences of "magic
numbers" have been detected for particles containing up to 2700 atoms, from the
observation of abundances in mass spectra. Extending the ideas orginally proposed by
Balian and Bloch [4], recent calculations show that such a shell structure must be
attributed to the electronic subshells of delocalized valence electrons moving in a central
potential [5,6]. The shell periodicity is proportionnal to the cube root of the number n of
valence electrons. Superimposed to this primary shell periodicity, a beat pattern is
observed. The corresponding supershell structure is explained in terms of interference of
amplitude associated with classical closed orbits of electrons.The beat mode has a
minimum found around 1100 valence electrons by Bjørnholm et al. [7], whereas Martin
et al. determined the transient region around $n=830$ [3,8]. Calculations are able to
reproduce both values, depending on the detailed shape of the radial potential [3,5,6].
We present here new results obtained on lithium clusters. We observed a shell structure
up to $n=2100$ with a beat mode minimum located near $n=900$. The temperature
dependence upon the observability of the shell structure is also investigated.

2. EXPERIMENT.

The Li clusters are produced by a gas-aggregation type source [3,9-11]. The metal vapor
is produced at a moderate pressure, a few of 0.1 torr, by heating a molybdenum oven.
Metal atoms effuse in a helium flow through a 4 mm hole (Fig.1). They are carried
along by 20-25 torr of helium into a 6cm long, liquid nitrogen cooled copper tube. The
metal clusters grow along the helium stream which progressively cools them and
dissipates their formation energy. The cluster growth stops at the end of the tube, where
the metal cluster-helium mixture enters the three successive chambers of a differential
pumping system.

P. Jena et al. (eds.), Physics and Chemistry of Finite Systems: From Clusters to Crystals, Vol. I, 369–374.

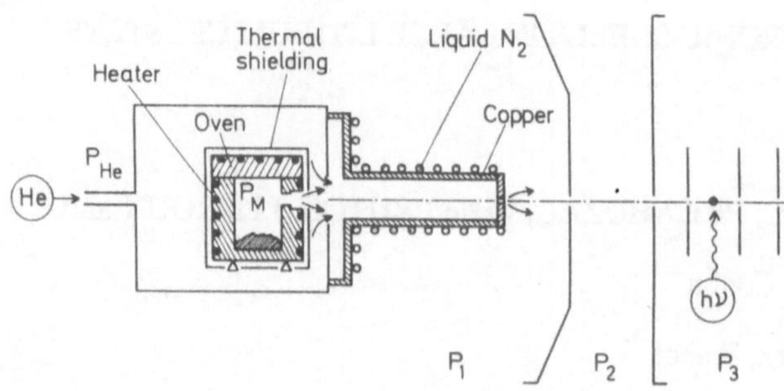

Figure 1 : Schematic of our gas-aggregation source. For 20 torr of He, the differential pressures $P_{1,2,3}$ are typically 0.5 torr, $2 \, 10^{-4}$ and $5 \, 10^{-6}$ torr respectively. The hole diameter in the liquid nitrogen cooled copper tube is 2mm.

The neutral particles are photoionized at a photon energy hν=3.50 eV provided by the third harmonic of a pulsed Nd-YAG laser. Ionization takes place in between a multiplate accelerating system of a Wiley-Mc Laren time-of-flight mass spectrometer [12]. The length of the drift tube is 120cm and typically Li_n clusters are accelerated to 2.8 kV. A secondary electron multiplier delivers ion signals averaged by a Le Croy 9400 digital oscilloscope. With this source a pseudo-gaussian distribution of neutral Li_n clusters is generated, the mean size of which is quite sensitive to the metal vapor pressure. Varying this latter from $5 \, 10^{-2}$ torr to $5 \, 10^{-1}$ torr, promotes a shift of the maximum from n=200 to n=900. Simultaneously the relative width $\Delta n/n$ is increasing from 0.6 to 1.2 respectively. Such a source gives medium-size clusters [11]. This differs from those obtained by other gas-aggregation sources which generate only large neutral clusters [3,10].

RESULTS.

At low laser fluence (1 mW/cm^2) one observes s-shaped irregularities superimposed on the envelope of the size distribution (Fig.2). When the metallic vapor pressure is varied, the envelope shifts, whereas these structures remain fixed in time, and so in size. They appear with a modulation which decreases at large size, beyond the maximum of the distribution. These features must be attributed to electronic shells of the valence electrons as observed by other authors with Na_n clusters [2,3]. The photon energy hν=3.50eV is choosen above the ionization potentials of large lithium clusters and in the expected range of the maximum

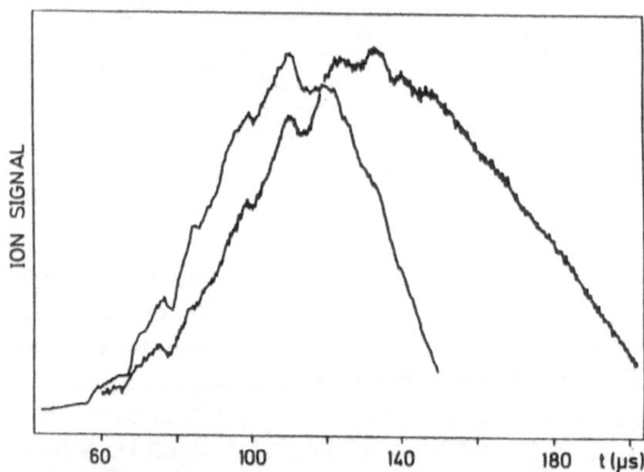

Figure 2 : Shift of the mass distribution vs the metallic vapor. Left curve : Lithium pressure 0.2 torr. Right curve : lithium pressure 0.7 torr. Successive shells are clearly visible as strong irregularities.

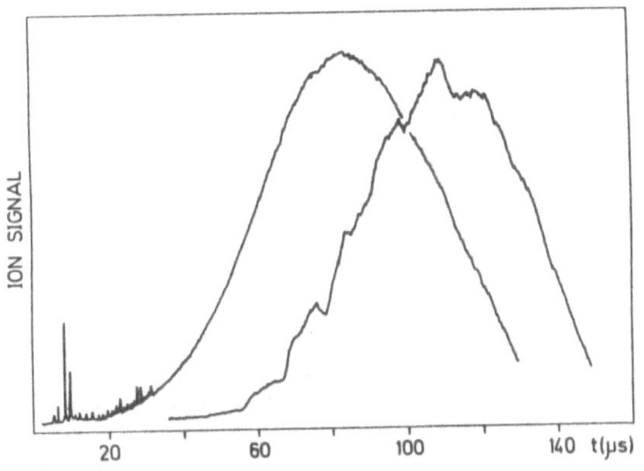

Figure 3 : Shift of the mass distribution vs the laser fluence. The profile peaked at large size is the same as Fig.2 (low metallic vapor pressure). The smooth profile shifted toward smaller time-of-flight is taken at the same lithium pressure but at a slightly higher laser fluence. Notice the occurence of mass resolved very small size fragments.

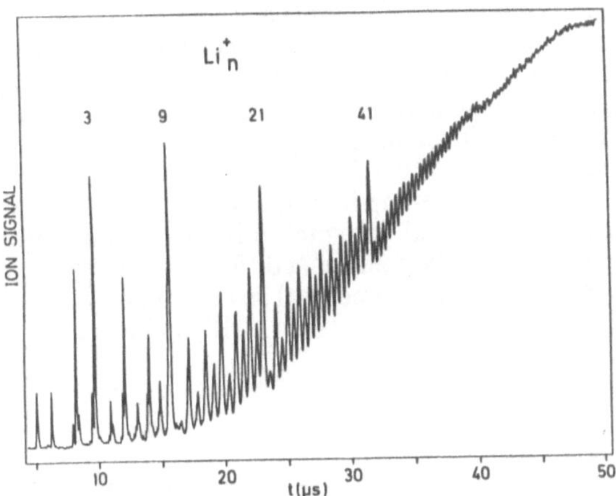

Figure 4 : Small size lithium clusters obtained by photoevaporation from large sizes. Closed shell, odd-even alternations [15] are visible. The first ion peak is due to ionized helium atoms. The rising bakground corresponds to the residual envelope observed at low laser fluence.

of the photo-absorption cross-section, so that the shells are revealed from evaporative cooling as shown in ref. [13,14].When the laser fluence is slightly increased (up to 5 mW/cm^2), the irregularities are washed-out and a smooth profile is observed (Fig.3). Under these conditions, several photons are absorbed during the ionizing laser pulse. The subsequent dissociation shifts noticeably the mean size of the mass distribution toward smaller sizes. Indead, each ion parent gives rise to a broad distribution of fragments [13] and the sharp structures are erased. When the laser fluence is still increased (up to few W/cm^2), the efficient multistep photoexcitation-dissociation process generates small clusters up to the monomer (Fig.4). In this case the well known shell closing effects [15] are clearly visible as an abrupt change in ion intensities after n=3,9,21,41,69 and 93. Mass spectra as shown on Fig.4 provide an accurate and absolute scaling of the time-of-flight mass spectrometer. However care must be taken to the use of natural lithium which contains two isotopes ^{7}Li (92.6%) and ^{6}Li (7.4%). A cluster of a given size n results on a binomial distribution of the isotopic mixture ^{6}Li$_x$ ^{7}Li$_{n-x}$ with the probability $C_n^x \, \alpha^x (1-\alpha)^{n-x}$ [16]. C_n^x is the binomial coefficient and α is the natural abundance of the ^{6}Li isotope. In the limit of large n, the isotopic distribution peaks at x=nα. This leads to a relative change in the mass position of the ion peaks

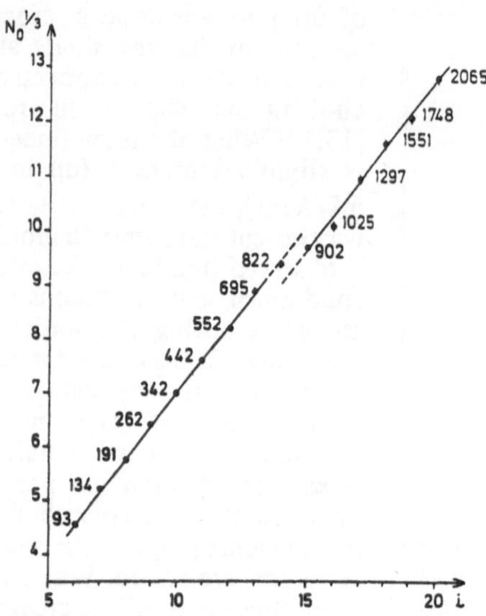

Figure 5 : Magic numbers $N_0^{1/3}$ vs the shell index and the corresponding number of valence electrons. Notice the abrupt phase-shift in the n=800-900 size domain

given by $\dfrac{\Delta n}{n} = \alpha(1\,\dfrac{6}{7})=0.01$. The determination of the relative maxima of the shell structure is obtained by using such a corrected size calibration.

The valence electron number corresponding to the successive relative maxima as seen on Fig.2 and 3, and averaged over several mass spectra, are easily determined. In a semiclassical picture one expects a periodicity of shells proportionnal to the linear dimension of the droplet $\approx n^{1/3}$. Plotting the "magic numbers" N_0, i.e the shell maxima, on a $n^{1/3}$ scale against a running index i, accounting for the shell number, one obtains two series of points at approximately equal intervals (Fig.5). The two straight lines are shifted relative to each other by one-half unit of i around n$\approx$900 as found in one of the experiments performed with Na_n clusters [8]. This characterizes the beat mode minimum of a supershell structure.

It is interesting to compare the slope of this straight line which should be independent of the nature of alkali-atoms in the cluster. We found $\Delta(N_0^{1/3})=(0.63\pm0.03)\Delta i$, a value in excellent agreement with the value found for sodium clusters [7,8]. The absolute values found for the N_0 numbers show noteworthy similarities with those either deduced from sodium experiments [3,7,8] or calculated by using a Wood-Saxon potential or a "wine-bottle" type potential [3,5,6]. However some differences exist. Mainly we found an extremum at n=308$\pm$5, not shown in the plot of the Fig.5. This shell is neither observed with Na_n clusters nor predicted by calculations. Keeping in mind the sensitivity of the subshell bunching to the detailed shape of the radial potential, this suggests that the potential to be considered for lithium clusters could be somewhat different with regard to the potential used for sodium clusters. Moreover, the geometrical arrangement of atoms in the clusters, can play a role in determining the magic numbers [17,18] and this cannot be totally ruled out.

CONCLUSION

This first study of shell effects in lithium clusters allows a comparison with results obtained with sodium clusters. Experimental values support the semi-classical interpretation of the shell and supershell structures. However the detailed analysis of the observations suggests the need to reconsider more accurately the radial potential shape of lithium clusters. This could be obtained from more precise measurements on isotopically enriched ^7Li.

ACKNOWLEDGEMENTS.
The authors would like to thank T.P.Martin for valuable and stimulating discussions.

*Permanent adress : Department of Chemistry, The Johns Hopkins University, Baltimore, Maryland 21218, USA.
K.B. wishes to acknowledge the partial support of the U.S. National Science Foundation under grant CHE-9007445.

REFERENCES.

[1] Göhlich.H, Lange.T, Bergmann.T, Martin.T.P., Electronic shell structure in large metallic clusters, Phys. Rev. Lett. 65, 748-751 (1990).

[2] Bjørnholm.S, Borggreen.J, Echt.O, Hansen.K, Pederson.J, Rasmussen.H.D., observation of electronic shells and shells of atoms in large Na clusters. Phys. Rev. Lett. 65, 1627-1630 (1990).

[3] Martin.T.P, Bergmann.T, Göhlich.H, Lange.T. Mean field quantization of several hundred electrons in sodium metal clusters. Chem. Phys. Lett. 176, 343-347 (1991).

[4] Balian.R, Bloch.C. Distribution of eigenfrequencies for the wave equation in a finite domain : III Eigenfrequency density oscillations. Ann. Phys. 69, 76-160 (1972).

[5] Nishioka.H, Hansen.K, Mottelson.B.R. Supershells in metal clusters. Phys. Rev. B 42, 9377-9386 (1990-II).

[6] Genzken.O, Brack.M. Temperature dependence of supershells in large sodium clusters. To be published and private communication.

[7] Pederson.J, Bjørnholm.S, Borggreen.J, Hansen.K, Martin.T.P, Rasmussen.H.D. A new periodic system with three thousand elements. To be published, Nature (1991).

[8] Martin.T.P, Bjørhnholm.S, Borgreen.J, Bréchignac.C, Cahuzac.Ph, Hansen.K, Pederson.J. Electronic shell structure of laser-warmed Na clusters. To be published, Chem. Phys. Lett. (1991).

[9] Sattler.K, Mühlbach.J, Recknagel.E. Generation of metal clusters containing from 2 to 500 atoms. Phys. Rev. Lett. 45, 821-824 (1980).

[10] Franck.F, Schulze.W, Tesche.B, Urban.J, Winter.B. Formation of metal clusters and molecules by means of the gas aggregation technique and characterization of size distribution. Surface Science 156, 90-99 (1985).

[11] Bréchignac.C, Cahuzac.ph, Carlier.F, de Frutos.M, Masson.A, Roux.J.Ph. Generation of rare earth metal clusters by means of the gas aggregation technique. Z.Phys. D - Atoms, Molecules and clusters 19, 195-197 (1991).

[12] Wiley.W.C, Mclaren.I.H. Time-of-flight mass spectrometer with improved resolution. The Review of Scient. Instrum. 26, 1150-1157 (1955).

[13] Bréchignac.C, Cahuzac.Ph, Carlier.F, Leygnier.J. Collective excitation in closed-shell potassium cluster ions. Chem. Phys. Lett. 164, 433-437 (1989).

[14] Bréchignac.C, Cahuzac.Ph, Carlier.F, de Frutos.M, Leygnier.J. Cohesive energies of K_n 5 n 200 from photoevaporation experiment. J.Chem. Phys. 93, 7449-7456 (1990).

[15] Knight.W.D, Clemenger.K, de Heer.W.A, Saunders.W.A, Chou.M.Y, Cohen.M.L. Electronic shell structure and abundances of sodium clusters. Phys. Rev. Lett. 52, 2141-2143 (1984).

[16] Bréchignac.C, Cahuzac.Ph. Evolution of photoionization spectra of metal clusters as a function of size. Z. Phys. D Atom, Molecules and clusters B, 121-129 (1986).

374

[17] Martin.T.P, Bergmann.T, Göhlich.H, Lange.T. Observation of electronic shells
     and shells of atoms in large Na clusters, Chem. Phys.Lett. $\underline{172}$, 209-213 (1990).
[18] Mansikka-aho.J, Suhonen.J, Valkealahti.S, Hammaren.E, Manninen.M.
     Electronic shell structure in icosahedral metal clusters, to be published (this issue).

STRUCTURE AND DYNAMICS OF INTERMEDIATE SIZE ALUMINIUM CLUSTERS

S. Debiaggi[#*] and A. Caro[#]
[#]Paul Scherrer Institute. 5232- Villigen. Switzerland.
[*]Universidad Nacional del Comahue. 8300 Neuquen. Argentine.

ABSTRACT. We use molecular dynamics with a semiempirical quantum chemistry model at a Hartree level of approximation to predict the structure of Al_n clusters. The relative simplicity of this model allows us to relax up to few hundred atoms. We find that icosahedral and cube-octahedral symmetries are almost degenerate, but that for the cases where we performed annealing, deeper minima exist without symmetry.

1- Introduction:

Due to their simple electronic structure, Al clusters have been a favourite element in experimental and theoretical works. Compared to alkali metal clusters, they have the additional interest to s-p hybridizes to form trivalent atoms becoming a bridge to transition metal clusters. The experimental data for Al cluster is very extensive; however, information on the cluster structures is still lacking and most of what is known comes from theoretical work. Density Functional - Simulated Annealing has been performed for Al_n [1], with n up to 10, leading to new stable structures (numerous buckled planar arrays are found), and spin multiplicities. Transitions from planar to non-planar structures are found at n = 5, and to states with minimum spin degeneracies at n = 6.

To our knowledge there is only one experimental report on dissociation energies of neutral clusters [2], which have been calculated through their measured dissociation energies for ionized clusters and the corresponding ionization potentials measured in photo-dissociation experiment. All calculations show poor agreement with these results. The SCF CI calculation of Petterson [3] and Upton [4] are substantially lower, while LSDA systematically overestimate the binding [1], (see Figure 1). For bigger Al clusters, ab-initio calculations are computationally very hard; restricted geometries [5], and full annealed 13 and 55 atom clusters have been studied [6]. For Na [7] it has been found that low symmetry structures have less energy than their corresponding high-symmetry partners.

Semiempirical approaches become relevant as they can help in the search of global minima, extrapolate to regions of larger size, and enable predictions of the effects of temperature. Such type of results have been obtained by Pacchioni et al. [8] with an INDO method computed according to a symmetry restricted (no-relaxation of the structures) and spin unrestricted iterative procedure. Their calculation of cohesion extends up to n = 49, showing an overall trend as a function of cluster size which agrees with the experimental data, although still slightly lower than the measured values, (see Figure 1).

In this work we present results concerning Al clusters in the medium size region (n < 256). We use a simple quantum chemistry model, that properly describes the angular dependence of

375

P. Jena et al. (eds.), Physics and Chemistry of Finite Systems: From Clusters to Crystals, Vol. I, 375–380.
© 1992 *Kluwer Academic Publishers.*

the chemical bonding, as it was proved on the predicted structures of small s-p metal clusters [9], and also give good results for bulk and surface properties [10]. We calculate dissociation energies as a function of cluster size and we analyze the effects of relaxations, comparing the relative stability of icosahedral, octahedral, and, in some cases, fully annealed configurations. These last simulations show that the global minimum has no symmetry.

2) The model:

We use a self-consistent spinless tight-binding or linear combination of atomic orbitals (LCAO) model for the description of the chemical bonding. For our study of Al we adopt a parametrization according to the so-called Extended Hückel model [11]. The Hamiltonian is a single-particle LCAO, namely:

$$H = \Sigma\, \varepsilon_{i,\alpha}\, |i,\alpha\rangle\langle i,\alpha| + \Sigma\, t_{i\alpha,j\beta}\, (\,|i,\alpha\rangle\langle j,\beta|\,) \tag{1}$$

$\varepsilon_{i,\alpha}$ is the self consistent diagonal element that can be obtained by iterations from an empirical expression, as for instance Gray's equations [12] (on-site correlation): $\varepsilon_{i,\alpha} = \varepsilon_{i,\alpha}{}^0 + U(q_i - z_i)$, where $\varepsilon_{i,\alpha}{}^0$ is the αth energy level of the free atom i. The second term represents the change in the intra-atomic Coulomb potential, q_i is the self-consistent total charge of atom i, z_i is the charge of the free atom i. U is an average of the intra-atomic Coulomb integral of the valence electrons; its value is chosen to reproduce charge transfer effects obtained in an ab-initio calculation on a 55 atom Al cluster [13]. $t_{i\alpha,j\beta} = K\, S_{i\alpha,j\beta}$, is the hopping integral from site j orbital β to site i, orbital α; K is an adjustable parameter fit to bulk band width, and $S_{i\alpha,j\beta}$ is the overlap between atomic Slater orbitals. $|i,\alpha\rangle$ represents the orbital α located at site i. Only valence 3s, 3p, and 3d electrons are taken into account in the sum over orbitals, giving a basis of nine orbitals per atom. The resulting overlap integrals are radial functions times an angular part depending on the orientation of the vector r_j-r_i. For practical reasons, relevant in the dynamics, a smooth spatial radial cut-off is imposed at a distance equal to 5.35 Å. The orthogonal LCAO formulation involves the self-consistent solution of:

$$(\,H - \lambda I\,)\,\phi = 0 \tag{2}$$

The total energy is calculated as $E = E_{coh} + E_{rep}$ where $E_{coh} = 2\, \Sigma\, \lambda_n\, f(\lambda_n)$ and $E_{rep} = 1/2\, \Sigma_{i,j}\, V_{i,j}$. E_{coh} is given in terms of the doubly occupied molecular levels λ_n, $f(\lambda_n)$ is the Fermi function at a finite low temperature, λ_n are the solutions of Eq. 2, and E_{rep} is an adequate repulsive pair potential chosen to reproduce some set of properties, namely: the equation of state of Rose et al. for bulk Al [14] (lattice parameter, cohesive energy, and bulk modulus), the interstitial formation enthalpy, the relative stability of fcc versus bcc phase, and the relaxation of some selected surface. For the relaxation procedure the forces are calculated using the Hellman-Feynman theorem. The transferability of this parametrization is excellent since it predicts the energy and distance in the singlet Al_2 in agreement with a much better calculation [15]

Solving these equations for the largest problem reported in this work (256 atoms) implies finding 400 eigenvalues and eigenvectors out of a 2304 x 2304 matrix, which is done either by direct diagonalization or using the Car - Parrinello algorithm [16] as implemented in [17]. For these size of the basis and fraction of eigenvalues, both procedures require similar computing time.

3- Results:

The determination of the parameters involved in the model is reported elsewhere [10]. In the present section we present the cohesive energy of the clusters as a function of the clusters size and the relative cohesive energy between the icosahedral (I_h), octahedral (O_h), and some fully annealed structures. For n = 13, 19, 43, 55, 79, 87, 135, 141, 147 and 177, the initial structure was relaxed by a steepest descend method until a convergence of 0.01 eV/at in the forces was reached. This procedure preserves the original symmetry. For 13, 55, and 147 a simulated annealing was first performed to search the absolute minima.

Only for 13, 55 and 147 atoms do the regular cube-octahedral and icosahedral structures exist. Cube-octahedral structures can be obtained by properly cutting a perfect fcc array parallel to the {1,1,1} and {1,0,0} planes. These cube-octahedra have 6 square {1,0,0} and 8 triangular {1,1,1} faces. The ideal icosahedral structure can be obtained by distortion of the regular octahedron of the same number of atoms, as pointed out by Mackay [18]. Icosahedral structure have five-fold symmetry, and all of its faces (20 in total) have distorted (1,1,1) directions. The inter-shell nearest neighbour distance is almost the same as in the cube-octahedron, but the intra-shell distances are stretched uniformly by 5%. The resulting geometry has then lower packing but higher coordination in the surface atoms (6 for I_h and 5 for O_h). In the case of Al_{19} the double icosahedra was generated by the addition of two capped Al_{13} icosahedra. Close packed structures with 19, 43, 79, 87, 153, 141 and 177 atoms have shell-like growth, composed of 2, 3, 5, 6, 7, 8, and 9 shells respectively around a central atom.

We have calculated the energy difference predicted by this semiempirical model for these symmetries. The results are summarized in Table 1. Al_{13} is O_h with a cohesive energy of approximately 40 meV/at less than its I_h partner. For the remainder clusters the more stable isomers are the I_h, with an energy difference decreasing with increasing size, being less than 10 meV/at for the biggest of the set, the Al_{147}. In all the cases, the small magnitude of the energy differences make both structures almost degenerate within the precision of this approximation.

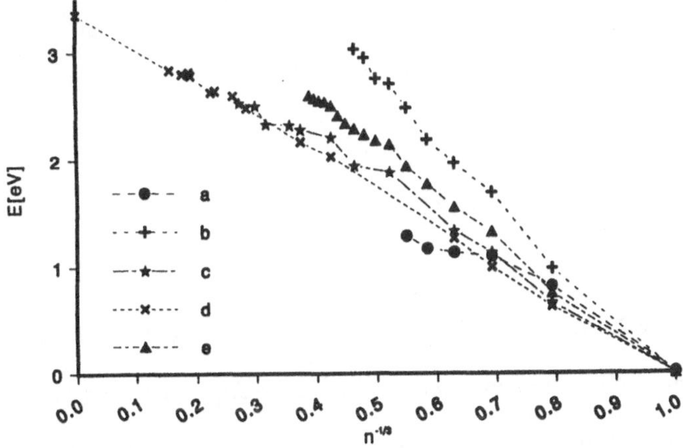

Figure 1. Calculated and measured cohesive energies of Al clusters. a- SCF - CI, Ref. [3]. b- LSD, Ref. [1]. c- INDO, Ref. [8]. d- Present work. e- Experimental results, Ref. [2].

The amplitude of the relaxations are significant in the O_h symmetry but have small influence on the energy. They can in some cases reverse the relative stability of the structures as in Al_{19} and Al_{147} (see Figure 2).

In I_h structures the relaxation of the shells is always inwards in order to decrease the surface nearest neighbour distances, which are larger than in O_h by 5%. The final distance between the central atom and its first shell is considerably reduced by this effect. The first shell radius is 2.79 Å in Al_{13}, 2.69 Å in Al_{55} and 2.61 Å in the Al_{147}. In the relaxed and no-relaxed configurations the average coordination number is conserved.

In the O_h structures an important reconstruction is observed. Interchange of shells are produced between 2nd. (1,0,0) and 3rd. (1,1/2,1/2) shells in Al_{55} and Al_{147}, and also for the outer 5th. (3/2,1/2,0) and 6th. (1,1,1) shells in the Al_{147} cluster. In this last case, the 7th. shell moves inwards almost at the same distance from the centre than the displaced 5th shell. The resulting aspect of the relaxed clusters is such that the {1,0,0} faces appear rounded, and in both cases the final configuration have more coordination than in the ideal unrelaxed cases.

Figure 2. Unrelaxed, relaxed and fully annealed cube-octahedral Al_{147}

Also breathing mode relaxations were performed leading in all the cases to higher energy minima, as can be seen in Table 1.

The global minima in the energy surface of Al_{13}, Al_{55} and Al_{147} have been studied by combining the LCAO model with molecular dynamics using a Langevin thermostat. An initial configuration with O_h or I_h symmetry was heated up to 2000 K for some equilibration time, and then different configurations are selected as starting structures for the quenching and relaxation process.

N	Symmetry	Unrelaxed	Relaxed	Breathing	$\varepsilon\%$	Annealed
2	$C_{\infty h}$	-	0.633	-		-
3	C_{2v}	-	1.003	-		-
4	C_{2v}	-	1.271	-		-
13	I_h O_h C_1	1.986 2.006 -	1.994 2.036 -	1.994 2.036 -	-2.4 -3.6 -	- - 2.140
19	I_h	2.172	2.188	-	-	-
19	O_h	2.157	2.167	-	-	-
43	O_h	2.470	2.487	-	-	-
55	I_h O_h C_1	2.565 2.539 -	2.627 2.603 -	2.617 2.553 -	-5.5 -2.3 -	- - 2.669
79	O_h	2.637	2.648	-	-	-
87	O_h	2.610	2.634	-	-	-
135	O_h	2.754	2.785	-	-	-
141	O_h	2.734	2.791	-	-	-
147	I_h O_h C_1	2.757 2.770 -	2.832 2.821 -	2.815 2.779 -	5.0 1. -	- - 2.851
177	O_h	2.784	2.804	-	-	-
256	O_h -	2.842	no	-	-	-

Table 1. Cohesive energy in eV/atoms of the unrelaxed, relaxed, breathing relaxed and fully annealed configurations.

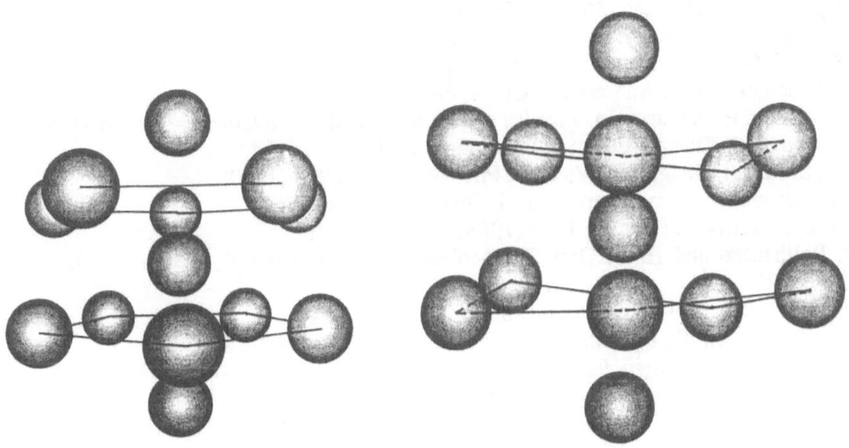

Figure 3. Left: Ideal icosahedral Al_{13}. Right: Geometry of the global minimum.

In this way we find that Al_{13} evolves to a unique deep minimum which is 0.1 eV/atom below the symmetric ones. The average coordination number of the surface atoms, which is evaluated considering a cutoff distance of 3.8 Å, is 5,67. The structure can be considered as a distortion of the I_h symmetry in the sense that the pentagonal rings rotate around their common axis (see Figure 3). For Al_{55} and Al_{147} the situation is more complex as several close minima appear. Preliminary results on a small number of samples show the gain in energy per atom is less important as the size increases; the cohesive energies are 2.67 eV/at for Al_{55} (C_1), and 2.85 eV/at for Al_{147} (C_1). The structures can be characterized by an average coordination of 7.1 and 7.4 respectively. The eccentricities, defined as the ratio of the maximum and minimum axis of the ellipsoid of inertia are 1.08 and 1.07 respectively.

4- Conclusions:

We presented results concerning medium size Al clusters obtained with a self-consistent LCAO model adjusted to reflect the bulk behaviour of the material. Our binding energies are in a reasonable agreement with the reported experimental values but a slower convergency to the bulk values is observed. We have found that for those clusters that were full annealed, namely Al_{13}, Al_{55} and Al_{147}, the I_h and O_h structures are definitely non stable.

Acknowledgements: We gratefully acknowledge F. Liu and P. Jena for fruitful discussions.

REFERENCES

[1] R. O. Jones. Phys. Rev. Lett. **67**, 224 (1991).

[2] M. F. Jarrold, U. Ray, J. E. Bower, and K. Creegan. J. Chem. Soc. Faraday Trans. **86**, 2537 (1990).

[3] L. G. M. Peterson, Ch. W. Bauschlicher, and T. Halicioglu. J. Chem. Phys. **87**, 2205 (1987).

[4] T. H. Upton. Phys. Rev. Lett. **56**, 2168 (1986).

[5] H. P. Cheng and R. S. Berry. Phys. Rev. B **43**, 10647 (1991).

[6] J. Y. Yi, D. J. Oh, and J. Bernholc. Chem. Phys. Lett. **174**, 461 (1990).

[7] U. Rothlisberger and W. Andreoni. J. Chem. Phys. **94**, 8129 (1991).

[8] G. Pachioni and P. Fantucci. In "Physics and Chemistry of Small Clusters", ed. by P. Jena, B. K. Rao and S. N. Khanna. Plennum Press, New York, 1987, p. 843.

[9] A. Caro, S. R. de Debiaggi, M. Victoria, Phys. Rev. B **41**, 913 (1990).

[10] S. Debiaggi and A. Caro. To appear in J. Phys. Cond. Matter.

[11] R. Hoffman, J. Chem. Phys **39**, 1397 (1963).

[12] C. J. Ballhausen and H. B. Gray, in "Molecular Orbital Theory", Benjamin, New York (1965), p. 125.

[13] F. Liu and P. Jena, private communication.

[14] J. H. Rose, J. R. Smith, F. Guinea and J. Ferrante, Phys. Rev B **30**, 2963 (1984).

[15] T. H. Upton. J. Phys. Chem. **90**, 754 (1986).

[16] R. Car and M. Parrinello, Phys. Rev. Lett. **55**, 2471 (1985).

[17] K. Laasonen and R. M. Nieminen, J. Phys.: Cond. Matter **2**, 1509 (1990).

[18] A. L. Mackay. Acta Cryst. **15**, 916 (1962).

Molecular Dynamics Studies of the Melting and Freezing of CCl$_4$ Clusters

JIAN CHEN and LAWRENCE S. BARTELL
Department of Chemistry
University of Michigan
Ann Arbor, MI 48109

ABSTRACT. Phase transitions in a 225-molecule cluster of carbon tetrachloride have been studied by molecular dynamics simulations. A 5-site model potential function was developed to reproduce the density and heat of vaporization of the bulk liquid. Computations began with molecules in a nearly spherical cluster distributed in an orientationally disordered fcc structure at 5 K. Clusters were heated to 200 K in 10 degree steps of 50 ps each, then cooled to 10 K. Translational and rotational transitions were followed by monitoring the translational and rotational diffusion and mean rotational entropies of individual molecules. Melting began at the surface and propagated inward as the temperature increased. Translational and rotational melting occurred together. The solidification of molten clusters upon cooling proceeded from the center to the surface. At the high cooling rates of the simulations, however, molecules were unable to organize into crystalline arrays and solidified into a glassy structure, instead. Despite this difference between melting and freezing, there was a striking similarity between the heating and cooling curves portraying the fraction of molecules in liquid-like motion at any temperature.

1. Introduction

The study of phase transitions in clusters has attracted considerable attention during the last decade. A large proportion of the research has been carried out by molecular dynamics methods and most of it has focused upon monatomic systems [1,2,3]. Experimental and computational research in our laboratory on phase transitions, by contrast, is directed towards the dynamics of transformations in clusters of polyatomic molecules. Such clusters have very different properties than those of monatomic molecules. In a recent electron diffraction investigation of CCl$_4$ in a supersonic jet, the kinetics of freezing of supercooled, liquid clusters was investigated [4]. The aim of the present investigation is to study phase changes in clusters of CCl$_4$ by molecular dynamics (MD) simulations. Clusters of carbon tetrachloride formed by condensation in a supersonic expansion are of a plastically crystalline phase. Indeed, orientationally disordered phases are commonly generated when clusters are condensed from the vapor. Molecular dynamics techniques are uniquely suited for their ability to expose for analysis the molecular behavior responsible for the special properties of such phases.

Molecular orientation, of course, introduces elements of interest absent from the conventional MD simulations of clusters of Lennard-Jones spheres. We have developed

P. Jena et al. (eds.), Physics and Chemistry of Finite Systems: From Clusters to Crystals, Vol. I, 381–386.
© 1992 *Kluwer Academic Publishers.*

procedures to analyze the rotational behavior of molecules during melting and freezing processes. Noteworthy among the properties sharply altered in such phase transitions are the rotational diffusion coefficients and mean rotational entropies, quantities which can be computed for individual molecules as functions of the temperature and position in a cluster.

In Section 2 the MD computations performed for the carbon tetrachloride cluster are described. In Section 3 the procedures for analyzing the simulations are outlined and illustrative examples of their application are presented. The principal conclusions are summarized in Section 4.

2. Details of Computation

The cluster system studied consisted of 225 molecules of carbon tetrachloride. Molecules were considered to be rigid tetrahedra with bondlengths of 1.769 Å. Interactions between molecules were represented by pairwise-additive atom-atom potential energies expressed by Lennard-Jones functions. Potential parameters were, for ε (kJ/mol), 0.209 (C-C) and 1.117 (Cl-Cl), and for σ (Å), 3.80 (C-C) and 3.472 (Cl-Cl). For C-Cl, ε was taken as the geometric mean and σ was taken as the arithmetic mean of the C-C and Cl-Cl parameters. These parameters were chosen to yield the enthalpy of vaporization and molar volume of the liquid in a Monte Carlo computation based on 128 molecules in a cubic volume with periodic boundary conditions imposed.

The initial configuration for molecular dynamics simulations was a spherical cluster containing 225 molecules that was carved out of crystal with randomly oriented molecules distributed in an fcc lattice at 5 K. The disordered fcc structure corresponds to the metastable phase Ia of crystalline CCl_4, the phase which tends to form spontaneously when the liquid freezes under normal conditions[5]. Because the structures of the stable solid forms of CCl_4 are complex and unknown, it seemed reasonable in this preliminary study to model the fcc phase.

Molecular dynamics simulations consisted of a series of heating (or cooling) steps, each nominally 10 degrees warmer (or cooler) than the previous step. At each temperature examined runs were carried out for 10,000 time steps (5 fs each). The first 1000 time steps rescaled velocities in a heat bath of the desired temperature. Configurational energies and temperatures were then averaged over the next 9,000 steps run at constant energy. The very first run at 5 K was equilibrated with a heat bath for 5,000 time steps, and run at constant energy for 10,000 time steps. This procedure was repeated to allow the initial configuration time to relax before heating to 10 K.

3. Analysis of Results

1. Translational Diffusion

The motion of a molecule in a cluster is considered to be a combination of diffusion and vibration in the cage of neighboring molecules. For pure diffusion the mean-square translational displacement of a molecule is related to the coefficient of diffusion, D_r, by

$$<(r(t)-r(t=0))^2> = 6D_r t.$$

The conventional method to calculate the diffusion coefficient in a cluster is to determine the time-dependence of the displacements by following the statistical average $<(r(t)-r(t=0))^2>$ computed by sampling over all molecules. At low temperatures, where motions are pure vibrations, the mean-square displacement quickly rises to a constant value and D_r, of course, is negligible. At high temperatures the cluster is molten, all molecules diffuse, and the statistical average increases linearly with time. At intermediate temperatures, molecules at different sites in the cluster behave differently, and it is instructive to study the diffusion of individual molecules. Here, the conventional method using the statistical average over the molecules obviously cannot work. Even if the mean-square displacements of individual molecules were tabulated, the statistical averages would be very poor. Hence, it is necessary to resort to a differential technique averaging the slopes of mean-square displacements over many different segments of a molecular trajectory[6] as explained in the following.

Because the diffusion processes of molecules are Markovian, the statistical average of the squared displacement is independent of the time and depends only on the time interval. This fact enables us to extract from the chaotic trajectory of a single molecule enough independent samples with different initial times to calculate the temporal evolution of well-averaged values of the desired quantity. In other words, we calculate the mean-square displacement from the expression

$$\frac{1}{N}\sum_{i=1}^{N} (\mathbf{r}(t+t_i)-\mathbf{r}(t_i))^2$$

where N is the number of different initial times t_i sampled. This sum approaches the statistical average $<(\mathbf{r}(t)-\mathbf{r}(t=0))^2>$ as the number of samples increases. The diffusion coefficient is calculated from the slope of the curve. It is convenient to adopt a plausible minimum value of the coefficient D_r to identify molecules which can be considered to be undergoing genuine translational diffusion. Considerations of signal-to-noise ratios in the present work suggest that a suitable value for D_r to serve as a criterion of "translational melting" is 3.3×10^{-6} cm^2/s.

Presented in Figure 1 are the numbers of translationally diffusing molecules at different temperatures, according to the above criterion. Below 50K the entire cluster is frozen. As the temperature increases, the number of diffusing molecules steadily increases and the cluster begins to melt. When the temperature exceeds 160K, a temperature 90 degrees below the bulk melting point, the cluster is completely molten. When the molten cluster is cooled, the number of translationally frozen molecules increases. Examination of the positions of the diffusing molecules reveals that, in general, the further a molecule is from the center of the cluster, the higher is its diffusion coefficient. That is, the surface melts before the interior as the cluster is heated and freezes later as the cluster is cooled.

A visual inspection of images of clusters cooled from fully melted or partly melted aggregates clearly reveals that the freezing process is not the reverse of the melting process at the high cooling rate characteristic of the MD method. Freezing leads to a glassy structure, not to a crystal. The modest hysteresis displayed by the diffusion coefficient in Figure 1 is similar to that seen in transformations that are induced by comparable heating and cooling rates in cases where cooling yields a crystalline phase, however[7].

2. Rotational Diffusion

Coefficients for rotational diffusion can be defined in a manner analogous to those for translational diffusion. Since all bonds of carbon tetrachloride are equivalent for a molecule in an isotropic medium (and, therefore, are nearly equivalent for a molecule in a cluster) it is natural to follow the mean-square displacement of bonds from their directions in an initial configuration. For small angular displacements, coefficients of diffusion can be defined by the relation

$$\langle \theta(t)^2 \rangle = 4D_\theta t$$

where the factor of 4 corresponds to the two-dimensional character of the bond orientation.

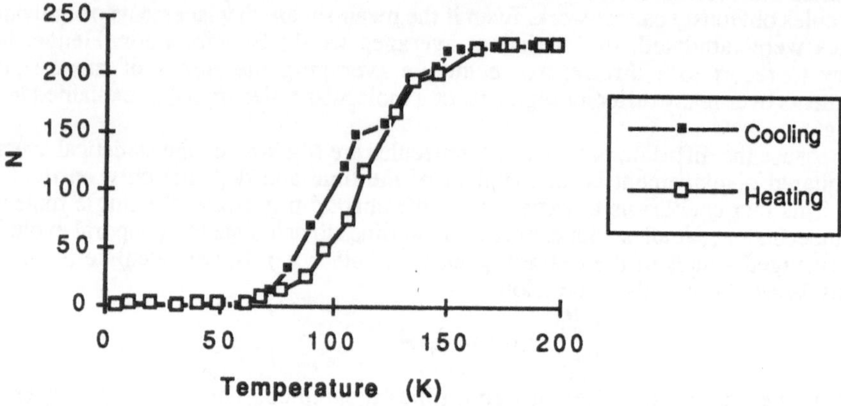

Figure 1. Number of molecules diffusing translationally during heating and cooling processes.

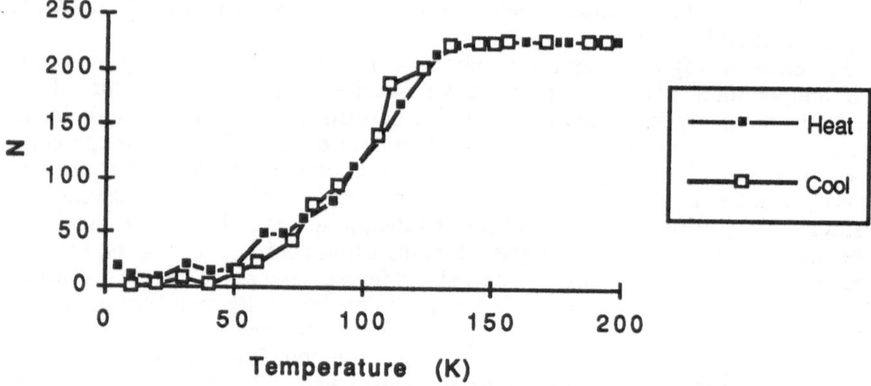

Figure 2. Number of molecules diffusing rotationally during heating and cooling processes.

Just as in the case of translational diffusion, it is desirable to monitor the diffusion coefficients for individual molecules. This requires, of course, that different segments of the rotational trajectories be averaged as they are for translational diffusion. For rotational diffusion, however, distinguishable angular displacements are confined to the range described by spherical polar coordinates. Therefore, the mean-square angular displacement can easily approach the limit $(\pi^2-4)/2$ in simulations of modest length. If the diffusion is diagnosed by taking the average over comparatively short segments of time, however, no problem arises. As for translational diffusion, it is expeditious to adopt a critical value for the coefficient of diffusion. Coefficients above this value are considered to indicate that the molecule in question is in a "rotationally melted" part of the cluster. A reasonable value in our simulations is 8×10^9 rad^2/s which corresponds to a displacement of Cl atoms of 2.5×10^{10} Å^2/s, a value of the same order of magnitude as the critical value 3.3×10^{10} Å^2/s adopted for translational diffusion. These critical values are both a comparable distance above the noise level of the simulations.

The number of molecules in rotationally melted portions of the cluster are plotted in Figure 2. Figure 2 closely resembles Figure 1 although, at a given temperature, a few more of the molecules appear to be diffusing rotationally than translationally. Rotational freedom is greater near the surface of the cluster just as it is for translation.

3. Rotational Entropy

It is well known that the Boltzmann H-function gives a numerical description of the disorder of a statistical system. The H-function decreases as the system becomes disordered. We introduce the orientational H-function to give a quantitative measure for orientational disordering.

The rotational Boltzmann H-function is defined as

$$H_R = \int f(\theta,\phi,\psi) \, log(f(\theta,\phi,\psi)) \, d^3\Omega$$

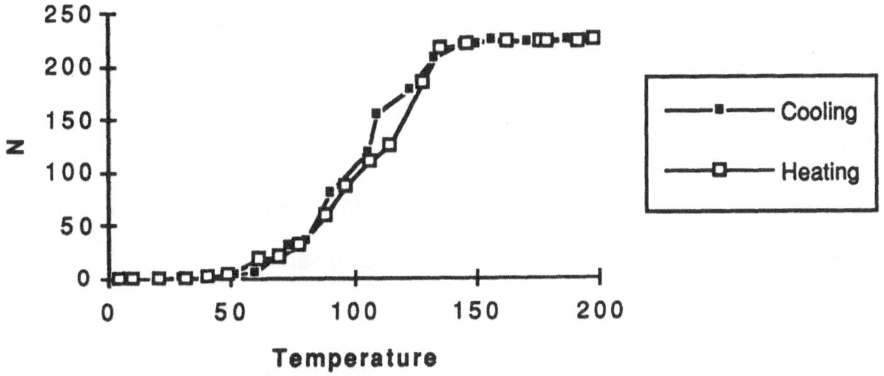

Figure 3. Number of rotationally mobile molecules during heating and cooling processes according to the rotational entropy.

where θ,ϕ,ψ are the Euler angles, $d^3\Omega$ is the volume element in the 3-dimensional space of Euler angles and $f(\theta,\phi,\psi)$ is a probability density. The rotational entropy $S_R = - k_B H_R$ is a part of the thermal entropy of the cluster. We use the rotational entropy, so defined, as an additional index of the rotational organization of clusters. As in the case of diffusion, the index can be calculated for an individual molecule if a histogram corresponding to $f(\theta,\phi,\psi)$ is accumulated over many time steps. The rotational entropy increases as the orientational disordering increases and approaches a limit when the probability density corresponds to a uniform distribution of molecule orientations. We choose a value slightly below the limit to be the critical value. If the rotational entropy of a molecule is greater than this value, the molecule is considered to be orientationally mobile.

The number of molecules in orientationally molten regions according the rotational entropy method is shown in Figure 3. We may tentatively conclude from the similarity of Figures 2 and 3 that the rotational entropy and rotational diffusion are comparable indicators of "rotational melting", even though the former samples the entire molecular orientation while the latter samples only the directions of bonds.

4. Discussion

The molecular dynamics simulations show that a 225-molecule cluster melts from the surface inwards when heated and freezes from the interior to the surface when cooled. Translational diffusion, rotational diffusion and molecular rotational entropy are effective gauges of phase change and indicate that clusters of CCl_4 melt translationally virtually as soon as they melt rotationally, unlike clusters of TeF_6 and benzene studied in parallel computations in our laboratory[8]. When CCl_4 clusters freeze at the high rates of cooling characteristic of MD simulations (2×10^{11}K/s in the present investigation) they become glassy rather than crystalline. This behavior is consistent with recently published experimental studies of the freezing of large clusters of CCl_4 which indicate that the maximum nucleation rate for freezing into the crystalline solid is $\sim 10^{29}$ $m^{-3}s^{-1}$, a peak rate which occurs in the vicinity of 170K. Therefore, because crystallization in the present studies of clusters whose volume is $3.6 \times 10^{-26}m^3$ would require a nucleation rate of $\sim 10^{36}$ $m^{-3}s^{-1}$, there is insufficient time for crystallization to occur. The near superposition of the heating and cooling curves in Figures 1-3, then, requires a certain correspondence between transitions between crystal and liquid, on the one hand, and between liquid and glass on the other. Further details about the molecular behavior occuring in the transitions will be described elsewhere[9].

5. Acknowledgement

Acknowledgement is made to the Donors of The Petroleum Research Fund, administered by the American Chemical Society, for the support of this research. We thank Messrs. S. Xu and F. Dulles for their expert assistance in computations.

6. References

1. Berry, R. S., Lecture at Varenna Summer School in Italy in 1988.
2. Cheng, H. P. and Berry, R. S. (1991) Mat. Res. Soc. Syp. Proc., **206**, 241
3. Jena, P., Rao, B. K. and Khanna, S. N. (eds.) (1987), Physics and Chemistry of Small Clusters, Plenum, New York.
4. Bartell, L. S. and Dibble, T. D. (1991), J. Phys. Chem., **95**, 1159.
5. Rudman, R. and Post, B. (1968) Molec. Cryst. **5**, 95.
6. Bartell, L. S., Dulles, F. J. and Chuko, B. (1991), J. Phys. Chem., **95**, 6481.
7. Bartell, L. S. and Xu, S., J. Phys. Chem., in press.
8. Dulles, F. J., Chuko, B. and Bartell, L. S., in this volume.
9. Chen, J. and Bartell, L. S., to be published.

MELTING AND SOLIDIFICATION OF ULTRAFINE PARTICLES INVESTIGATED BY OPTICAL AND ELECTRON MICROSCOPY TECHNIQUES

P. CHEYSSAC, R. KOFMAN and A. AOUAJ
Laboratoire de Physique de la Matière Condensée, URA 190
Université de Nice - Sophia Antipolis, 06034 NICE Cedex, FRANCE

ABSTRACT. The melting and freezing behaviours of lead inclusions, size ranging from 10nm to 100nm have been studied by optical and TEM measurements.The melting process of non spherical particles, influenced by local curvature effects, is compared to the melting of spherical particles. Inclusions exhibit a deep undercooling. Nucleation processes are controlled by various parameters: the influences of the nature of the matrix or the electron beam on nucleation are discussed.

1. Introduction

Theoretical as well as experimental efforts are in progress in order to advance our understanding of the physical processes which are involved in melting and freezing phenomena. Experimental results obtained by two complementary techniques, optics and transmission electron microscopy (TEM), on distributions of small metallic particles embedded in an amorphous dielectric matrix, are reported.

Small metallic particles are known to exhibit a thermodynamical size effect: the melting temperature Tm depends on the size [1]. A spherical particle is described by a constant radius of curvature which governs, among others, Tm. As a non spherical particle is described by local radii of curvature, influence of the geometry of the particle on Tm is expected. Experiments on spherical and non spherical particles are reported and a schematic representation of the melting of non spherical particles is proposed. In the liquid state, we show that a decrease in T leaves the particles in an undercooled state over a wide interval. The freezing temperature Ts may be influenced by various parameters such as the nature of the matrix or the experimental technique; Ts has been determined for individual spherical particles by electron microscopy. An unexpected influence of the electron beam on the freezing temperature has been observed.

2. Experimental procedures

2.1 THE SAMPLES

Distributions of small lead particles have been prepared by evaporation-condensation in UHV onto substrates suited for optical (sapphire) and TEM measurements (carbon copper grids).

Prior to the evaporation of lead, a thin layer of silicon monoxyde SiO is evaporated onto the substrates. Lead is evaporated and a suited thermal treatment is applied before covering the particles with another layer of SiO.This treatment enables a control of the mean radius of the size distribution and of the shape of the particles. For low temperature substrates (173K) this leads to non spherical particles which are not in their equilibrium shape. At higher temperature (473K) particles are obtained in the liquid state so with their equilibrium shape: a truncated sphere, due to the contact

387

P. Jena et al. (eds.), Physics and Chemistry of Finite Systems: From Clusters to Crystals, Vol. I, 387–392.
© 1992 *Kluwer Academic Publishers.*

angle characteristic of the wetting of liquid lead on SiO.

The second layer of SiO insures protection against oxydization. Such samples composed of lead inclusions in SiO show excellent reproducibility of the optical results: the geometrical parameters (obtained by image analysis) of the size distribution, as well as the shape of the particles, are retained as many cycles of melting and freezing are performed.

2.2 OPTICAL EXPERIMENT.

Let us remind here only its basic features; more details are given in [2].

Usually, the phase transitions of a material produce a change in a physical parameter. Depending upon the parameter under study, such a variation may be sharp. In particular, the loss of crystalline order during melting leads to a variation of the dielectric function of the material. Then, when T evolves, a jump in the reflectance R can be associated with a phase transition. High sensitive reflectance measurements are a very useful non disturbing method for the study of phase transitions. The results can be interpreted within the frame of a mean field theory, hence. the physical informations obtained are averaged values over sizes and states of the particles in the distribution.

2.3 TRANSMISSION ELECTRON MICROSCOPY EXPERIMENTS.

Conventional TEM and dark field electron microscopy (DFEM) masurements have been performed on a Philips CM12 microscope fitted with a low light level video camera. Contrary to optics, electron microscopy gives individual results for each particle, but it may be a disturbing technique.

TEM gives images suited for the determination of the geometrical parameters of the size distribution. Unfortunately, such a technique suffers a lack, namely the height of the particle along the direction of the electron beam remains unknown, at least for non spherical particles. For spherical particles, the contact angle being known, a measurement of the diameter gives its height.

DFEM has been described in details in [3] ; essentially, it consists in forming a true image from the diffracted beam. Depending upon the crystalline or amorphous state of the sample, diffraction imaging will result in variation of contrast. Such a technique enables to determine the state (solid or liquid) of a particle, and its evolution with T.

In TEM, with a heating holder, the sample temperature depends upon various parameters yielding errors in its determination. Heating of the sample by the electron beam is partially minimized by inserting the grids between two small apertures which form an oven; nevertheless, the measured temperature is the oven one.The temperature of the sample may either be lower or higher than the oven one, because radiation losses, limited thermal conductivity of the copper grid and carbon film and heating of the sample by the beam are competing.

3. Experimental results

3-1 REFLECTANCE

Experimental results are given (Fig.1) for two distributions of ultrafine spherical and non spherical particles; they have the same (projected) equivalent mean radius with a similar standard deviation, the number of particle per unit area (surface density) being slightly different. A comparison of their mass thicknesses dm shows that dm is higher (7nm) for spherical particles than for non spherical ones (4nm); taking account of the different surface densities increases this difference.This means that non spherical particles are flat, i.e. their "height", perpendicular to the substrate and usually non measurable with TEM, is smaller than their equivalent diameter. This has been checked by TEM measurements on cross-sections of ion thinned samples [4].

Let us remark that: when T varies, the variations of the optical constants of solid or liquid lead, SiO and sapphire, are weak (10^{-4} K^{-1} to 10^{-5} K^{-1}) so it can be demonstrated [5] that, for a single phase distribution, R(T) varies linearly. On the contrary, the optical constants of lead change sharply in a 10% range when the metal undergoes the solid to liquid transition.

Hence, any deviation of R(T) from a linear behaviour can be interpreted as a change in the filling factors of the solid or liquid lead.

3-1-1 *Melting* A comparison of Fig.1a and 1b shows that the temperature interval in which the whole distribution is solid extends to 546K for spherical particles, and only 484K for non spherical ones. Melting of non spherical particles begins at a lower temperature than for spherical ones, the end of melting being nearly the same (~ 589K) for the two distributions.

3-1-2 *Freezing* In both cases, a very large undercooled region appears. Spherical particles freeze below 464K, non spherical ones below 446K. In the two cases, the reflectance returns to a linear behaviour at T_L = 410K. Remark that, for smaller particles, this value can still be lower T_L ~ 392K [2], showing a very deep $(\Delta T_S/T_B ~ 0.347)$ undercooled state.

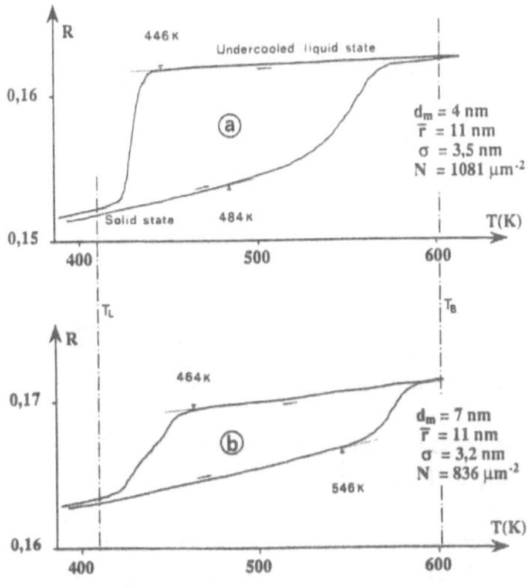

Fig. 1: Reflectance for two distributions of lead inclusions in SiO.
a : non-spherical particles   b : spherical particles

3-2 ELECTRON MICROSCOPY

3-2-1 Melting The melting behaviour of non spherical particles has been recently described [3]; the main result is: at temperatures just below Tm, a liquid layer appears at the surface of the particle (Fig.2), whose local thickness depends on curvature and temperature.

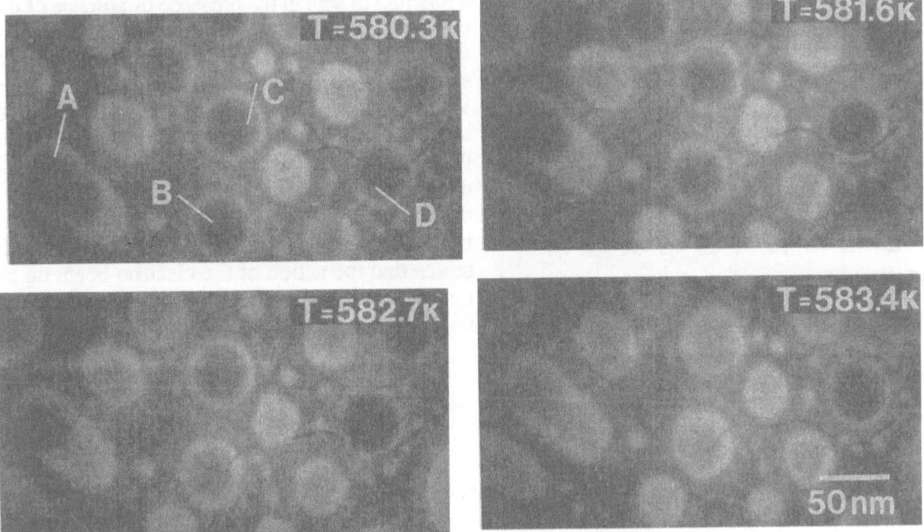

Fig. 2: Melting of non-spherical and spherical particles as shown by DFEM measurements. Particle A exhibits continuous melting. At T = 582.7 K particle B is melted. At T = 583.4 K particles A, B, C are melted while particle D is still solid.

This situation goes on over an interval of temperature of a few degrees until the remaining solid core is circular shaped. At this stage, the solid core melts suddenly (the melting time is lower than the time interval between two video frames: 40ms). So, melting of non spherical particles can be described as a continuous process followed by a first-order transition.

The same experiment has been performed on quasi-spherical particles; no continuous process can be observed; melting of spherical particles appears as a first-order transition (Fig.2). As pictures do not reflect the whole features of the phenomenon, it must be outlined that dynamical observation allow an easy determination of the sharp transitions.

3-2-2-*Freezing* For a given distribution, the melting and freezing intervals of temperature are known by optical measurements. Previous observations relative to undercooling have been confirmed; furthermore the influence of the electron beam on the freezing temperature of a distribution of particles has been studied with a two-steps procedure outlined below: each step begins when all particles are melted. The first step is realized with the electron beam switched on all the time. As T decreases, images of a given field are recorded. Some particles freeze and sharp variations in contrast determine Ts for each particle of the field; this corroborates the optically determined temperatures, taking into account the heating of the beam.

The second step is the following: all particles being melted, the beam is switched off and T is decreased down to a value T1 where some particles freeze. Then, one leaves the critical domain of temperature by increasing T up to T2 where particles which are not freezed remain liquid, while the solid ones are below their melting point and remain solid. The electron beam is switched on and images of the same field are recorded. This procedure is repeated by varying T1 by an amount $\Delta T = 10K$ and the sizes of the solid particles are determined as a function of T1. For the same particles, Ts has been determined as a function of size in the presence or absence of the electron beam. When the beam is on, Ts is known with the error discussed above. This error is taken symmetrical compared to the experimental data. When the beam is off, the only known temperature is T1, so Ts is in the interval T1+ ΔT. On Fig.3, the points are the experimental data, error bars are drawn in the worst case. One can clearly see that the average freezing temperature of the distribution is lower in the absence of the electron beam. It seems that the action of the electron beam on an undercooled material increases the density of nucleation centers.

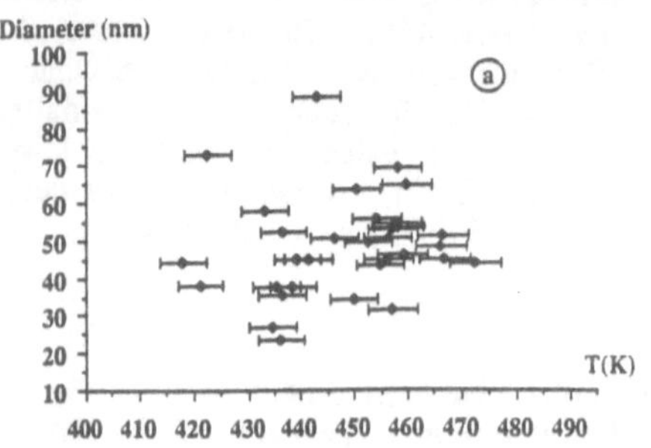

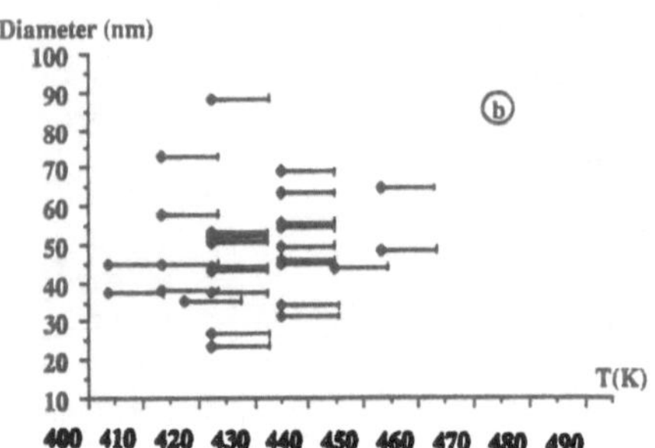

Fig. 3 : Freezing temperatures of a distribution of quasi-spherical particles. a: under electron beam, b: without electron beam

4-Discussion

4-1 MELTING

For a distribution of non spherical particles, as shown by reflectance measurements, the melting interval is larger than for a distribution of spherical particles having the same mean radius. A model of the melting process of non spherical particles can be elaborated from DFEM observations (see Fig.4). As a fairly good approximation, the particle is an oblate ellipsoid. A cross section along a plane perpendicular to the substrate reveals zones where the curvature is larger than the curvature of a spherical particle having the same equivalent radius.

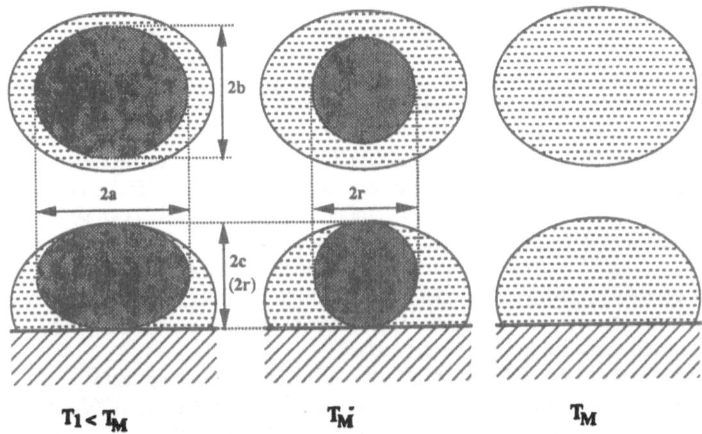

Fig. 4: Schematic representation of the melting of a non-spherical particle as it appears in DFEM (up) and as we assume it is in a plane normal to the substrate (down). T_M is the melting temperature of the final core.

Experiments show that the transition begins by a local nucleation of the liquid located at the higher curvature zone for the lower temperature. This transition evolves in a continuous and also reversible way until the solid core reaches a spherical shape. At this point, the core melts suddenly [3] in the same way than a spherical particle. Surface melting, as it has been presented by other workers [6] in the case of plane interfaces, certainly exists for curved interfaces but cannot be experimentally distinguished with our methods from the local melting associated with local curvature. The reason is that the surface melting liquid layer is very small (a few atomic layers).

For spherical particles, R(T) curves represent the thermodynamical size effect [1] alone, the width of the transition is only related to the polydispersion of the particle size. For non spherical particles, in addition to polydispersion exists a local melting associated with local curvature, this local melting increases the width of the transition on the low temperature side.

The role of the embedding matrix is questionnable: two different amorphous materials, SiO and Al2O3 have been used. For equivalent distributions of lead particles, the experimental results do not show significant differences.The wetting of such matrices by lead is low (the contact angle is large, > 100°), so the interaction between the inclusion and the matrix is very weak. Furthermore, even when the surface energies between lead and the matrix are not known, the rise of a liquid layer at the surface of the inclusion means an energetically favourable situation which is described by the following relation: $\gamma_{sl} + \gamma_{lm} < \gamma_{sm}$ (s for solid, l for liquid, m for matrix).Therefore, it seems that, on one hand, an amorphous dielectric matrix does not act upon the nucleation process of the liquid, unless by retaining the geometrical shape of the inclusion which insures the reproducibility of the experiment. On the other hand, experiments [7] performed on similar distributions of lead inclusions, but in a crystalline (Al) material, have revealed another type of behaviour: melting as shown by X-rays occurs at temperatures higher than T_B (till 67K above); a size effect still exists, as proven by the broadening of the transition region, but is superimposed to this superheating which means a strong interaction between Pb and Al. Lead is in epitaxy with Al, the surface of the inclusion being clamped to the matrix, nucleation of the liquid at the surface is delayed [8].

The existence of a phenomenon which may be influenced by a possible pressure of the matrix on the inclusion must also be discussed. It has been shown [9] that, in the case of an amorphous matrix, an increase of pressure due to the increase in temperature or to the partial melting would lead to results contradictory to experiments. One concludes that an amorphous matrix whose thickness is of the order of 10nm relaxes and does not apply a pressure which has to be taken into account for the discussion of experimental results. On the contrary, for a crystalline matrix, the effect of such a pressure must be discussed.

4-2 FREEZING

We have shown that lead inclusions in an amorphous matrix can be driven in a deep undercooled state ($\Delta Ts/TB \sim 0.347$). This must be compared to the recent 48m high drop-tube experiments performed in Grenoble [10] on millimeter sized particles of W and Re. Such particles are in containerless and weightlessness conditions, $\Delta Ts/TB \sim 0.144$ for W and $\Delta Ts/TB \sim 0.283$ for Re. The use of ultrafine particles embedded in an amorphous matrix under UHV conditions, by increasing $\Delta Ts/TB$, certainly enables a limitation in the number of impurities which can act as nucleation centers for freezing. When it is amorphous, the matrix does not seem to influence the freezing. On the contrary, using a crystalline matrix such Al lowers the undercooling to the order of $\Delta Ts/TB \sim 0.035$ [7], this can be interpreted by the existence of a high number of nucleation centers inside the inclusions. For an amorphous matrix, the high undercooling suggests that, for some particles, the limit of homogeneous nucleation has been reached.

The apparent symmetry between melting and freezing, as it appears on the optical results, suggests the existence of a size effect on freezing [2]. On the contrary, DFEM measurements seem to deny any correlation between size and freezing temperatures (see Fig.3). From the present state of our results, one can only say that, if a size effect exists for freezing, its interpretation cannot be based on an analogy with the thermodynamical size effect observed for melting. An effect related to the density of impurities which can act as nucleation centers, this density depending on the size of the particles, may rather be searched for.

Such an assumption is consistent with DFEM results showing that freezing takes place without being related to the size, because the density of impurities must fluctuate around a mean value.

At evidence, homogeneous and heterogeneous nucleation processes are both involved in the observed phenomenon; further experiments with various matrices appear still necessary to obtain a better understanding of these different regimes.

ACKNOWLEDGMENTS. The authors thank Y.Lereah and G.Deutscher from Tel Aviv University for valuable collaboration on part of this work and the Center for electron microscopy (CCMA) of the University of Nice-Sophia Antipolis where present experiments have been done.

REFERENCES:
1- Buffat., Ph. and Borel, J.P.(1976) 'Size effect on the melting of gold particles', Phys. Rev. A.13, 2287-2298.
2- Cheyssac, P., Kofman, R. and Garrigos, R. (1988) 'Solid-liquid phase transitions optically investigated for distributions of metallic aggregates. Absence of hysteresis for the smallest sizes', Physica Scripta 38,164-168.
3- Kofman, R., Cheyssac, P., Lereah, Y. and Deutscher, G. (1991) 'Melting of non-spherical ultrafine particles', Zeitschrift für Physik D, 20, 267-271.
4- Kofman, R., Cheyssac, P., Lereah,Y. and Deutscher, G. To be published
5- Cheyssac, P.(1987) 'Du monocristal à l'agrégat: leurs propriétés optiques appliquées à l'étude de la transition de phase solide-liquide', Thesis, Université de Nice.
6- Pluis, B., Denier van der Gon, A.W., Frenken, J.W.M. and Veen, J.F. van der (1987) 'Crystal-face dependence of surface melting', Phys. Rev. Lett. 59, 2678-2681
7- Bohr, J.(1991). 'Epitaxial clusters in single crystal hots', Zeitschrift für Physik D, 20, 215-218.
8- Daeges, J., Gleiter, H. and Perepezko, J.H. (1986) 'Superheating of metal crystals', Physics Letters A, 79-82
9-Lereah,Y., Deutscher, G., Cheyssac, P.and Kofman, R. (1991) 'Comment on "A direct observation of low-dimensional effects on melting of small lead particles" by J.Bohr', Europhys. Letters.14, 87-90
10- Vinet, B., Cortella, L., Favier, J.J.and Desre, P. (1991) 'Highly undercooled W and Re drops in an ultrahigh vacuum drop tube', Appl. Phys. Lett. 58, 97-99

MONTE CARLO STUDIES OF PHASE TRANSITIONS IN CLUSTERS OF TELLURIUM HEXAFLUORIDE AND BENZENE

FREDERIC J. DULLES, BOR-JING CHUKO, AND LAWRENCE S. BARTELL
Department of Chemistry
University of Michigan
Ann Arbor, Michigan 48109-1055

ABSTRACT. The melting, and in some cases, the freezing transitions of clusters of tellurium hexafluoride and benzene containing 12, 13 and 14 molecules, have been investigated by means of isothermal Monte Carlo simulations, using rigid model molecules with pairwise additive potentials between atomic sites (7 for TeF_6, 12 for benzene.) Various diagnostic methods were used to analyze results, including caloric curves, relative root mean square distance fluctuations (δ) and indices of translational and rotational diffusion. Results show transitions analogous to melting and freezing ocuring over a range of temperatures. Consequences of the magic number 13, so prominently displayed by Lennard-Jones spheres, were found to be more conspicuous in benzene clusters than in the case of TeF_6. Vibrational and librational motions of the molecules in cold, solid-like clusters were found to be much freer for TeF_6, but not for benzene, than in the bulk. Rotational diffusion in both cases began at lower temperatures than translational diffusion. As the small clusters were heated they fragmented more readily than larger clusters at the same temperature, and evaporated rapidly upon melting.

1. Introduction

Computer simulations of the properties of clusters have provided an intimate view of the behavior of molecules during phase changes. This information has led to an increased understanding of the dynamics and thermodynamics of transitions in condensed matter. Because only a small proportion of the simulations to date have been carried out on systems of polyatomic molecules, it seemed worthwhile to study several such systems in order to examine behavior that monatomic systems cannot exhibit, and to help clarify observations made in experimental investigations of clusters. We present here initial results of Monte Carlo computations performed on small clusters of benzene and tellurium hexafluoride.

2. Methods

Simulations were carried out with a Monte Carlo (MC) program differing only in minor details from one described in detail by Jorgensen and coworkers [1]. Interaction potential functions were those tabulated elsewhere [2,3]. In the case of benzene with 12 interaction sites, the pairwise additive intermolecular potential energies adopted were more complex than Lennard-Jones functions [2] while, for TeF_6 with 7 sites, the Lennard-Jones form was used [3]. Approximately spherical clusters of 12, 13, and 14 molecules were

P. Jena et al. (eds.), Physics and Chemistry of Finite Systems: From Clusters to Crystals, Vol. I, 393–398.
© *1992 Kluwer Academic Publishers.*

constructed to have the molecular packing observed in bulk crystals, orthorhombic for benzene and bcc for TeF_6. Simulations began at low temperatures where the clusters quickly relaxed to structures of lower energy. Temperatures were increased in steps of 5 K every million trial moves. In some cases warm clusters were cooled back to low temperatures at the same rate. Move sizes were increased as the temperature was increased to maintain an acceptance rate for moves of approximately 45%. Molecular coordinates were saved every 1000 moves for benzene and every 2000 for TeF6, and thermodynamic averages were accumulated.

Analyses of the runs included the conventional indicators of melting, namely caloric curves and the Lindemann index δ, and new indices to diagnose translational and rotational diffusion. The latter indices, R_D, based on a random walk model, were constructed to remove the arbitrariness of the Monte Carlo move size. Amplitudes of translational and rotational oscillations of the molecules (as distinguished from their migration by diffusion) were also determined.

3. Results

All clusters relaxed to structures with polytetrahedral motifs at low temperature, making the 13- molecule clusters more or less icosahedral. As they warmed they became increasingly labile and deformed. Although some of the benzene clusters suffered fragmentation at temperatures lower than those reached in the data presented, clusters remained intact in all of the simulation results reported herein.

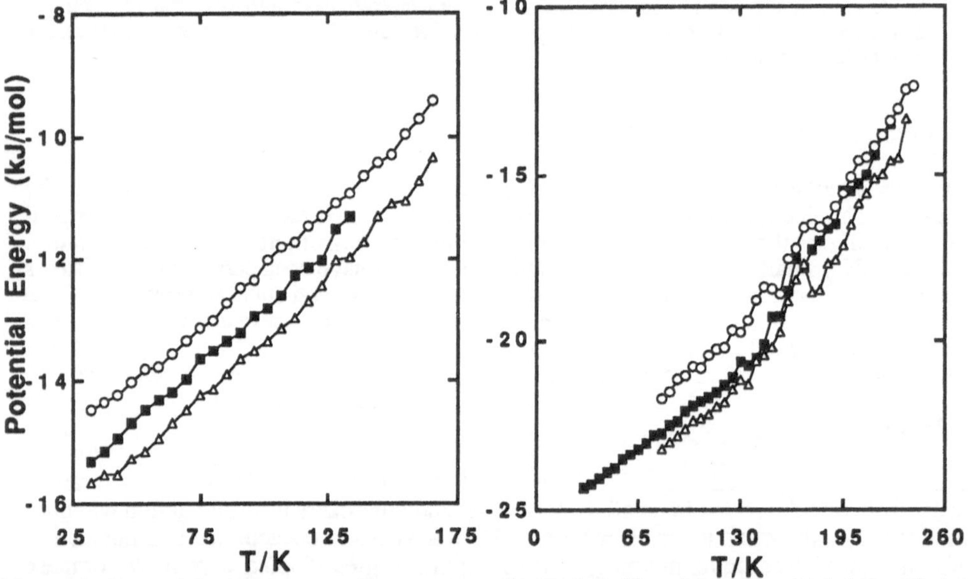

Figure 1. Potential energy versus Temperature for TeF_6(left) and benzene (right) for 12-(O), 13-(■) and 14-(Δ) molecule clusters

Plotted in Figure 1 are caloric curves for clusters of the two compounds. Several aspects deserve comment. Although analyses to be described later indicate that TeF_6 clusters are solid-like at the lower temperatures shown and molten at the higher temperatures, there are no conspicuous breaks in the curves of the type typical of phase changes observed in systems of Lennard-Jones spheres [4] and in larger clusters of TeF_6.[5] Apparently, the transition from a fairly rigid to a very loose cluster is spread out over a large range of temperature. In addition, the relatively even spacing between the curves for the 12-, 13-, and 14- molecule clusters indicates that 13-molecule clusters enjoy no special stability of the sort characteristic of van der Waals clusters of monatomic molecules. Benzene is strikingly different from TeF_6 as can be seen in Figure 1. Its clusters do exhibit a special stability at the "magic number" of 13, and a melting transition is more evident in the caloric curves. Runs of only one million moves, however, do not provide sufficient statistical averaging to yield smooth caloric curves. The oscillations are indicative of chaotic excursions between solid-like and liquid-like structures in the transition region.

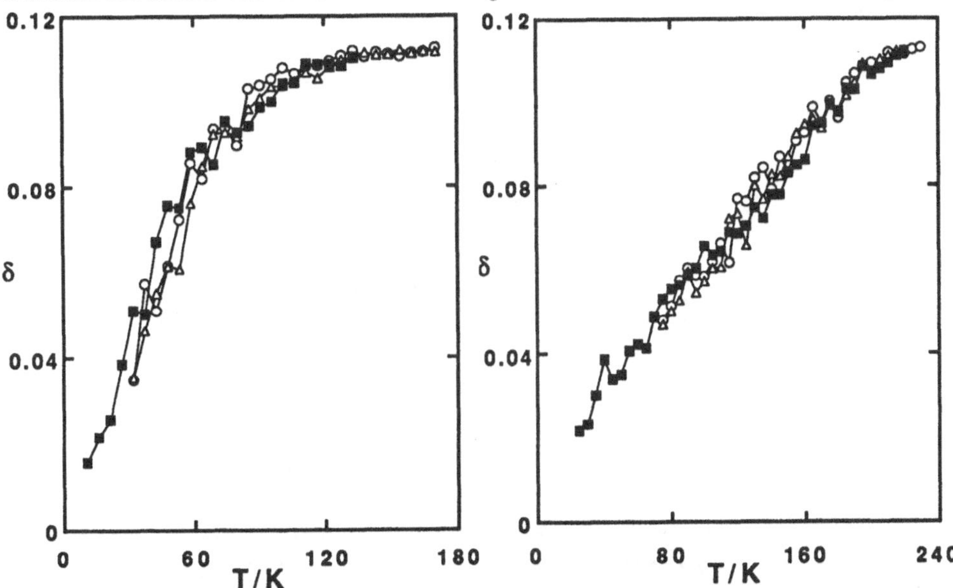

Figure 2. Lindemann index (δ) for TeF_6(left) and benzene (right) for 12-($\circ$), 13-($\blacksquare$) and 14-($\triangle$) molecule clusters.

That the clusters are undergoing a transition over the range of temperature studied is shown more sensitively in the Lindemann index depicted in Figure 2 than in the caloric curves. The Lindemann index δ represents the root-mean-square vibrational amplitude of centers of mass of pairs of neighboring molecules relative to the mean distance between them. This manifestation of thermal motion has been demonstrated to change markedly during melting transitions of atomic clusters, the liquid-like phase being attained when the index exceeds 0.1 [4]. As has been explained elsewhere [5], the critical value for melting is rather less than 0.1 for clusters of polyatomic molecules. By this criterion both the TeF_6 and benzene clusters melt well before the simulations are terminated, according to Figure 2.

On a molecular scale, the property that most essentially discriminates between liquid and solid phases is the freedom of molecules to move among each other. This property of self-diffusion is most naturally defined in terms of the time-rate of increase of the mean-square displacements of molecules from their initial positions. In Monte Carlo computations the increase with time must be replaced by the increase with cycle number. A dimensionless index, R_D, has been devised to express the self diffusion in Monte Carlo runs [5]. It is unity when molecules execute completely free random walks, and is smaller, the more constrained the motions are. For translational diffusion in bulk liquids it may be 0.2 or somewhat less, and for bulk crystals it is far smaller than 0.1. From the results presented in Figure 3 it appears that the clusters are indeed becoming liquid-like as the temperature is raised. On the other hand, whereas the indices for large clusters level off when clusters have melted fully (see Figure 4), they have not leveled off for the present very small clusters by the time temperatures are reached at which evaporation becomes a serious problem. Therefore, Figures 2 and 3 are somewhat at variance with each other in implications about the degree of completion of melting over the temperature range indicated.

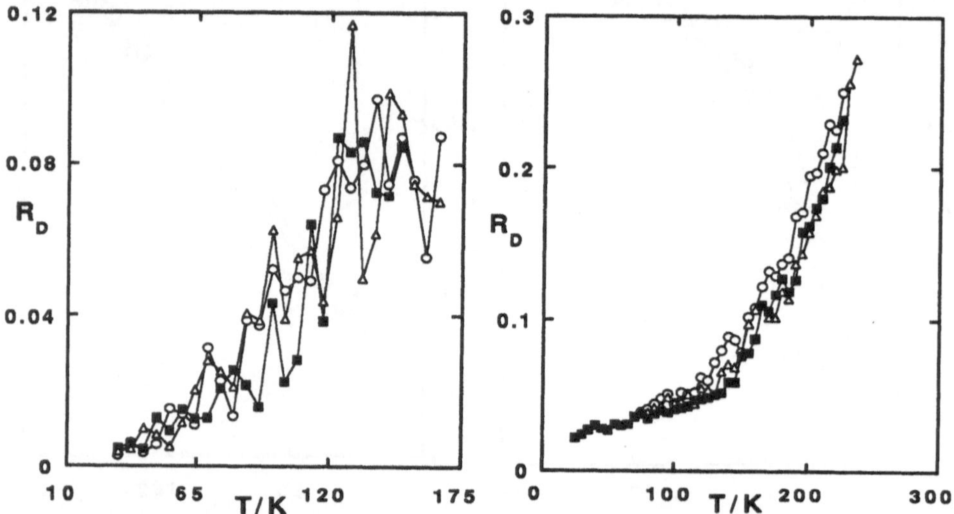

Figure 3. Index R_D of translational self diffusion versus temperature for TeF_6 (left) and benzene (right) clusters of 12-(O), 13-(■) and 14-(Δ) molecules.

An additional complicating factor in melting is the occurrence of rotational diffusion. In the example of a 225-molecule cluster of CCl_4 described elsewhere in this volume [6], rotational melting occurred virtually in step with translational melting. The present extremely small clusters are quite different. Tellurium hexafluoride molecules rotate with little hindrance at the lowest temperatures, well before they diffuse translationally. Benzene molecules are more restrained in this respect but still begin to tumble markedly before they are warm enough to diffuse away from their initial sites. Curiously, when they are cold they flip about their two-fold axes more readily than they spin about their six-fold axes.

4. Discussion

Very small clusters of polyatomic molecules exhibit unmistakable signs of transforming from solid-like to liquid-like entities as they are warmed. The transitions are not as clearly defined as they are for larger clusters or even as they are for clusters of Lennard-Jones spheres of the same size [4]. This may be a consequence of the rotations of the quasispherical molecules before translational motions become free. One element of interest is the fact that benzene, which is less spherical than TeF_6, exhibits special stability in 13-molecule clusters (just as do clusters of spherical atoms) whereas TeF_6 does not.

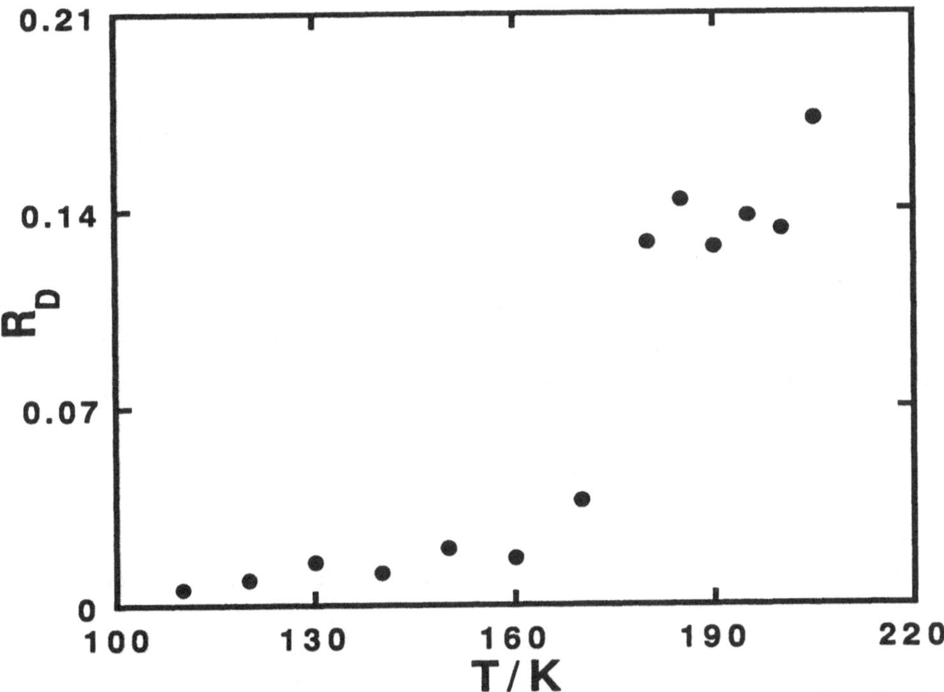

Figure 4. Index R_D as applied to translational diffusion of TeF_6 clusters of 128 molecules. Results from reference 5.

A plausible rationale for this behavior is that benzene packs naturally in an orthorhombic structure which is as close to a cubic closest packing arrangement (fcc) as is allowed by the molecular shape. Argon crystals are fcc, and 13-molecule fragments (cuboctahedra) readily transform to the favored icosahedral structure by undergoing small deformations. TeF_6, by contrast, naturally packs into bcc crystals (or the very closely related monoclinic form under some conditions) where the coordination is entirely different. Fragments of TeF_6 with 13 molecules, therefore, are not especially stable relative to those with 12 or 14 if the packing geometry tends toward that in the crystal. In addition, the fact that the compact quasi-icosahedral form that the molecules arrange themselves into at low temperature is not closely related to a natural crystal structure of TeF_6 probably accounts for the high degree of rotation in even the coldest clusters. The absence of a natural intermeshing of the molecules corresponds to a low barrier to rotation.

The more favorable fitting together of benzene molecules is indicated by their restrained oscillatory motions. At temperatures below the softening transition of the clusters the amplitudes of translational and rotational vibrations of molecules are little different from their values in crystalline benzene at the same temperature [7] The fact that molecules in a cluster are not as fully surrounded as they would be in the crystal seems to be compensated for by the extra number of attractive contacts in an icosahedron relative to a cuboctahedron. Further results of computations for clusters of benzene and TeF_6 will appear elsewhere [8].

5. Acknowledgements

This research was supported by a grant from the National Science Foundation. We gratefully acknowledge the awarding of a Regents´ Fellowship to F.J.D. and a generous allocation of computing time from the University of Michigan Computing Center.

6. References

1. Jorgensen, W.L., Binning, R.C., Jr. Bigot, B. (1984) J. Am. Chem. Soc. 103, 4393.
2. Bartell, L.S., Sharkey, L.R., Shi, X. (1988) J. Am. Chem. Soc. 110, 7006
3. Bartell, L.S., Powell, B.M., (1991) Mol. Phys., in press Bartell, L.S., Xu, S. (1991) J. Phys. Chem., in press
4. See e.g.: Honeycutt, J.D., Andersen, H.E. (1987) J. Phys. Chem. 91, 4950
   Etters, R.D., Kaelberer, J.B. (1977) J. Chem. Phys.. 66, 3233
   Davis, H.L., Jellinek, J., Berry, R.S. (1987) J. Chem. Phys., 86, 6456
   Beck, T.L., Jellinek, J., Berry, R.S.,(1987) J. Chem. Phys., 87, 545
   Quirke, N. (1988) Mol. Simulation, 1, 249
5. Bartell, L.S., Dulles, F.J., Chuko, B. (1991) J. Phys Chem. 95, 6481
6. Chen, J., Bartell, L.S., this volume
7. Jeffrey, G.A., Ruble, J.R., McMullan, R.K., Pople, J.A. (1987) Proc. R. Soc. Lond. A, 414, 47
8. Dulles, F.J., Chuko, B., Bartell, L.S., to be published

Vapor Phase Homogeneous Nucleation of Polar Molecules

M. Samy El-Shall, D. Wright and R. Caldwell
Department of Chemistry
Virginia Commonwealth University
Richmond, Va. 23284-2006, USA

ABSTRACT. Homogeneous nucleation in supersaturated vapors of highly polar substances such as CH_3CN, CH_3NO_2, C_6H_5CN and $C_6H_5NO_2$ has been studied by use of a diffusion cloud chamber. The results are in serious disagreement with the predictions of the classical nucleation theory. The discrepancy has been explained in terms of dipole-dipole interaction within the curved surface of the embryonic droplets.

I. Introduction

There is currently an enormous growth of interest in studying the kinetics and thermodynamics of homogeneous nucleation in supersaturated vapors [1,2]. Vapor phase homogeneous nucleation involves the decay of a metastable state, i.e., a supersaturated vapor, by the spontaneous occurence of large thermal fluctuations, that is, through the formation of droplets of the liquid phase. Droplets that are larger than a critical size grow, and thus the stable phase results. The supersaturation at which the metastable state collapses and the nucleation rate increases explosively is known as the critical supersaturation. The experimental studies of homogeneous nucleation are usually concerned with two fundamental measurements. These are the determination of the temperature dependence of the critical supersaturation for the onset of nucleation and the determination of the temperature and supersaturation dependences of the rate of nucleation.
The most popular nucleation theory, the so-called classical nucleation theory (CNT), is based on the "capillarity approximation" [3,4]. This amounts to assuming that the clusters of the new phase are liquid drops, having all of the features of macroscopic drops,

399

P. Jena et al. (eds.), Physics and Chemistry of Finite Systems: From Clusters to Crystals, Vol. I, 399–404.
© 1992 *Kluwer Academic Publishers.*

including a well defined interfacial tension. This theory possesses reasonable predictive capability for the critical supersaturation required for the onset of nucleation of many substances [5]. Most of these substances are either non-polar or weakly to moderately polar (dipole moment less than 2 Debye) and it appears that the nucleation of highly polar liquids has not been investigated. The goal of our study is to measure the homogeneous nucleation rates of polar substances and compare the results with the predictions of several nucleation theories. These measurements can provide important correlations between the homogeneous nucleation process and the molecular properties of the system under investigation such as dipole moment and polarizability. In a previous report we presented results for the homogeneous nucleation of acetonitrile [6]. Those results showed that the critical supersaturation required for the onset of nucleation, as measured in a diffusion cloud chamber, is higher than the prediction of the CNT by more than 40%. We attributed this discrepancy to an oriented dipole-dipole interaction within the curved surface of the embryonic droplets which causes the free energy barrier for nucleation to increase relative to the randomly oriented dipoles in weakly polar molecules. In the present study we extend that investigation to other polar molecules containing a polarizable phenyl group such as benzonitrile and nitrobenzene. This enables the mutual effects of dipole moment and polarizability on the nucleation threshold to be studied.

2. Experimental

The homogeneous nucleation data were obtained by use of an upward thermal diffusion cloud chamber [5,6]. A typical chamber is illustrated in Figure 1. The chamber is designed so that to a high degree of approximation, one dimensional diffusion takes place through a carrier gas from a lower heated pool of liquid to an upper metal plate at which the vapor molecules are condensed. A solution of the one dimensional equations for heat conduction and binary diffusion then provides the variation of temperature and partial pressure in the chamber. As vapor pressure is approximately an exponential function of temperature while temperature and partial pressure profiles in the chamber are almost linear, the vapor is supersaturated throughout. The supersaturation can be made as large as desired by increasing the temperature gradient. The critical chamber state is defined as the point at which the temperature difference across the chamber is sufficient to yield a regular rain of drops (~1 drop/cm^3/sec). The onset of nucleation is determined by observing the forward scattering of light from drops falling through a horizontal He-Ne laser beam positioned at an elevation lower than 0.5 (reduced height) of the cloud chamber. These drops originate near the elevation at which the maximum (peak) supersaturation occurs (0.7 reduced height). Critical supersaturations (S_C) determined in the chamber are plotted against temperature as shown by the individual peaks in Figure 2. Each of these peaks represents several points around the maximum supersaturation achieved in the chamber at the temperature of the experiment. The "envelope" of these many peaks constitutes the curve of S_C vs. T, obtained from experiment. The dashed line in Figure 2 represents the theoretical predictions of S_C vs. T as obtained from the classical nucleation theory.

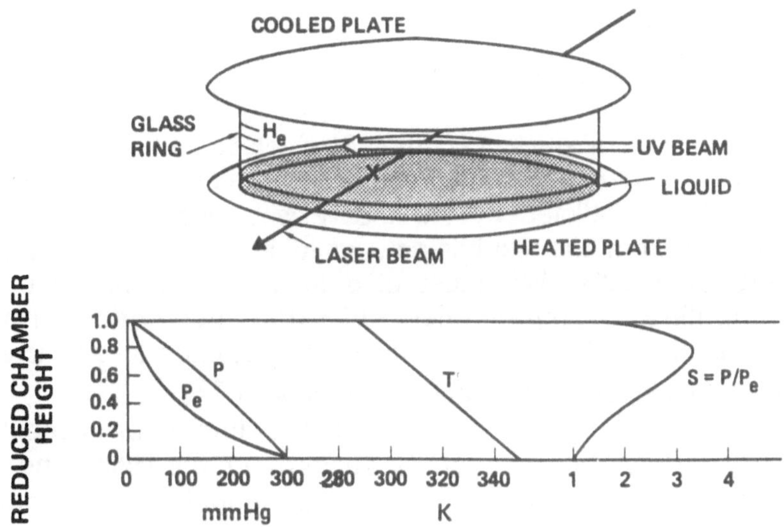

Figure 1 : (a) Sketch of diffusion cloud chamber with relevant components indicated.
(b) Typical courses of diffusant partial pressure and supersaturation.

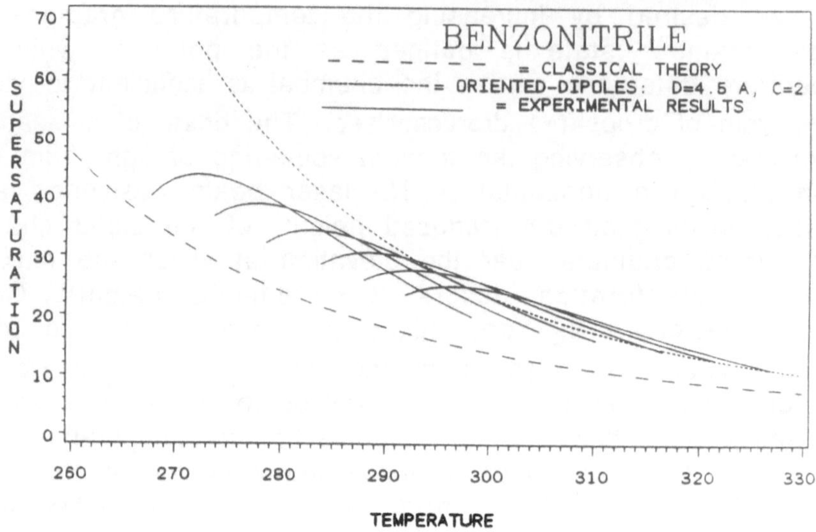

Figure 2: Critical supersaturation vs. temperature for benzonitrile. Dashed line represents theory. Dotted line represents oriented dipole model with d = 4.5 Å and c = 2 .

Results and Discussion

It is clear from Figure (2) that the classical theory deviates strongly from the experimental results. For example, at 280 K, S_C for benzonitrile predicted by the theory is at least 44% lower than that obtained from the experiment, while at 310 K the deviation is still more than 38%. We must also indicate that because of the sharp nonlinear dependence of the nucleation rate on supersaturation a 30% reduction in S_C corresponds to an increase in the rate of nucleation by a factor of nearly 10^{12}.

As indicated earlier, the classical theory has been accepted as essentially correct because of its success in predicting the critical supersaturations (within 5%) for many nonpolar and moderately polar substances over a wide temperature range. Therefore, the discrepancy in Figure (2) appears to indicate a significant failure of the classical theory in predicting the homogeneous nucleation of highly polar liquids. Similar discrepancies for nitromethane (μ=3.5D) and nitrobenzene (μ=4.3D) have been found. These observations can be explained by considering the effect of the dipole-dipole interaction of the oriented surface molecules on the surface free energy of the critical clusters. It can be shown that dipole-dipole interaction of the highly oriented surface molecules in a spherical surface will result in an increase in the surface free energy of the droplet. The deviation from parallel orientation involves an increase in the surface potential energy which is given by

$$U_s = (1/2) \; [\mu^2 (1-\cos f) \; c \; j_s] \; / \; 2d^3$$

where μ is the magnitude of the dipole moment, f is the angle between any two nearest neighbor surface dipoles, c is the number of surface nearest neighbors of a surface molecule, d is the distance between nearest neighbors and j_s is the total number of molecules at the surface of the droplet. Therefore, the total surface

$$F_n(s) = \sigma \, (r) \; A = \sigma^0 \, A + U_s$$

where $\sigma \, (r)$ is a surface tension dependent on the droplet radius due to the effect of surface curvature on dipole-dipole interaction and σ^0 is the flat surface tension used in the classical theory. Thus, a corrected surface tension can be expressed as

$$\sigma \, (r) = \sigma^0 + 2\mu^2 c \; / \; [4\pi d^3 \; (3v/4\pi)^{2/3} \; n^{2/3}]$$

where v is the volume per molecule in the bulk liquid. As indicated in Fig.(2), the critical supersaturation calculated from the classical theory by using the corrected surface tension compares rather well with the experimental results. This result supports the assumption of oriented dipoles within the curved surface. With the corrected surface tension, it is now possible to estimate the size of the critical nuclei involved in the homogeneous nucleation of these polar liquids. Figure 3 exhibits a plot of the critical size as a function of temperature. The critical size was calculated using the experimental S_c and the corrected surface tension, $\sigma \, (r)$ obtained from the dipole orientation model.

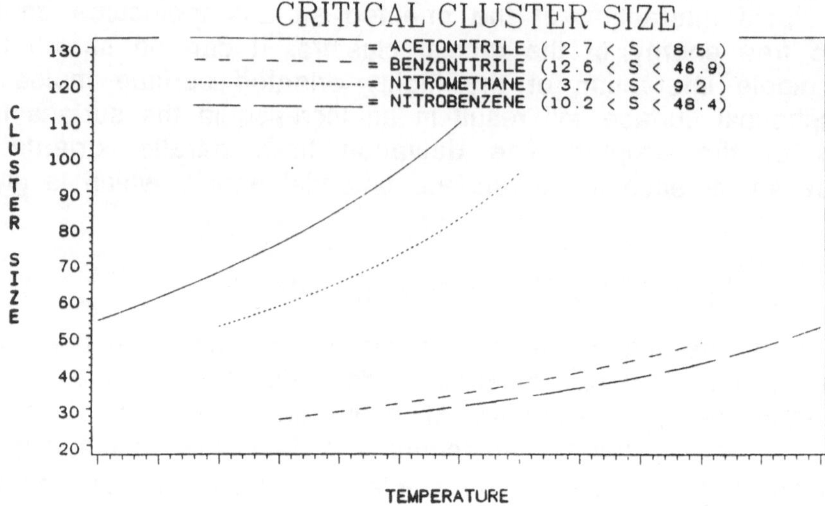

Figure 3: Critical size corresponding to the onset of nucleation as a function of temperature.

Conclusions

A significant disagreement between the experimental results and the predictions of the classical nucleation theory has been found for highly polar liquids ($\mu > 3D$). The failure of the classical theory to explain the nucleation of highly polar liquids has been attributed to the dipole-dipole interaction within the curved surfaces of the critical clusters.

Acknowledgement

Acknowledgement is made to the donors of the Petroleum Research Fund, administered by the American Chemical Society and to the Thomas F. and Kate Miller Jeffress Memorial Trust for the partial support of this research.

References

1."Atmospheric Aerosols and Nucleation" edited by P.E.Wagner and G.Vali, Lecture Notes in Physics, vol.309 (Springer, Berlin, 1988).
2.F.F.Abraham,"Homogeneous Nucleation Theory" (Academic Press, New York,1974).
3.See for example: "Nucleation", edited by A.C.Zettlemoyer (Marcel Dekker, New,York,1969).
4. J. Frenkel, "Kinetic Theory of Liquids" (Dover, New York, 1955), Chap.7.
5. M.S. El-Shall, J Chem. Phys. 90. 6533 (1989) and references therein.
6. D. Wright, R. Caldwell and M.S. El-Shall, Chem. Phys. Lett. 176, 46 (1991).

MELTING OF NICKEL CLUSTERS

I. L. GARZÓN[1] AND J. JELLINEK[2]

[1]*Instituto de Física, Universidad Nacional Autónoma de México, Apartado Postal 2681, 22800 Ensenada,
Baja California, México*
[2]*Chemistry Division, Argonne National Laboratory, Argonne, Illinois 60439 USA*

ABSTRACT.   The meltinglike phenomenon in Ni_n, n = 19,20,55, clusters is studied using microcanonical molecular dynamics simulations. The interaction between the atoms in the clusters is modelled by a size-dependent Gupta-like potential that incorporates many-body effects. The clusters display the "usual" stages in their meltinglike transition, which characterize also Lennard-Jones (e.g., noble gas) and ionic clusters. In addition, Ni_{20} passes through a so-called premelting stage found earlier also for Ni_{14}.

1.   Introduction

Phase changes, especially the meltinglike phenomenon, is one of the challenging and ever more active areas of cluster and small particle research [1]. Until recently, the bulk of the theoretical work concentrated primarily on Van der Waals (Lennard-Jones) systems. Most of the experimental studies, however, were performed on metal microparticles [2]. Different microscopy techniques were utilized to detect the melting (or the solidification) transition and to characterize the size dependence of the melting temperature of metal particles in the micrometer and nanometer size ranges. Only in a few theoretical studies has the phenomenon of a solid-to-liquid like transition in small metal clusters been considered. Sawada and Sugano [3] used a Gupta-like potential to characterize the meltinglike behavior of 6- and 7-atom transition metal clusters. They identified a so-called fluctuating state in the 6-atom cluster in which structural rearrangements can take place in a diffusionless manner. Ercolessi et al [4] employed the so-called "glue" potential to study the melting behavior of the Au_n, n = 219, 477, and 879. Surface melting as a precursor to complete melting has been found for n = 477 and 879. Surface melting in such a small cluster as Cu_{55} has also been detected in simulation studies of Cheng and Berry [5]. We have performed a molecular dynamics investigation of the meltinglike transition in 12- 13-, and 14-atom transition metal clusters [6] and 6-, 7-, and 13-atom gold clusters [7] using a size-dependent Gupta-like potential. For Ni_{13}, only a small reduction (~6%) in the melting temperature, as compared to the bulk value, is predicted. The corresponding reduction in the melting temperature of Au_{13} is estimated at ~50%. A novel intermediate stage in the meltinglike transition in clusters, which we called "premelting" [6] and which is, in general, different from the so-called "coexistence" [1], has been found for Ni_{14}. Both the qualitative and the quantitative features of the nickel clusters mentioned have been reproduced recently using an embedded-atom potential [8].

Here we report results on the meltinglike behavior of Ni_n, n = 19, 20, 55, derived from the size-dependent Gupta-like potential. The theoretical background and the computational

405

P. Jena et al. (eds.), Physics and Chemistry of Finite Systems: From Clusters to Crystals, Vol. I, 405–410.
© *1992 Kluwer Academic Publishers.*

procedure are sketched in the next section. The results are presented and discussed in section 3. A summary is given in section 4.

2. Theoretical Background and Computational Procedure

The dynamical behavior of the clusters at different fixed total energies was obtained by solving numerically Newton's equations of motion for all the degrees of freedom. The forces acting on the atoms were calculated from a size-dependent many-body Gupta-like interaction potential, which is written in reduced units of the potential energy V^* and interatomic distances r_{ij}^* [6] as

$$V^* = \frac{1}{2} \sum_{j=1}^{n} \left\{ A \sum_{i=1}^{n} \exp(-p(r_{ij}^*-1)) - \left[\sum_{i=1}^{n} \exp(-2q(r_{ij}^*-1)) \right]^{1/2} \right\} \qquad (1)$$

The values of the parameters for nickel are: $A = 0.101036$, $p=9$, $q=3$ [3,6]. The absolute potential energy V and distances r_{ij} are obtained from their corresponding reduced values V^* and r_{ij}^* using material- and size-dependent units of energy U_n and distance r_{on}: $V = V^* \cdot U_n$, $r_{ij} = r_{ij}^* \cdot r_{on}$ [6]. The clusters were prepared initially with zero total linear and angular momenta. The Verlet algorithm [9] was utilized to integrate the equations of motion. Trajectories of length of 10^5 - 10^6 steps, with a step size of $7.8 \cdot 10^{-16}$s, were generated on a grid of total energies (per atom) large enough to observe the solid-to-liquid like transition in each cluster. The total energy in the individual runs was conserved within 0.75%.

To detect and characterize the meltinglike transition, a variety of quantities was calculated along the trajectories. Of these we present in the next section the following: (1) Short-time (over 500 steps) average of the kinetic energy per atom as a function of time. The goal of this averaging is to attenuate the fluctuations in the kinetic energy caused by the vibrational motions. (2) Caloric curve, i.e., long-time (along the entire trajectory) average of the kinetic energy per atom as a function of the total energy per atom. (3) Root-mean-square (rms) bond length fluctuation δ,

$$\delta = \frac{2}{n(n-1)} \sum_{i<j}^{n} \frac{(\langle r_{ij}^2 \rangle - \langle r_{ij} \rangle^2)^{1/2}}{\langle r_{ij} \rangle} \;, \qquad (2)$$

as a function of temperature κT (in energy units),

$$\kappa T = \frac{2 \langle E_\kappa \rangle}{3_n - 6} \;, \qquad (3)$$

where E_κ is the total kinetic energy of the atoms in a cluster, κ is the Boltzmann constant, and $< >$ stands for the long-time average.

3. Results and Discussion

We have determined the minimum energy structures of the clusters performing numerical thermal quenching. The double icosahedron is the most stable geometry of Ni_{19} ($V^* = -18.641$). The most stable configuration of Ni_{20} ($V^* = -19.646$) is the 19-atom double icosahedron with an additional atom placed over a threefold face between two fivefold rings. The most stable structure of Ni_{55} ($V^* = -57.400$) is a two-shell icosahedron.

Figure 1 displays the short-time averaged kinetic energy per atom as a function of time for different total energies per atom. As the total energy increases, the pattern of the graphs for Ni_{19} and Ni_{55} changes from an essentially constant, to moderately fluctuating but single-branched, to multi-branched, and eventually to highly fluctuating with no individual branches resolvable. This evolution of the pattern is similar to that observed in certain size Lennard-Jones and ionic clusters [1], as well as in Ni_{12} and Ni_{13} [6,8]. Graphs (a) correspond to low-energy rigid clusters, graphs (b) - to "softer" but still preserving their structure clusters, graphs (c) - to spontaneous isomerization transitions (the different branches represent different isomers), and graphs (d) - to high-energy "melted" or liquid-like clusters. The changes in the pattern for Ni_{20} are, however, different. First the short-time averaged kinetic energy converts from a nearly constant (a) to a two-branched (b), then again to a single-branched albeit with larger fluctuations (c), then to multi-branched (d), and eventually to strongly fluctuating with no separable branches (e). A similar evolution of the graphs with the total energy is characteristic also for Ni_{14} [6,8]. It has not, however, been found in 14- and 20-atom Lennard-Jones or ionic clusters. As in the case of Ni_{14}, the two-branched curve (b) corresponds to the "premelting" stage, which involves only a limited topological rearrangement in the vicinity of the additional to the double icosahedron "surface" atom. Graph (d) represents isomerization transitions involving the entire Ni_{20} cluster. Graph (c) corresponds to a stage that is intermediate between the premelting and the full blown isomerizations; the two-branched pattern of premelting is obscured by the fluctuations due to large-amplitude oscillations in the underlying Ni_{19}. The (a) and (e) graphs correspond to the solidlike and liquidlike forms, respectively, of the Ni_{20} cluster.

In Figure 2, the caloric curves for the three clusters are shown. The meltinglike transition is associated with the variations in the slope of the curves. These variations are much larger in complete shell icosahedra Ni_{19} and Ni_{55} than in Ni_{20}.

Finally, the rms bond length fluctuations δ as a function of κT are depicted in Figure 3. The sharp rise in δ for Ni_{19} and Ni_{55} signifies the meltinglike transition in these clusters (cf. the Lindemann criterion [10]). The graph for Ni_{20} displays two abrupt changes: the first one, at lower temperature, corresponds to premelting, while the second one, at higher temperature, - to complete melting. To find the absolute temperatures at which the δ's change sharply, one has to convert the corresponding κT values from reduced to absolute units. The needed for this values of U_n (see section 2) were obtained in accordance with $U_n = V_n/V_n^*$; the calculated in the framework of an embedded-atom approach energies of the most stable structures [11] were used as V_n and the listed above corresponding energies in reduced units as V_n^* (incidentally, by comparing the characteristic distances in these structures in absolute and reduced units of length, one obtains the values of r_{on}). The magnitudes of $U_{19} = 3.380$ eV and $U_{55} = 3.537$ eV put the melting temperature (as specified by the value $\delta = 0.18$) of the Ni_{19} and the Ni_{55} clusters at ~1080K and ~1300K, respectively. We have not determined the exact value of U_{20}, but estimated it to be very

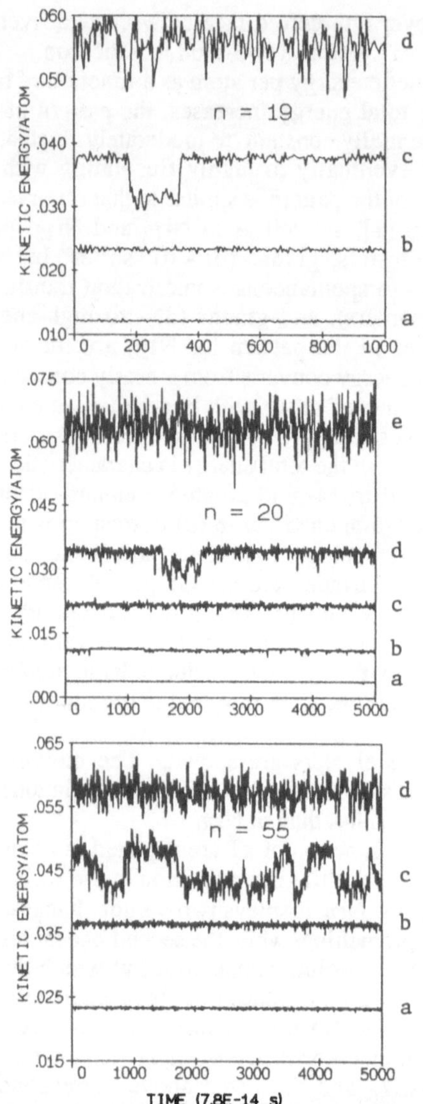

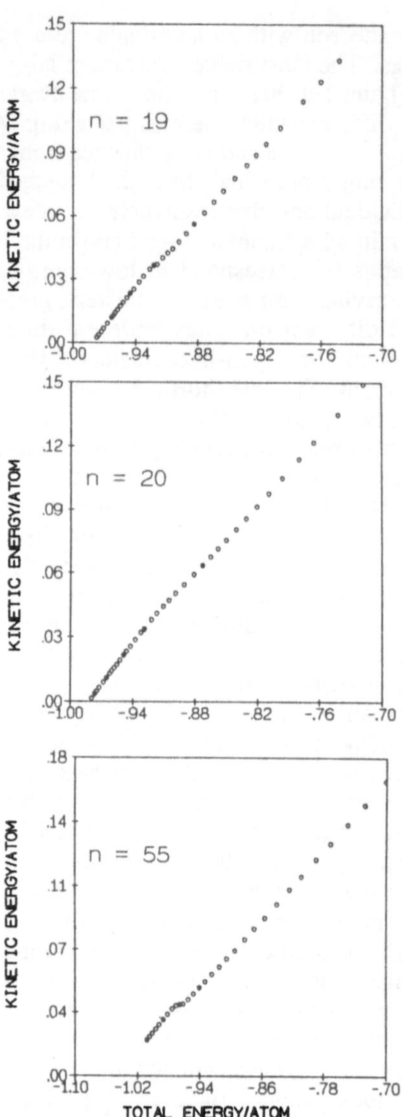

Figure 1. Short-time averaged kinetic energies per atom (in reduced units) as a function of time. The total energies per atom (in reduced units) are: n=19: a) -0.963, b) -0.945, c) -0.923, d), -0.884; n -20: a) -0.977, b) -0.965, c) -0.949, d) -0.929, e) -0.873; n = 55: a) -1.008, b) -0.987, c) -0.966, d) -0.941.

Figure 2. Caloric curves in reduced units of energy. The full circles correspond to total energies per atom considered in Fig. 1.

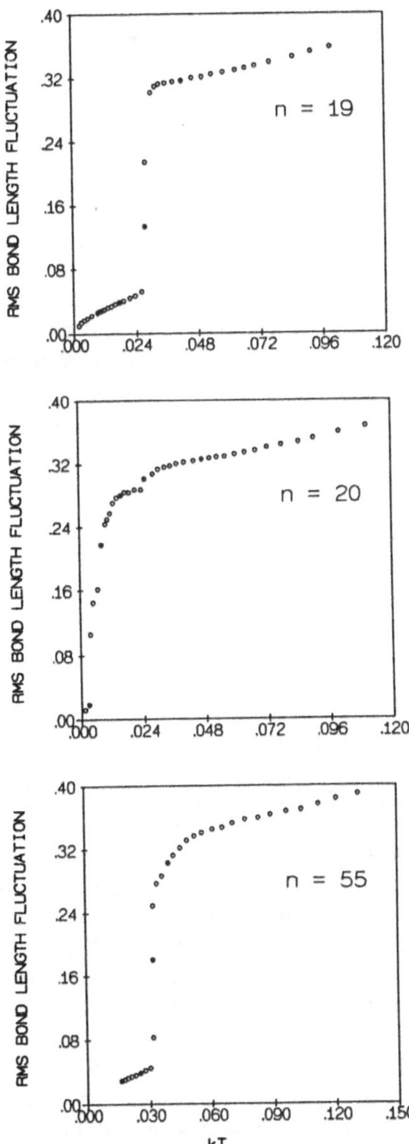

Figure 3. Rms bond-length fluctuations as a function of κT (in reduced units). The full circles correspond to total energies per atom considered in Figure 1.

close to that of U_{19}, although slightly smaller. This puts the premelting and melting temperatures of Ni_{20} at ~210±50K and ~930±50K, respectively.

4. Summary

A brief discussion of the meltinglike behavior of Ni_n, n=19,20,55, clusters described by a size-dependent Gupta-like potential has been presented. As the internal energy of the clusters is increased, they undergo a solid-to-liquid like transition which, similar to the case of Van der Waals and ionic clusters, includes an isomerization ("coexistence") stage. The Ni_{20} cluster exhibits also a low-temperature premelting state similar to that found earlier for Ni_{14}. It appears that the premelting phenomenon is characteristic of clusters with complete shell icosahedral geometries plus one atom, provided the potential describing the interaction between the atoms incorporates many-body effects. It is of interest to test this proposition on Ni_{56}, Ni_{148},... clusters.

Acknowledgments

This work was supported by DGAPA-UNAM Project 104289 (ILG) and by the Office of Basic Energy Sciences, Division of Chemical Science, US-DOE under contract number W-31-109-ENG-38 (JJ). We thank the supercomputing time provided by DGSCA-UNAM.

References

1. See, for example, *Physics and Chemistry of Small Clusters*, P. Jena, B. K. Rao, and S. N. Khanna, Eds., Plenum Press: New York, 1987; Adv. Chem. Phys., I. Prigogine and S. A. Rice, Eds., 10, Part 2, Wiley-Interscience: New York, 1988; *Elemental and Molecular Clusters*, G. Benedek, T. P. Martin, and G. Pacchioni, Eds., Springer: Berlin, 1988; and references therein.

2. See, for example, Ph. Buffat and J. P. Borel, Phys. Rev. A 13, 2287 (1976); J. P. Borel, Surf. Sci. 106, 1 (1981); S. Iyima and T. Ichihashi, Phys. Rev. Lett. 56, 616 (1986); D. J. Smith, A. K. Petford-Long, R. L. Wallenberg, and J.-O. Bovin, Science 233, 872 (1986); P. M. Ajayan and L. D. Marks, Phys. Rev. Lett. 60, 585 (1988); T. Castro, R. Reifenberger, E. Choi and R. P. Andres, Phys. Rev. B 42, 8568 (1990); M. E. Lin, A. Ramachandra, R. P. Andres and R. Reifenberger, this volume; and references therein.

3. S. Sawada and S. Sugano, Z. Phys. D 14, 247 (1989).

4. F. Ercolessi, W. Andreoni, and E. Tossatti, Phys. Rev. Lett. 66, 911 (1991).

5. H.-P. Cheng and R. S. Berry, Mat. Res. Soc. Symp. Proc. 205, 241 (1991).

6. J. Jellinek and I. L. Garzon, Z. Phys. D. 20, 239 (1991).

7. I. L. Garzon and J. Jellinek, Z. Phys. D 20, 235 (1991)

8. Z. B.Güvenç, J. Jellinek and A. F. Voter, this volume.

9. L. Verlet, Phys. Rev. 159, 98 (1967).

10. I. Z. Fisher, *Statistical Theory of Liquids*, University of Chicago: Chicago, 1966.

11. J. Jellinek, unpublished results.

PHASE CHANGES IN NICKEL CLUSTERS FROM AN EMBEDDED-ATOM POTENTIAL

Z. B. GÜVENÇ AND J. JELLINEK
Chemistry Division, Argonne National Laboratory, Argonne, Illinois 60439 USA

A. F. VOTER
Theoretical Division, MSJ569, Los Alamos National Laboratory, Los Alamos, New Mexico 87565 USA

ABSTRACT. The meltinglike behavior of Ni_n, n=12,13,14,19 clusters is studied using molecular dynamics simulations. The cohesion in clusters is modelled by an embedded-atom potential incorporating many-body effects. The features of the phase change transition derived are compared to those obtained earlier from pairwise interactions and a different many-body (Gupta-like) potential.

1. Introduction

Structural, phase and phase change properties of atomic clusters have been and remain in the focus of theoretical studies of these systems [1]. The question of how the molecularlike rigidity and floppiness of clusters evolves into bulk phases and phase transitions, as the clusters grow in size, is one of the many intriguing questions yet to be answered. It appears that the structural and phase properties of clusters are intimately related to and often play a decisive role in determining the other features of these systems, for example, their chemical reactivity [2].

In the framework of cluster studies, the phenomena of phases and phase changes, traditionally viewed as thermodynamic in nature, can be described and understood, especially for smaller clusters, in purely dynamical terms. A meltinglike transition in a cluster can be viewed as a transition from a highly restricted and correlated motion of the constituent particles, characteristic of a solid, to their uncorrelated large-amplitude or even diffusive motion, characteristic of a liquid. Since the type of the motion of the particles at a given total energy is defined by the forces acting on them, the details of a phase change transition in a cluster depend, in general, on the nature of its cohesion. Van der Waals (in particular, Lennard-Jones) clusters have been studied extensively, and the melting properties of ionic and covalent clusters have also been examined [1]. Only very recently, however, have the studies been extended to metallic clusters [3,4]. One of the reasons for this is that mimicking the metallic cohesion, where the many-body effects play an important role, is a nontrivial task. This is particularly true for transition metal clusters, in which the d-electrons may be involved in bonding.

In this communication, we give a brief account of the results of our studies of the melting behavior of Ni_n, n=12,13,14,19, clusters based on an embedded-atom potential. The theoretical background is given in the next section. In section 3 the results are presented and discussed. A summary of the findings is given in Section 4.

2. Theoretical Background

We extracted the structural and dynamical properties of clusters from microcanonical

P. Jena et al. (eds.), Physics and Chemistry of Finite Systems: From Clusters to Crystals, Vol. I, 411–416.
© *1992 Kluwer Academic Publishers.*

molecular dynamics simulations. Hamilton's equations of motion were solved for all the atoms in a cluster on a grid of total energies using Hamming's modified predictor-corrector propagator with a step size of 10^{-15} s. The clusters were prepared with zero initial total linear and angular momenta. The lengths of the runs ranged from 2.1 to $3.0 \cdot 10^6$ steps. The total energy was conserved within 0.03%. The forces experienced by the atoms were calculated from an embedded-atom potential [5] fitted to reproduce simultaneously the measured equilibrium lattice constant, the equilibrium cohesive energy, the bulk modulus, the three cubic elastic constants and the vacancy formation energy of the bulk nickel, as well as the equilibrium bond length and bond energy of the nickel dimer. The general form of the potential is

$$V = \sum_{i<j}^{n} \phi(r_{ij}) + \sum_{i}^{n} F(\bar{\rho}_i) , \tag{1}$$

where ϕ is a pairwise (Morse) interaction, F is the so-called embedding energy, r_{ij} is the distance between atoms i and j, and $\bar{\rho}_i$ is the local electron density experienced by atom i. This density is supplied by all the atoms j surrounding atom i:

$$\bar{\rho}_i = \sum_{j(\neq i)}^{n} \rho_j(r_{ij}) , \tag{2}$$

where ρ_j is the electron density created by atom j at the position of atom i. For details of the parameterization and fitting of the potential see Ref. 5.

To characterize the phases and phase changes in the clusters, a number of time-dependent and equilibrium quantities has been calculated and monitored as a function of the total energy (per atom). Of these we present here results for the following:
1) "Instantaneous temperature" T_i of the cluster as a function of time,

$$T_i = \frac{2<E_k>_s}{(3n-6)k} , \tag{3}$$

where E_k is the kinetic energy of the cluster, k is the Boltzmann constant, and $<>_s$ stands for a short-time average. This average was calculated over time intervals of $5 \cdot 10^{-13}$s to attenuate the fluctuations in the kinetic energy due to the vibrational motions in the cluster;
2) Relative root-mean-square (rms) bond length fluctuation δ as a function of the temperature T of the cluster, where

$$\delta = \frac{2}{n(n-1)} \sum_{i<j} \frac{(<r_{ij}^2> - <r_{ij}>^2)^{1/2}}{<r_{ij}>} , \tag{4}$$

$<>$ stands for the long-time (over the entire trajectory) average, and T is calculated in accordance with Eq. (3) with $<>_s$ replaced by $<>$;
(3) power spectra $I(\omega)$,

$$I(\omega) = 2 \int_{0}^{\infty} C(t) \cos\omega t \, dt, \tag{5}$$

calculated from the velocity autocorrelation function C(t),

$$C(t) = \frac{\sum_{j=1}^{n_t} \sum_{i=1}^{n} \vec{v}_i(t_{0j}+t) \cdot \vec{v}_i(t_{0j})}{\sum_{j=1}^{n_t} \sum_{i=1}^{n} \vec{v}_i^2(t_{0j})} \,, \tag{6}$$

where $\vec{v}_i$ is the velocity of atom i and n_t is the number of the different time origins t_{0j}. Up to 8000 time origins along a trajectory were used to suppress the effect of the initial conditions and to assure convergence of C(t).

3. Results and Discussion

The time dependence of the instantaneous temperature of a cluster, which is a measure of its short-time averaged kinetic energy, is an informative characteristic of the state of this cluster at a given total energy. At low energies the cluster oscillates around its most stable structure. An averaging over a few oscillations along the trajectory yields the short-time averaged kinetic and potential energies as functions of time; these remain almost constant, and their sum is, of course, the fixed total energy of the cluster. As its total energy increases, the cluster begins to sample its other isomeric forms as well. These latter have, in general, different potential energies and therefore the short-time averages of the potential, and thus of its complement - the kinetic energy, begin to substantially deviate from a constant. The instantaneous temperature T_i of a cluster considered as a function of time is, thus, a sensitive indicator of its structural rearrangements, i.e., of its phase changes. (The possible exceptions are the transitions between permutational isomers or isomers with accidental degeneracy. But even in these cases, the passage over a barrier is reflected in a very short time decrease of the value of T_i. The situation is more complicated in the case of rotating clusters [6], where even nondegenerate isomers may have both equal, or at least similar, and different T_i's).

In Fig. 1, the patterns of the time dependence of T_i at different total energies are shown for each of the Ni_n, n = 12,13,14,19, clusters. One notices that indeed, as the total energy increases, the instantaneous temperature evolves from being essentially constant to fluctuating, first around a single value and then around two or more values, which form distinguishable branches. Eventually, at high energies, the fluctuations become so large that no separate branches can be resolved. This general evolution of the patterns of T_i with the total energy is similar to what has been found earlier for nonmetallic clusters [1]. Quenching thermally the trajectories from points corresponding to different branches of T_i, we have confirmed that these branches correspond to different isomers of a cluster; a detailed characterization of these isomers will be given elsewhere. Thus, as in the case of nonmetallic clusters, the nickel clusters considered here pass, when energized, through a stage of spontaneous isomerizations with the different isomers surviving long enough on the time scale of their characteristic vibrational periods. And, as found earlier, the isomerization stage persists over a finite range of total energies.

A careful examination of the panel of Fig. 1 corresponding to Ni_{14} reveals also some differences. As the energy increases, the pattern of T_i changes from being essentially constant, to two-branched, to moderately fluctuating with no branches resolvable, then back to fluctuations around distinct branches, and eventually to large fluctuations with no distinguishable branches. Such behavior has not been observed in nonmetallic clusters.

414

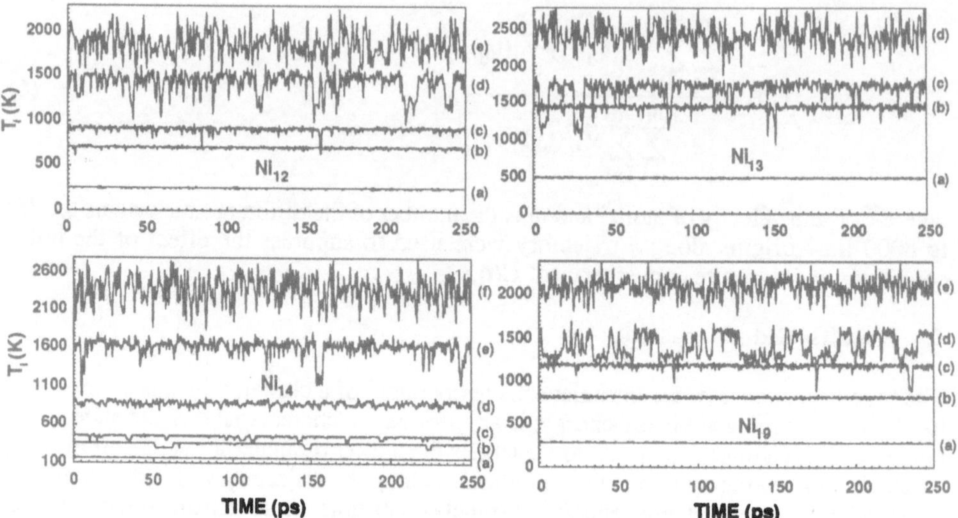

Figure 1. Instantaneous temperatures T_i as functions of time. The graphs correspond to the following total energies per atom: Ni_{12}: (a) -2.98 eV, (b) -2.88 eV, (c) -2.83 eV, (d) -2.67 eV, (e) -2.52 eV; Ni_{13}: (a) -3.05 eV, (b) -2.82 eV, (c) -2.73 eV, (d) -2.40 eV; Ni_{14}: (a) -3.11 eV, (b) -3.06 eV, (c) -3.05 eV, (d) -2.93 eV, (e) -2.73 eV, (f) -2.47 eV; Ni_{19}: (a) -3.25 eV, (b) -3.12 eV, (c) -3.02 eV, (d) -2.91 eV, (e) -2.66 eV.

Fig. 2, which displays the relative rms bond length fluctuations δ as functions of the temperature T, clearly indicates that the change in the state of the clusters is a meltinglike transition. The signature of this transition is a sharp increase in the value of δ; cf. the Lindemann criterion for bulk melting [7]. The most resistant to melting is Ni_{13} with the transition occurring close to the bulk melting temperature (1726 K). This can be rationalized in terms of the high stability of the 13-atom icosahedral equilibrium structure. The reference to the structure explains, however, only the relative stability of this cluster. The meltinglike transition in an initially icosahedral Ar_{13}, described by a Lennard-Jones potential, takes place far below the triple point of argon. It is the difference in the nature of bonding in Ni_{13} and Ar_{13} that is ultimately responsible for the large difference in the degree of reduction in the melting temperature of these clusters. The resistivity to melting decreases in the order Ni_{19}, Ni_{12} and Ni_{14}. The most stable equilibrium structures of these clusters are: Ni_{19} - double icosahedron, Ni_{12} and Ni_{14} - relaxed structures obtained from an icosahedral Ni_{13} by removing a surface atom and adding an atom over a threefold face, respectively. The T-dependence of δ clearly displays the peculiarities of the melting behavior of Ni_{14}: it exhibits more stages in its meltinglike transition than the other clusters. The first sharp increase in δ at T<500 K corresponds to intermittent isomerizations between two topologically similar structures, one with the 14th atom over a threefold face and the other over an edge. The graphs (b) and (c) for Ni_{14} in Fig. 1 correspond to this "premelting" [4b] stage. The next, almost horizontal, part of the δ curve represents the stage in which groups of atoms of the 12-atom surface of the underlying 13-atom cluster "rock" around the icosahedral geometry in a fairly symmetrical fashion. The 14th atom occasionally inserts itself into the surface of the 13-atom cluster, and another atom is subsequently

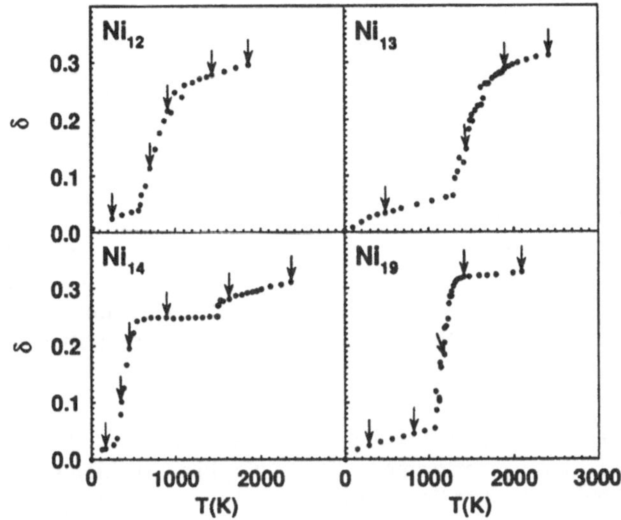

Figure 2. Relative rms bond length fluctuations δ as functions of the temperature T. The points indicated by the arrows correspond to the total energies represented in Fig. 1.

expelled onto the surface. A representative pattern of the T_i's for this stage is given by the graph (d) of the corresponding panel in Fig. 1. The second sharp rise of δ in the vicinity of T = 1500 K signifies initiation of isomerization transitions and eventually melting of the cluster; cf. the graphs (e) and (f), respectively, for Ni_{14} in Fig. 1. Comparison of the δ curves for the Ni_{13} and Ni_{14} clusters indicates that the major effect of the 14th atom is adding the premelting stage. The meltinglike transition proper takes place in the two clusters over essentially the same temperature range and is described by very similar values of δ.

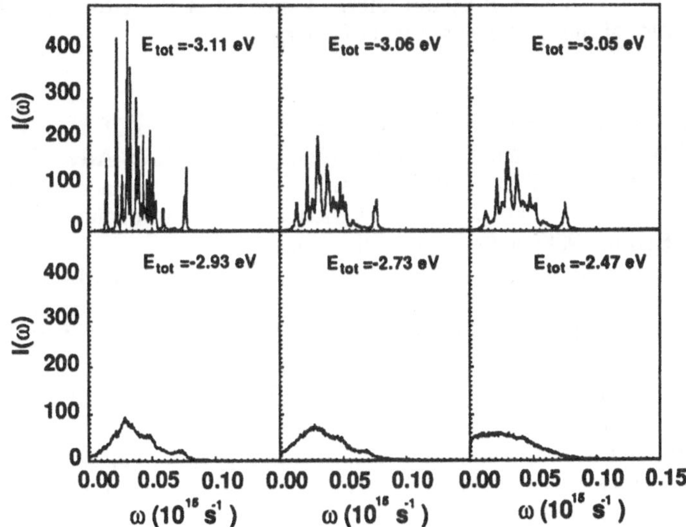

Figure 3. Power spectra of the Ni_{14} cluster at different total energies per atom. $I(\omega)$ is in units of 10^{-15} s.

The distributions of the vibrational frequencies for Ni_{14} at energies corresponding to the cases (a)-(f) of Fig. 1 are shown in Fig. 3. These also confirm the meltinglike nature of the transition taking place in the cluster. At low energies the motion of atoms is characterized by oscillations with well-resolvable individual frequencies. No diffusive motion ($\omega = 0$), characteristic of a liquid, is present. As the energy increases, a continuum of frequencies, including $\omega = 0$, sets in.

Finally, we note that all the features described above, including the premelting phenomenon, are in agreement with what has been found for the melting behavior of transition metal clusters using a size-dependent Gupta-like potential [4b]. The common feature of the embedded-atom and the Gupta-like potentials is that both incorporate many-body effects. It is these effects which are probably responsible for the specific phase change features of nickel (or even more generally, metal) clusters above.

4. Summary

A brief report on the meltinglike behavior of small nickel clusters derived from an embedded-atom potential has been presented. The meltinglike transition in Ni_n, n=12,13,14,19 is characterized by a finite-size (in terms of the total energy) isomerization range similar to that found earlier in studies of nonmetallic clusters. The premelting phenomenon in Ni_{14} and the only very moderate reduction in the melting temperature of Ni_{13} are, however, novel features. It is reassuring that all our findings agree, even in detail, with the melting properties derived for transition metal clusters using a different, so-called Gupta-like potential. Both potentials incorporate many-body effects, which seem to have a substantial effect on the mechanism of the meltinglike transition in clusters of at least some sizes.

Acknowledgments

Work at ANL is performed under the auspices of the Office of Basic Energy Sciences, Division of Chemical Science, US-DOE under contract number W-31-109-ENG-38. Work at LANL was supported by the Energy Conversion and Utilization Technologies (ECUT) Program of the U. S. Department of Energy.

References

1. See for example: *Physics and Chemistry of Small Clusters*, P. Jena, B. K. Rao, and S. N. Khanna, Eds., Plenum Press: New York, 1987; Adv. Chem. Phys. I. Prigogine, and S. A. Rice, Eds., 10, Part 2, Wiley-Interscience: New York, 1988; *Elemental and Molecular Clusters*, G. Benedek, T. P. Martin, and G. Pacchioni, Eds., Springer: Berlin, 1988; and references therein.
2. J. Jellinek and Z. B. Güvenç, this volume; S. J. Riley and E. Parks, this volume.
3. S. Sawada and S. Sugano, Z. Phys. D. 14, 247 (1989); F. Ercolessi, W. Andreoni, and E. Tossatti, Phys. Rev. Lett. 66, 911 (1991); A. Bulgac and D. Kusnezov, Phys. Rev. B (in press).
4. (a) I. L. Garzon and J. Jellinek, Z. Phys. D 20, 235 (1991); (b) J. Jellinek and I. L. Garzon, *ibid*, 20, 239 (1991).
5. A. F. Voter and S. P. Chen, Mater. Res. Soc. Symp. Proc. 82, 175 (1987).
6. J. Jellinek and D. H. Li, Phys. Rev. Lett. 62, 241 (1989); D. H. Li and J. Jellinek, Z. Phys. D. 12, 177 (1989).
7. I. Z. Fisher, *Statistical Theory of Liquids*, University of Chicago: Chicago, 1966.

STRUCTURE AND DYNAMICS OF ARGON CLUSTER IONS

T. IKEGAMI, T. NAGATA, and T. KONDOW
Department of Chemistry, Faculty of Science, The University of Tokyo
Bunkyo-ku, Tokyo 113
Japan

ABSTRACT. The geometrical and electronic structures of Ar_n^+ ($n = 3 \sim 27$) were calculated on the basis of a diatomic-in-molecules (DIM) model. In the electronic ground state, the positive hole was found to be localized on the central three atoms forming a trimeric ion core. The calculated transition wavelength shifts from 520 nm to 580 nm as the cluster size increases from 14 to 20, in accord with experimental observations. The origin of the shift was interpreted in terms of the interaction between the ion core and the surrounding solvent atoms in the framework of the solvated core model.

1 Introduction

It is well accepted that the argon cluster cations consist of a small ion core surrounded by neutral atoms [1–9]. In various theoretical works, the ion core has been postulated to be Ar_2^+ [1–3], because of its large binding energy (1.33 eV): Photodissociation dynamics [4,5] and the photoabsorption band [6] of Ar_3^+ have been interpreted in terms of the dimeric ion core. However, recent diatomics-in-molecules (DIM) calculations have indicated that the positive hole is localized on the three atoms forming a trimeric ion core [7]. In addition, several experimental data have been interpreted on the basis of the trimeric ion core [7–11]. The visible absorption spectra of Ar_n^+ show a broad and intense band peaking near 520 nm irrespective of the cluster size in the range between $n = 3$ and 14 [8]. This behavior supports the trimeric ion core model. As the cluster size increases further, the peak shifts toward 600 nm in the range of $n = 15 \sim 20$, but does not shift any more beyond $n \simeq 20$ [8]. In the photodissociation of Ar_n^+, ionic and neutral fragments are produced: The kinetic and angular distributions of these photofragments [9] have also indicated the presence of the trimeric ion core in Ar_n^+.

These observation are closely related to the geometrical and electronic structures of Ar_n^+. In the present work, the DIM model developed by Kuntz and Valldorf [7]

P. Jena et al. (eds.), Physics and Chemistry of Finite Systems: From Clusters to Crystals, Vol. I, 417–422.
© 1992 *Kluwer Academic Publishers.*

was employed to determine the geometrical and electronic structures of Ar_n^+ as well as dipole transition moments between the electronic ground state and the excited states. The calculated spectral shift for $n = 14 \sim 20$ agreed well with the experimental observation. A solvated core model was proposed to explain the spectral shift.

2 Calculation

2.1 BASIS SET AND HAMILTONIAN MATRIX

In the DIM calculation, a valence-bond type basis set, $\{\ \Xi_A^w : 1 \leq A \leq n,\ w = x,y,z\ \}$, was used [7]. The subscript A represents an Ar atom having a hole, and superscript w indicates the atomic orbital where the hole is located. For example, Ξ_3^z represents a wave function of a valence-bond structure in which the third Ar atom has a hole in the $3p_z$ orbital. The construction of the DIM hamiltonian matrix in this basis set was described in ref. [7].

2.2 GEOMETRY OPTIMIZATION

The geometrical structures of Ar_n^+ ($n = 3 \sim 27$) were optimized by the steepest descent method by using analytical derivatives of the potential energy. The initial geometries of Ar_n^+ were generated by adding one atom randomly to the optimized configuration of Ar_{n-1}^+. The number of the initial geometries was $30 \sim 100$. The analytical derivatives were calculated as follows: The expected value of the force exerting on an atom C, $\langle F_C \rangle$, is written as

$$\langle F_C \rangle = \langle \Phi_g | \nabla_C \mathcal{H} | \Phi_g \rangle, \tag{1}$$

where Φ_g is the wave function of the electronic ground state, and expanded as

$$\Phi_g = \sum_{A,w} g_A^w \Xi_A^w, \tag{2}$$

with g_A^w as expansion coefficients. On the basis of the diabaticity of the basis set employed, the right hand side of eq. (1) is rewritten [12] as

$$\langle \Phi_g | \nabla_C \mathcal{H} | \Phi_g \rangle = \sum_{A,w} \sum_{B,v} g_A^{w^*} g_B^v \nabla_C \langle \Xi_A^w | \mathcal{H} | \Xi_B^v \rangle. \tag{3}$$

The gradient of the potential energy is readily obtained from the analytical derivatives of the hamiltonian matrix elements with respect to the atomic coordinates.

2.3 TRANSITION DIPOLE MOMENT

The matrix elements of the electronic dipole operator, $\mathcal{M}$, were estimated by using the localized point charge model:

$$\langle \Xi_A^w | \mathcal{M} | \Xi_B^v \rangle = r_A \delta_{AB} \delta_{wv}, \tag{4}$$

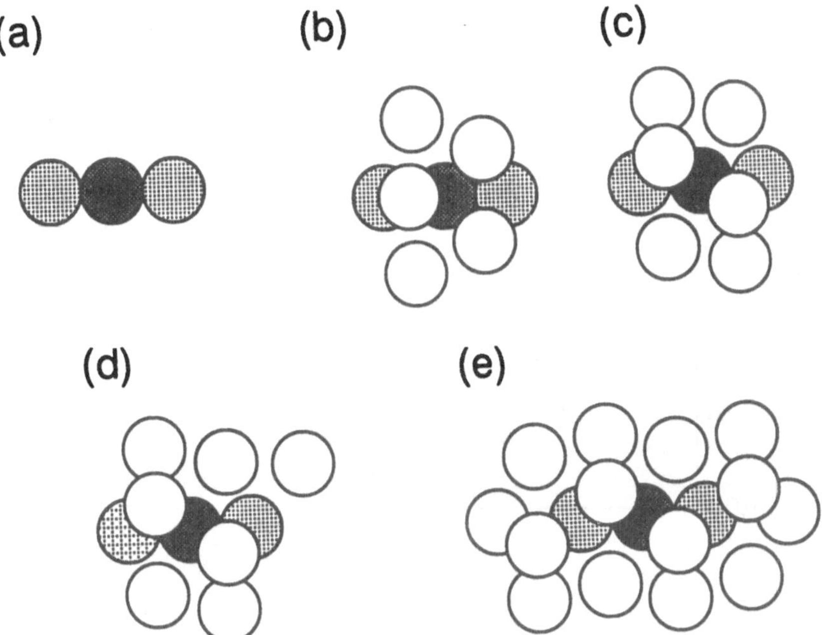

Figure 1: The optimized structures of (a) Ar_3^+, (b) Ar_8^+, (c) Ar_{13}^+, (d) Ar_{14}^+, (e) Ar_{25}^+.

where r_A is the position vector of atom A. In this model, the contribution from the transition within a single atom is neglected. This approximation was verified by the fact that the oscillator strength for the Ar_3^+ system calculated by this model was in excellent agreement with those obtained by POL-CI calculations.

3 Results

3.1 GEOMETRICAL STRUCTURES

The optimized structures of Ar_3^+, Ar_8^+, Ar_{13}^+, Ar_{14}^+, and Ar_{25}^+ are shown in fig. 1. The circles show the Ar atoms and depth of the blackness in the circles is proportional to the charge density on the atoms. The hole is localized on the central three atoms forming a trimeric ion core for $n = 3 \sim 27$. The structure evolves from the bare ion core, Ar_3^+, to the ion core with the completed first solvation shell in Ar_{25}^+ as follows: Neutral atoms envelop the ion core along its axis as the cluster size increases from $n = 4$ to 12, and a cylindrical shell is completed in Ar_{13}^+ (fig. 1c). Neutral atoms start to attach on one end of the ion core at $n = 14$, and finally

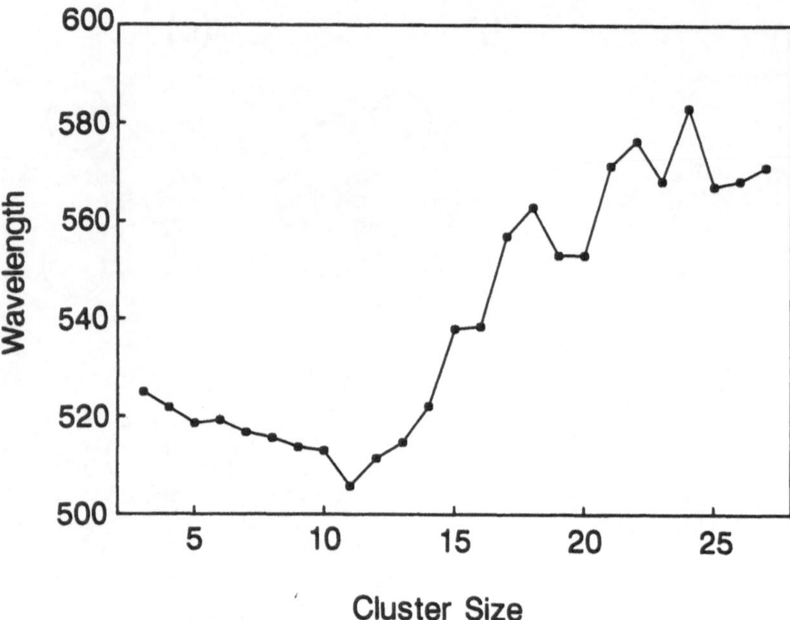

Figure 2: The expected values of the transition wavelength.

form a pentagonally pyramidal cap composed of 6 atoms at $n = 19$. In the range of $n = 20 \sim 25$, another pentagonally pyramidal cap is constructed on the other end of the ion core.

3.2 TRANSITION WAVELENGTH

The *averaged* transition energy from the electronic ground state (denoted by 0) to the excited states is defined as

$$\Delta \bar{E}_n = \frac{\sum_{i=1}^{3n-1} f_{i0}(E_n^i - E_n^0)}{\sum_{i=1}^{3n-1} f_{i0}}, \tag{5}$$

where n is the cluster size, E_n^i is the energy of the i-th excited state, and f_{i0} is the oscillator strength between 0 and i-th states. The *averaged* transition wavelength was calculated from $\Delta \bar{E}_n$, and plotted as a function of the cluster size in fig. 2. The wavelength decreases slightly from 520 to 510 nm, turns to increase at $n = 11$, and then tends to level off around 580 nm above $n = 20$. This tendency accords with the experimental observation [8].

4 Discussion

As shown in fig. 1, the positive hole is localized on the ion core in the electronic ground state. Hence, the electronic ground state is described by a linear combination of 9 states where the hole is confined in the ion core. These core states (C-states) span the subspace in the electronic configuration space. The rest of the space is, by definition, spanned by $3n - 9$ surrounding states (S-states) where the hole is delocalized among the solvent atoms. Since the hole has a tendency to be delocalized in the excited states, the wave function of the excited states should be expressed by a linear combination of the *excited* C-states and the S-states. Both states have comparable energies and interact strongly with each other. On the contrary, the *ground* C-state, which is the primary component of the ground state, is energetically isolated from all the S-states, so that the energy of the electronic ground state is independent of the cluster size.

Presumably, the third *excited* C-state (Φ_3^C) is responsible for the absorption band near 520 nm, since the corresponding band arises mainly from the transition to the third excited state in Ar_3^+. If the interactions between Φ_3^C and the S-states were completely neglected, the energy of Φ_3^C, $E(\Phi_3^C)$, would not change with the cluster size. As shown in fig. 2, however, the energy of Φ_3^C increases (blue shift) from $n = 3$ to 11 and then decreases (red shift) for $n > 11$, as a consequence of the interactions with the S-states. This behavior can be illustrated by considering an effective S-state (Φ_{eff}^S). The energy of the effective S-state, $E(\Phi_{eff}^S)$, and the interaction energy between Φ_{eff}^S and Φ_3^C are expected to increase with the cluster size. We assume here that $E(\Phi_{eff}^S)$ is lower than $E(\Phi_3^C)$ at $n = 4$ and increases to exceed $E(\Phi_3^C)$ at $n = 11$. Under these circumstances, the energy level of Φ_3^C shifts toward higher energy (blue shift) as the cluster size increases from 4 to 11 and then turns to be lowered for $n > 11$ (red shift). Significant red-shift starting at $n = 14$ indicates that the ion core begins to interact more strongly with the solvent atoms at this size where the Ar atom begins to cap one end of the ion core. Since the hole orbital in Φ_3^C is extended along the axis of the ion core, the Ar atoms attached to the ends of the axis can interact strongly with the core.

In the photodissociation of Ar_n^+ at 532 nm, the fast neutral photofragments were detected up to $n = 13$. These fast fragments are likely to be produced from the direct dissociation of the ion core, because Φ_3^C associated with the direct dissociation interacts weakly with the S-states. For $n \geq 14$, on the other hand, the strong interactions between Φ_3^C and the S-states cause to suppress the production of the fast fragments in the photodissociation.

Acknowledgement

The authors gratefully acknowledge Prof. Suehiro Iwata for valuable discussion on all aspects of this work. We used the TKYVAX computer at the University of Tokyo meson science laboratory (UTMSL).

References

[1] (a) Sáenz, J. J., Soler, J. M., and García, N. (1985) 'Evaporation of small clusters of noble gases by ionization', Surf. Sci. **156**, 121–125. (b) Haberland, H. (1985) 'A model for the processes happening in a rare-gas cluster after ionization', Surf. Sci. **156**, 305–312. (c) Polymeropoulos, E. E. and Brickmann, J. (1985) 'Magic numbers in ionized rare-gas clusters', Surf. Sci. **156**, 563–571.

[2] Lethbridge, P. G., Del Mistro, G., and Stace, A. J. (1990) 'Hard-sphere cluster-ion structures', J. Chem. Phys. **93**, 1995–2003.

[3] Böhmer, H. -U. and Peyerimhoff, S. D. (1989) 'Stability and structure of singly-charged argon clusters Ar_n^+, $n = 3 - 27$. A Monte-Carlo simulation', Z. Phys. D–Atoms, Molecules and Clusters **11**, 239–248.

[4] Woodward, C. A., Upham, J. E., Stace, A. J., and Murrell, J. N. (1989) 'The photofragmentation of Ar_3^+', J. Chem. Phys. **91**, 7612–7620.

[5] (a) Snodgrass, J. T., Roehl, C. M., and Bowers, M. T. (1989) 'Photodissociation dynamics of Ar_3^+', Chem. Phys. Lett. **159**, 10–16. (b) Bowers, M. T., Palke, W. E., Robins, K., Roehl, C., and Walsh, S. (1991) 'On the structure and photodissociation dynamics of Ar_3^+', Chem. Phys. Lett. **180**, 235–240.

[6] DeLuca, M. J. and Johnson, M. A. (1989) 'Observation of a UV absorption band in Ar_3^+ near 300 nm', Chem. Phys. Lett. **162**, 445–448.

[7] Kuntz, P. J. and Valldorf, J. (1988) 'A DIM model for homogeneous noble gas ionic clusters', Z. Phys. D–Atoms, Molecules and Clusters **8**, 195–208.

[8] Levinger, N. E., Ray, D., Alexander, M. L., and Lineberger, W. C. (1988) 'Photoabsorption and photofragmentation studies of Ar_n^+ cluster ions', J. Chem. Phys. **89**, 5654–5662.

[9] Nagata, T., Hirokawa, J., and Kondow, T. (1990) 'Photodissociation of Ar_n^+ cluster ions', Chem. Phys. Lett. **176**, 526–530.

[10] Chen, Z. Y., Albertoni, C. R., Hasegawa, M., Kuhn, R., and Castleman Jr., A. W. (1989) 'Ar_3^+ photodissociation and its mechanisms', J. Chem. Phys. **91**, 4019–4025.

[11] Nagata, T., Hirokawa, J., Ikegami, T., and Kondow, T. (1990) 'Photodissociation of Ar_3^+ cluster ion', Chem. Phys. Lett. **171**, 433–438.

[12] Amarouche, M., Gadea, F. X., Durup, J. (1989) 'A proposal for the theoretical treatment of multi-electronic-state molecular dynamics: Hemiquantal dynamics with the whole DIM basis (HWD). A test on the evolution of excited Ar_3^+ cluster ions', Chem. Phys. **130**, 145–157.

MOLECULAR DYNAMICS SIMULATION OF INFREQUENT EVENTS: EVAPORATION FROM COLD METALLIC CLUSTERS

Z.A.INSEPOV and E.M.KARATAJEV
Kazakh Polytechnical Institute
Satpaev Str. 22
SU-480013 Alma-Ata
Kazakhstan

ABSTRACT. For calculation of evaporation rate of cold metallic clusters the theory which combined transition state theory (TST) and molecular dynamics (MD) simulation is proposed. We have factored the evaporation rate into two terms: $k_d = k_{TST} * F$, where the equilibrium factor k_{TST} is the usual transition state theory rate constant and F is the dynamical factor. For k_{TST} calculation MD method and compensating potential method are used. F is calculated by MD method. The evaporation rate calculations of 14-atoms iron cluster for temperatures T=1210-2800 K are carried out. The calculated value of F is about 0.8. The obtained evaporation rates lie in the range $10^{-5} - 10^1$ s^{-1} and are fitted well by Arrhenius dependence $k_d = A\exp(-E/k_B T)$, where $A=4.104*10^6 s^{-1}$, $E=4.97*10^{-12}$ erg, k_B is Boltzmann constant.

1. Introduction

The MD method was used earlier for the calculation of evaporation rate from Lennard-Jones (LJ) clusters for temperature range $k_B T/\varepsilon = 0.3-0.7$ (ε is LJ potential depth) and cluster sizes n=3-100 [1,2]. Direct calculation by MD method of evaporation rates for lesser temperatures is nonrational because of enormous need of computation time. Consequently an expansion of the possibility of MD method for the infrequent events calculation become vital.

2. Model and Results

In present paper the model of evaporation rate from cold clusters is explored that the TST and MD methods are combined. The evaporation rate can be expressed [3] : kd=kTST*F , where kTST is equilibrium TST term, F is dynamical factor. For calculation of kTST the compensating potential method [3] is used. In this case an evaporation occurs from the shallow potential well $W(q,s)=V(q,s)-U(s)$, where $V(q,s)$ is exact atom-cluster interaction potential, $U(s)$ is compensating potential, q is relative coordinates of evaporating atom in cluster center-of-mass coordinate system, s is radial reaction coordinate. Contribution to kTST from the deep potential U(s) takes into account analitically. For kTST we get

P. Jena et al. (eds.), Physics and Chemistry of Finite Systems: From Clusters to Crystals, Vol. I, 423–427.
© 1992 *Kluwer Academic Publishers.*

424

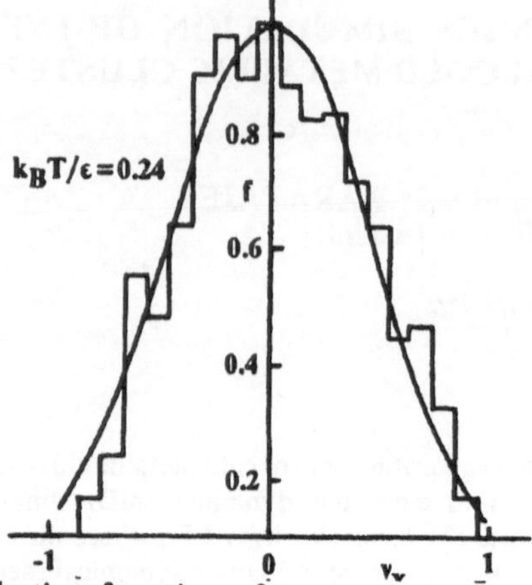

Figure 1. Velocity distribution function of atoms.

$$k_{TST}=L_s(2k_BT/\pi m)^{1/2}g(s_0)\exp(-U(s_0)/k_BT)/\int dsg(s)\exp(-U(s)/k_BT) \qquad (1)$$

where L_s is a number of probable ways of evaporation, m is a reduced mass of atom and cluster, s_0 is a radial reaction coordinate value at which there is an escape of atom from cluster, $g(s)=\int dq\exp(-W/k_BT)$ is calculated by MD method. If detail equilibrium of cluster with gas medium is proposed for F one can get [3]: $F=\alpha\exp(-V(s_0)/k_BT)$, where α is the sticking coefficient of atom on cluster which can be calculated by MD method.

The calculations of evaporation from cluster which consists of 14 Fe atoms (one atom on the surface of 13-atom icosahedron) are carried out. All atoms interact via LJ potential with parameters for Fe: $\varepsilon=1.064*10^{-12}$ erg, $\sigma=2.95$ A. The atomic initial coordinates in cluster are chosen at low temperatures and then strained in a linear temperature expansion approach. The velocities of atoms have been chosen from Maxwellian distribution. By means of velocities rescaling the cluster temperature was kept up. In this paper we have got the interaction potential between atom and cluster in continium approach:

$$U(s)=4\pi n_0[A/(s-R)^9-B/(s-R)^3]$$

where $A=(1-1/C^9)/45-(1-1/C^8)/40D$, $B=(1-1/C^3)/6-(1-1/C^2)/4D$, $C=(s+R)/(s-R)$, $D=s/(s-R)$, R is a cluster radius, n_0 is adjustable parameter. If R approaches infinity we can get a Mie (9-3) potential and when R approaches σ U(s) is very close to LJ potential.

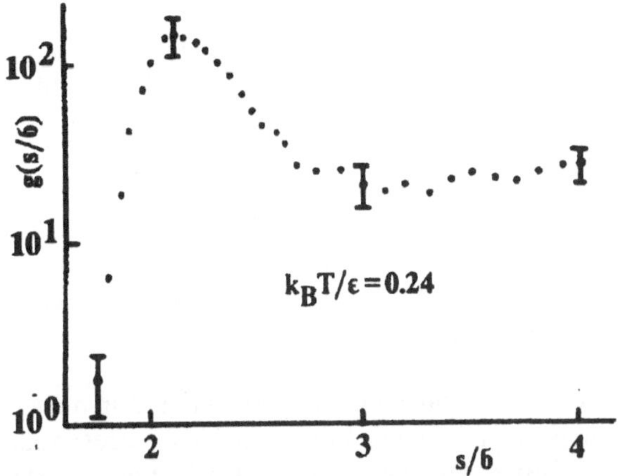

Figure 2. Variation with radial reaction coordinate of the g(s)

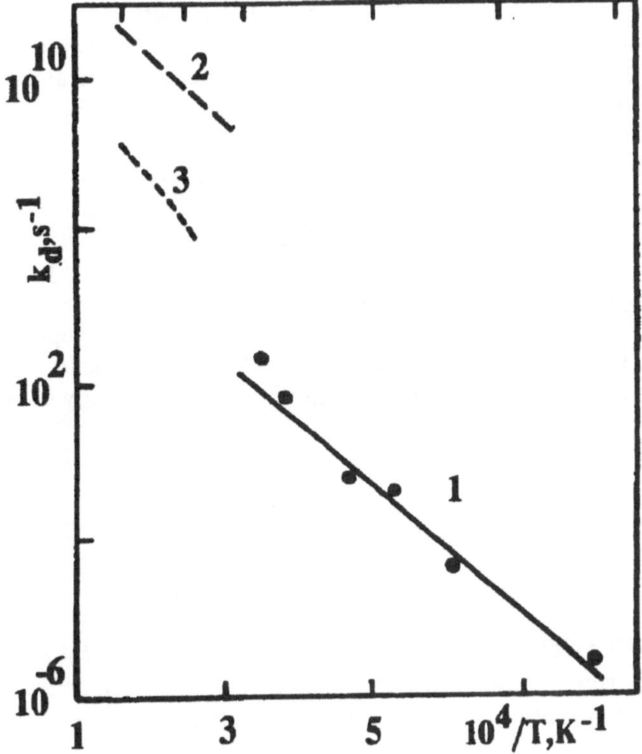

Figure 3. The evaporation rates as a function of temperature.

Cluster was placed in the spherical cavity with radius s0 centered at the center-of-mass of 13-atom icosahedron cluster . The statistical independent configurations and velocities of cluster were chosen for formation of canonical ensemble. Histogram and Maxwellian theoretical distribution of atoms velocities are plotted in Fig.1.

The function $g(s)$ was calculated as a fraction of time which trajectories spend in the region from s to s+ds in a field of potential $W(q,s)$ averaged over a canonical ensemble. The calculated values of $g(s)$ as a function of s are plotted in Fig.2. The value $s_0=4\sigma$ corresponds to precision and availability for $g(s)$ calculation. By means of $g(s)$ substituting in (1) and further numerical integrating one can find k_{TST}.

The sticking coefficient α was calculated by MD method in a such way. The gas atoms with different impact parameters and velocities corresponding the cluster temperature collide with cluster. The gas atom results sticking to cluster if relative full energy of atom-cluster system became less than zero. The sticking coefficient was calculated as $\alpha=N_s/N_0$, where N_s is number of trajectories which lead to sticking, N_0 is a full number of trajectories. N_0 was chosen approximately equal to 200. In the range of temperatures $k_BT/\varepsilon=0.16-0.37$ α can be set to 0.8. This value of α slightly more than one's in [4] where α was calculated in frame of more complicated MD model. The evaporation rates as a function of a temperature are shown in Fig.3.

One can see our calculated points (curve 1) are fitted good by Arrhenius dependence $k_d=A\exp(-E/k_BT)$, where $A=4.104*10^6\,s^{-1}$, $E=4.97*10^{-12}$ erg. The evaporation rate which was calculated directly by MD method [1] for 14-atom Fe cluster (curve 2) and theoretical evaporation rate for 15-atom ionic copper cluster [5] (curve 3) are shown in Fig.3 too. The values of A and E obtained in paper [1] were $A=1.453*10^{14}\,s^{-1}$, $E=5.545*10^{-12}$ erg. It is obvious that k_d has a jump in the temperature range T=2970-3350 K. The observed difference in preexponential factor A eventually probable originate from considerable increasing of activation entropy of cluster [6].

Acknowledgments

We thank Professor G.E.Norman for helpful discussions.

References

1. Bedanov,V.M., Vaganov,V.C., Gadiak,G.V., Kodenev,G.G. (1988) 'Chislennoe modelirovanie isparenia lennard-djonnsovskich klusterov i raschet scorosti zarodischeobrazovania v peresischennom pare',Chim.Fizika (in Russian) 7,412-419.

2. Insepov,Z.A.,Karatajev,E.M., Norman,G.E. (1991) 'The kinetics of condensation behind the shock front', Z.Phys.D 20,449-451.

3. Grimmelmann,E.K., Tully,J.C., Helfand,E. (1981) Molecular dynamics of infrequent events:Thermal desorption of xenon from a platinium surface',J.Chem.Phys. 74,5300-5310.

4. Bedanov,V.M. (1989)'Koefficient kondensacii malich klusterov i ego vlianie na scorost zarodischeobrazovania.Raschet metodom molecularnoi dinamiki',Chim. Fizika (in Russian) 8,117-121.

5. Klots,C.E. (1987) 'The Evaporative Ensemble',Z.Phys.D 5,83-89.

6. Robinson,P.J.,Holbrook,K.A. (1972), Unimolecular Reactions,Wiley-Interscience,London.

COMPUTER SIMULATION OF DYNAMICS OF METAL CLUSTER IMPACT WITH CRYSTAL SURFACE

Z.A.INSEPOV and B.Z.KABDIEV
Kazakh Polytechnical Institute
Satpaev Str. 22
SU-480013 Alma-Ata
Kazakhstan

ABSTRACT. The model of initial stage of cluster epitaxy is developed. In frame of this model dynamical behaviour of 55 Al atoms cluster impact with own surface and structural properties of formed deposit are explored using conventional and stochastic molecular dynamics simulation. Even for small initial cluster translational energies the deposit has a plate form due to significant acceleration in substrate potential. Cohesive energy increases monotonically in the range of cluster translational energy 1-70 eV.

1. Introduction

Ionized cluster beam deposition is used to obtain metal, semiconductor and insulator films at low temperature substrates [1-3]. Up to now some details of this process are isufficiently clear. Therefore computer simulation has important role to explore initial stage of cluster epitaxy. Previously molecular dynamics (MD) technique was used to study the cluster deposition on crystal surface [4-6]. Two-dimensional [4] and three-dimensional silicon [5,6] systems were modelled. The limitation of these studies was the application of direct velocity renormalization technique to remove excess energy resulting from the cluster collision with surface.

The aim of this paper is to develop a model and to explore the dynamics of 55 atoms Al cluster impact of its (100) FCC surface by MD and stochastic molecular dynamics (SMD) simulation.

2. Model

The pair additive Morse potential is used to simulate the interactions between atoms of the system. Initial positions of 55 cluster atoms are chosen in accordance with the icosahedron structure. To achieve the minimum potential energy configuration the structural relaxation technique is used. Then to keep up cluster temperature T the SMD technique [7] is employed. Initally 357 substrate atoms were placed in sites of a cylindrical sample cut from two layers of (100) FCC crystal surface. Top layer consists of 177 atoms and only 101 atoms can are movable in accordance with MD algorithm.

429

P. Jena et al. (eds.), Physics and Chemistry of Finite Systems: From Clusters to Crystals, Vol. I, 429–433.
© 1992 *Kluwer Academic Publishers.*

The boundary MD atoms can move in accordance with SMD technique. Others 76 atoms are fixed in the initial positions. Usage of SMD allows to carry out thermal stabilization for the desirable substrate temperature. In addition this technique allows to account adequately the response of system due to strong short-lived perturbation. To simulate momentum and energy transfer into substrate the 180 bottom layer atoms block can be movable (as whole) in normal direction using SMD technique. Continious potential was used to take into account the other part of the substrate. Initial velocities of the substrate atoms are chosen from Maxwellian distribution. Translational velocity of the cluster was chosen normaly to the substrate surface.

In this simulation following properties of system were calculated: translational and internal kinetic energies of the cluster, number of atoms placed in different deposited layers, number of cluster atoms penetrated into the substrate, average cohesive energy. In addition the maximum two-dimensional diffusion coefficient and order parameter of deposit are estimated.

3. Results and Discussion

Following parameters of Morse potential for Al are used: potential well depth $D=0.2703$ eV, minimum potential position $r_0=3.253$ A, increment parameter is equal 1.1646 A^{-1}. Two values of cluster temperature T are used: 300 K and 900 K. Substrate temperature is equal to 300 K. Four values of the initial cluster translational energy E_0 are used: 1.86, 7.43, 29.73, 66.90 eV. The value of time step in MD algorithm is varied from $8.2*10^{-16}$ s to $3.3*10^{-15}$ s for different values of E_0 to keep up desired accuracy of calculations. The total time of every MD run is about 10^{-12} s. The value of time unit t_0 is equal to 0.33 ps.

Time evolution of translational and internal kinetic energies of cluster at $E_0=7.43$ eV and T=300 K are shown in Fig.1. When cluster approaches to the surface the significant acceleration in substrate potential is observed. The greater part of translational energy transforms into internal kinetic energy of cluster. Estimated cooling time of deposited cluster is about $(20-25)*3.3*10^{-13}$ s. Analysis of same curves for other values of E_0 and T shows that cooling time is changed slightly with translational energy and cluster temperature. At the moment of maximum heating of cluster at impact with surface the value of two-dimensional diffusion coefficient is estimated about 10^{-3} cm^2/s. The final temperatures of the cluster and of the substrate are equal to the initial substrate temperature. Our model allows to transform the initial center-of-mass cluster kinetic energy into the thermal energy of the top layer atoms and to take into account the momentum transfer into the substrate.

The normalized (to number of atoms in cluster) number of atoms placed in different deposited layers at T=900 K is shown in Fig.2. There is a tendency to the one-layer plate shape of deposit when E_0 increases. The number of atoms penetrating into the

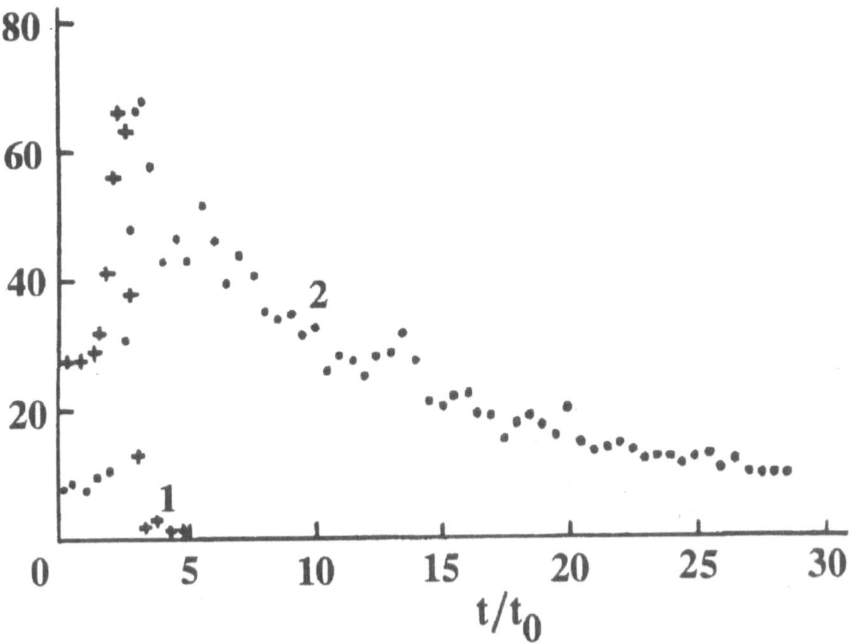

Figure 1. Time evolution of cluster kinetic energy , 1 - translational , 2 - internal

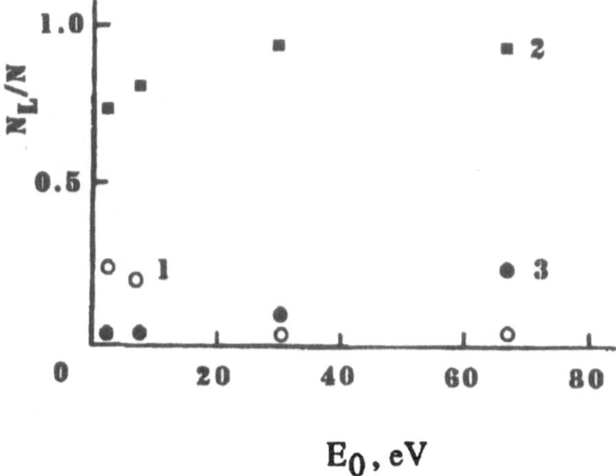

Figure 2. Normalized number of atoms , 1- top deposited layer, 2- bottom deposited layer , 3 - penetrated into substrate.

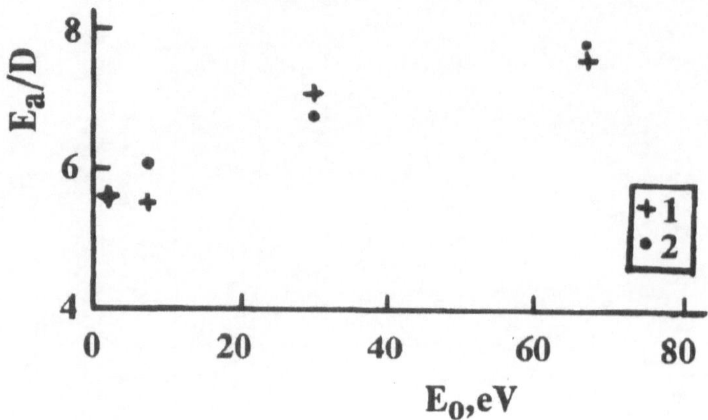

Figure 3. Cohesive energy,1 - at T=900 K, 2 - āt T=300 K.

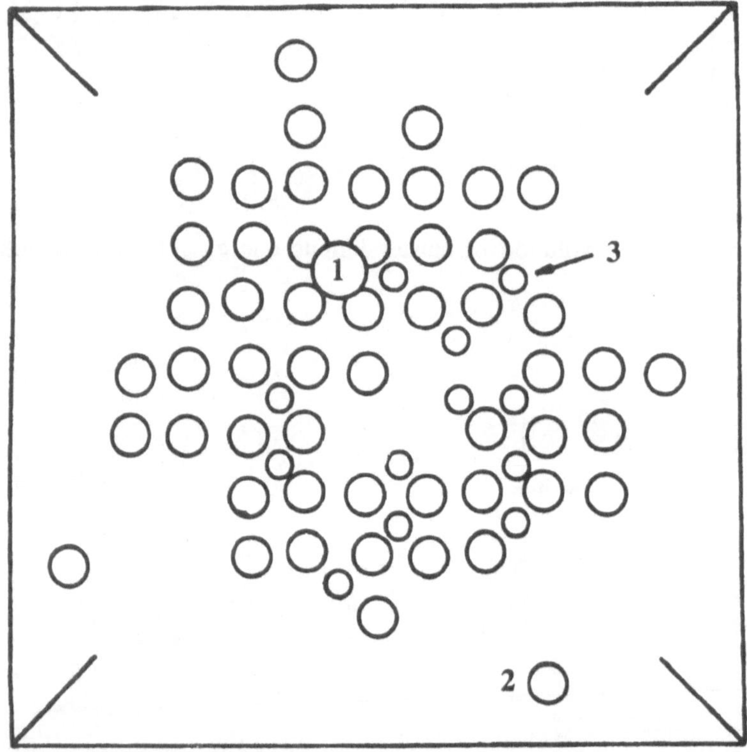

Figure 4. Schematical final top view, signs 1,2,3 - as in Figure 2.

substrate is increased with E_0. Note that in previous MD studies [4-6] the mixing phenomena at interface cluster-substrate is not observed. Even at small E_0 values the great part of atoms are placed in the bottom deposited layer. The area of contact between deposit and substrate is increased with E_0. Two-dimensional order parameter calculated for the bottom deposited layer is about 0.8 and depends on E_0 slightly.

One of the important properties of deposited films is cohesive energy (Fig.3). This value is defined as the average potential energy of cluster atoms. In accordance with experimental results [2,3] the value of cohesive energy increases with E_0. This fact obviously follows from the plate shape of deposit as well as from the penetration of the cluster atoms into the substrate.

The schematic final top view of deposit at E_0=66.90 eV and T=900 K is shown in Fig.4. We can see that almost all atoms are placed in one deposited layer. In the centre of deposit we can see an empty area. This picture is specific for this range of E_0 value. In our opinion this is the first observation of a new "ring spreading" phenomena for the model using pair additive potential.

Acknowledgments

We thank Prof. H.Haberland who suggested the subject of this paper and for providing copies of his works prior to the publication and Prof. G.E.Norman for stimulating discussions.

References

1. Takagi, T. (1986) 'Ionized cluster beam technique',Vacuum 36,27-31.
2. Haberland, H., Karrais, M., Mall, M. (1991) 'A new type of cluster and cluster ion source',Z.Phys.D 20.
3. Haberland, H., Karrais, M., Mall, M. (1991) 'A new type of cluster ion source for thin film deposition',to be published in: Proceedings of the 1990 Fall Meeting of the MATERIALS RESEARCH SOCIETY.
4. Muller, K.-H. (1987) 'Cluster-beam deposition of thin films: a molecular dynamics simulation',J.Appl.Phys. 61,2516-2521.
5. Biswas, R., Grest, G.S., Soukoulis, C.M. (1988) 'Molecular-dynamics simulation of cluster and atom deposition on silicon (111) ', Phys.Rev.B 38,8154-8162.
6. Kwon, I., Biswas, R., Grest, G.S., Soukoulis, C.M. (1990) 'Molecular- dynamics simulation of amorphous and epitaxial Si film growth on Si (111)', Phys.Rev.B 41,3678-3687.
7. Heerman, D.W. (1986) Computer Simulation Methods in Theoretical Physics, Springer-Verlag, New-York.

Space/Time spectroscopic Analysis of Laser Vaporization Process
from Si Target under Gas Environment

A. Kasuya and Y. Nishina
Institute for Materials Research
Tohoku University
Sendai 980
Japan

ABSTRACT.   Laser vaporization processes in Si are analyzed in vacuum
and in gas environment by space/time resolved optical spectrometry.
The results show that the collisions of inert gas atoms with a
vaporized Si cluster enhances its fragmentation in the pressure range
up to 20 Torr, whereas in higher pressures, they suppress its decomposition into highly excited atoms. Both processes may be understood
as energy relaxations of highly excited Si clusters. The decomposition
of highly charged clusters are discussed in terms of electrostatic
Coulomb effect.

1. Introduction

The laser vaporization technique is widely used for forming microclusters and thin films in various gas environments [1-5]. Its dynamical processes, however, have not been analyzed in situ in details of
real time measurements. This paper reports our space/time resolved
analysis on the interaction of vaporized particles with the environmental gas during the initial stage of target vaporization.

2. Experimental results

A KrF excimer laser beam of its photon energy, 5.0 eV, peak
intensity, 50-100 MW/cm^2, pulse width, 10 ns is incident normal (z-
direction) to the target of Si at room temperature. The luminescence
from the plume of vaporized particles is measured from the tangential
x-direction by a conventional video camera and by a streak camera.
Figure 1a shows the result in Ar atmosphere by the video camera.
It represents luminescence patterns of the plume in the z-y plane. The
results show that both intensity and size of the plume increase with
increasing pressure up to 5 Torr of Ar, and then decrease with further

P. Jena et al. (eds.), Physics and Chemistry of Finite Systems: From Clusters to Crystals, Vol. I, 435–440.

increase in the gas pressure. Figure 1b shows the pattern in Xe
atmosphere. The enhancement of luminescence is much more pronounced in
Xe than in Ar. Our result in He atmosphere, on the other hand, shows
much weaker enhancement as compared to Ar and Xe.

Figures 2a and 2b show spectral distributions of the luminescence
in the range from 380 to 420 nm in vacuum (2a) and in 10 Torr of Xe
(2b) measured by the time-integrated mode of our streak camera. The
line at 390 nm is from monomers of neutral Si atom and 412 and 386 nm
from those of ions. The ionic lines are broadened due to the Stark
effect as discussed in [4]. Figures. 2a and 2b show that the overall
feature of spectral distribution in Xe atmosphere is similar to that
in vacuum. Measurements in other photon energy range between 200 nm
and 800 nm also show no additional spectral lines or bands due to the
presence of Ar or Xe. The essential difference lies in the spatial
distribution. The z-position of the maximum intensity is near 0.1 mm
in vacuum, whereas it is shifted to about 0.3 mm in Xe.

The space/time variation of the plume pattern in 10 Torr Xe is
measured by our streak camera system synchronized with the laser pulse.
Figures 3a and 3b show result for the neutral Si line at 390 nm, and
Figs. 4a and 4b for the ionic Si line at 386 nm, respectively. The
patterns in Xe exhibit a delay of over 15 ns for their maximum
intensities with respect to that of laser pulse. The patterns indicate
also that the vaporized particles are ejected with a nearly equal
initial velocities both in vacuum and in Xe gas but they are decele-
rated by collision with Xe as they expand away from the target surface
in Xe gas. The initial velocity is found higher than 10^6 cm/s.

3. Discussion

It is clear from our experimental results that the enhancement of
luminescence in an inert gas atmosphere is due to collision of vaporized particles with the inert gas. The most characteristic feature
found in our measurement is that Si atoms are excited over their first
ionization level of 8.15 eV by colliding with the inert gas at room
temperature. Many collision induced phenomena are caused by a transfer
of energy from the energetic incident atom to the target particle. In
our case, however, the target particles are in motion with their
velocity over 10^6 cm/s, whereas the inert gas is moving with the
thermal velocity at room temperature. Our experimental results also
show that there is not much energy transfer from vaporized particles to
the rare gas since no radiative transition is observed from the excited
state of rare gas. The energy for ionization and the enhancement of
luminescence, therefore must come from the internal energy of vaporized
particles themselves. It is very hard, however, to realize that the
energy of the order 10 eV is stored for a period longer than 10 ns in
the vaporized particles consisting of monomers or dimers of Si atom.

In our model of laser vaporization[2-4], the target atoms leave
the surface in the form of microcluster of various diameters comparable

to the skin depth of the target solid. Under an intense laser beam
well above the threshold for the cluster emission, these clusters are
formed on the surface in the early part of 10 ns laser pulse and
continue to absorb the latter part of the same pulse even after they
leave the surface. They absorb laser energy to the extent to become
unstable and start fragmentation. The laser energy is absorbed both in
the electronic system and vibronic system, but their relaxation times
are much less than 1 ns. The decomposition times found in our results
in Figs. 3a 3b, 4a and 4b exceed 10 ns. Furthermore, the vibronic
system can not store energy well above the binding energy of the system
which is much less than the ionization energy in Si. The only system
that a cluster can store such a high energy is in its ionic state.
Cluster can be multiple-charged up to a critical value at which
it breaks up by Coulomb explosion. The multiple-charged cluster is,
therefore, inherently unstable. It can disintegrate by any external
disturbance causing fluctuation in the distribution of excess charge.
A collision with an atom causes a local charge fluctuation via
electronic interaction directly between the atom and the cluster or
through vibronic excitation of the cluster. Our experimental results
may be explained in terms of this type of collision-induced process.

A possible mechanism to produce multiple-charged clusters is
their photoionization by the incident laser beam. Our previous investigation in bulk Si by nitrogen laser excitation of photon energy
3.7 eV shows that the two-photon photoemission is the dominant
mechanism of electron emission[2]. Assuming that the efficiency of
two-photon photoemission is about the same for bulk Si and cluster Si,
one finds that 10^7 electrons are emitted from the surface of a
spherical cluster of diameter 3,000 A by a laser pulse width of 10 ns
and intensity 100 MW/cm^2. This calculation shows that the cluster can
easily be multiple-charged by multi-photon photoemission processes. If
an intense laser beam is incident on such a cluster it can be photoionized to reach a critical state of instability so that its collision
with a gas atom would lead to local fluctuations of charge distribution
over the surface of the particle. Then the energy build-up would
induce fragmentation of the particle into monomers/dimers.

According to the above interpretation of our experimental results,
the luminescense would be enhanced monotonically with increasing
pressure of the inert gas. The results in Fig. 2, however, show that
the enhancement exhibits its maximum 'around 10 Torr and diminishes with
further increase in the gas pressure. This result can be explained as
follows: The collision of a hot cluster with a rare gas atom at room
temperature gives rise to de-excitation of the cluster. If a large
number of collisions take place under a high gas pressure over 100 Torr
during 10 ns of laser excitation, the loss of internal energy in the
cluster through collisions quenches the energy accumulation due to
absorption of laser energy. Hence the cluster will be cooled down
rather than excited to the critical state of instability.

The relaxation process for an excited cluster is by evaporation of atoms from its surface or by radiative transitions. For multiple-charged clusters, the evaporation of a neutral atom results in an increase in its charge-to-mass ratio, and tend to make them more unstable. Our model of laser vaporization predicts an efficient production of highly charged clusters to cause possible Coulomb explosion.

4. Conclusion

The observation of laser vaporization in the inert gas predicts presence of interesting features of excited cluster. The cluster is a good absorber and reservoir of laser energy. It gives off energetic atoms and ions efficiently by its decomposition induced by gas collision at room temperature. These energetic particles may be used for synthesis of exotic materials.

The authors would like to thank Professor C. Horie and Professor Y. Sasaki for valuable discussions. This work is supported by Grant-in-Aid for Scientific research from the Ministry of Education in Japan.

References

[1] See for example, P. Jena, B.K. Rao and S.N. Kanna (eds.), Physics and Chemistry of Small Clusters, NATO ASI series B: Physics, vol. 158 (1987).

[2] Kasuya, A., and Nishina, Y. (1987) 'Dynamical characteristics of laser vaporization process in group IV elements", in S. Sugano, Y. Nishina and S. Ohnishi (eds.), Microclusters, Springer Series in Materials Science 4, Springer-Verlag, Berlin, pp.249-256.

[3] Kasuya, A., Nishina, Y. (1989) 'Nanosecond emission spectra in us space of Si particles sputtered by laser irradiation', Z. Phys. D12, 493-496.

[4] Kasuya, A., Nishina, Y. (1990) 'Spectroscopic evidence for cluster emission from laser ablated surface in group IV elements', D.C. Paine and John C. Bravman (eds.), MRS Symposia Proc. vol. 191, pp.73-78.

[5] Eryu, O., Murakami, K., Masuda, K., Kasuya, K., Nishina, Y. (1989) 'Dynamics of laser-ablated particles from high T_c superconductor $YBa_2Cu_3O_y$', Appl. Phys. Lett., 54, 2716-2718.

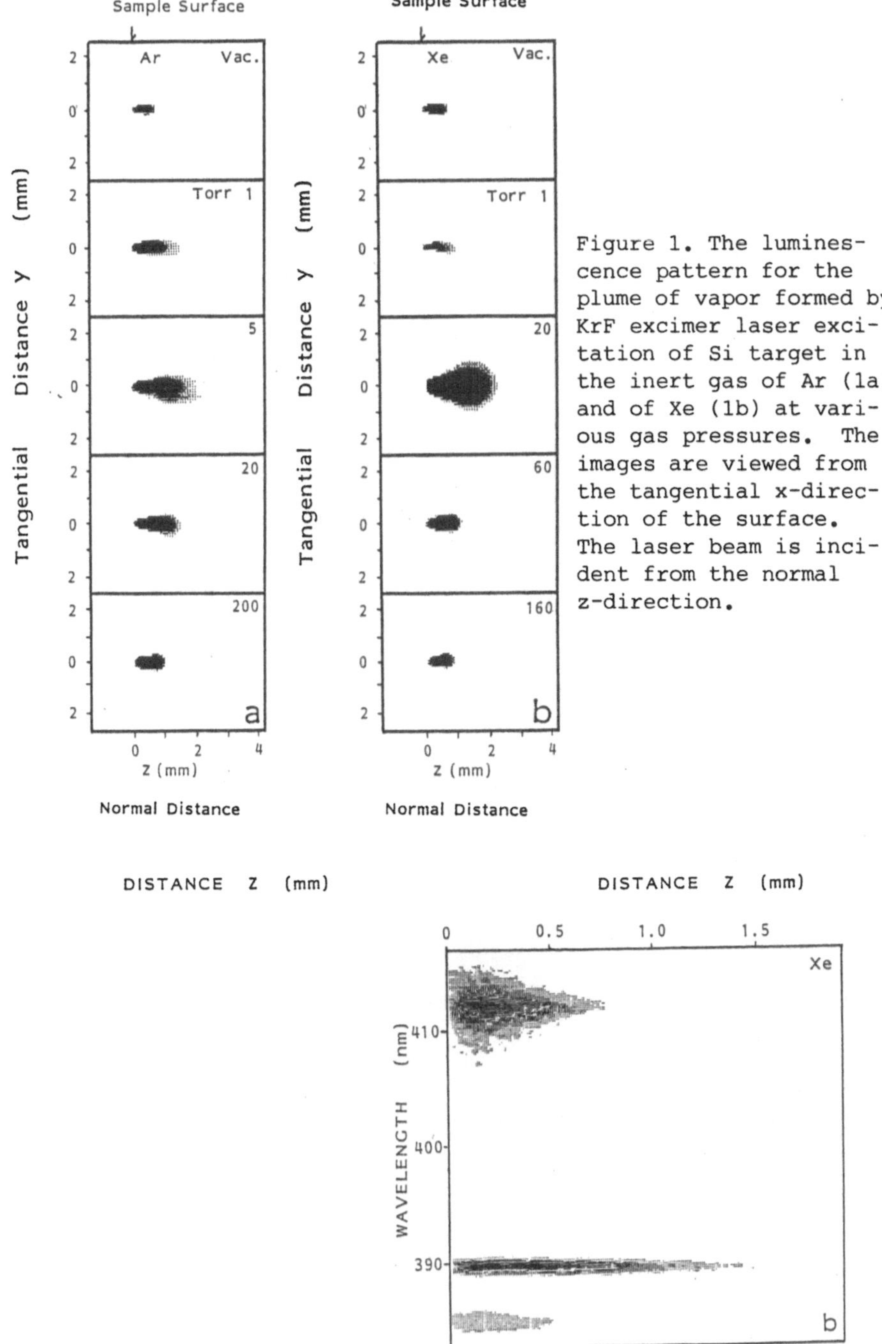

Figure 1. The lumines-
cence pattern for the
plume of vapor formed by
KrF excimer laser exci-
tation of Si target in
the inert gas of Ar (1a)
and of Xe (1b) at vari-
ous gas pressures. The
images are viewed from
the tangential x-direc-
tion of the surface.
The laser beam is inci-
dent from the normal
z-direction.

Figure 2. The spectral distribution of luminescence from the vaporized
Si particles in vacuum (2a) and in Xe gas of 10 Torr (2b).

440

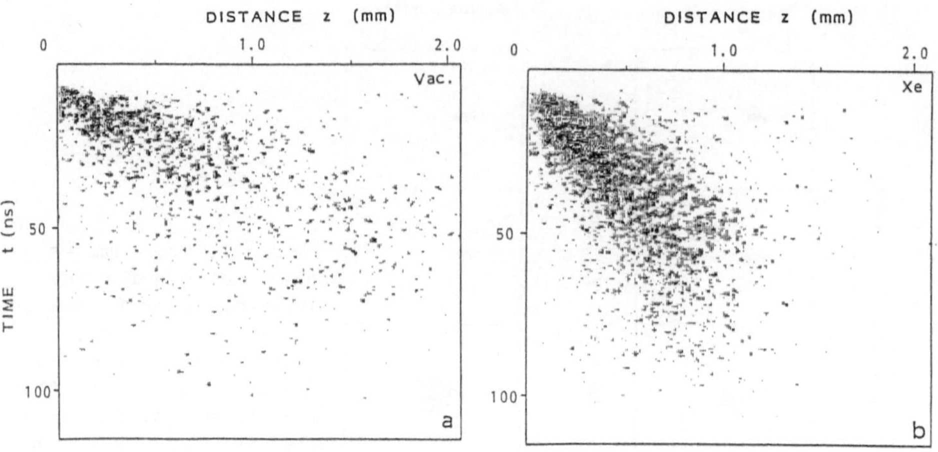

Figure 3. The space/time intensity distribution of luminescence for the neutral Si line at 390 nm in vacuum (3a) and in Xe gas of 10 Torr (3b).

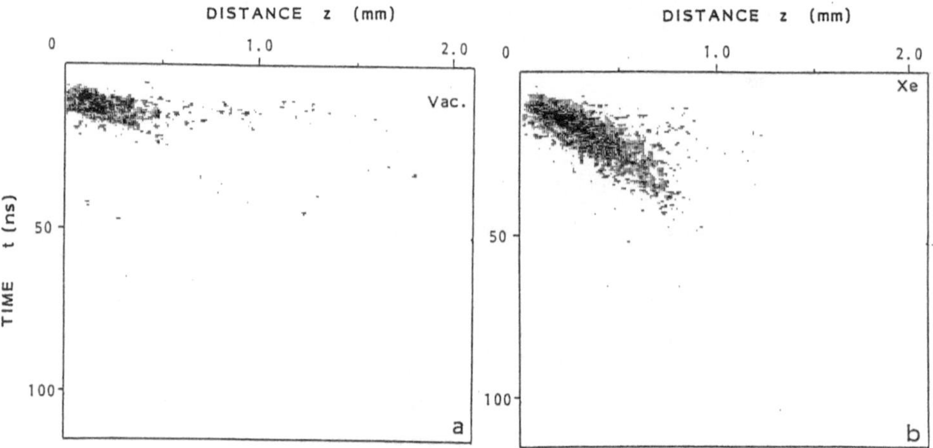

Figure 4. The space/time intensity distribution of luminescence for the ionic Si line at 386 nm in vacuum (4a) and in Xe gas of 10 Torr (4b).

Structural Transition from Small Clusters to Bulk-like Structures
- *Ab initio* Molecular Dynamics Study of Li, Be, and B clusters -

R. KAWAI, M. W. SUNG[†] and J. H. WEARE[†]

Department of Physics, University of Alabama at Birmingham
UAB Station, Birmingham, AL 35294, USA

[†]*Department of Chemistry, University of California, San Diego*
La Jolla, CA 92093-0340, USA

ABSTRUCT: The structural transitions of Li_N, Be_N, and B_N clusters toward the bulk-like structures have been systematically studied using *ab initio* molecular dynamics simulations. Lithium clusters become close-packed structures based on icosahedral packing above $N \approx 25$. Beryllium clusters begin to develop metallic binding at $N \approx 6$. Boron clusters below $N=13$ are quite different from its crystal structure. The icosahedral cage becomes stable at $N=14$.

1. Introduction

The structural transition of clusters to the bulk-like structures has been one of major interests in cluster science. For example, small rage gas clusters, e.g., Ar_{13}, are believed to form icosahedral structures whereas the crystal form of these elements is FCC. The transition from the icosahedral to FCC structure is expected to occur at $N \leq 1,000$[1,2]. Since rare gas atoms are chemically inert and spherical, their structures are closely related to close-packed structures of hard spheres. Therefore, because of the similar packing fraction in FCC and icosahedral structures, the transition from icosahedral packing to FCC is mainly caused by the weak long range interaction and the existence of surface. On the other hand, the structures of small clusters of metallic and covalent elements are quite different from their bulk structures because their electronic structures differ significantly. Therefore, simple forms of empirical interaction can not describe the structural transitions of these elements. *Ab initio* calculations are strongly desired.

Moreover, in recent years the chemical properties of clusters such as chemical reactivity[3] are of increasing interest. Such properties strongly depend upon the size of clusters and their geometrical structures. However, it is not possible at the present to determine the structure of small clusters directly from experimental data. Theoretical investigation is again necessary. Searching for the lowest energy structures of large clusters was even theoretically formidable. Simulated annealing using *ab initio* molecular dynamics (AIMD)[4] is know to be the best mean at the present. This approach has been successfully applied to many systems. The AIMD simulation also can be used to simulate chemical reaction from the first principles.

In this Proceedings, the structures of Li_N ($N \leq 40$), Be_N ($N \leq 20$) and B_N ($N \leq 14$) clusters are systematically investigated using the AIMD simulations.

P. Jena et al. (eds.), Physics and Chemistry of Finite Systems: From Clusters to Crystals, Vol. I, 441–446.
© *1992 Kluwer Academic Publishers.*

442

2. Methods

The lowest energy structures are searched by the simulated annealing method using the AIMD simulations. Electronic density is calculated within the local density approximation of density functional theory and the density is updated every time step in the molecular dynamics simulation using the Car-Parrinello method for beryllium and boron clusters[5,6] and the conjugate gradient method for lithium clusters[7].The detailed computational method is written elsewhere[6,7].

3. Results

Li clusters

Since metallic electrons are almost uniformly distributed and screen out the core charges, the ion cores behave as a hard sphere in the uniform electron gas and tend to form close-packed structures such as FCC. In analogy to the rare gas systems, clusters of such metallic elements are expected to form icosahedral packing. For example, Al_{13} cluster, which is an important building block in the quasicrystals, is believed to be icosahedral[8]. Similarly, alkali metal clusters are also predicted to form icosahedral packing despite the BCC crystal structures at a room temperature[9]. On the other hand, small metallic clusters are often far from close-packed structures. Structural transition from a low packing fraction to a high packing fraction is an interesting problem because this transition is a consequence of change in the nature of valence electrons.

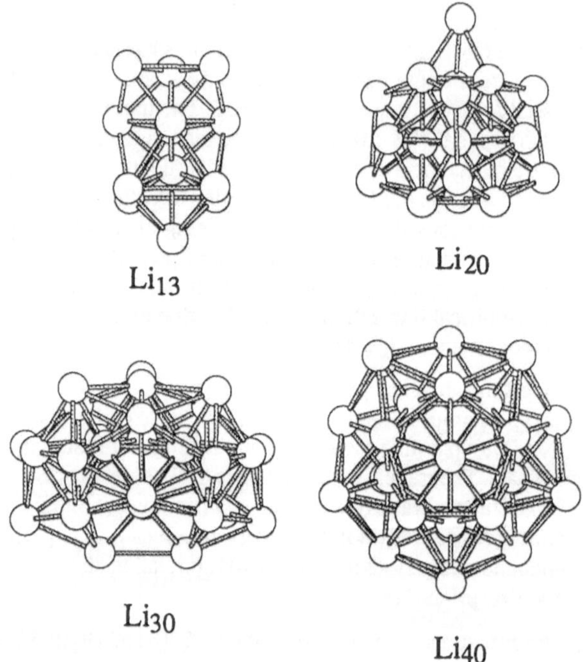

FIG. 1. Structures of Li_{13}, Li_{20}, Li_{30}, and Li_{40} clusters. Li_{30} and Li_{40} contains many pentagonal pyramids.

Pentagonal pyramids appear very often even in the small lithium clusters as pointed out in the recent papers[10,11]. However, these clusters are quite different from the rare gas clusters except for Li_7, indicating that the small lithium clusters are not close-packed. On the other hand, all the clusters above N=25 consist of pentagonal pyramids as shown in FIG. 1. Therefore, the transition from non-icosahedral packing to icosahedral packing takes places below N=25. Furthermore, we found that there are three stages before reaching to the close-packed structure based on the icosahedron. Such stages are illustrated by the highest coordination number (N_c) plotted in FIG. 2. Since small clusters do not have enough atoms to make sufficient coordinations, the highest coordination number grows linearly (N_c=N-1, N$\leq$9) as the size of cluster increases. At N=11, the system reaches to the point where at least one atom has enough coordination to stabilize the structure. The local coordination around this atom does not change until N=16 at which a new coordination number (N_c=10) begins. The twelve coordination, which is as high as that of FCC, HCP and icosahedral packing, begins between N=22 and 24 and bulk-like metallic cohesion becomes dominant above these sizes. In terms of packing fraction, lithium clusters reach to the bulk limit about N=25.

Beryllium clusters

Even more interesting transition takes place in the group II systems. Beryllium diatomic molecule are weakly bound (D_e=0.11eV) by the van der Waals interaction. Therefore, small Be clusters are expected to form similar structures to the rare gas clusters. For examples, Be_{13} was predicted to be an icosahedron[12]. On the other hand, beryllium forms a HCP metallic crystal. Since the transition from non-metallic to metallic state changes binding properties such as cohesive energy and bond distance significantly, dramatic transition is also expected in the geometric structures.

The binding of beryllium clusters grows very quickly as the size increases[5]. By N=4, the gap between the s-band and the p-band vanishes. Metallic cohesion becomes dominate at N$\approx$6[5,13]. Although geometric structures below N=7 are similar to those of the rare gas systems, the Jahn-Teller distortion is increasingly large. Above this size, the structures are no longer resemble to the rare gas clusters. The packing fraction is rather low even at N=20 in comparison to the bulk HCP-

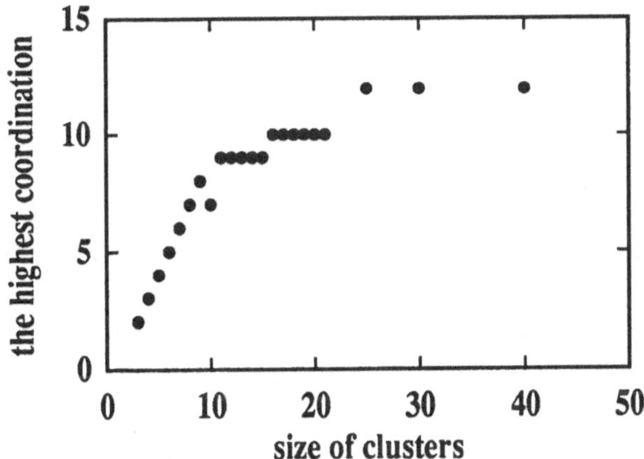

FIG. 2. The highest coordination numbers of lithium clusters.

444

structure.(see FIG. 3).

Although three-dimensional packing fraction is low, many of small beryllium clusters have layered structures and these layers consist of two dimensional close-packing, i.e., hexagonal packing (FIG. 3a), suggesting anisotropic binding. This anisotropic nature is consistent with the bulk crystal structure which has a smaller c/a ratio than the ideal HCP structure.

Boron clusters

The icosahedral cage of B_{12} is a common building unit in the boron-rich materials such as α rhombohedral boron crystal. A borane, $(B_{12}H_{12})^{-2}$ is also known to be iacosahedral. Becuase of this ubiquity, the structure of a bare B_{12} cluster is proposed to be icosahedral. However, the icosahedral unit in the boron-rich crystals strongly interact with its neighbors and the interactions are even stronger than the internal bonds of the icosahedral cage. Therefore, isolation of a B_{12} icosahedron from the crystals creates many dangling bonds. Such a structure is normally unstable and the icosahedral cage is not the lowest energy structure[6]. It is interesting how many boron atoms are needed to create a structure based on the icosahedral cage.

Boron clusters below N=6 are planar and a pentagonal pyramid of B_6 is the first three dimensional structure. (see FIG. 4) However, unlike Li_7 and Be_7, B_7 is not a pentagonal bipyramid but a hexagonal pyramid. A hexagonal bipyramid appears to be the lowest energy structure of B_8. Inter-

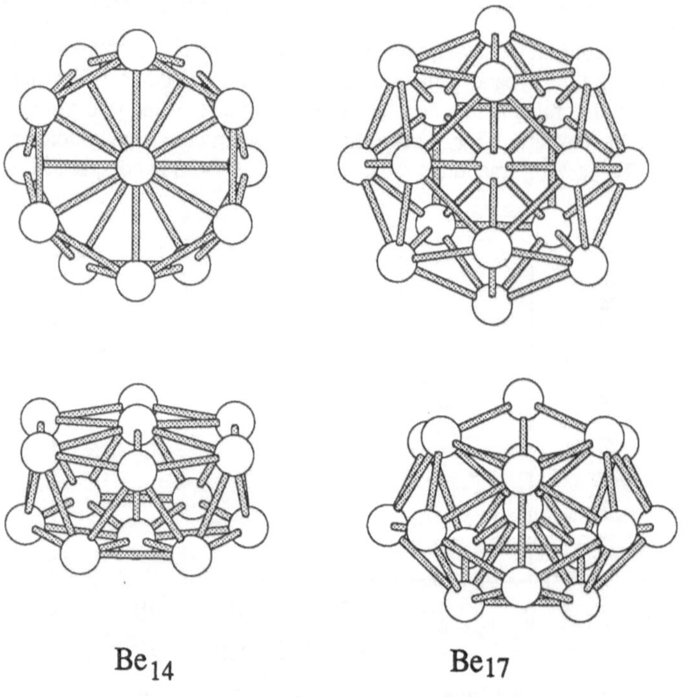

FIG. 3. Structures of Be_{14} and Be_{17} clusters.

estingly, the similar development continues from B_9 to B_{12} as shown in FIG. 5. These structures suggest that the boron clusters prefer the hexagonal ring to the pentagonal ring. This growth pattern stops at N=13 and the C_{6v} and C_{3v} structures (FIG. 6) are found to be degenerate ground states for B_{13}. Both structures still contains hexagonal rings. However, B_{14} is an icosahedral cage with two capping atoms and does not have hexagonal patterns as shown in FIG. 6. It is surprising that only two capping atoms stabilize the icosahedral cage.

This work was supported by the U. S. Office of Naval Research (ONR-N00014-87-K0675) and the U. S. National Science Foundation (ASC-9014996). Calculations were carried out at the Naval

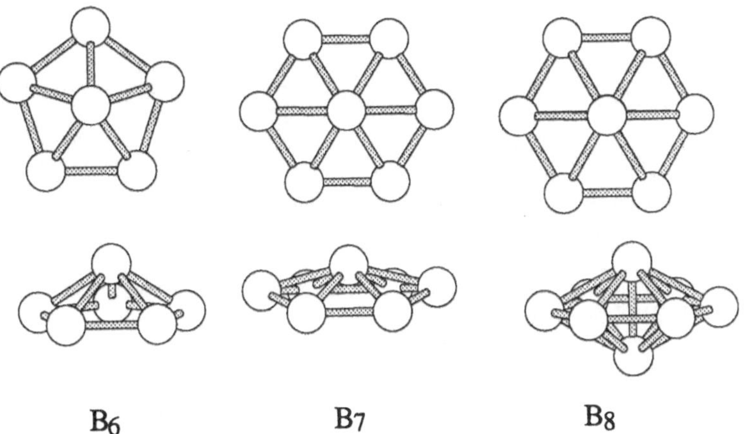

FIG. 4. Structures of B_6, B_7, and B_8 clusters. Two different views are shown for each cluster.

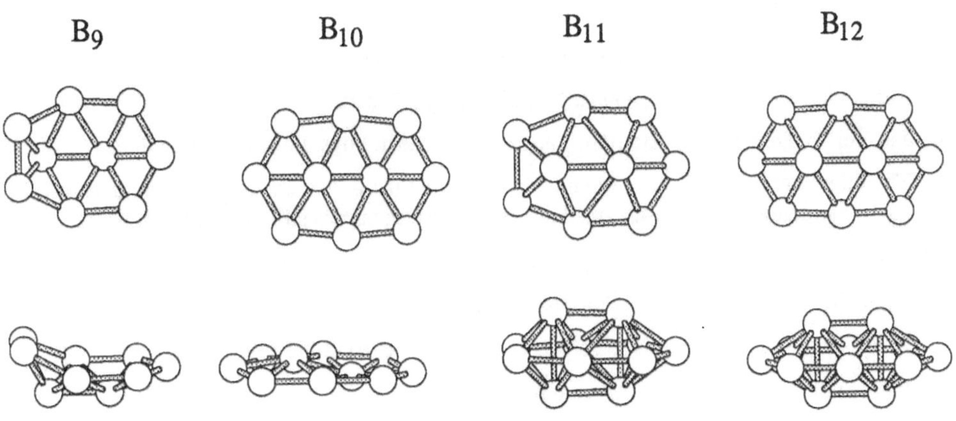

FIG. 5. Structures of B_9, B_{10}, B_{11}, and B_{12} clusters. The structures of B_{11} and B_{12} can be considered as bi-pyramids of B_9 and B_{10}, respectively.

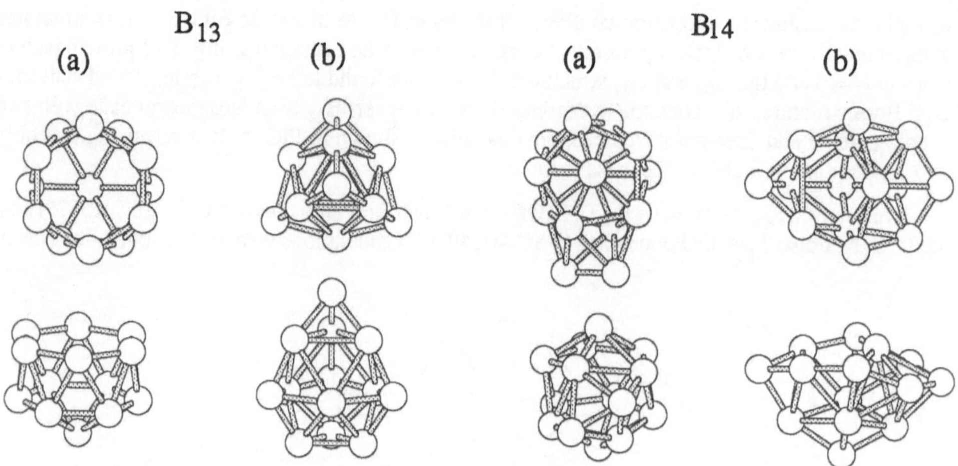

FIG. 6. Structures of B_{13} and B_{14} clusters. For B_{13}, the structures (a) and (b) have the same energy. For B_{14}, the structure (a) is slightly more stable than (b)

Research Laboratory, Naval Oceanographic Office Superconductor Center, San Diego Supercomputer Center and IBM Los Angeles Scientific Center.

References

1. B. Raoult, J. Farges, M. F. Deferaudy, and G. Torchet, Philos. Mag. **B60**, 881 (1989).

2. W. Miehle, O. Kandler, T. Leisner, and O. Echt, J. Chem. Phys. **91**, 5940 (1989).

3   For example, P. Fayet, A. Kaldor, and D. M. Cox, J. Chem.Phys.**92**, 254 (1990).

4. R. Car and M. Parrinello, Phys. Rev. Lett.**65**, 80 (1985).

5. R. Kawai. and J. Weare, Phys. Rev. Lett. **65**, 80 (1990).

6. R. Kawai and J. Weare, J. Chem. Phys. **95**, 1151 (1991).

7. Ming-Wen Sung, J. Weare, and R. Kawai, to be published.

8. Jae-Yel Yi, Dirk J. Oh, J. Bernholc, and R. Car, Chem. Phys. Lett. **174**, 461 (1990).

9. The lithium crystal is BCC at room tempertaure but has a martensitic transition to a close-packed structure at near 70K. This low-teperature structure is still under investigation.

10. P. Fantucci, J. Koutecky, and G. Pacchioni, J. Chem. Phys. **80**, 325 (1984).

11. O. Sugino and H. Kamimura, Phys. Rev. Lett. **65**, 2696 (1990).

12. E. Blaisten-Barojas and S. N. Khanna, Phys. Rev. Lett. **61**, 1477 (1988).

13. P. V. Sudhakar and K. Lammertsma, preprint.

14. L. Hanley, J. L. Whitten, and S. L. Anderson, J. Phys. Chem. **92**, 5803 (1988).

DYNAMICS OF THE CLUSTER-SURFACE INTERACTION: COLLISIONAL
DETACHMENT OF SIZE SELECTED CARBON NEGATIVE CLUSTER IONS

T. MORIWAKI, H. SHIROMARU AND Y. ACHIBA
Department of Chemistry, Tokyo Metropolitan University
Hachioji, 192-03 Tokyo, Japan

ABSTRACT. Size selected carbon negative cluster ions (n = 5-22, 60, and 70) were examined on the collisional electron emission at solid surface as a function of the incident kinetic energy (0-1keV). Two different mechanisms were found below 1 KeV collision energy. Electron emission occurring at high energy (> 300 eV) is totally governed by velocity of the incoming cluster ions. The electron emission at the low collision energy, on the other hand, is considerably dependent on the molecular structures of the carbon clusters, i.e., linear chain, monocyclic ring, and hollowclosed-cage structures. The latter process is a new phenomenon which may be closely related with a collisional electron detachment of negative cluster ions.

1. Introduction

Cluster-surface interactions are very interesting subjects from view points of not only understanding nature of clusters but also application of clusters to new materials. Of particular interest is that whether such interactions are of size specific or not. However, so far little have been known on the size specific subjects even for energy region below 1keV[1,2]. Very recently Beck et al.[3] have first demonstrated the size specific phenomena induced by collision of clusters with solid surface, where impact induced cleaving and melting of alkali halide clusters were investigated.

In the present work, we have studied collision induced processes of mass selected carbon negative ions with MoS_2 single crystal, paying a special attention on the process of electron emissions as a function of the cluster size and the collision energy. The reason why we measure here electrons is that the detection of electrons is relatively easy and highly sensitive comparing with other particles. In addition to such a technical advantage, the emitted electrons induced by collision of negative ions would have unique and useful microscopic information on the cluster-surface interactions.

In the course of present work, we found the following interesting

P. Jena et al. (eds.), Physics and Chemistry of Finite Systems: From Clusters to Crystals, Vol. I, 447–452.

phenomena on the electron emission processes. 1) Upon collision of carbon negative cluster ions with MoS_2 surface, there appears two different electron emission processes in the the energy region lower than 1keV. 2) One of which dominantly occurs at rather higher collision energy (>300eV), and is well characterized by the fact that electron emission efficiency is totally dependent on the velocity of the clusters. 3) The other one occurs at low collision energy with narrow energy window. The electron emission efficiency depends on rather microscopic properties of the incident clusters such as a geometric structure and an electron affinity. The plots of electron emission efficiency as a function of collision energy show maxima at a certain collision energy.

2. Experimental

Figure 1 shows the experimental setup used in the present work. Carbon clusters were produced by usual laser vaporization combined with a pulsed He flow [4]. Negative ion clusters were accelerated and mass selection was performed by TOF method[2]. The target of MoS_2 single crystal was placed on the bottom of a coaxial reflectron with which the incident kinetic energy of the clusters was controlled by applying negative voltage on a bottom plate. Emitted electrons were detected by a multichannel plate with a hole ($\emptyset$ 8.0mm) at center, through which the negative ion beam passes into the reflectron. The coaxial reflectron has an advantage in comparison with an angled reflectron such that the incident angle of the incoming cluster beam is kept constant while the incident kinetic energy is changed. Therefore, the interaction area on the target does not change.

The sample of fullerenes such as C_{60} and C_{70} was prepared by arc heating of graphite and isolated by a preparative HPLC described elsewhere[5]. Pure C_{60} and C_{70} powder was first dissolved in benzene and dropped onto a Cu rod for the laser vaporization.

Estimation of the numbers of cluster ions has ambiguity if we use an electron multiplier as an ion detector without paying attention on mass dependences of secondary electron emission efficiency (γ). Actually, the previous our experiments on the size dependent γ values determined for n = 5-22 negative carbon cluster ions seems to have large systematic errors[2]. Therefore, in the present work, first, we determined γ values by means of a Faraday cage for each size of cluster ions. The absolute values of the number of incident cluster ions (I_{ion}) as well as emitted electrons (I_e) were successfully

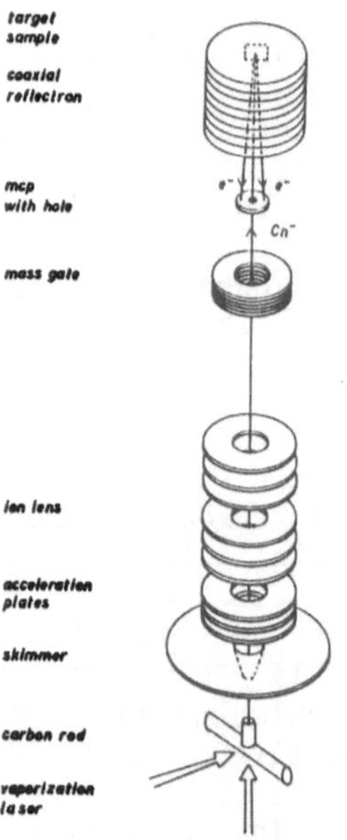

Fig.1 A schematic drawing of apparatus for mass selected cluster beam-solid surface collision experiments.

obtained as a pulsed current at 640eV. The γ values for other incident kinetic energies were determined as relative values to those at 640 eV with an electron multiplier.

3. Results & Discussion

According to the previous works[6], carbon negative cluster ions have three forms, i.e., a linear chain, a monocyclic ring and a hollowclosed-cage structure, and the preferential form much depends on their sizes and internal energies. Concerning the carbon clusters smaller than n=30, both linear chain (for n<10) and monocyclic ring (for $10 \leq n < 30$) forms are dominant[7], while the carbon clusters of C_{60} and C_{70} are certainly of hollowclosed-cage[8]. Furthermore, as described in our previous paper[9], under certain laser vaporization condition, negative carbon

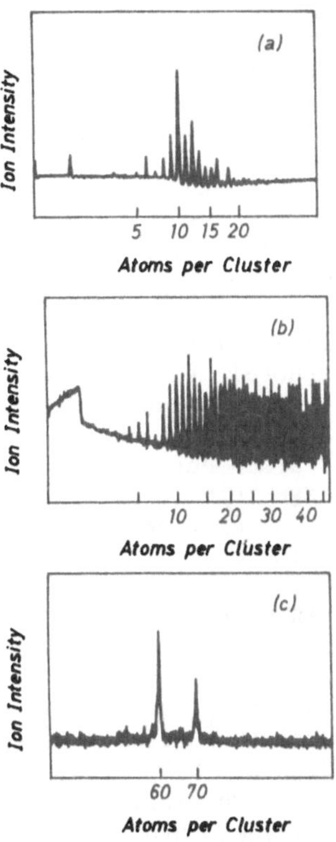

Fig.2 TOF mass spectra of carbon negative cluster ions. The condition of the laser vaporization was optimized so as to obtain (a) highly abundance of the ring form, and (b) the chain form. Chemically isolated C_{60} and C_{70} powder was used for the spectrum (c).

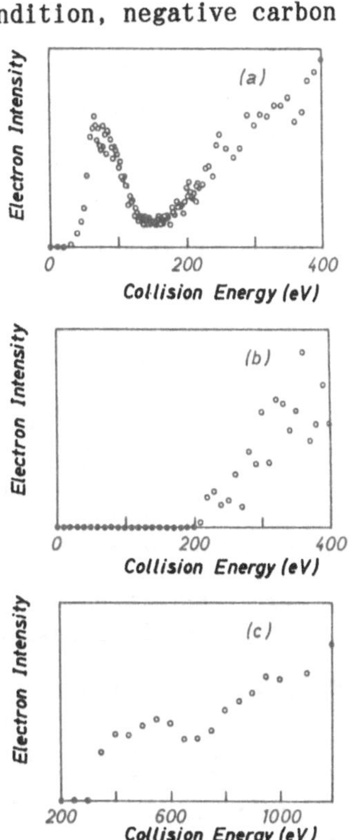

Fig.3 Variations of electron emission intensities as a function of the collision energy for (a) C_{10}^{-} (ring), (b) C_9^{-} (chain), and (c) C_{60}^{-} (hollowclosed cage).

450

cluster ions exist as a linear chain up to n=15-20. These experimental facts suggest that we can perform the cluster collision experiments not only as functions of incident energy and size, but also as a geometrical structure.

Figures 2(a)-(c) display typical mass distributions of carbon negative ions revealing the presence of three forms which were produced by the "controlled" laser vaporization. Note here that we used an electron multiplier as an ion detector. The spectrum (a) is well characterized by the presence of several strong magic numbers, reflecting that the carbon clusters larger than $n = 10$ exist as a monocyclic ring. On the other hand, the spectrum (b) reflects the "chain dominant" mass distributions for which we can see no magic like behaviors. Mass abundance smoothly distributes over whole range. Figure 2 (c) is a mass spectrum of fullerenes, for which chemically isolated C_{60} and C_{70} powders were used.

Figures 3 (a)-(c) show the plots of electron emission efficiencies for the carbon negative clusters of n=10 (ring), n=9 (chain) and n=60 (hollowclosed cage), respectively, as a function of incident kinetic energies. Here, the electrons emitted by collision of negative cluster ions with the target were detected after traveling through the electric field of the reflectron. The electric field applied to the reflectron acts as a decceleration for the incident negative cluster ions, whereas it works as an acceleration for the electrons at this time. Most of the electrons emitted at the target go back to the same way with a small divergence, because the intrinsic kinetic energies of the electrons initially generated by collision are negligibly small (<3eV)[2].

As can be seen from these curves, the general features commonly observed for all of these carbon forms are that electron emission efficiencies gradually increase by increasing kinetic energy in the higher energy region (>300eV). However, it is also certainly distinct that for the ring (C_{10}) and the hollowclosed-cage form (C_{60}), the curves have a peak or a hump at a certain kinetic energy lower than that mentioned above. From these experimental evidences, it is readily suggested that there are two different mechanisms in electron emission processes. Let us discuss these two processes in more detail.

High Energy Collision

In order to clarify what governs the electron emission efficiency in the high collision energy (>300eV), we measured absolute values of electron emission efficiency, γ, for the carbon clusters of n=5,10,16,18. For this purpose, we used a Faraday cage with keeping the incident kinetic energy constant at 640eV. The resulting values are 0.39, 0.31, 0.24, 0.18 for n=5,10,16,18,respectively. Comparing these observed values, it was found that the γ values have a linear relationship with the velocity of the corresponding carbon clusters as shown in fig.4. Here, the γ for C_1^- was the values reported by Tsuji et al.[10].

The presence of the linear relationship strongly suggests that the electrons produced in the higher energy region are closely related with a kind of kinetic emission[11]. This conclusion is also supported by the fact that almost the same values are obtained for the corresponding neutral carbon clusters with the same size and same kinetic energy

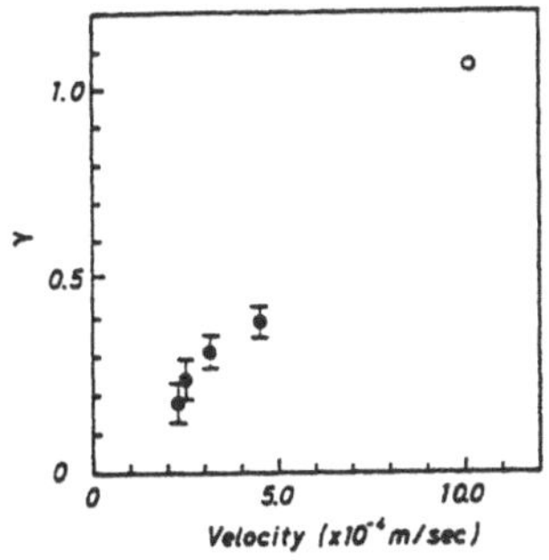

Fig.4 A plot of the values obtained for C_5^-, C_{10}^-, C_{16}^-, and C_{18}^- against velocities of the clusters. The value for C_1^- used here is the one reported in reference [10].

(here, the neutral carbon clusters were produced by photodetachment after acceleration.)[2].

Low Energy Collision

Apparently, the plots of the γ values show a peak at around 50-200 eV kinetic energy for all the ring carbon clusters examined here. The peak intensity varies within the same order from n = 10 to 18. For the fullerenes, the peak positions shift to the larger kinetic energy at around 550 eV. However, no peak-like shape appears in the γ plots for the chain-form clusters (n = 5-9). The γ values almost monotonically increase in this case.

No peaks appearance for C_5^- - C_9^- chain carbons was further supported by the experimental evidences that even in the case of carbon clusters larger than n = 10, if they exist as a chain form, they show little trace of the electron emission in the low energy region[12]. This fact may correspond to absence of collisional electron emission for C_5^- - C_9^- below 200 eV.

Figure 5 shows the relationship of the appearance energy of the electron emission at the low energy and EA values[7] for the ring-form clusters. The clusters with the higher electron affinity tend to require the larger collision energy to emit electrons. Such behaviors strongly suggest that the collision with the solid surface involves an excitation of an uppermost electron of the negative cluster ions through conversions of the incident kinetic energy into the internal excitation energy, and thus the electrons are ejected into vacuum from the clusters. Therefore, the process is considered as a collisional electron detachment induced by a surface potential, possibly by a repulsive potential. Here, the question may arise why such an electron

Fig.5 A plot of electron emission appearance energies of the low energy peaks against the electron affinities (EA) obtained for the ring-form carbon clusters; [●] C_{10}^-, [△] C_{12}^-, [▼] C_{14}^-, [□] C_{16}^-, [◆] C_{18}^-. The EA values are from reference [7].

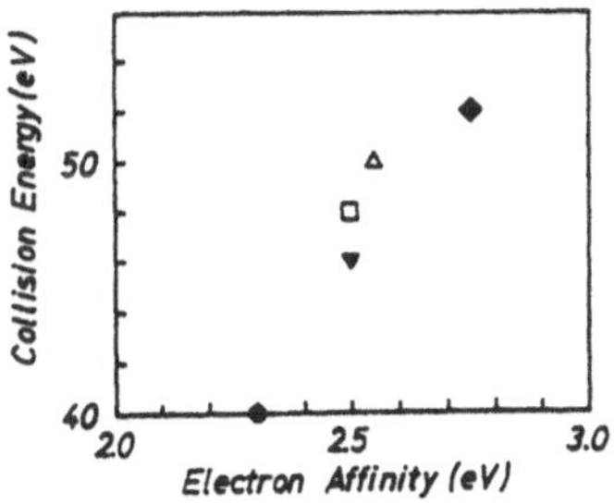

detachment does not take place on the linear chain clusters. The most plausible interpretation at moment is that since the outermost electrons in the chain forms are very much localized at the edges of the chain, it may dramatically reduce the cross section of the electron detachment process.

Acknowledgments

The authors thank Mr. Ken-ichi Aihara and Mr. Hiroyuki Matsuura for their experimental assistance.

References

[1] Hsieh,H.,   and   Averback,R.S.   (1990)   'Molecular-dynamics investigation of cluster-beam deposition' Phys. Rev. B. 42, 5365-5368.

[2] Shiromaru,H.,   Moriwaki,T., Kittaka,C., and Achiba,Y.   (1991) 'Collisional electron emission at solid surfaces induced by mass selected negative cluster ions', Z. Phys. D - Atoms, Molecules and Clusters 20, 141-143.

[3] Beck,R.D., John,P.S., Homer,M.L., and Whetten,R.L. 'Impact-induced cleaving and melting of alkali-halide nanocrystals', Phys. Rev., submitted.

[4] Dietz,T.G., Duncan,M.A., Powers,D.E., and Smalley,R.E. (1981) 'Laser production of supersonic metal cluster beams', J. Chem. Phys. 74, 6511-6512.

[5] Kikuchi,K.,   Nakahara,N.,   Honda,M.,   Suzuki,S.,   Saito,K., Shiromaru,H., Yamauchi,K., Ikemoto,I., Kuramochi,T., Hino,S., and Achiba,Y.   (1991) 'Separation,   detection,   and uv/visible absorption spectra of fullerenes; C_{76}, C_{78}, and C_{84}', Chem. Lett., 1607-1610.

[6] Cheshnovsky,O., Pettiette,C.L., and Smalley,R.E. (1989) 'UPS of metal and semiconductor clusters', in J.P.Maier (ed.), Ion and cluster ion spectroscopy and structure, Elsevier, Amsterdam, 373-415.

[7] Yang,S., Taylor,K.J., Craycraft,M.J., Conceicao,J., Pettiette,C.L., Cheshnovsky,O., and Smalley,R.E. (1988) 'UPS of 2-30-atom carbon clusters: chains and rings', Chem. Phys. Lett. 144, 431-436.

[8] Krätschmer,W., Lamb,L.D., Fostiropoulos,K., and Huffman,D.R. (1990) 'Solid C_{60}: a new form of carbon', Nature 347, 354-358.

[9] Achiba,Y., Kittaka,C., Moriwaki,T., and Shiromaru,H.   (1991) 'Evidence of linear chain for n $\geq$10 carbon negative ion clusters revealed by mass selected photodetachment spectroscopy', Z. Phys. D - Atoms, Molecules and Clusters 19, 427-429.

[10] Tsuji,H., Miyata,K., Taya,T., Ishikawa,J., and Takagi,T. (1985) 'Negative ion implanter and characteristics of negative ion beam', Proc. 9th symposium on ISAT '85 Tokyo, 195-202.

[11] Kaminsky,M. (1965) Atomic and ionic impact phenomena on metal surfaces, Springer, Berlin.

[12] Shiromaru,H., Moriwaki,T., Aihara,K., Matsuura,H., and Achiba,Y. to be published.

ON THE SPECTRA, STRUCTURE AND DYNAMICS OF ANILINE-AR$_n$ CLUSTERS: SIMULATION STUDIES COMPARED WITH THE RESULTS OF R2P2CI SPECTROSCOPY

P. PARNEIX, P. HERMINE, F.G. AMAR* and Ph. BRECHIGNAC
*Laboratoire de Photophysique Moléculaire, Bât 213,
Université de Paris-Sud, 91405 Orsay, FRANCE.
*Permanent address: Dept. of Chemistry, University
of Maine, Orono, ME 04469, USA.*

ABSTRACT. Molecular dynamics simulations of Aniline-(Ar)$_n$ Van der Waals clusters have been performed. Isomerisation studies for An-(Ar)$_3$ are described. Statistical evaporation rates for An-(Ar)$_3$ are given and the results discussed in relation to isomerisation. Simulated spectra of An-(Ar)$_n$ with n=1-3 obtained with the spectral density method are also presented.

1. INTRODUCTION

We have previously reported resonant two-photon photoionization (R2P2CI) spectra of a series of aniline-Ar$_n$ clusters[1]. The use of two different colors permitted us to obtain the absorption spectra of mass-selected clusters while minimizing fragmentation. The region near the $S_0 \dashrightarrow S_1$ transition of aniline is scanned by the first laser photon. It is the resonant absorption of this beam which constitutes the spectrum. In a second step, molecules are ionized to their ground ionic state by photoionization with a second laser whose frequency is adjusted to just above the ionization threshold. Figure 1 shows a series of spectra taken with this technique.

The aniline-Ar spectrum (Figure 1a) is characterized by a relatively intense 0^9_0 band, shifted by 53 cm^{-1} to the red of the pure An 0^0_0 band . A series of smaller peaks extending to higher energy are attributed to Van der Waals modes of the complex. This pattern is roughly repeated 53 cm^{-1} further to the red for the An-(Ar)$_2$ spectrum (Figure 1b). Thus, aniline-(Ar)$_2$ clusters obey the same kind of red shift additivity law observed for complexes of other aromatic molecules[2,3] and which can be explained by the existence of two equivalent binding sites on either side of the aromatic ring.

There are two groups of bands in the An-Ar$_3$ spectrum (Figure 1c,e): the first group has a band origin which is red-shifted by -74 cm^{-1} relative to the bare molecule; the second band origin is blue-shifted by 28 cm^{-1} relative to the bare molecule. A similar spectrum has also been observed by Bieske et al[4].

We have found in our structural modelling[1] that there are two isomers which we call the "3+0" (most stable) and the "2+1" which is less stable by about 94 cm^{-1}. These structures are shown near the relevant spectra. We associate the red-shifted band with the "2+1" structure, under the assumption that a single Ar atom found near the nitrogen atom contributes a blue-shift to the spectrum so that the overall red-shift of "2+1" is less than that of "1+1". The "3+0" is quite reasonably associated with the band that has a net blue-shift. Additional evidence that these two bands really correspond to two different isomers is found in figure 1c,d,e which show a dramatic pressure dependence of the relative intensities of the two groups of bands in the An-Ar$_3$ spectrum. As the Ar pressure increases, the red-shifted band (attributed to the "2+1" isomer) diminishes and the blue-shifted band (attributed to the "3+0" isomer) grows in intensity. It is interesting to note that an extended Franck-Condon progression becomes evident for the blue-shifted band ("3+0"). These pressure changes in the expansion are naturally accompanied by cluster temperature changes[1]. This suggests to study the dynamical behaviour of this cluster. In Section 2 below we shall address several important questions: How does the dynamical behaviour of the system evolve as its internal energy is

P. Jena et al. (eds.), Physics and Chemistry of Finite Systems: From Clusters to Crystals, Vol. I, 453–458.
© 1992 *Kluwer Academic Publishers.*

454

increased? What is the magnitude of the potential barrier for the atom side-crossing transition (allowing the interchange between isomers "2+1" <------> "3+0")? Do the experimental results reflect some of this dynamics? Section 3 is devoted to semi-classical spectral simulations which have been performed for An-Ar, An-Ar$_2$ and An-Ar$_3$.

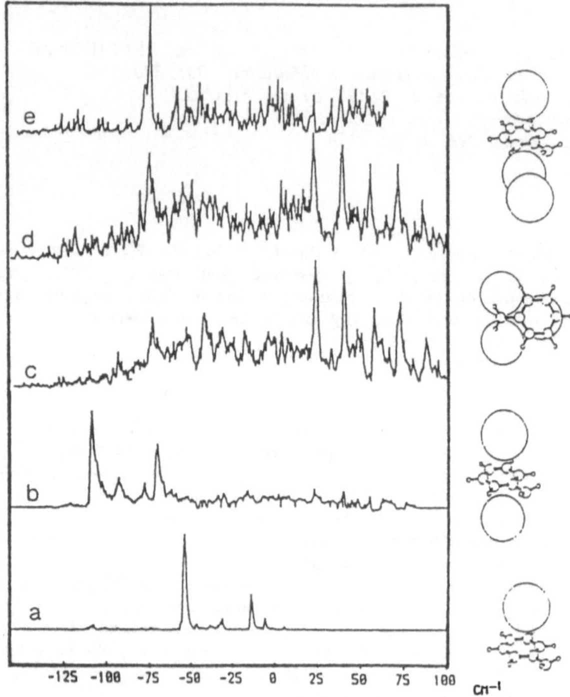

Figure 1. Spectra of Aniline-(Argon)$_n$ for n=1 to 3.

2. DYNAMICS OF AN-AR$_3$

As in our previous structural study[1] the intramolecular potential energy surface (PES) was built from the simple atom-atom potential forms similar to those first proposed by Ondrechen et al.[5] supplemented by a dipole-induced dipole interaction to partially account for electrostatic terms. The Ar-Ar interaction was described by a fit of Aziz and Chen[6] potential to a sum of even powers of 1/r for reasons of computational simplicity[7]. The An molecule is considered as a rigid body whose orientation is described by quaternions parameters[8]. Classical equations of motion are numerically solved by standard molecular dynamics techniques to sample the cluster phase space. Additional details are given in ref. 1.

2.1. ISOMERISATION

For this study, we have run An-Ar$_3$ trajectories at energies ranging from the absolute minimum of the PES, E_{min}, up to around the dissociation threshold, $E_{min} + E_{diss}$. From each trajectory (of duration 710 ps), we extract the kinetic temperature $T = 2<\varepsilon> / 3nk_B$ and the normalized r.m.s. Ar-Ar bond fluctuation, δ_{Ar-Ar} which characterizes the rigidity of the cluster. The values obtained for the temperature and the δ_{Ar-Ar} parameter as a function of energy, are reported in Figure 2.

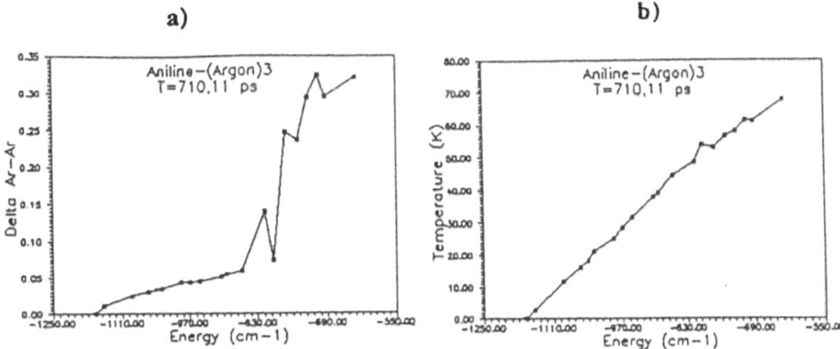

Figure 2. δ_{Ar-Ar} and the temperature as a function of the energy E.

A rapid increase of δ_{Ar-Ar} (figure 1a) is observed when the energy is equal to about E_{iso}= -830 cm^{-1}. It corresponds approximatively to a temperature of 40 K. This parameter rises dramatically from values less than 0.1 to values of 0.2 or higher in accord with the Lindemann criterion for melting which is now well proven for a variety of cluster systems[9]. During the trajectories, we have monitored the evolution (as a function of time) of the cluster structure. In fact, the rapid increase of δ_{Ar-Ar} is the result of the passage of one Ar atom from one side of the substrate to the other side (transition "2+1"<----->"3+0"). On the caloric curve (figure 2b) the temperature remains proportionnal to the internal energy at low energy. When the energy is approximatively superior to E_{iso}, we can note a net inflection whith respect to this straight-line corresponding to the passage over the potential barrier between the two isomers ("2+1" and "3+0"). The fluctuations of δ_{Ar-Ar} in the transition region can be explained by the occurence of sub-trajectories where the system remains trapped in a given configuration (so that in this band of energy the system can have a solid-like or a liquid-like behaviour). The first side crossing transition has been observed for E=-814 cm^{-1}. It provides an upper limit for the value of the barrier potential height which is then less than 438 cm^{-1}.

From these results it seems that the temperature characteristic of the side to side transition is too high ($\cong$ 40 K) to explain the observed growing of the "3+0" isomer with Ar partial pressure as a result of intramolecular isomerization. This confirms the dominant role of the nucleation processes, pointed out in our previous work[1], in the spectral modifications of Figures 1c,d,e.

2.2. EVAPORATION

We have restricted ourselves, so far, to energies less than the binding energies E_{diss} of An-Ar$_3$. It is also interesting to investigate the cluster dynamics when the internal energy E becomes larger than $E_{min} + E_{diss}$, necessary to remove one Ar atom, so that the unimolecular reaction:

$$An\text{-}Ar_3 \; \text{-----> } \; An\text{-}Ar_2 + Ar$$

can be observed.

The determination of evaporation rates by M.D. requires that one evaluate the number of non-evaporated clusters as a function of time for an initial ensemble of trajectories, since for unimolecular decay:

$$k\,t = -\ln(N(t)/N(0))$$

This number can be drawn from the determination of the "evaporation time" on each "evaporative" trajectory among a statistical set of 100 trajectories.

At each energy considered, we start with an ensemble of trajectories, already equilibrated at some low energy. E < E_{diss} and then instantaneously add kinetic energy to the cluster by scaling the velocities, subject to the constraint that J = 0. The final energy E after scaling will, of course, be larger than E_{diss}. The evaporative trajectories are identified by a distance criterion but time of evaporation is defined by means of a dynamical criterion (last time at wich the evaporated atom had a negative radial velocity)[10].

It is interesting to note that, as shown in Figure 3, the added energy is redistributed among the internal modes of the cluster in a very short time ($\cong$ 10 ps). Since the idea of evaporation presupposes thermal equilibrium, we eliminate those trajectories which evaporate in less than 5 picoseconds.

456

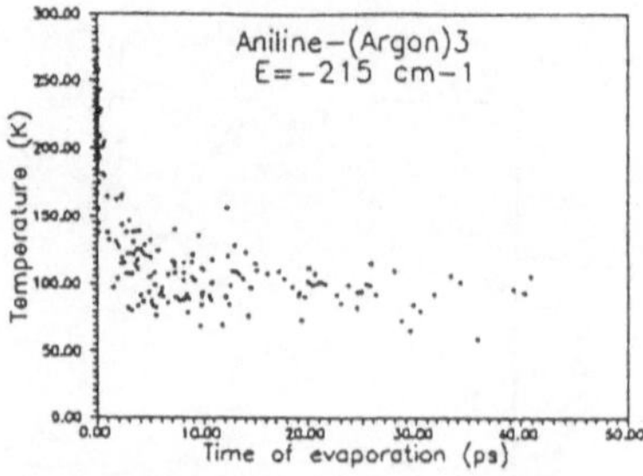

Figure 3. Cluster temperature as a function of evaporation time for E=-215 cm⁻¹.

The behaviour of ln(N(t)/N(0)) as a function of time is plotted in Figure 4a for two different energies. The evaporation rates are derived from the slopes of such plots. The results are displayed in Figure 4b as a function of E.

a)   b)

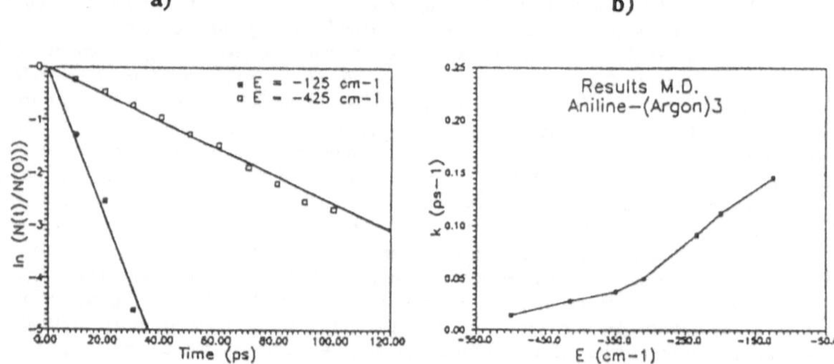

Figure 4. a) ln(N(t)/N(0)) as a function of time for 2 different energies.
b) Evaporation constant as a function of the energy E.

In order to get a better insight into the dynamical processes occurring during these trajectories, we have examined two parameters, chosen to reflect the instantaneous configuration of the system, at the exact time defined by the dynamical evaporation criterion. We label the ejected atom with an index of 3, the other two atoms within the remaining sub-cluster being 1 and 2. The z-axis being orthogonal to the substrate plane, these parameters are z_1+z_2 and $z_1+z_2+z_3$. They are plotted against each other in Figure 5 for a given set of 100 trajectories. It is striking that, although the $z_1+z_2+z_3$ data for the whole cluster are spread over all accessible values, the corresponding z_1+z_2 for the sub-cluster is rather constrained around discrete values, which is consistent with the fact that it is left "colder". Indeed evaporation is known to be a cooling mechanism, as a result of the removal of an energy equal to E_{diss}. Then, at the time determined by the dynamical criterion, the _evaporative_ configuration of the cluster really shows one Ar atom leaving a cold sub-cluster.

In fact, during the trajectory, a transfer of energy occurs between the sub-cluster and the ejected atom. At the short times, all the Ar atoms can undergo side-crossing transitions, and the dynamics shows that _the average residence times in each conformer are about the same._ However we always observe a larger number of _evaporative_ configurations "2+1" than of "3+0", the net yield (number of evaporations from each conformer) being: "3+0"/"2+1" = 0.61. On the other hand the average time spent in the last configuration before the dynamical criterion is shorter for the _evaporative configuration_ "3+0" than for the _evaporative_ configuration "2+1" (see Table I). At first sight these two findings seem contradictory, but a strict steric reasoning , considering the cold nature of the sub-cluster, can account for them.

Just before the evaporation time , two Ar atoms get trapped in a given configuration while the third atom can explore very freely the configuration space. Figure 5 shows that the probability to get a "1+1" sub-cluster is twice smaller than for a "2+0". For obvious steric reasons, if the ejected Ar atom flies over a "1+1" sub-cluster, the evaporative configuration will always be a "2+1" structure, but if it flies over a "2+0" sub-cluster, it can be equally in a "2+1" or "3+0" configuration. Then, within this statistical picture, the expected yield "3+0"/"2+1" is 0.5. In fact, we can see it overestimates the number of "2+1" isomers. This difference is the result of the dynamical behaviour of the cluster in its final phase, which shows (Table 1) that the cluster evaporates more quickly when it is in a "3+0" structure. This is consistent with the simple mechanical picture in which when the system explores a "3+0" configuration, the probability for the third Ar atom to have a energy-transferring collision with an other Ar atom is greater than in a "2+1" configuration.

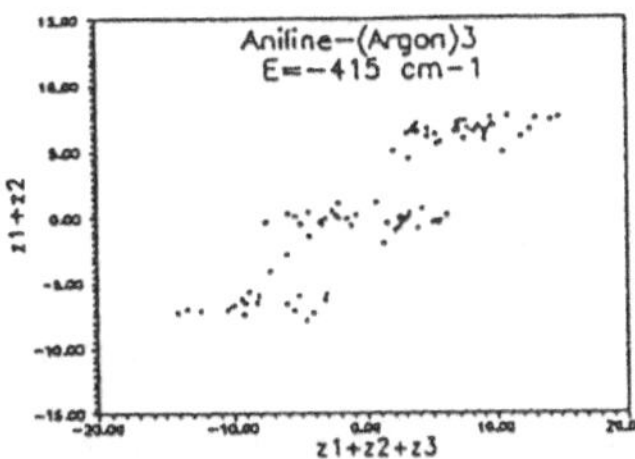

Figure 5. z1+z2 as a function of z1+z2+z3 at E=-415 cm^{-1}. The values are in angstroem.

Table 1.   Duration of the last configuration before evaporation.

Energy (cm^{-1})	Time "2+1" (ps)	Time "3+0" (ps)
- 415	11.20	4.22
- 315	6.23	4.45
- 215	3.13	2.34
- 115	2.11	1.25

3. SEMICLASSICAL SIMULATION OF SPECTRA

In this last section, we present spectra for An-Ar$_n$ (with n=1 to 3) which have been simulated using the spectral density method proposed by Mukamel[11]. The spectrum is generated from the Fourier transform of the function, U(t), which is the difference between the excited and ground electronic states of the cluster along a trajectory propagated in the ground state S$_0$. The spectra shown in Figure 6 are averages over several trajectories from a statistical ensemble, generated in the following way: we first run a trajectory using Nosé dynamics at a chosen temperature, T. Since Nosé dynamics thermostats the system, the total internal energy of the cluster can fluctuate; we accept a configuration from this trajectory only if its total energy is within 1 cm^{-1} of the most probable energy associated with the given T.

An empirical term, of the form $\Delta E\, e^{-\beta r}$ (r being the distance between each Ar atom and the Nitrogen atom) was finally introduced and its average along the trajectories added to the spectral shift. This allows us to reproduce all of the experimental frequency shifts, as apparent in figure 6.

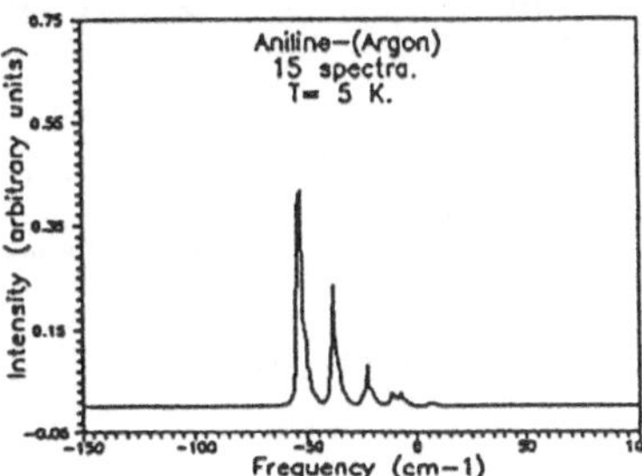

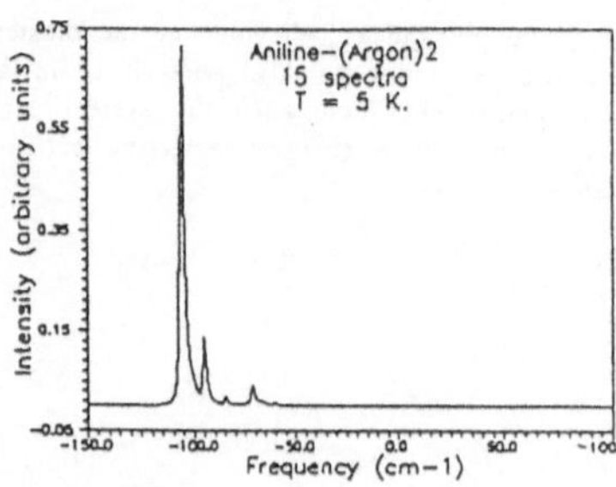

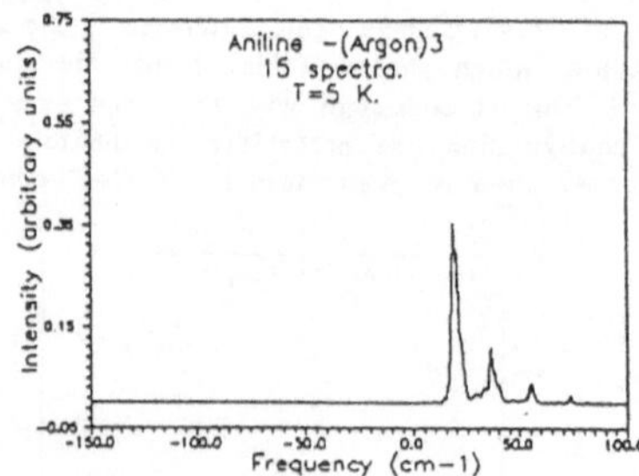

Figure 6 . Simulated spectra at T= 5 K for An-Ar, An-Ar$_2$, An-Ar$_3$ ("3+0").

It is quite remarkable that the frequencies of all the observed Van der Waals modes are in essential agreement with experiment, the b_x frequency (bending motion along the long axis) being slighty too small. The relative intensities of the bands differ somewhat from the experimental values, the major discrepency being a too large Franck-Condon activity of the b_x mode, reflecting the difference between the PES for S_0 and S_1 states. Nonetheless it is very satisfactory to see that the characteristic Van der Waals progression (18 cm^{-1}) in "3+0" An-Ar$_3$ isomer is well predicted.

In general, the spectral density method seems to be more suitable for studying the bigger clusters. Indeed the spectral shift and the Van der Waals frequencies depend on the temperature of the cluster. This is unphysical in small clusters where mode spacing is relatively large. For the bigger system, the excitation of hot bands should be able to explain this behaviour.

4. CONCLUSION

The aniline-(Ar)$_n$ clusters, because of their large spectral shifts, appear to be a most interesting system for the study of isomerisation and evaporation dynamics. We have observed isomer-specific evaporation phenomena in our simulations. As our understanding of these processes and the associated reverse nucleation and condensation phenomena increases, we hope to be able to say more about how cluster formation conditions might be used to control the formation of specific isomers as well as specific cluster sizes.

ACKNOWLEDGMENTS

We thank S. Leutwyler for a useful discussion concerning the spectral density method.

REFERENCES

1. Hermine, P., Parneix, P., Coutant, B., Amar, F.G. and Bréchignac, Ph. (1991). Z. Phys. D. in press.
2. Hayman, Ch. A., Brumbauch, D.V. and Levy, D.H. (1984). J. Chem. Phys. 80, 2256-2264.
3. Even, U., Amirav, A., Leutwyler, S., Ondrechen, M.J., Berkovitch-Yellin, Z. and Jortner, J. (1982). Faraday Discuss. Chem. Soc., 73, 153.
4. Bieske, E.J., Vichanco, A.S., Rainbird, M.W. and Knight, A.E.W. (1991).J. Chem. Phys. 94, 7029-7037.
5. Ondrechen, M.J., Berkovitch-Yellin, Z. and Jortner, J. (1981). J. Am. Chem. Soc. 103, 6586.
6. Aziz R.A. and Chen H.H. (1977). J. Chem. Phys. 67, 5719-5726.
7. Weerasinghe S. and Amar, F.G., unpublished work.
8. Evans, D.J. and Murad, S. (1977). Mol. Phys. 34, 327-331.
9. Berry, R.S., Beck, T.L., Davis, H.L. and Jellinek, J. Advances in Chemical Physics, edited by I. Prigogine and S.A. Rice (Wiley, New-York, 1988), Vol LXX, Part 2.
10. Weerasinghe, S. and Amar, F.G. (1991). Z. Phys. D 20, 167-171; Amar, F.G. and Weerasinghe, S. Proceedings of the 24 th Jerusalem Quantum Chemistry Conference, Kluwer Academic.
11. Fried, L.E. and Mukamel, S. (1991). Phys. Rev. Lett. 66, 2340-2343.
12. Nosé, S. (1984). Mol. Phys. 52, 255-268; (1984). J. Chem. Phys. 81, 511-519.

COMPUTER SIMULATION STUDY OF THE EVAPORATION MECHANISMS
IN LENNARD-JONES CLUSTERS

C.E. Román and I.L. Garzón
Instituto de Física, Universidad Nacional Autónoma de México,
Apartado Postal 2681, 22800 Ensenada, Baja California, México

ABSTRACT. The evaporation mechanisms of small Lennard-Jones clusters
have been studied using molecular dynamics method. Heating up processes
were applied to clusters with n=7,12,13, and 14 atoms in a step-like
manner. At temperatures around the cluster melting point, the cluster
started to evaporate monomers. In this temperature region we have used a
classical trajectory analysis to calculate the survival probability, the
evaporation rate and the average kinetic energy release as a function of
the cluster total energy. These results have been interpreted in terms
of statistical theories of molecular dissociation.

1. Introduction

The understanding of the evaporation mechanisms in small charged
clusters is of fundamental importantance in the interpretation of mass
spectra obtained in nozzle expansion experiments. Theoretical [1] and
experimental [2] approaches have been applied to study these aspects of
cluster dynamics. In the last few years, the experimental study of the
dynamics in cluster evaporation, together with statistical models of
energy transfer, allow to make estimations of the binding energy of the
evaporated species from the measurement of the cluster evaporative
lifetime and the average kinetic energy release [3,4].

From a theoretical point of view the evaporation dynamics of
clusters have been studied mainly through molecular dynamics (MD)
simulations. For charged van der Waals clusters MD simulations predicted
multiple decay, reactions after ionization, in a picosecond time scale
[5] and provided the time evolution of cluster size distributions which
mimic the real experiments for long times [6].

Evaporation of neutral van der Waals clusters also has been studied
via MD method [7-10]. The main interest of these studies has been
directed to the analysis of the applicability of various theoretical
models to the description of the cluster evaporation dynamics.

We have used MD simulations to study the thermal decay of

459

P. Jena et al. (eds.), Physics and Chemistry of Finite Systems: From Clusters to Crystals, Vol. I, 459–464.
© 1992 *Kluwer Academic Publishers.*

Lennard-Jones clusters [7] and to calculate the evaporation rate and the average kinetic energy release as a function of the cluster total energy [8]. In both cases we have done the study for the 13-atom cluster, due to its special structural stability.

Weerasinghe and Amar [9] have shown that the phase space theory together with accurate anharmonic densities-of-states are adequate to describe the evaporation of the 13-atom argon cluster. The decay behavior of neutral Lennard-Jones clusters with 12-14 atoms was studied by Smith [10]. He obtained the cluster half-lives and decay energies for the three clusters studied.

In the present work we extend our calculations of the evaporation rate and the average kinetic energy release as a function of the cluster total energy for Lennard-Jones (LJ) clusters with 7,12 and 14 atoms. In section II we present the theoretical background and describe the computational procedure. Section III is dedicated to show the results and their discussion and a brief summary is given in section IV.

2. Theory and Computational Procedure

Microcanonical MD simulations have been done on the 7-,12- and 14-atom neutral clusters using a Lennard-Jones potential in reduced units [8]. The Verlet algorithm [11] with a time step of 0.01τ (τ is the reduced time unit) was used to solve the equations of motion. The calculations started from the most stable configuration for each cluster taking care that the linear and angular momenta of the whole cluster was equal to zero. Next, each cluster was heated up in a step-like process to obtain the caloric curve: long time average of the kinetic energy per particle as a function of the total energy per particle. These time averages involve at least 10^6 time steps. From the caloric curves we select several cluster total energies corresponding to a hot cluster which were used to make the evaporation studies. Following the same procedure as in reference 8, we generate an ensemble of 100 different initial conditions for each total energy previously selected, and for each cluster size. Then, we used each one of these initial conditions to make long MD runs looking for the time at wich the cluster start to evaporate. We say that evaporation is taking place when the evaporated monomer is at a distance larger than five reduced units of distance from the cluster center of mass. A limit of 10^7 time steps was imposed as the maximum propagation time. This means that a trajectory which has no evaporation within this limit was considered a bound one. With the information about the time at which the evaporation occurs, for each initial condition, we calculate the cluster survival probability P(t) as a function of time. This quantity is the probability that a cluster trajectory has not led to evaporation and is calculated from [12]

$$P(t) = (1/N) \sum_{1}^{N} \theta_1(t) \qquad (1)$$

where $\theta_1(t)=1$ if by the time t the trajectory generated by the initial condition i has not led to evaporation and $\theta_1(t)=0$ if it has. N is the total number of trajectories (100 in these cases). The distribution of lifetimes obtained by this procedure was fitted (by least squares) to ln P(t) = -Kt, where K is the evaporation rate. The average cluster lifetime is given by K^{-1}.

3. Results and Discussion

For the 7-atom cluster we have calculated the survival probability as a function of time for 8 total energies. They are showed at the top panels of figure 1. The left hand side panel shows P(t) for the four lowest total energies, whereas the rigth hand side panel presents P(t) for the four highest total energies considered. The middle panels show the corresponding P(t) for n=12. In this case, six different total energies were used. The three lowest are to the left and the three highest are to the rigth. The bottom panel present the same sequence for n=14 with seven different total energies. From all these figures we observe that in general they follow an exponential decay behavior for P(t) as a function of time. This trend motivate us to plot the logarithm of P(t) vs. time and fit by least squares a linear function. The slope of this function is taken to be the evaporation rate K. During the fitting procedure we observe that the long time data can not be adjusted with a linear function. The same effect has been pointed out by Smith [10] but at the present time there is not a conclusive interpretation of this non-linearity.

Figure 2a shows the evaporation rate as a function of the cluster total energy. In this figure we include our previous results for n=13 [8], together with the new data for n=12 and 14. We also report error bars wich represent an estimation of the influence of statistical fluctuations in the calculation of the evaporation rates. The average lifetime of the cluster as a function of its total energy per particle can be obtained directly from this curve. For the energy range investigated in this study the lifetimes are going fron about 100 ns for the lowest energies to 10 ns at higher energies, for the three clusters investigated.

To study the relative stability of the 12-, 13-, and 14-atom clusters at a given temperature we have plotted the logarithm of K as a function of temperature. This plot is shown if figure 2b. In this figure we observe a wide range of temperatures at which ln K, for n=13, is lower than the corresponding value for n=12 and 14. This result allow us to say that there is a range of temperatures at which the 13-atom LJ cluster is the most stable of the three clusters investigated.

It is important to point out that similar results to those presented above were obtained by Smith [10], at higher cluster total energies. This fact is a good test for our calculations and indicate that the stability of the 13-atom cluster extends over a wider range of cluster total energies or temperatures.

From our simulations we also have extracted data for the calculation of the average kinetic energy release of the evaporated monomers. This quantity was calculated as the average of the kinetic energy of the particle which is ejected from the cluster at the

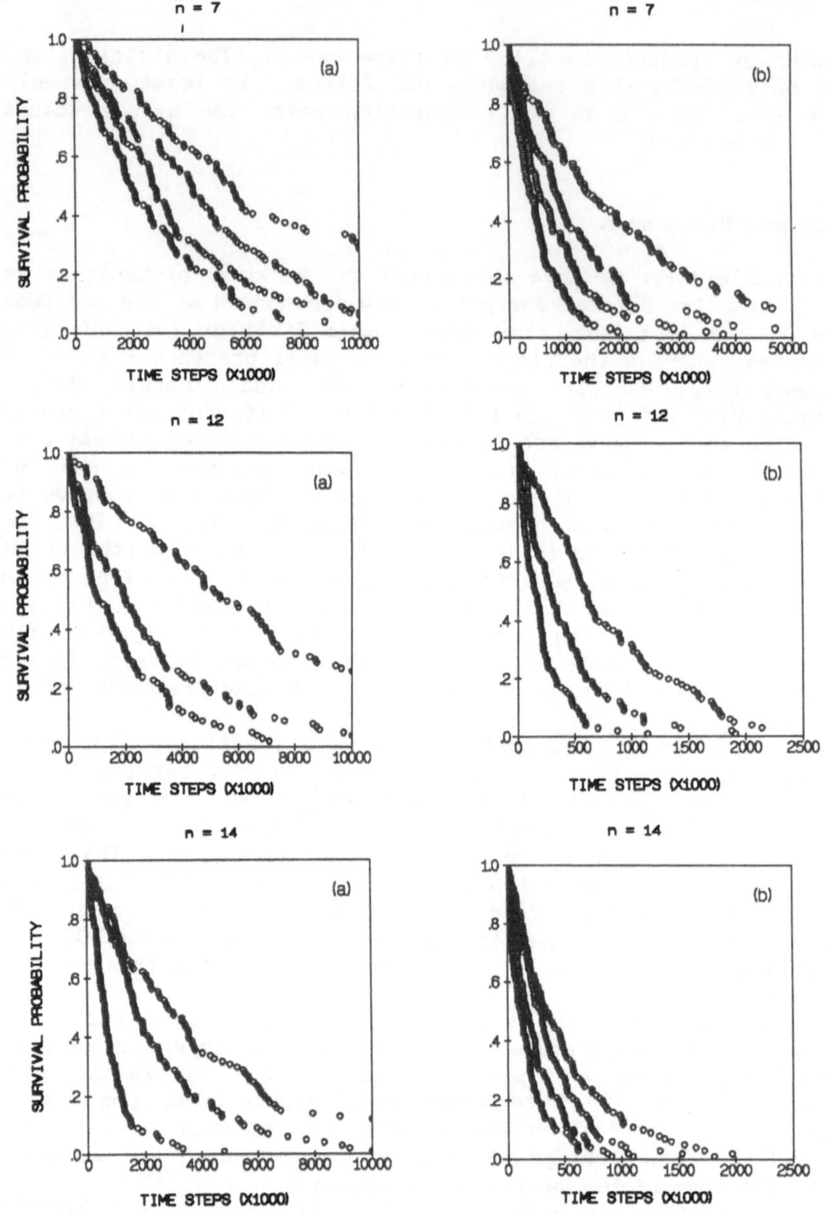

Figure 1. Survival probability as a function of time for different total energies per particle. The curves correspond to clusters with n = 7, 12 and 14 atoms.

evaporation time. For n=7 the kinetic energy release is almost constant as a function of the cluster total energy. Its value is around 0.13ε, where ε is the LJ energy parameter. The kinetic energy release for n=12,13 and 14 varies slightly as a function of the cluster total energy. For n=12, it changes between 0.20ε at low total energies and 0.30ε at higher total energies. For n=13 and 14 the corresponding variation is from 0.25ε at low total energies to 0.35ε at higher total energies.

The numerical results for the evaporation rate as well as the average kinetic energy release obtained through MD simulations can be compared with the predictions given by theoretical models of molecular dissociation like RRK theory [13], and other theories like the Engelking model [3]. Using reasonable values for the parameters which enter in the expressions for the evaporation rate and the kintetic energy release coming from that models, we have calculated them, as a function of the cluster total energy. The comparisson of our numerical results with those coming from the theoretical models indicate that the RRK theory adjust better the simulation data than those coming from the Engelking model. The average kinetic energy release in terms of the cluster total energy is well reproduced by the RRK theory in contrast to what was reported by Franklin [14], when he described molecular dissociation experiments.

The failure of the Engelking model to predict the evaporation rates and the average kinetic energy for LJ clusters was not expected because this approach takes into account more details of the evaporation process and was used with success by Engelking [3] and Brechignac, *et al.* [4] to describe the evaporation dynamics of various clusters.

Recently, Weerasinghe and Amar [9] have proposed to use the phase space theory with anaharmonic densities-of-states to characterize the evaporation of clusters. They have showed that, for argon clusters with 13 atoms, the evaporation rate and the average kinetic energy release obtained via MD simulations, can be recovered using such theory. This

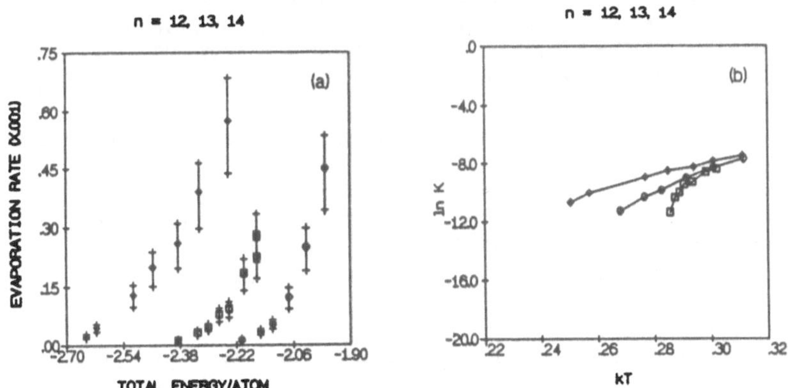

Figure 2. a) Evaporation rates vs. total energy per particle. The bars show the statistical errors. b) Temperature dependence of the evaporation rate. Circles are for n=12, squares for n=13 and diamonds for n=14.

fact offer new avenues for a theoretical interpretation of the evaporation mechanisms of clusters obtained through computer simulations.

4. SUMMARY

We have used constant energy molecular dynamics method with a Lennard-Jones potential to make a systematic simulation study of the evaporation process in 7-, 12-, and 14-atom van der Waals clusters. We have calculated the evaporation rates as a function of the cluster total energy from a linear fit of the logarithm of the cluster survival probability as a function of time. The average kinetic energy release were calculated for the same clusters for a variety of total energies. The results from these studies indicate that the cluster average lifetime depends strongly on its total energy or temperature. The calculated cluster lifetimes change from about 10 to 100 ns with a decrease of the cluster total energy, for the three sizes investigated. The averaged kinetic energy release fluctuates around 0.10ε for n=7 and changes between 0.20ε and 0.40ε for n=12 and 14.

The computational results were compared with data coming from theoretical models like RRK theory and the Engelking model. Our results indicate the RRK theory adjust better the simulation data.

ACKNOWLEDGEMENTS

This work was supported by DGAPA-UNAM under Project IN-103189.

REFERENCES

1) C.E. Klots, J.Chem.Phys. **83**,5854 (1985); Z.Phys. **D5**, 83 (1987).
2) T.D. Märk, Int. J. Mass Spectrom. Ion Phys. **79**, 1 (1987).
3) P.C. Engelking, J.Chem.Phys. **85**, 3103 (1986); ibid. **87**, 936 (1987).
4) C. Brechignac, C. Cahuzac, Ph. Leygnier, J. Weiner, J.Chem.Phys. **90**, 1492 (1989).
5) J.M. Soler, J.J. Saenz, N. Garcia, O. Echt, Chem.Phys.Lett. **109**, 71 (1984).
6) R. Casero, J.M. Soler, J. Chem. Phys. **95**, 2927 (1991).
7) I.L. Garzón and M. Avalos-Borja, Z.Phys. **D12**, 185 (1989).
8) C.E. Román and I.L. Garzón, Z.Phys. **D20**, 163 (1991).
9) S. Weerasinghe and F.G. Amar Z.Phys. **D20**, 167 (1991).
10) R.W. Smith, Z.Phys. **D21**, 57 (1991).
11) L. Verlet, Phys.Rev. **159**, 98 (1967).
12) S.K. Gray, S.A. Rice, D.W. Noid, J.Chem.Phys. **84**, 3745 (1986).
13) O.K. Rice, H.C. Ramsberger, J.Am.Chem.Soc. **49**, 1617 (1927); S.L. Kassel, J.Chem.Phys. **32**, 225 (1928), ibid. **32**, 1065 (1928).
14) J.L. Franklin, Science **193**, 725 (1976).

MONTE-CARLO GEOMETRY OPTIMIZATION OF SODIUM CLUSTERS WITH A DISTANCE-DEPENDENT MONOELECTRONIC HAMILTONIAN

F. SPIEGELMANN and R. POTEAU
Laboratoire de Physique Quantique (URA 505 du CNRS)
IRSAMC, Université Paul Sabatier
118 route de Narbonne
31062 Toulouse Cedex
France

ABSTRACT. The formulation of a distance-dependent tight-binding hamiltonian expressed in the basis of s-orbitals and including the effects of p-orbitals via perturbation theory is shown to provide a simple and efficient calculation of the potential energy surfaces for sodium clusters. A few applications illustrate the ability of such a model to provide both qualitative and quantitative results for the purpose of optimizing geometries of small microclusters. Thus geometries and stabilities of Na_n clusters (n=2-21) are obtained via Monte-Carlo simulated annealing, in good agreement with more sophisticated calculations when available. Larger clusters with constrained symetries are investigated in the range Na_{55}-Na_{2057}.

1. Introduction

Despite the progress in computational techniques for studying molecular properties, sophisticated methods such as Configuration Interaction (CI) or Density Functional Theory (DFT), the determination of structural properties and stabilities of clusters remains a difficult task beyond 10-atom clusters. In the recent years, the combination of DFT and the Car-Parinello (CP) Molecular Dynamics [1] provided a powerful algorithm for the study of energetical and dynamical properties of small clusters, particularly silicon [2] and sodium clusters [3,4] until approximately 20 atoms. Nevertheless, in order to overpass this limit, simplified hamiltonians are certainly needed, and should help to bridge the gap between additive potentials which may prove unadequate to treat metallic or covalent bonds and fully ab-initio methods. The tight-binding method (or Hückel theory in chemistry) constitutes a possible alternative solution. However, even in the so-called "extended Hückel" version, an important drawback is the usual failure in performing geometry optimization, which for cluster systems is a serious difficulty. In the field of condensed matter, Chadi [5] and Harrison [6] originally proposed extended versions of the tight-binding theory as an attempt to provide a better distance dependence. Such formulations explicitly add the ion-ion repulsion and sometimes the on-site electron-electron repulsion and were widely used particularly in order to study surface or defects in silicon. Sophisticated versions, including explicitly the electrostatic interactions were recently proposed [7]. In this paper, we report the application of a distance-dependent tight-binding hamiltonian for different studies of sodium clusters, namely geometry

P. Jena et al. (eds.), Physics and Chemistry of Finite Systems: From Clusters to Crystals, Vol. I, 465–470.
© *1992 Kluwer Academic Publishers.*

optimization concerning Na_n clusters (n=2-21), and calculations of symmetry constrained structures of larger sodium clusters in the range n=55-2057.

2. The Model

The model consists of a monoelectronic hamiltonian [8] :

$$\hat{h} = \sum_{i,j} h_{ij}\, a_i^\dagger\, a_j$$

where $a_i^\dagger$ and a_i are creation and anihilation operators corresponding to an s orbital on site i. Although $\hat{h}$ is expressed in the model space of s-orbitals only, the matrix elements, which are distance-dependent, include perturbatively the effect of p orbitals :

$$h_{ii} = h_{ii}^{(0)} + h_{ii}^{(2)} = \sum_{k\neq i}^{atoms} \left[\rho_{ss}(R_{ik}) - \frac{t_{s\sigma}^2(R_{ik})}{\varepsilon_{3p} - \varepsilon_{3s}} \right] = \sum_{k\neq i}^{atoms} \rho^{eff}(R_{ik})$$

$$h_{ij} = h_{ij}^{(0)} + h_{ij}^{(2)} = t_{ss}(R_{ij}) - \sum_{k\neq i,j}^{atoms} \left[\frac{t_{s\sigma}(R_{ik})\, t_{s\sigma}(R_{jk})}{\varepsilon_{3p} - \varepsilon_{3s}} \times \frac{\vec{R}_{ik} \cdot \vec{R}_{jk}}{|R_{ik}||R_{jk}|} \right]$$

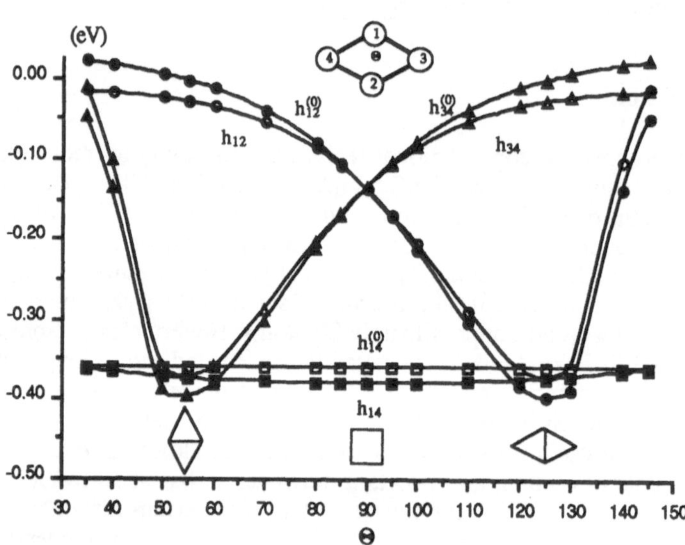

Figure 1. Dependence of Na_4 transfer integrals with the geometry.

The three functions $\rho_{ss}(R)$, $t_{ss}(R)$ and $t_{s\sigma}(R)$ represent respectively the ion-ion repulsion, the s-s and the s-p_σ transfer integrals at distance R. The total energy is calculated as the sum of the one-electron energies ε_i, namely the eigenvalues of $\hat{h}$. A proportionality dependence with respect to the orbital overlaps $S_{ss}(R)$ and $S_{s\sigma}(R)$ was imposed to the transfer integrals, namely $t_{s\sigma}(R)/t_{ss}(R)=\lambda(S_{s\sigma}(R)/S_{ss}(R))$. $\rho_{ss}(R)$, $t_{ss}(R)$ and the constant λ (providing thus $t_{s\sigma}(R)$) were determined in such a way that the energy of the ground state $E(^1\Sigma_g^+)=2\varepsilon(\sigma_g)$ and the energy of the first excited state $E(^3\Sigma_u^+)=\varepsilon(\sigma_g)+\varepsilon(\sigma_u)$ reproduce at any distance accurate *ab initio* potentials for the Na_2 diatomics, an extra condition being the fitting of the binding energy of the Na_4 cluster. An originality of the present formulation with respect to former ones is the perturbative treatment of the unoccupied 3p shell, which yields effective matrix elements in the model space as the sum of zeroth order contributions $h_{ij}^{(0)}$ (s-only) and second order ones $h_{ij}^{(2)}$ (s-p interactions). As an illustration, *fig* 1 shows the angle dependence of some off-diagonal matrix elements for Na_4 in the rhombus geometry (the sides being kept constant). The effect of s-p terms is to increase the magnitude of the transfer integral on the small diagonal and decrease it on the long diagonal, thus reducing the antibonding character of the highest occupied orbital $1/\sqrt{2}(s_3-s_4)$. In general, the global effect of second order s-p terms is to increase the stability of the clusters. It is of particular importance in the case of Na_7 pentagonal bipyramid where the magnitude of the transfer integral between the two apex atoms is significantly increased, due to the three-body terms involving atoms of the pentagonal basis.

3. Geometries and stabilities of clusters from Na_2 to Na_{21}

A non-local optimization method was used, namely the Monte-Carlo Simulated Annealing method and the Metropolis algorithm. From an initial arbitrary structure, the clusters are heated up to 800K, and slowly cooled down to 500K, again heated up to 800K, and the temperature is then continuously lowered down to 0K (by steps of -30K). For each temperature, the number of random moves is proportional to the number of atoms in the cluster. The most stable geometries of clusters obtained in this work are illustrated in *fig* 2a. For the smaller clusters Na_3-Na_8, the geometries are the same as those found in *ab-initio* CI calculations [9,10] which also agree with the DFT-CP results, except for the lowest structure of Na_8. As concerns larger clusters, the only *ab-initio* results are the DFT-CP results which concern n=9,10,13,18 and 20. For all these clusters, the ground state structures obtained with the present model are in almost complete agreement with the DFT-CP structures (the only exception is Na_{10}, for which the best DFT-CP structure is our second one, 5 meV above the absolute minimum). From the present study, one may stress some dominant trends. Up to n=14, the clusters seem to reconstruct for each size, and no continuous evolution is apparent, although some pieces may appear as recurrent patterns (pentagonal bipyramid, tetracapped tetrahedron, fully-capped trigonal prism, and the bicapped square antiprism) in the lowest isomers of clusters between Na_9-Na_{14}, defining various families in a very narrow energy-range. Oppositely, all lowest energy clusters in the range Na_{15}-Na_{21} are built by capping an icosahedral core, which is remarkable since no stable isomer for Na_{13} or Na_{14} was found close to the icosahedron (even Jahn-Teller distorted). Na_{19}, although an anti-magic number with respect to stability, exhibits a rather compact structure, namely the D_{5h} double icosahedron. The structure obtained for Na_{20} (the same as DFT-CP [4]) is close to be spherical, although the three axial ratios are not identical. The dissociation energies with respect to the lowest channel are depicted in *fig* 2b and it can be seen that the dissociation energies of Na_8 and Na_{20} are particularly high with respect to their neighbours. Beyond n=14, the clusters systematically evaporate a monomer, as far as energy criterions are considered.

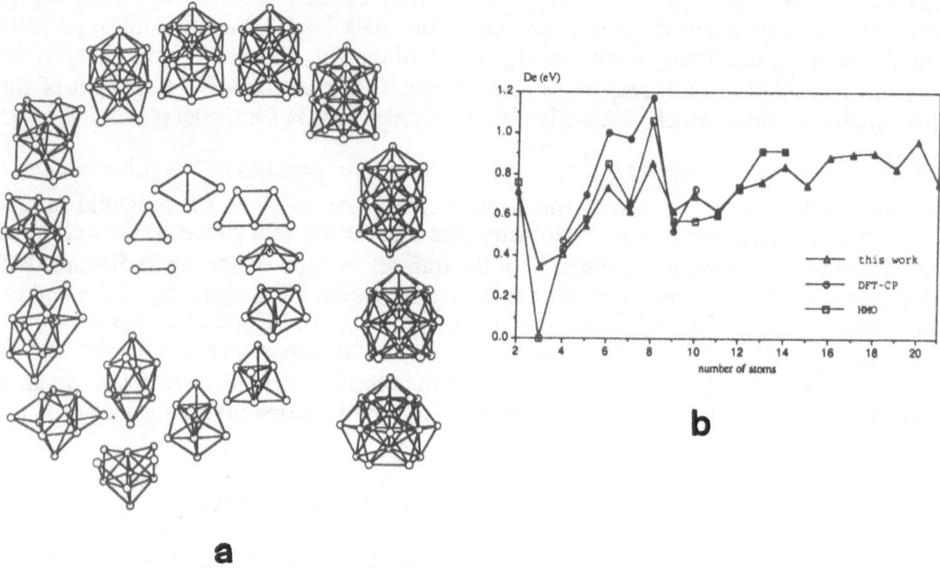

Figure 2. *a*/Lowest energy geometrical structures of small sodium clusters Na$_2$-Na$_{21}$. *b*/Dissociation energies of small sodium clusters with respect to the lowest dissociation channel. Filled or empty symbols correspond to the evaporation of a monomer or a dimer respectively. DFT-CP = ref 4. HMO = ref 11.

4. Stability Of Large Icosahedral And Cuboctahedral Clusters

We have also studied larger clusters in the range n=55-2057 with constrained I$_h$ symmetry (icosahedron), cuboctahedral pieces of the fcc lattice (O$_h$ symmetry), and also regular pieces of the centered cubic (cc) lattice (with NxNxN elementary cells with N ranging from 2 to 9). The I$_h$ and O$_h$ structures are indeed of interest, since Martin *et al.* [12] recently observed transitions from electronic shell magic numbers to compact shell magic numbers. We only achieved partial relaxation optimizing the intershell distance of O$_h$ and I$_h$ structures, and the lattice parameter of the cubic clusters. The least stable structures are the cubic ones (see *f*ig 2), illustrating the unfavorable surface energies provided by these clusters. As concerns Na$_{55}$, the icosahedron is more stable than the cuboctahedron. However, an inversion takes place for Na$_{147}$ and for larger clusters, cuboctahedral structures seem to be the most stable. The present results obtained on sodium clusters seem to be different from some conclusions of Martin *et al.* [13], obtained from observations on Mg$_n$ and Ca$_n$ clusters, where intermediate peaks in the mass spectra are interpreted as corresponding to the partial filling of the twenty external faces of the icosahedron. However, accurate experimental results are not yet available on sodium clusters. One cannot either exclude misleading results due to incomplete optimization.

From a general point of view, convergency to the solid cohesion energy (1.13 eV) is slower than in DFT calculations.

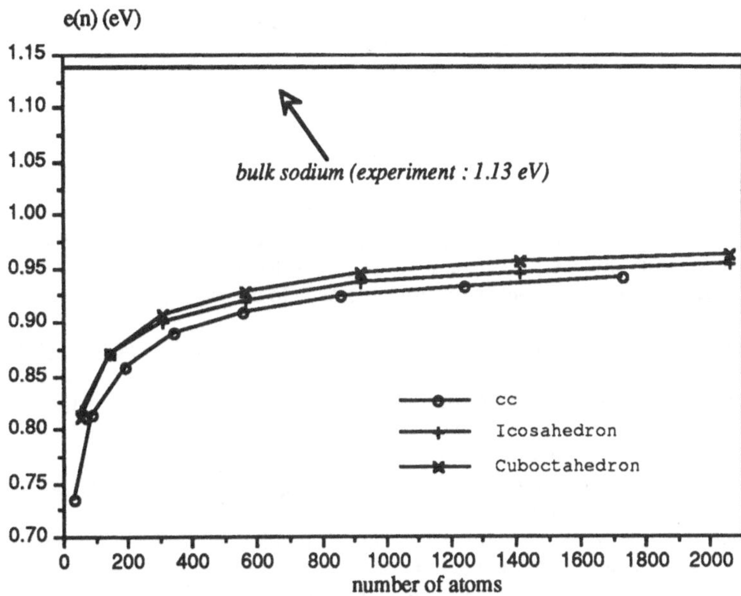

Figure 3. Binding energy per atom for large clusters in the range 55-2057.

5. Conclusion

Simple hamiltonians such as improved tight-binding hamiltonians are able to provide not only a qualitative but even a quantitative insight in the geometrical structures and stabilities of some particular clusters. Although correlation does not appear explicitly, a part of it is implicitly included since the integrals are fitted in order to reproduce almost exact correlated results on Na_2, which means that a part of the electronic correlation preexisting in Na_2 is transferred to larger clusters [14]. As an improvement, an explicit electron-electron repulsion term could be added to the present model. From a more practical point of view, the model provides a very efficient calculation of the potential energy surface, yielding reliable results for small clusters whenever comparison with *ab-initio* methods can be achieved. The sequence of low energy structures thus obtained shows that although the stability of small alkali clusters seems to be related with the electronic structure more than with the geometric structure, pentagonal and icosahedral arrangements may play a role in this growth. The conclusions obtained here are fully consistent with DFT calculations. Moreover, a precise knowledge of the ground state geometrical structures should be helpful to get some further insight in the vertical optical spectra observed for small alkali clusters and understand wether fine structure in those spectra are due to cluster shapes or to quantum effects [15]. Beyond optimized static structures of small clusters, the simplicity of the model also makes possible calculations of larger clusters with only partial relaxation in the range 100-1000 which is not much explored theoretically up to now [16]. It should also enable the study of dynamics and thermodynamics of small clusters, isomerization processes, and dependence on temperature.

[1] Car, R. and Parinello, M. (1985) 'Unified approach for molecular dynamics and density functional theory', Phys. Rev. Lett. 55, 2471-2474.
[2] Andreoni, W. (1991) 'Computer simulations of small semi-conductor and metal clusters', Zeitschrift für Physik D 20, 31-36.
[3] Röthlisberger, U. and Andreoni, W. (1991) 'Temperature-induced structural transitions and vibrational properties of microclusters. *Ab initio* molecular dynamic studies', Zeitschrift für Physik D 20, 243-246.
[4] Röthlisberger, U. and Andreoni, W. (1991) 'Structural and electronic properties of small sodium microclusters (n=2-20) at low and high temperature : New insights from *ab-initio* molecular dynamics studies', J. Chem. Phys. 94, 8129-8151.
[5] Chadi, D.J. (1978) 'Energy-minimization approach to the atomic geometry of semi conductor surfaces', Phys. Rev. Lett. 41, 1062-1065.
[6] Harrison, W.A. (1980) 'Electronic Structure and the Properties of Solids', Freeman, San Francisco.
[7] Sutton, A.P., Finnis, M.W., Pettifor, D.G. and Ohta,Y. (1988) 'The tight binding bond model', J. Phys. C 21, 35-66. Majewski, J.A. and Vogl, P. (1987) 'Simple model for structural properties and crystal stability of sp-bonded solids', Phys. Rev. B, 9666-9682.
[8] Poteau, R. and Spiegelmann, F. (1991) 'Distance-dependent Hückel-type model for the study of sodium clusters', Phys. Rev. B, to be published.
[9] Bonacic-Koutecky, V., Fantucci, P. and Koutecky, J. (1988) 'Systematic *ab-initio* Configuration Interaction study of alkali metal clusters. II. Relation between electronic structure and geometry of small sodium clusters', Phys. Rev. B 37, 4369-4374.
[10] Spiegelmann, F. and Pavolini, D. (1988) '*Ab initio* calculation of the electronic structure of small Na_n, Na_n^+, K_n and K_n^+ clusters ($n \leq 6$) including core-valence interaction', J. Chem. Phys. 89, 4954-4964.
[11] Lindsay, D.M., Wang, Y., George, T.F. (1987) 'The Hückel model for small metal clusters. II. Orbital energies, shell structures, ionization potentials and extrapolation to the bulk limit', J. Chem. Phys. 86, 3500-3511.
[12] Martin, T.P., Bergman, T., Göhlich, H. and Lange, T. (1990) 'Observation of electronic shells and shells of atoms in large Na clusters', Chem. Phys. Lett. 172, 209-213.
[13] Martin, T.P., Bergman, T., Göhlich, H. and Lange, T. (1991) 'Evidence for icosahedral shell structure in large magnesium clusters', Chem. Phys. Lett. 176, 343-347.
[14] Blaise, P., Spiegelmann, F., Maynau, D. and Malrieu J.P. (1990) 'Alkali-metal clusters : an s-band uncorrelated versus (s+p) highly correlated problem', Phys. Rev. B 41, 5566-5577.
[15] Pollack, S., Wang, C.R.C., and Kappes, M.M. (1991) 'On the optical response of Na_{20} and its relation to computational prediction', J. Chem. Phys. 94, 2496-2501.
[16] Lindsay, D.M., Wang, Y., George, T.F. (1990) 'The Hückel model for small metal clusters. IV. Orbital properties and cohesive energies for model clusters up to several hundred atoms', J. Clust. Science 1, 107-126.

INVESTIGATION OF THE STABLE STRUCTURES OF Ni_x, Fe_x, $Ni_{x-y}P_y$, $Ni_{x-y}B_y$, $Fe_{x-y}P_y$, AND $Fe_{x-y}B_y$ ($x=13, y<13$) MICROCLUSTERS

V.S.STEPANYUK, B.L.GRIGORENKO, A.A.KATSNELSON
Moscow State University, Leninskie Gori, 119899
Moscow, Russia

A. SZASZ
Eötvos University, Muzeum krt. 6-8, H-1088
Budapest, Hungary

ABSTRACT. Computer simulation of the structure of Ni_x, Fe_x, $Ni_{x-y}P_y$, $Ni_{x-y}B_y$, $Fe_{x-y}P_y$, $Fe_{x-y}B_y$ clusters has been carried out by molecular dynamics simulation using empirical potentials of interaction. Results indicate that the symmentry of the clusters changes with increasing of number of metalloid atoms (y). A surprising result is that when the Stillinger-Weber potential is used the great majority of the Morse minima are not supported.

INTRODUCTION

The investigation of physical and chemical properties of small particles becomes more and more interesting lately. It is because there are wide possibilities for application of these particles in catalysis, laser technology, biology, etc. Some recent works concerning metal clusters [1,2], inert gases [3,4] and covalent crystals [5,6] made it possible to understand many structural and electronic properties of small clusters. Inert gases clusters are formed mainly as icosahedral structures with the number of atoms of 13,55,... For simple metals clusters the stable structures are relised with the number of atoms of 8,20,.... Obviously, this can be explained by the model of so called spherical shells.

There are many works concerning cluster physics. Nevertheless, the number of investigated objects remains limited. For example, metal-metalloid systems (which are very improtant from practical point of view) have not been investigated yet in the form of clusters. The exposure of stable configurations of these systems will make it possible to understand the processes of growth of metal-metalloid systems and their chemical activity. Since at present there exists an opportunity to get the clusters of required size and composition, so the information about the stable configurations may be used to regulate the physical

P. Jena et al. (eds.), Physics and Chemistry of Finite Systems: From Clusters to Crystals, Vol. I, 471–477.
© 1992 *Kluwer Academic Publishers.*

472

properties.

This article concerns the investigation of structrue of small cluster systems Ni-P, Ni-B, Fe-P and Fe-B and also structural changes influenced by the clusters composition changes.   Section   II   offers   the   basic   method   of investigation, section III gives the results and their discussion, section IV presents the main conclusions.

METHOD OF CALCULATION

Molecular dynamics simulation was used to define stable geometrical configurations of small clusters of Ni-P, Ni-B, Fe-P, Fe-B.   In the calculations empirical potentials of interatomic interaction were used.

Stillinger-Weber potentials [7] and Morse potentials [8] were used for Ni-P. Morse potentials [9-11] were used for the other systems.

Stillinger-Weber potentials are given by

$$V(r) = \begin{cases} A\epsilon[(\alpha r/\sigma)^{-p}-(\alpha r/\sigma)^{-q}]\, [\exp(\alpha r/\sigma-a)^{-1}] \\ 0, \quad \alpha r/\sigma > a \end{cases}$$

The parameters chosen were p=12, q=0, a=1.652194, σ=2.2183 Å, and ϵ=0.08 eV. For the case of Ni-Ni, A=1.0x8.805977, α=1.0, for Ni-P, A=1.5x8.805977, α=2.49/2.2(=1.1318), and for P-P, A=0.5x8.805977, α=2.49/2.2(=1.1318).

Morse potential is given by

$$V(r) = \epsilon\{\exp[-2\alpha(r/r_0-1)]-2\,\exp[-\alpha(r/r_0-1)]\}f(r/r_0)$$

$$f(z) = \begin{cases} 1 & z<1 \\ 3z^4-8z^3+6z^2 & 1<z<r_c/r_0 \\ 0 & z>r_c/r_0 \end{cases}$$

$$x = r/r_0; \quad z = (x-r_c/r_0)/(1-r_c/r_0); \quad \alpha =3.76; \quad r_c/r_0 = 1.4$$

The parameters for Morse potentials in the investigated system are shown in Table 1.

Trajectories of particles in real space are found by means of integration of Newton equation.   In this article algorithm [12] for integration of movement equation was used. The initial conditions R(t=0) and particles velocity V(t=0) were chosen at will, so that it had no influence on the final result.

The potential V(R) determines the surface in 3N-dimensional space.   This surface is characterized by the existence of the number of minima that correspond to various stable and metastable structures. The evolution of clusters geometry with final temperature looks like trajectories on constant energy surface in 6N-dimensional phase space. The projection of this trajectory on 3N-dimensional configurational space will pass through all possible minima of potential energy, corresponding to possible structures. Each point of this configurational space can be connected to local (or global) minimum. Mathematically this procedure is performed by means of steepest descent minimization.

TABLE 1. The parameters for Morse potentials

	metal metal		metal metalloid		metalloid metalloid	
	$r(\text{Å})$	$\epsilon(\text{eV})$	$r(\text{Å})$	$\epsilon(\text{eV})$	$r(\text{Å})$	$\epsilon(\text{eV})$
Fe-B	2.585	0.32	2.057	0.342	3.515	0.015
Ni-B	2.532	0.52	2.015	0.55	2.28	0.015
Fe-P	2.620	0.51	2.253	0.848	3.406	0.194
Ni-P	2.53	0.19	2.24	0.32	2.10	0.06

Molecular dynamics simulation determines the positions $\{R_i\}$ and $\{V_i\}$ at each time step. The steepest descent minimization method allows one to find the instant position of local minimum, where the interatomic forces and velocities are equal to zero.

RESULTS AND DISCUSSIONS

$Ni_{13-y}P_y(y=0-7)$. $Ni_{13}(y-0)$ cluster in the most stable condition is realised in icosahedral structure. While using Stillinger-Weber potential for $Ni_{12}P$ cluster the atom of phosphorus is "pushed out" on the cluster surface. Note that the 6 axes of the fifth order five disappear. The remaining axis of the fifth order pass through the centre of the atom of phosphorus and through the centre of the cluster. In Morse potential case $Ni_{12}P$ (M) the atom of phosphorus is located in the centre of the cluster.

In Stillinger-Weber cluster $Ni_{11}P_2$ the atoms of phosphorus form P-P pair which is located on the cluster surface. While in this cluster the axis of the second order can be observed (see Fig. 1.1), in the cluster $Ni_{11}P_2$ (Morse) the second atom of phosphorus is located on the axis of the fifth order.

When y=3 in Stillinger Weber cluster $Ni_{10}P_3$ the atoms of phosphorus form an equilateral triangle that is located on the cluster surface nearest to the centre. The cluster possesses only the axis of the third order (see Fig.1.2). The Morse cluster $Ni_{10}P_3$ possesses the axis of the second order (see Fig. 2.1).

The study of the structure of the cluster Ni_9P_4 shows that in this Stillinger-Weber cluster the atoms of phosphorus

form tetrahedra, one atom of which enters the volume of the
cluster and the axis of symmetry of the third order is
observed (Fig. 1.3). In Ni_9P_4 (Morse) the clsuter composed of
atoms of phosphorus also possesses the geometry of tetrahedra
and, consequently, the axis of the third order (Fig. 2.2).

In Stillinger-Weber Ni P case one can observe the axis
of the second order, and in Morse Ni_8P_5 and Morse Ni_7P_6 -
planes of symmetry.

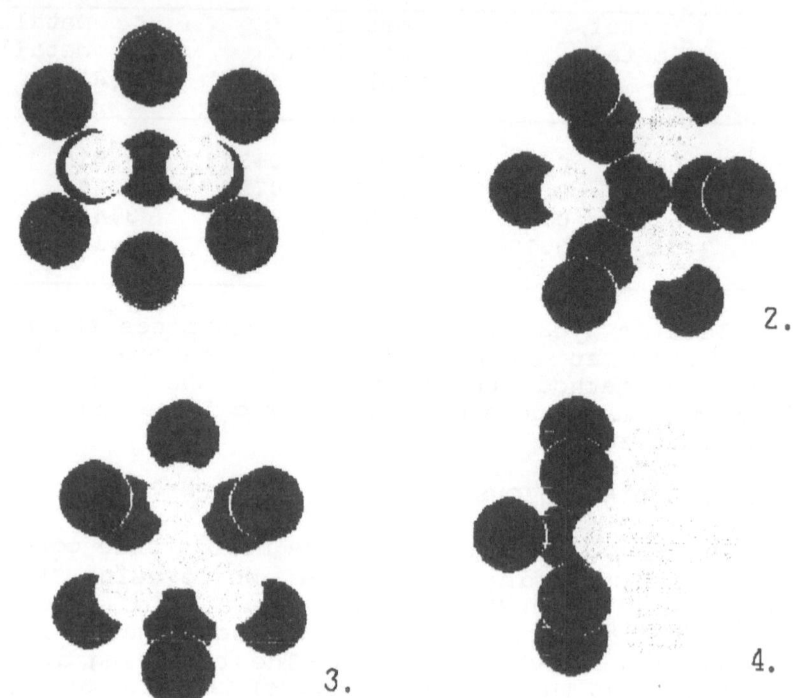

Fig. 1.   The Stillinger-Weber clusters (1) $Ni_{11}P_2$,
(2) $Ni_{10}P_3$, (3) Ni_9P_4, (4) Ni_6P_7.

In Stillinger-Weber Ni_6P_7 the atoms of phosphorus form
pentagonal bipyramid (Fig. 1.4). Note that the increase rise
of concentration of the atoms of phosphorus in Stillinger-
Weber cluster $Ni_{13-y}P_y$ more distinctly distinguishes the
isolation of Ni and P atoms, i.e. phosphorus and nickel
clusters form "independent" structures.
$Ni_{13-y}P_y$, $Ni_{13-y}B_y$, $Fe_{13-y}P_y$, $Fe_{13-y}B_y$. This section represents the results of
calculation using Morse potential.

When y=1 in the clusters $Ni_{12}P$ and $Fe_{12}P$ the atoms of
phosphorus are located in the centre of the cluster. Quite
a different picture can be found in $Ni_{12}B$ and $Fe_{12}B$. The boron
is "pushed out" to the surface, from the 6 axes of the fifth

order only one remains and it pass through that atom and the centre of the cluster.

If the number of metalloid atoms is increased (y=2) the process will go in such a way: the atoms of boron in $Ni_{11}B_2$ and $Fe_{11}B_2$ will form pairs B-B, that are located on the surface, like we had it in Stillinger-Weber cluster $Ni_{11}P_2$ case (Fig. 1.1); the atoms of phosphorus in $Ni_{11}P_2$ and $Fe_{11}P_2$ are located on the axis of the fifth order.

Fig. 2.   The Morse clusters (1) $Ni_{10}P_3$, (2) Ni_9P_4.

When y=3 in $Ni_{10}P_3$, $Fe_{10}P_3$ the axis of symmetry is of the second order (Fig. 2.1). In $Ni_{10}B_3$ and $Fe_{10}B_3$ the cluster seems to be the syntheses of two structures with the axes of symmetry of different orders - the fourth and the fifth. Fig. 3.1 and 3.2 show the cluster $Ni_{10}B_3$ ($Fe_{10}B_3$) from two opposite views.

If the next atom of metal is replaced by metalloid in the cluster Ni_9P_4 the axis is of symmetry of the third order (Fig. 2.2); in Ni_9B_4 and Fe_9B_4 (Fig. 3.3) the axis of symmetry is of the fourth order which is absent in icosahedral structures possessing point groups of symmetry 532. That means that the structural changes which appeared with the changes in the clusters composition may be non-trivial.

If y is increased (y=5,6) in $Ni_{13-y}P_y$(M) and $Fe_{13-y}P_y$ there are only the planes of symmetry. Note that in Fe_8P_5 the cluster of atoms of Fe represents two squares that have the angle of 45° between their diagonals.

In Ni_8B_5 (Fe_8B_5) and Ni_7B_6(Fe_7B_6) (Fig. 3.4) there are the elements of cubical cell (their "genesis").

The above mentioned results show that the behaviour of clusters $Ni_{13-y}B_y$ and $Fe_{13-y}B_y$ is the same, which is due to the used potentials.

476

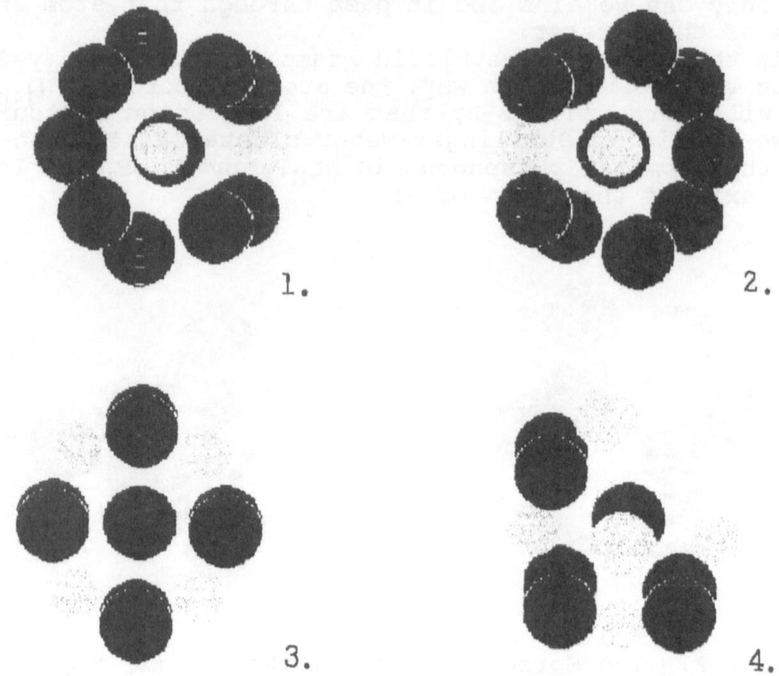

Fig. 3.   (1) and (2) $Fe_{10}B_3$, (3) Fe_9B_4, (4) Fe_7B_6.

CONCLUSIONS

1. The empirical potentials of interatomic interaction which have some common features in small ensembles lead to quite different basic results.
2. In the clusters Ni-B and Fe-B the atoms of metalloid are located in the surface of the clusters, while in Ni-P and Fe-P the atoms of metalloid are also introduced into "the volume" of the clusters.
3. The structures of the clusters $Ni_{13-y}B_y$ and $Fe_{13-y}B_y$ is the same while the configurations of $Ni_{13-y}P_y$ and $Fe_{13-y}P_y$ are different.
4. For the clusters $Ni_{13-y}B_y$ and $Fe_{13-y}B_y$ non-trivial transitions between the structures with different symmetry types were discovered.

REFERENCES

1.   Reuse, F., Khanna, S.N., de Coulon, V. and Buttet, J. (1990) "Pseudopotential local-spin-density

studies of neutral and charged Mg_n ($n \le 7$) clusters", Phys. Rev. B, 41, 11743-11759.

2. Shillady, D.D., Jena, P., Rao, B.K., Press, M.R. (1988) "Theoretical study of geometry of Mn_5" Int. J. Quantum Chem. Quantum Chem. Symp. (USA), N22, 231-236.

3. Northby, J.A. (1987) "Structure and binding of L.-J. clusters: $13 \le N \le 147$", J. Chem. Phys. 87, 6166-6177.

4. Beck, T.L. and Berry, R.S. (1988) "The interplay of structure and dynamics in the melting of small clusters", J. Chem. Phys. 88, 3910-3922.

5. Feuston, B.P., Kalia, R.K. and Vashishta, P. (1987) "Fragmentation of silicon microclusters: A molecular-dynamics study", Phys. Rev. B, 35, 6222-6239.

6. Antonio, G.A., Feuston, B.P. and Kalia, R.K. (1988) "Fragmentation of germanium microclusters: A molecular-dynamics study", J. Chem. Phys., 88, 7671-7686.

7. Weber, T.A. and Stillinger, F.N. (1985) "Interactions, local order and atomic rearrangement kinetics in amorphous nickel phosphorous alloys", Phys. Rev. B, 32, 5402-5411.

8. Ji-Chem Li and Cowlam, N. (1987) "Interatomic pair potentials for met. alloy glasses", Phys. and Chem. Liq. 17, 29-44.

9. Bratkovsky, A.M. and Smirnov, A.V. (1990). "The geometry of short-range order in Fe-B, Ni-B metallic glasses", Reports of the seventh USSR Conference "Structure and properties of metallic liquid and amorphous alloys".

10. Fujiwara, T. and Ishii, Y. (1980) "Structural analysis of models for the amorphous metallic alloy $Fe_{100-x}P_x$", J. Phys. F. 10, 1901-1911.

11. Fujiwara, T., Chen, H.S. and Waseda, Y. (1981) "On the structure of Fe-B metallic glasses of hypereutectic concentration", J. Phys. F. 11, 1327-1333.

12. Rahman, A. and Stillinger, F.H. (1971) "Molecular study of liquid water", J. Chem. Phys. 55, 3336-3359.

COMPUTER SIMULATION OF FREEZING AND MELTING IN TRANSITION METAL CLUSTERS.

L. T. WILLE[†] AND H. DREYSSÉ
Laboratoire de Physique du Solide
Université de Nancy-I
54506 Vandoeuvre-les-Nancy
France

† *permanent address:*
Department of Physics
Florida Atlantic University
Boca Raton, FL 33431
USA

ABSTRACT. We report on Monte Carlo simulations of the liquid-solid transformation in transition metal clusters. The energetics are expressed in terms of a many-body potential derived from the tight-binding model with Born-Mayer repulsion. By monitoring the radial fluctuation function and specific heat, we observe for certain cluster sizes, distinct freezing and melting temperatures, separated by a coexistence region. These depend in a non-monotonic fashion on cluster size and are found to correlate well with the magic numbers.

1. Introduction

As this symposium testifies, the study of microclusters continues to thrive. Great progress has been made in recent years, both in calculational and experimental techniques, and numerous applications have been suggested. Of particular interest has been the realization that small particles may exhibit phase transformations that are very similar to those observed in bulk materials, yet possess a number of remarkable, distinct features. It was first argued by Berry and co-workers [1] based on general grounds that small clusters should have unequal freezing (T_f) and melting (T_m) temperatures with a coexistence region in between. While most of the computational evidence for this phenomenon has been for rare gas clusters (with a Lennard-Jones or Morse pair potential), the same behavior has also been observed in computer simulations of transition metal clusters [2-5] and in Na-clusters [6]. In the present work we discuss new simulation results for transition metal clusters.

The possibility of phase transitions in small clusters was first suggested in computer simulations and later treated in a more rigorous framework [1,7,8]. There is now overwhelming evidence, from theory, experiment, and simulation, that the two-state picture is correct, although the width of the temperature interval $T_m - T_f$ depends sensitively and in a non-monotonic fashion on cluster size. Within the coexistence region an ensemble of clusters will be a mixture of 'solid' and 'liquid' forms as evidenced, for example, by a

P. Jena et al. (eds.), Physics and Chemistry of Finite Systems: From Clusters to Crystals, Vol. I, 479–484.
© 1992 *Kluwer Academic Publishers.*

480

bimodal distribution of the short-time averaged kinetic energies. Any given form, however, persists for time periods long compared to a typical vibrational period. The transition from one form to the other can be interpreted as a passage between local minima on the potential energy surface, with the saddle point crossings driven by a collective motion of the particles. The nature and location of these extrema is closely connected with the thermodynamic properties of the system. Finding the global minimum is an NP-complete problem [9], because of the exponential growth of the number of minima with cluster size, but the precise topology depends on the range of the pair interaction and this may be used to guide the minimization problem and to clarify the thermodynamics [10].

2. Model

Because of the strongly localized d-electrons, the electronic structure and bonding in transition metals cannot be adequately described in terms of a simple pair potential. Various approaches have been used to study transition metal clusters, for example, based on the embedded-atom method [11-12], effective medium theories [13], and Finnis-Sinclair potentials [14]. In the present work we use a potential first proposed by Gupta [15] to study surfaces and later employed by Tománek *et al.* [16] to investigate small clusters. Very recently a potential of the same form has been used to study the liquid-solid transformation in transition metal clusters [2-5].

In the Gupta potential the total energy of the cluster is taken to be of the form:

$$E = \sum_i \left[-\sqrt{Z_i} + \frac{q}{p\sqrt{Z_i}} \sum_j e^{-p(r_{ij} - 1)} \right],$$ \hfill (1a)

where

$$Z_i = \sum_j e^{-2q(r_{ij} - 1)}$$ \hfill (1b)

is the effective coordination number of atom i. In (1) all energies are expressed in reduced units and the distances r_{ij} between atoms i and j are in units of the bulk lattice constant. This expression has been derived by a Friedel-type argument, within the tight-binding model under the assumption of a repulsive Born-Mayer potential (of the form $\exp(-qr)$) at small distances, with the first three moments of the density of states fit exactly and taking the distance dependence of the hopping integrals to be of the form $\exp(-pr)$. For a detailed discussion of this type of potential see Ref. [17]. A list of parameter values appropriate for various transition metals is given in Ref. [18]. In the present work no attempt is made to fit any specific transition metal, and, as it is the aim to study the qualitative behavior of this type of potential, the usual choice p = 9, q = 3 is made[16].

In the Monte Carlo method a Markov chain of states is generated with the state transition probability determined by a Boltzmann factor. Starting from an initial configuration at temperature T, atoms are displaced in a random direction over a random distance, if the change in energy ΔE leads to a Boltzmann factor $\exp(-\Delta E/kT)$ that exceeds a random number in the interval [0,1]. The distribution of step lengths is adjusted to yield an acceptance rate of approximately 50%. Typically 5×10^5 updates per atom (MCS, Monte Carlo Steps) are attempted at each temperature, where the system is allowed to equilibrate for 2×10^5 attempts before averages are calculated. The presence of a phase change at a certain temperature can be detected in the simulations as a maximum in the specific heat, which can be expressed in terms of fluctuations of the average energy and energy-squared (averages denoted by angular brackets):

$$C_v = \frac{1}{T} (<E^2> - <E>^2).$$ (2)

Another useful quantity to monitor phase changes is the bond length fluctuation function:

$$\delta = \frac{2}{N(N-1)} \sum_{i<j} \frac{\sqrt{<r_{ij}^2> - <r_{ij}>^2}}{<r_{ij}>},$$ (3)

which shows a characteristic jump near the upper bound of the coexistence region.

3. Results and discussion

Monte Carlo simulations have been performed for a range of cluster sizes, but to be specific the present paper will concentrate mainly on the six-atom cluster. In that case, the potential (1) with $p = 9$, $q = 3$, possesses only two minima, corresponding to the octahedron (OCT) and the tripyramid (TP), with the former being the ground-state, with total energy -9.30903, compared to -9.07686 for the latter. To estimate the relative sizes of the catchment regions for these two isomers, 100 quenches (following the path of steepest descent on the potential energy hypersurface) were performed, starting from a random arrangement of the six atoms in a $4\times4\times4$ volume: 69 of these ended up in the TP state and 31 in the OCT configuration. The existence of those two minima, their ordering, and the relative abundances of the isomers are in good agreement with the calculations of Diep, Sawada, and Sugano [2-5], who used a similar potential. However, those authors did not consider the explicit dependence of the effective coordination number on the atomic distances and as a consequence our numerical values are different.

Ten independent Monte Carlo simulations were performed with the parameters mentioned earlier and all quantities were averaged over those 10 runs. Fig. 1 shows results for the specific heat and bond-length fluctuations where the temperature is expressed in energy units with Boltzmann's constant k equal to unity. To be noted is the broad maximum in C_v, which peaks near $T_m \approx 0.044$ and the jump in δ which occurs near $T_f \approx 0.038$. These values delineate the coexistence region: below T_f the cluster is in the 'solid' form and confined to one minimum, above T_m the cluster is in the 'liquid' state. To further clarify this behavior, we perform an analysis first proposed by Stillinger and Weber [19]. After the cluster has reached thermal equilibrium, we periodically interrupt the simulation and quench the configuration obtained at that moment to determine in which catchment region the system has arrived. The simulation is then continued from the configuration attained before quenching. This allows us to estimate the frequency of barrier crossings between the two minima. The results for two temperatures near the boundaries of the coexistence region are shown in Fig. 2. We observe that near T_f, saddle point crossings are rather infrequent and that the system quickly returns to the ground-state, while near T_m rapid oscillations from one minimum to the other occur. This is in perfect agreement with the expected general behavior.

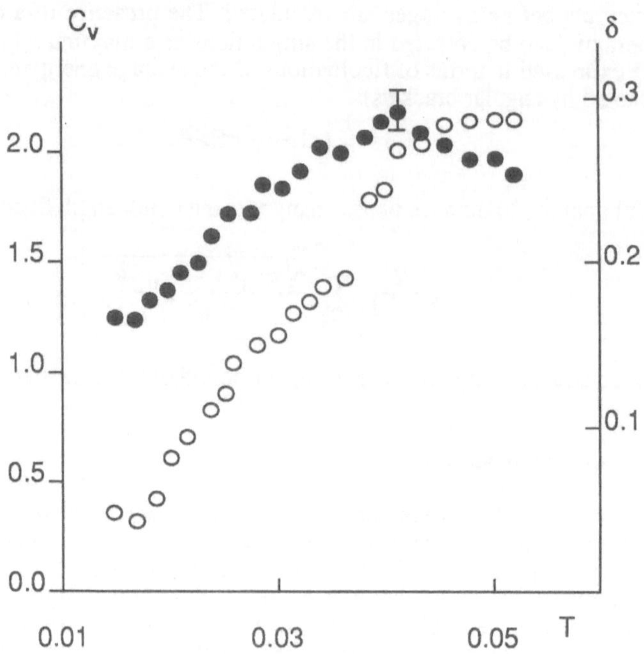

Figure 1. Specific heat C_v (solid circles) and bond length fluctuation function δ (open circles) as a function of temperature T for N=6 cluster. The broad maximum in C_v and the monotonic increase, followed by a jump and a plateau in δ are characteristic of coexistence of liquid and solid forms. A typical error bar is indicated.

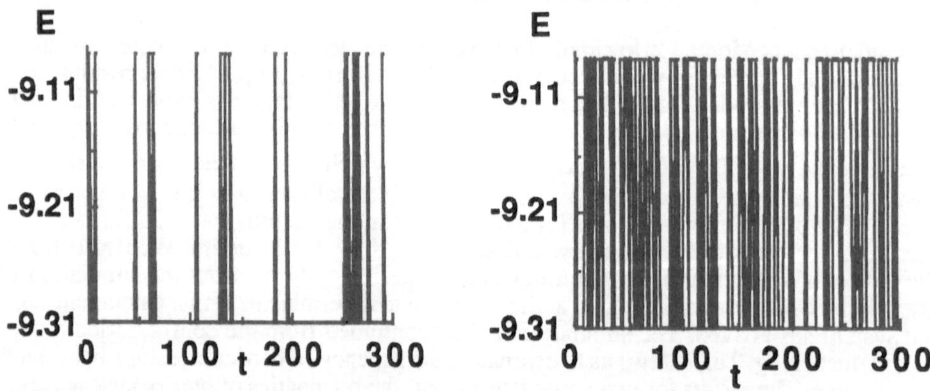

Figure 2: Distribution of quenched states as a function of time (in 10^3 MCS, after equilibration), oscillates between octahedron (E = -9.31) and tripyramid (E = -9.08). Leftmost graph corresponds to temperature T = 0.038 near T_f (lower limit of coexistence region) and rightmost graph to T = 0.044 near T_m (upper limit of coexistence region).

Simulations for clusters with N = 13 atoms (for which the ground-state is an icosahedron) and with N = 19 atoms (for which the ground-state is a double icosahedron) have also been performed and will be discussed in more detail in a forthcoming publication. Both cluster sizes correspond to magic numbers for a Lennard-Jones interaction and also for the present potential. As expected [1] they consequently exhibit a relatively broad coexistence region.

In conclusion, we have established the coexistence of liquid and solid states in a realistic model for transition metal clusters. Our general results are in good agreement with those of Diep, Sawada, and Sugano [2-5], who used a similar potential but without taking into account the explicit distance dependence of the effective coordination number Z_i (1b). As a consequence bond lengths are somewhat shorter in our model and freezing and melting temperatures are higher. The cluster sizes corresponding to a magic number (N = 13, 19, etc.) have a particularly broad coexistence region. For N = 6 the relation between the solid-liquid transformation and the potential energy hypersurface has been clarified.

Acknowledgements.

LTW was supported by grant No. DE-FG-05-89ER45392 from the U. S. Department of Energy and gratefully acknowledges the hospitality of the members of the Laboratoire de Physique du Solide at the University of Nancy. HD acknowledges financial support from NATO and CNRS/NSF.

References

[1] Berry, R. S., Beck, T. L., Davis, H. L., and Jellinek, J. (1988) 'Solid-liquid phase behavior in microclusters', Adv. Chem. Phys. **70**, 75-138, and references therein.

[2] Sawada, S. and Sugano, S. (1989) 'Dynamics of transition-metal clusters', Z. Phys. D **12**, 189-191.

[3] Sawada, S. and Sugano, S. (1989) 'Structural fluctuation and atom-permutation in transition-metal clusters', Z. Phys. D **14**, 247-261.

[4] Diep, H. T., Sawada, S., and Sugano, S. (1989) 'Melting and magnetic ordering in transition-metal microclusters', Phys. Rev. B **39**, 9252-9259.

[5] Sugano, S. (1990) 'New developments in the theory of microclusters', Phys. Scr. **T31**, 72-77.

[6] Chen, J., Brink, D. M., and Wille, L. T. (1991) 'Electronic structure and stability of small sodium clusters', Mat. Res. Soc. Symp. Proc. **206**, 139-144.

[7] Berry, R. S. and Wales, D. J. (1989) 'Freezing, melting, spinodals, and clusters', Phys. Rev. Lett. **63**, 1156-1159.

[8] Labastie, P. and Whetten, R. L. (1990) 'Statistical thermodynamics of the cluster solid-liquid transition', Phys. Rev. Lett. **65**, 1567-1570.

[9] Wille, L. T. and Vennik, J. (1985) 'Computational complexity of the ground-state determination of atomic clusters', J. Phys. A: Math. Gen. **18**, L419-L422.

[10] Braier, P. A., Berry, R. S., and Wales, D. J. (1990) 'How the range of pair interactions governs features of multidimensional potentials', J. Chem. Phys. **93**, 8745-8756.

[11] Cleveland, C. L. and Landman, U. (1991) 'The energetics and structure of nickel clusters: Size dependence', J. Chem. Phys. **94**, 7376-7396.

[12] Foiles, S. M., Baskes, M. I., and Daw, M. S. (1986) ' Embedded-atom-method functions for the fcc metals Cu, Ag, Au, Ni, Pd, Pt, and their alloys', Phys. Rev. B **33**, 7983-7991.

[13] Christensen, O. B., Jacobsen, K. W., Nørskov, J. K., and Manninen, M. (1991) 'Cu cluster shell structure at elevated temperatures', Phys. Rev. Lett. **66**, 2219-2222.

[14] Marville, L. and Andreoni, W. (1987) 'Size dependence of the structural properties of transition-metal aggregates from an empirical interatomic potential scheme' J. Phys. Chem. **91**, 2645-2659.

[15] Gupta, R. P. (1981) 'Lattice relaxation at a metal surface' Phys. Rev. B **23**, 6265-6270.

[16] Tománek, D., Mukherjee, S., and Bennemann, K. H. (1983) 'Simple theory for the electronic and atomic structure of small clusters', Phys. Rev. B **28**, 665-673.

[17] Dreyssé, H., Tománek, D., and Bennemann, K. H. (1986) 'Multi-adatom interactions on metal surfaces', Surf. Sci. **173**, 538-554.

[18] Rosato, V., Guillope, M., and Legrand, B. (1989) 'Thermodynamical and structural properties of f.c.c. transition metals using a simple tight-binding model', Phil. Mag. A **59**, 321-336.

[19] Stillinger, F. H. and Weber, T. A. (1984) 'Packing structures and transitions in liquids and solids', Science **225**, 983-989.